Springer Complexity

Springer Complexity is an interdisciplinary program publishing the best researc-hand academic-level teaching on both fundamental and applied aspects of complex systems – cutting across all traditional disciplines of the natural and life sciences, engineering, economics, medicine, neuroscience, social and computer science.

Complex Systems are systems that comprise many interacting parts with the ability to generate a new quality of macroscopic collective behavior the manifestations of which are the spontaneous formation of distinctive temporal, spatial or functional structures. Models of such systems can be successfully mapped onto quite diverse "real-life" situations like the climate, the coherent emission of light from lasers, chemical reaction-diffusion systems, biological cellular networks, the dynamics of stock markets and of the internet, earthquake statistics and prediction, freeway traffic, the human brain, or the formation of opinions in social systems, to name just some of the popular applications.

Although their scope and methodologies overlap somewhat, one can distinguish the following main concepts and tools: self-organization, nonlinear dynamics, synergetics, turbulence, dynamical systems, catastrophes, instabilities, stochastic processes, chaos, graphs and networks, cellular automata, adaptive systems, genetic algorithms and computational intelligence.

The two major book publication platforms of the Springer Complexity program are the monograph series "Understanding Complex Systems" focusing on the various applications of complexity, and the "Springer Series in Synergetics", which is devoted to the quantitative theoretical and methodological foundations. In addition to the books in these two core series, the program also incorporates individual titles ranging from textbooks to major reference works.

Springer Series in Synergetics

Founding Editor: H. Haken

The Springer Series in Synergetics was founded by Herman Haken in 1977. Since then, the series has evolved into a substantial reference library for the quantitative, theoretical and methodological foundations of the science of complex systems.

Through many enduring classic texts, such as Haken's Synergetics and Information and Self-Organization, Gardiner's Handbook of Stochastic Methods, Risken's The Fokker Planck-Equation or Haake's Quantum Signatures of Chaos, the series has made, and continues to make, important contributions to shaping the foundations of the field.

The series publishes monographs and graduate-level textbooks of broad and general interest, with a pronounced emphasis on the physico-mathematical approach.

Editorial and Programme Advisory Board

Frank Schweitzer

Browning Agents and Active Particles

Collective Dynamics in the Natural and Social Sciences

With a Foreword by J. Doyne Farmer

With 192 Figures and 3 Tables

Springer

Frank Schweitzer
ETH Zürich
Professur für Systemgestaltung
Kreuzplatz 5
8032 Zürich
Switzerland
E-mail: fschweitzer@ethz.ch

1st ed. 2003, 2nd Printing

Library of Congress Control Number: 2007932745

ISSN 0172-7389

ISBN 978-3-540-73844-2 Springer Berlin Heidelberg New York

Springer is a part of Springer Science+Business Media

springer.com

Typesetting: supplied by the author
Production: LE-TEX Jelonek, Schmidt & Vöckler GbR, Leipzig, Germany
Cover: WMXDesign, Heidelberg

SPIN 12081387 54/3180/YL - 5 4 3 2 1 0 Printed on acid-free paper

Foreword

When we contemplate phenomena as diverse as electrochemical deposition or the spatial patterns of urban development, it is natural to assume that they have nothing in common. After all, there are many levels in the hierarchy that builds up from atoms to human society, and the rules that govern atoms are quite different from those that govern the geographical emergence of a city. The common view among many, if not most, biologists and social scientists is that the devil is entirely in the details. This school of thought asserts that social science and biology have little or nothing in common, and indeed many biologists claim that even different fields of biology have little in common. If they are right, then science can only proceed by recording vast lists of details that no common principles will ever link together.

Physics, in contrast, has achieved a parsimonious description for a broad range of phenomena based on only a few general principles. The phenomena that physics addresses are unquestionably much simpler than those of biology or social science, and on the surface appear entirely dissimilar. A cell is far more complicated than a pendulum or an atom, and human society, being built out of a great many cells, is far more complicated still. Cells and societies have many layers of hierarchical organization, with complex functional and computational properties; they have identities, idiosyncracies stemming from an accumulation of historical contingency that makes them impossible to characterize in simple mathematical terms. Their complexity is far beyond that of the simple systems usually studied in physics. So, how can methods from physics conceivably be of any use?

The answer, as demonstrated by Frank Schweitzer in this book, lies in the fact that the essence of many phenomena do not depend on all of their details. From the study of complex systems we have now accumulated a wealth of examples that demonstrate how simple components with simple interaction rules can give rise to complex emergent behaviors, even when, as this book illustrates, the components are themselves quite complicated. This is because, for some purposes, only a few of their features are relevant and the complexity of the collective behavior emerges from the interactions of these few simple features alone. So although individual people are very complicated, and their decisions about where to live may be based on complex, idiosyncratic factors, it may nonetheless be possible to understand certain statistical properties of the geographic layout of a city in terms of simple models based on a few simple rules. Furthermore, with only minor modifications of these rules, the same explanatory framework can be used to understand the dendritic patterns for zinc deposits in an electric field. It is particularly striking that such disparate phenomena can be explained using the same theoretical tools. We have long known in physics that many different phenomena can be explained with similar mathematics. For example, the equations that describe simple electric circuits consisting of resistors, inductors, and capacitors are precisely the same as those describing a system of masses and springs. This work shows that such mathematical analogies apply even more broadly than one might have suspected.

In the middle of the 20th century, John von Neumann said that "science and technology will shift from a past emphasis on motion, force, and energy to communication, organization, programming and control". This is already happening, but as we enter the 21st century, the scientific program for understanding complex systems is still in its infancy as we continue experimenting to find the right theoretical tools. One of the obvious starting point candidates is statistical mechanics. This is a natural suggestion because statistical mechanics is the branch of physics that deals with organization and disorganization. One example where this has already succeeded is information theory. Claude Shannon showed how entropy, which was originally conceived for understanding the relationship between heat, work, and temperature, could be generalized to measure information in an abstract setting and used for practical purposes such as the construction of an efficient communication channel. So perhaps there are other extensions of statistical mechanics that can be used to understand the remarkable range of emergent behaviors exhibited by many diverse and complex systems.

But the reality is that classical statistical mechanics is mostly about disorganization. A typical model in statistical mechanics treats atoms as structureless Ping-Pong balls. A classic example is Brownian motion: When a particle is suspended in a fluid, it is randomly kicked by the atoms of the fluid and makes a random walk. This example played a pivotal role in proving that the world was made of atoms and helped make it possible to quantitatively understand the relationship between macroscopic properties such as friction and microscopic properties such as molecular collisions. It led to the development of the theory of random processes, which has proved to be extremely useful in many other settings.

The framework Schweitzer develops here goes beyond that of Brownian motion by making the particles suspended in the fluid just a little more complicated. The particles become *Brownian agents* with internal states. They can store energy and information and they can sense their environment and respond to it. They can change their internal states contingently depending on their environment or based on their interactions with each other. These extra features endow them with simple computational capabilities. They are smarter than Ping-Pong balls, but no smarter than they need to be. By adjusting parameters, the behavior can vary from purely stochastic at one extreme to purely deterministic at the other. Schweitzer shows that even when the Brownian agents are quite simple, through their direct and indirect interactions with each other, they can exhibit quite complex behaviors.

Brownian agents can be used in many different contexts, ranging from atomic physics to macroeconomics. In this book, Schweitzer systematically develops the power of the Brownian agent model and shows how it can be applied to problems ranging from molecule to mind. At the lowest level they can be simple atoms or molecules with internal states, such as excitation, and simple rules of interaction corresponding to chemical reactions. They can be used to describe the properties of molecular motors and ratchets. Or they can be cells or single-celled organisms responding to stimuli, such as electric fields, light, or chemical gradients. They can be used to study the group feeding properties of bark beetle larvae, or the trail formation of ants creating and responding to pheromone signals. With just a few changes in the model, they can be pedestrians forming trails based on visual queues, or automobile drivers stuck in traffic. Or they can be voters forming opinions by talking to their neighbors, or workers deciding where to work in a factory.

Agent-based modeling is one of the basic tools that has emerged in recent years for the study of complex systems. The basic idea is to encapsulate the behavior of the interacting units of a complex system in simple programs that constitute self-contained interacting modules. Unfortunately, however, agent-based modelers often lack self-restraint, and create agents that are excessively complicated. This results in models whose behavior can be as difficult to understand as the systems they are intended to study. One ends up not knowing what properties are generic and which properties are unwanted side-effects.

This work takes just the opposite approach by including only features that are absolutely necessary. It demonstrates that agent-based modeling is not just for computer simulation. By keeping the agents sufficiently simple, it is also possible to develop a theoretical framework that sometimes gives rise to analytic results and provides a mental framework for modeling and interpreting the results of simulations when analytic methods fail. By insisting on parsimony, it adheres to the modeling style that has been the key to the success of physics (and that originally motivated the rational expectations equilibrium model in economics).

This book lays out a vision for a coherent framework for understanding complex systems. However, it should be viewed as a beginning rather than an end. There is still a great deal to be done in making more detailed connections to real problems and in making quantitative, falsifiable predictions. Despite the wide range of problems discussed here, I suspect that this is only a small subset of the possibilities where the Brownian agent method can be applied. While I don't think that Brownian agents will gain a monopoly on theoretical modeling in complex systems, this book does a major service by introducing this new tool and demonstrating its generality and power in a wide range of diverse applications. This work will likely have an important influence on complex systems modeling in the future. And perhaps most importantly, it adds to the unity of knowledge by showing how phenomena in widely different areas can be addressed within a common mathematical framework and, for some purposes, most of the details can be ignored.

Santa Fe, NM, USA
February 2003

J. Doyne Farmer

Preface

The emergence of complex behavior in a system consisting of interacting elements is among the most fascinating phenomena of our world. Examples can be found in almost every field of today's scientific interest, ranging from coherent pattern formation in physical and chemical systems to the motion of swarms of animals in biology and the behavior of social groups. The question of how system properties on the macroscopic level depend on microscopic interactions is one of the major challenges in complex systems and, despite a number of different attempts, is still far from being solved.

To gain insight into the interplay between microscopic interactions and macroscopic features, it is important to find a level of description that, on the one hand, considers specific features of the system and is suitable for reflecting the origination of new qualities, but, on the other hand, is not flooded with microscopic details. In this respect, *agent models* have become a very promising tool mainly for *simulating* complex systems. A commonly accepted *theory* of agent systems that also allows analytical investigation is, however, still pending because of the diversity of the various models invented for particular applications. It will be a multidisciplinary challenge to improve this situation, in which statistical physics also needs to play its part by contributing concepts and formal methods.

This book wants to contribute to this development. First, we introduce a particular class of agent models denoted as *Brownian agents* and show its applicability to a variety of problems ranging from physicochemistry to biology, economy, and the social sciences. As we will demonstrate, the Brownian agent approach provides a stable and efficient method for *computer simulations* of large ensembles of agents. Second, we do not want just to present simulation results but also want to use the methods of statistical physics to analyze the dynamics and the properties of systems with large numbers of Brownian agents, in this way contributing pieces for a formal approach to *multiagent systems.*

Similar to Brownian particles, *Brownian agents* are subject to both deterministic and stochastic influences, which will allow us to derive a generalized Langevin dynamics for their activities. Different from physical particles, however, Brownian agents have individual degrees of freedom that allow them to respond differently to external signals, to interact with other agents, to change their environment, or to perform active motion. Because all kinds of activities need energy, an important internal degree of freedom is the agent's energy depot. In this book, we will extensively investigate the influence of the internal energy depot on the active, self-driven motion of the agents that in this respect are denoted as *active particles*. But other activities, such as the interaction via an adaptive "landscape" – a field generated by agents that feeds back to their further behavior – will also be widely discussed.

Because of the many examples from very different application areas, the book will be of interest not only to *physicists* interested in self-driven motion or physicochemical structure formation, but also to *biologists* who want to model the collective behavior of bacteria or insect societies, or to engineers looking for effective algorithms for the self-assembly and optimization of networks. Major parts of the book are also devoted to Brownian agent models in *social*, *urban*, and *economic* problems, to inspire scientists from these fields to apply the concept. Among the examples are models for urban and economic agglomeration, and also models of human behavior, such as the motion of pedestrians and the formation of collective opinions. The book will show that within the framework provided by statistical physics, nonlinear dynamics, and the theory of stochastic processes, a formal description of Brownian multiagent systems can be achieved, which also allows us to derive critical parameters for computer simulations and to predict the outcome of a collective dynamics.

The investigations presented in this book have been carried out in the years 1992 to 2001 mainly at the Institute of Physics at Humboldt University, Berlin, in close collaboration with the Sonderforschungsbereich (SFB) 230 "Natural Constructions" (Stuttgart) (1992–1995), further at the Department of Physics at Emory University, Atlanta, Georgia (1993), and the Department of City and Regional Planning at Cornell University, Ithaca, New York (1997). New light was shed on the results obtained during these years, when I started to work at the GMD Institute for Autonomous Intelligent Systems (now part of the Fraunhofer Society) in Sankt Augustin in 1999. The challenge to combine concepts from distributed artificial intelligence and new simulation methods for multiagent systems with the approaches of many-particle physics eventually gave this book a new direction.

I want to express my sincere thanks to Werner Ebeling (Berlin) whose fundamental work on the physics of self-organization and evolution in complex systems is one of the methodological bases of my investigations. He was actively involved in developing the ideas of active Brownian particles. Over the years, he promoted my interdisciplinary engagement, which thrived in the broad-minded scientific atmosphere of his research group. Further, I am very indebted to Lutz Schimansky-Geier (Berlin). Many of the ideas about interacting Brownian particles presented in this book were developed in close collaboration with him. His suggestions and critical remarks always gave new impulses for improving and extending the concept. Finally, I would like to thank Heinz Mühlenbein (Sankt Augustin) for his enduring support of my work and for very many stimulating discussions in a superb working atmosphere.

For collaboration, for discussions, suggestions, critical remarks, for various forms of encouragement, for invitations and support, I would like to thank (in alphabetical order) Wolfgang Alt (Bonn), Torsten Asselmeyer (Berlin), Jörn Bartels (Rostock), Vera Calenbuhr (Brussels), Andreas Deutsch (Bonn), Jean-Louis Deneubourg (Brussels), Fereydoon Family (Atlanta), Hermann Haken (Stuttgart), Dirk Helbing (Stuttgart), Janusz Hołyst (Warsaw), Klaus Humpert (Stuttgart), Peter Fleissner (Vienna), Jan Freund (Berlin), Hans-Jürgen Krug (Berlin), Lui Lam (San Jose), Kenneth Lao (Atlanta), José Lobo (Ithaca), Thilo Mahnig (Sankt Augustin), Péter Molnár (Atlanta), Ludwig Pohlmann (Berlin), Steen Rasmussen (Los Alamos), Gernot Richter (Sankt Augustin), Rupert Riedl (Altenberg), Gerd Röpke (Rostock), Helge Rosé (Berlin), Andrea Scharnhorst (Berlin), Eda Schaur (Innsbruck), Hans-Joachim Schellnhuber (Potsdam), Richard Schuler (Ithaca), Gerald Silverberg (Maastricht), Jens Steinbrink (Berlin), Angela Stevens (Heidelberg), Benno Tilch (Stuttgart), Rüdiger Wehner (Zürich), Wolfgang Weidlich (Stuttgart), Olaf Weiss (Berlin), Jörg Zimmermann (Bonn), and all colleagues and friends not mentioned here.

This work was made possible by two personal grants from the *Deutscher Akademischer Austauschdienst* (DAAD) (1993) and the *Deutsche Forschungsgemeinschaft* (DFG) (1996–1998) which I would like to thank very much for their financial support.

Indispensable, and yet free of charge were the tools for completing this book, namely LaTeX, the GNU Emacs, and the LINUX operating system with its various useful applications. So, finally, I want to express my thanks to all the developers and maintainers of these wonderful programs.

Sankt Augustin and Berlin,
September 2002

Frank Schweitzer

Contents

1. Complex Systems and Agent Models

1.1 Introduction to Agent-Based Modeling

1.1.1 The Micro–Macro Link

The emergence of complex behavior in a system consisting of interacting elements is among the most fascinating phenomena of our world. Examples can be found in almost every field of today's scientific interest, ranging from coherent pattern formation in physical and chemical systems [86, 142, 263], to the motion of swarms of animals in biology [98, 383, 561] and the behavior of social groups [449, 518, 536].

Heuristic approaches to complex systems basically focus on the interaction between "microscopic" subsystems and the *emergence* of new qualities at the "macroscopic" system level (see Fig. 1.1) which cannot be easily derived from microscopic properties.

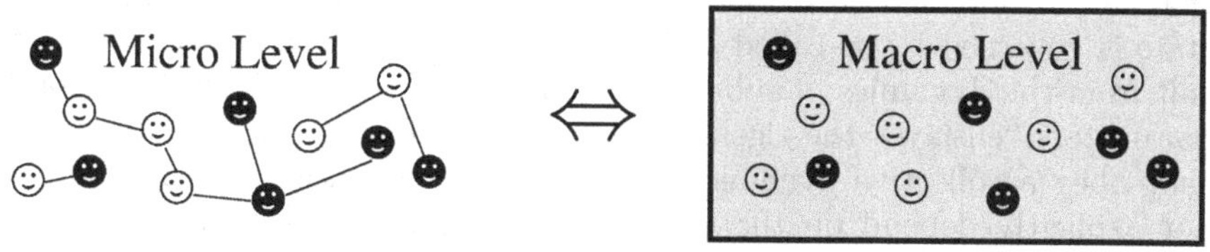

Fig. 1.1. The micro–macro link : how are the properties of the elements and their interactions (*"microscopic" level*) related to the dynamics and properties of the whole system (*"macroscopic" level*)?

Although there is no commonly accepted definition of a complex system [121], this is considered one of the eminent features in heuristic approaches, for instance, "Complex systems are systems with multiple interacting components whose behavior cannot be simply inferred from the behavior of the components."[1]

For specific physical systems, such as gases or semiconductors, statistical physics has provided different methods for deriving macroscopic properties

[1] New England Complex Systems Institute, `http://necsi.org/`

from known microscopic interactions. In particular, it was shown that systems consisting of a large number of elements (atoms, molecules, etc.) may display new properties that do not exist in single elements, but arise from their *collective* interaction. This could be *material* properties, such as conductivity or hardness, and also *dynamic* properties such as irreversibility.

A major advancement in the conceptual understanding and quantitative modeling of emergent properties in interacting systems was initiated by Hermann Haken in the 1970's, when he developed his concept of *Synergetics* [198, 199].[2] Here, again the system is composed of a large number of subsystems. They do not need to be elementary in the sense of "atoms" or "molecules," but can already have their own complexity.

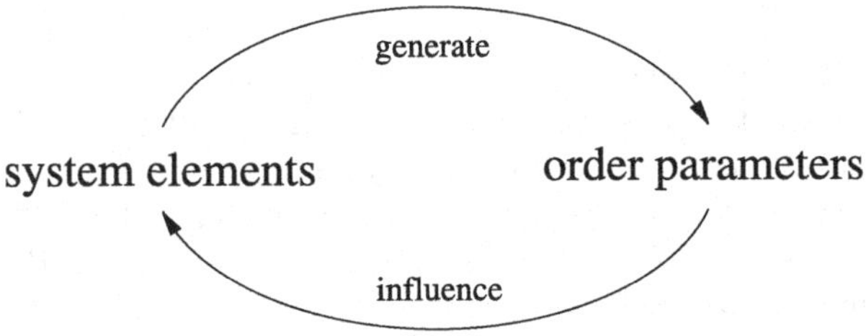

Fig. 1.2. Circular causation in synergetics: the system elements commonly generate one or more order parameters which in turn "enslave" their dynamics

Due to the nonlinear interactions of subsystems, new structures may arise that can be observed on the macroscopic level. The transition to the ordered state is governed by so-called *order parameters–collective* variables that result from the dynamics of subsystems. After they are established, the order parameters "enslave" the eigendynamics of the different subsystems so that they obey a collective dynamics. Noteworthy, the collective dynamics does not explicitly depend on the degrees of freedom of all subsystems, which could be of the order of 10^{23} for physical systems, but depend only on a few "aggregated" variables. This considerable reduction is also referred to as *information compression*, i.e., it is not necessary to know explicitly the values of all dynamical variables to describe collective dynamics. These paradigmatic insights are, of course, important also for the multiagent systems discussed in this book.

The appearence of emergent phenomena is often denoted as *self-organization*, i.e., "the process by which individual subunits achieve, through their cooperative interactions, states characterized by new, emergent properties transcending the properties of their constitutive parts" [58]. However, whether, or not these emergent properties occur depends, of course, not only on the

[2] The reader may take a look at the *Springer Series in Synergetics* for an overview of the vast applications of synergetics, ranging from laser physics to biology and psychology.

properties of subsystems and their interactions, but also on suitable external conditions and such as global boundary conditions and the in/outflux of resources (free energy, matter, or information). These are denoted as *control parameters* in synergetics. A description that tries to include these conditions is given by the following heuristic definition: "Self-organization is defined as spontaneous formation, evolution and differentiation of complex order structures forming in nonlinear dynamic systems by way of feedback mechanisms involving the elements of the systems, when these systems have passed a critical distance from the statistical equilibrium as a result of the influx of unspecific energy, matter or information" [502].

In this sense, self-organized structure formation can be considered the opposite of a hierarchical design of structures which basically proceeds *from top down to bottom*: here, structures *originate* bottom up, leading to an emerging *hierarchy*, where the structures of the "higher" level appear as a new quality of the system [94, 198]. To predict these global qualities from local interactions, fundamental limitations exist which are discussed, e.g., in chaos theory. Moreover, stochastic fluctuations also give unlikely events a certain chance to occur, which in turn affects the real history of the system. This means that the properties of complex systems cannot be determined by a hierarchy of conditions; the system creates its complexity in the course of evolution with respect to its global constraints. Considering that the boundary conditions may also evolve and new degrees of freedom may appear, *co-evolutionary processes* become important, and the evolution may occur on a qualitatively new level.

1.1.2 The Role of Computer Simulations

To gain insight into the interplay between microscopic interactions and macroscopic features in complex systems, it is important to find a modeling level, which, on one hand, considers specific features of the system and is suitable for reflecting the origination of new qualities but, on the other hand, is not flooded with microscopic details.

This raises the question of new scientific methodologies or tools suitable for investigating complex systems: "By complex system, it is meant a system comprised of a (usually large) number of (usually strongly) interacting entities, processes, or agents, the understanding of which requires the development, or the use of, new scientific tools, nonlinear models, out-of equilibrium descriptions and computer simulations."[3]

The "classical" methodological approach of theoretical physics aims to describe a system in terms of a set of equations, which can be (hopefully) solved analytically or at least numerically. The advanced techniques developed in this respect are still the backbone of physical science. But during the last

---

[3] Journal *Advances in Complex Systems*, `http://journals.wspc.com.sg/acs/acs.html`

20 years, *computer simulations* became more and more relevant as a third methodology in addition to formal theories and empirical studies (experiments). *Simulation* is not meant to be just another tricky algorithm to solve, e.g., difficult partial differential equations, it means that the basic interaction dynamics on the microscopic level is used to model the macroscopic system. *Molecular dynamics* is one of the prominent examples in physics.

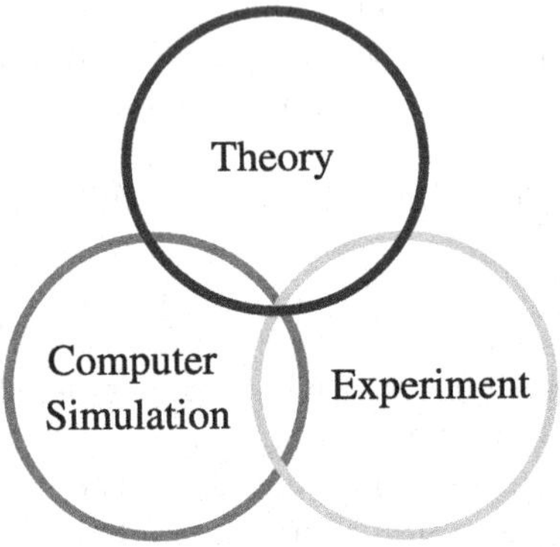

Fig. 1.3. Physics today – a methodological merger of theory, experiment, and computer simulations. Is a transfer of this approach to nonphysical complex systems feasible?

Thanks to the increasing power and the accessibility of affordable computers, microscopic or *particle-based* simulation techniques have become even more attractive recently. In combination with computer architectures that allow massive parallel computing, large complex systems can be tackled this way – for instance, in the field of *granular material* or *traffic systems*. Thus, the painstaking way of getting firm analytical results is cut short in many cases by computer simulations that try to sketch "microscopic reality" somehow. In physical systems, the microscopic properties of the elements and their interaction potentials may be known to a certain extent, but in biological or in socioeconomic systems, this becomes a rather weak point. Therefore, before summarizing all of the *advantanges* that we might expect from a particle-based or an agent-based approach to complex systems, we should first discuss some of the *problems* of this methodological approach.

In principle, there are two ways to bind computer simulations to reality. First, *experiments* may produce data that can be used, for instance, to calibrate the simulation. This may be feasible in physics or molecular biology, but it becomes rather difficult in complex systems in life or social sciences. Moreover, observable (or obtainable) data in these systems are mostly *aggregated* data that do not allow us to deduce microscopic actions. For example, city traffic simulations have to count on measured net fluxes of cars – the underlying OD (origin–destination) matrices that reflect individual driver decisions are not available (or census data). Another problem results from the difficulties in deducing appropriate dynamics from measured data – this is sometimes referred to as *reverse engineering.*

A second way to prevent computer simulations from just being computer games is to link them to an *analytical model* that makes certain *predictions* about the system's behavior. This should avoid a methodology, where the impact of a certain variable q characterizing the "microscopic entity" (or agent) on the macroscopic dynamics is studied by just varying q and then only comparing the runs of different simulations. By an analytical theory, the impact of q, e.g., on bifurcations (the appearance or disappearance of new solutions for the macrodynamics) or on phase transitions (the emergence of new collective states) should be *predicted*, ideally. Such a challenge – that will also be the aim of this book – has to face the problem that the link between a microscopic computer model and an analytical description of a macrosystem can be achieved only under certain modeling *reductions* that need to be carefully discussed. There is certainly a *trade-off* between a most realistic computer simulation that includes as much microscopic detail as possible and the possibility of matching such a system with a tractable analytical model. Although the latter is most desirable, it would need some compromises in the design of the microscopic simulations.

This raises the question of how much (or how detailed) microscopic information is needed to obtain a particular macroscopic behavior. As discussed in the previous section, the emergence of collective phenomena is in many cases connected only to a few variables, the order parameters. Thus not all microscopic details must be known; this has been referred to as information compression. On the other hand, the order parameters themselves are generated only from microscopic interactions and are not known at the outset in most cases.

To escape from this dilemma, we would like to propose the following procedure for designing particle-based computer simulations in this book:

- Instead of incorporating as much detail as *possible*, we want to consider only as much detail as is *necessary* to produce a certain emergent behavior. This rather "minimalistic" or reductionistic approach will, on the other hand, allow us in many cases to derive some analytical models for further theoretical investigation, and thus gives the chance to understand *how* emergent behavior occurs and *what* it depends on.
- To find the simplest "set of rules" for interaction dynamics, we will start from very simple models that *purposefully* stretch the analogies to physical systems. Similar to a construction set, we may then add, step-by-step, more complexity to the elements on the microlevel, and in this way gradually explore the extended dynamic possibilities of the whole model, while still having a certain theoretical background.

Reality is complex, but models do not have to be as complex as reality, to provide some insight into it. We know from various examples, i.e., the *logistic map* in nonlinear dynamics and its relation to population dynamics, that even very simple models can display complex behavior that sketches certain features of reality. The *art* of deriving such simple models that capture complex features of reality has resulted in many intriguing *paradigmatic examples* in natural and social sciences, ranging from the Ising system in physics [166, 168, 254, 277, 334, 353, 526] to the *minority game* in economics [81, 82, 324, 479].

These general remarks should have prepared the discussion of "entities" on the microlevel, which, in general, shall be denoted as *agents*.

1.1.3 Agents and Multiagent Systems

Agent-based systems are regarded as a *new paradigm* enabling an important step forward in empirical sciences, technology, and theory [517, 547]. But similar to other terms such as *complexity*, there is no strict or commonly accepted definition of an *agent*. However, some common attributes can be specified for particular agent concepts. To elucidate the differences among these concepts, we will first have a short look at the history of agent systems in *informatics*.

The simulation of human thinking and human behavior on the computer is one of the most ambitious efforts of informatics. The concept of *artificial intelligence* (AI) has traditionally focused on the *individual* and on rule-based paradigms inspired by psychology. However, in the 1980s, the opinion prevailed that *interaction* and *distribution* have to be considered as basic concepts of intelligence. A prominent representative, Marvin Minsky declared that intelligence is the result of numerous interacting modules, each solving primitive tasks [346, 347]. One of his followers, Carl Hewitt, became the father of *distributed* artificial intelligence (DAI), when he first created a model of interactive objects called *actors*. These actors carried different internal states, worked in parallel, and responded to messages from other objects. For the interaction of actors, social societies became the leading paradigm – with all the problems that arise when sociological concepts need to be transferred into technical concepts, and vice versa. Nevertheless, *socionics*,[4] the merger of concepts from distributed artificial intelligence and sociology – or *infonomics*,[5] a combination of concepts from informatics and economics – are challenging research fields today.

The advancement of actors is today called *agents* in informatics. In a so-called "weak notion of agency" that is also used in emerging disciplines such as *agent-based software engineering*, the agent is simply seen as "a self-contained, concurrently executing software process, that encapsulates some state and can communicate with other agents via message passing processes" [565]. If the focus is on "software technologies" used for the Internet or for

---

[4] `http://www.tu-harburg.de/tbg/SPP/Start_SPP.html`
[5] `http://www.infonomics.nl/`

business administration, agents can be subroutines, but also larger entities of a software program with some sort of persistent control.

A "stronger notion of agency" [565], on the other hand, applies to fields such as artificial intelligence, where in addition to the definition above, concepts that are more typical of humans are used. For example, knowledge, belief, intention, obligation – or even emotions – may be attributed to agents to a certain degree. In distributed artificial intelligence, therefore a distinction is made between the *reflexive* and the *reactive* agent. The first has an (internal) model or at least some knowledge about its environment that allows it to draw conclusions about some certain actions in a given situation (see also Sect. 1.1.4). The reflexive agent may act on either knowledge-based or behavior-based rules [315]; its deliberative capabilities may allow it to perform complex actions, such as BDI, Believe, Desire, Intention, based on what the agent thinks is true, what it would like to achieve, and how it expects to do it [357]. Other features are learning either in terms of adaptation or specialization, or genetic evolution of the internal program, etc. The reactive agent, on the other hand, simply "reacts" to signals from the environment without referring to internal knowledge. This signal-response type of action implies the same reaction to the same signal.

Agent power results mainly from their *interaction* in a larger environment. In particular, in fields such as economics, social science, and population biology, the major focus is rather on *cooperative interaction* instead of autonomous action. A *multiagent system* (MAS) then may consist of a *large number* of agents, which can be also of *different types.* That means, in general, that agents may represent very different components of the system; each already has its own complexity. Basically, an agent can be *any* entity in the system that affects other agents (including itself).

In molecular biology, agents may represent different types of enzymes acting together in a regulatory network. Taking the example of an ecosystem, agents may represent individual creatures interacting, e.g., symbiotically or in a predator–prey type of interaction. But, depending on the modeling level, agents may also represent the different species of a food chain, which is already an aggregated description. In a financial market, as another example, some agents may represent stockbrokers acting on certain price changes, analysts evaluating companies, or shareholders or banks or institutional investors, which do not interfere directly with the market, but through other traders or analysts, etc.

The interactions between agents may usually occur on *different spatial and temporal scales.* That means, in addition to *local* or spatially restricted interactions which may occur only at specific locations or if agents are closer, we also have to consider *global* interactions, where all agents are involved. Further, the timescale of interactions is of significant importance. Whereas some interactions occur rather frequently, i.e., on a shorter timescale, others become effective only over a long time. A third distinction to be mentioned is between direct and indirect interactions. The latter occurs, e.g., if agents anonymously interact via a "market" or use a common resource that can be exhausted in the course of time. In this way, the actions of all agents are indirectly coupled via the resource, and its current availability further provides some information about the cumulative activity of others.

In the following, we will only list some features of agents (see also [143, 382, 517, 547, 565]) and multiagent systems that are of particular relevance for the models and examples discussed in this book.

Internal agent structure:

- *Internal degrees of freedom:* Agents are characterized by different internal parameters. These are individual (continuous or discrete) variables that may also change in time (due to eigendynamics to be specified or due to impacts from the environment).
- *Autonomy:* The actions of agents are determined by their internal states ("inherent dynamism") and/or environmental conditions ("induced dynamism"). They are *not* controlled by other instances (i.e. system administrator); that means agents operate on their own and have some "control" over their actions and internal states.

Agent activities:

- *Spatial mobility:* Agents can move in space, i.e., they can migrate. Physical space is considered here a two-dimensional continuous space but can also be represented by a 2-D lattice. Agents that do not move but rather act at their current positions, are assumed as a special case of the mobile agent.
- *Reactivity:* Agents can *perceive* their environment and *respond* to it. Although their action may be triggered by actions of other agents or by environmental conditions ("induced dynamics"), they act autonomously.
- *Proactivity:* Agents take the initiative, i.e., they can start interaction with other agents or *change their environment*, e.g., by creating new quantities (information) or by depleting resources. This may induce further actions.
- *Locality:* Agents are small compared to the system size, and individual agent actions, it is assumed, do not change the system as a whole at once. Global effects therefore arise from accumulated agent actions. In general, agents are mainly affected by and in turn affect their local neighborhoods, but there are, of course, situations where all agents interact on a global level (i.e., via a mean field).

Multiagent features:

- *Modularity*: In MAS, a logical distinction is made between modules and their interactions. Particular modules (entities, subsystems) of a system are represented by respective agents. Depending on the granularity of the model, each of these modules may be composed of smaller modules.[6] Different from a *monolithic* view that treats the system as a whole, the modular view allows reconfigurability and extensibility of the MAS in an easier way.

- *Redundancy:* MAS consist of large numbers of agents, many of them similar in function and design. This means, on one hand, that critical instances are not represented by just one agent, and on the other hand, that the system does not break down if an agent fails to act in an expected manner. Because agents are "small," the excess capacity provided by redundant agents does not cost too much.
- *Decentralization:* MAS are not ruled *top-down* by centralized system control (which is the paradigm of the monolithic approach). Instead, competence, capacity, etc. are *distributed* across several agents. This in turn allows them to "create" a control *bottom-up*, in a self-organized manner, as the result of interaction between different agents.
- *Emergent Behavior:* In MAS, interaction between agents can produce new (and stable) behavior on the global level of the whole system. This represents a new quality that results from the aggregated behavior of the agents and therefore cannot be reduced to individual agents. Further, due to nonlinear effects, it is often hard to predict the emergent properties of the system from individual properties.
- *Functionality:* Vulnerable features of the MAS such as functionality are *not* associated with central instances or single agents but are *emergent* features of the MAS. Although each agent may have its own functions (or "behavior"), the functionality of the system as a whole, for instance, in problem solving, is *not* assigned to specific agents but results from the interaction of different agents.
- *Adaptation:* Modularity, decentralization, and emergent functionality are the basis for the MAS to adapt to changing situations. Here, also the excess capacity provided by redundant agents may play an important role. As in natural evolution, it ensures a reserve that can be utilized in unforeseen situations, i.e., for exploration of new possibilities, without losing the functionality of the system. Adaptation (sometimes also called *collective learning*) also needs a limited or *bounded in time memory* of the system, i.e., the system has to *forget*/unlearn its old states, interactions, etc. to adapt to the new ones.

---

[6] In the field of *classifier systems*, the so-called *Michigan approach* considers systems of "individuals" that act as modules rather being modular themselves [95]. The so-called *Pittsburgh approach* considers systems where "individuals" each have or develop a modular structure but do not interact besides some competitive effects [525]. The approach considered here is between these two extremes, i.e., it considers modules that *may* be considered modular themselves.

1.1.4 Complex Versus Minimalistic Agents

The complex behavior of a multiagent system as a whole basically depends (i) on the complexity of the agent (i.e., the range of possible actions) and (ii) on the complexity of the interaction. In Sect. 1.1.2, we already raised the question of how many microscopic details or how much agent complexity is needed to obtain a certain emergent behavior and have favored a rather "minimalistic" approach that includes only as much detail as *necessary* – not as possible. On the other hand, many of the currently discussed MAS seem to follow just the other route; therefore some remarks will apply here.

In addition to the properties summarized in Sect. 1.1.3, agents can be further equipped with different attributes – which, however, do not all appear necessary or useful in every application. For example, complex agents can develop an *individual world view*, i.e., each agent has its own model (sometimes called conceptual model) of the external world that surrounds it. Here, the manner in which the agent builds up its model of the world on the basis of the information it receives from its environment is of particular interest.

Besides "intelligent" features such as logical deduction or the capacity to learn, complex agents may also be able to develop individual *strategies* to pursue a certain goal. This goal could be either an *individual* goal such as maximizing its own utility or a *common* goal such as catching all agents of another kind at a certain place. The evolution of such strategies based on previous experiences plays an important role, e.g., in evolutionary game theory, and also in areas such as economics or trading in financial markets. Evolutionary games also recently attracted the interest of physicists [253, 302, 330, 498, 499]. One example of particular interest is the so-called *minority game*[7] [81, 82, 324, 479], where agents can develop a strategy, i.e., to choose either 0 or 1, based on information about the past record of winners. This seemingly "simple" model shows a rich variety of phenomena which are, on one hand, closely related to observations of real markets, but, on the other hand, can be analytically treated by advanced methods of statistical physics.

Other examples of quite complex agents can be found in *agent-based computational economics* (ACE)[8], which is meant to describe and to simulate the economic interaction between "agents" which could be, for example, individuals, firms, etc. [14, 135, 147, 236, 274, 294, 326]. Among other goals, economic agents may tend to maximize their utility based on a cost–benefit calculation. Here, a notable difference between the *rational* and the *bounded rational* agent is made. As one of the standard paradigms of neoclassical economic theory, the *rational agent* model is based on the assumption of the agent's *complete knowledge* of all possible actions and their outcomes or a known probability distribution over outcomes, and the *common knowledge*

---

[7] `http://www.unifr.ch/econophysics/minority`

[8] `http://www.econ.iastate.edu/tesfatsi/ace.htm`

assumption, i.e., that the agent knows that all other agents know exactly what he/she knows and are equally rational [419, 475]. Based on such complete knowledge, the agent then can make a *rational choice* of its next action.

Such an approach has been widely used for theoretical investigations of economic models such as for the calculation of equilibrium states, but it clearly involves a number of drawbacks and practical problems. Besides the methodological complications of, e.g., defining utility functions (that are assumed fixed, i.e., nonevolving), the rational agent must have an *infinite computing capacity* to calculate all possible actions [419]. And to assure that all agents are perfectly informed, this implicitly requires an *infinitely fast, loss-free* and *error-free dissemination of information* in the whole system.

A more realistic assumption would be based on the *bounded rationality* of agents [15, 422, 476], where decisions are not taken upon complete a priori information, but on incomplete, limited knowledge that involves uncertainties and is distributed at finite velocity. This, however, would require modeling the *information flow* between the agents explicitly as a spatially heterogeneous, time-delayed and noise-affected process. This will also be discussed in Sect. 1.2.2.

The *exchange of information* among agents and between agents and their environment is one of the preconditions of *common action* and cooperative capacity in MAS. As the different examples in this book show, such functionality can nevertheless also be achieved with rather simplex or minimalistic agents that do not have all the advanced features of complex agents. It has already been noticed that the *complexity of interaction* in MAS is even more important than the complexity of the *agent* itself. This is rather similar to physical or physicochemical systems, where a rich variety of structures can emerge from the interaction of *simple* entities.

Agents, in general, have a set of different rules to interact with each other. Which of the rules applies for a specific case may also depend on local variables, which in turn can be influenced by the (inter)action of the agents. Commonly, the freedom to define rules or interactions for the agents is much appreciated. However, each of these additional rules expands the state space of possible solutions for the agent system, which is known as *combinatoric explosion* of the state space. For 1000 agents with 10 rules, the state space already contains about 10^{13} possibilities. Hence, almost every desirable result could be produced from such a simulation model, and the freedom could very soon turn out to be a pitfall.[9] In fact, due to their rather complex facilities

[9] Parunak [382] suggests avoiding this situation by limiting the agent interaction *ex post*: "effective systems will restrict communication in some way." He quotes as an example Kauffmans Boolean networks that show self-organization only when sparsely connected [268]. This conclusion, however, cannot be generalized. In particular, *synergetics* has shown how self-organization also occurs in densely connected systems, via generating order parameters. Moreover, many examples in this book show that self-organization also works in systems with broadcasted information.

to set up rules and interactions, many of the currently available simulation tools lack the possibility of investigating the influence of specific interactions and parameters systematically and in depth.

This in turn raises the question of an alternative agent design which we call *minimalistic* here, to distinguish it from the *complex* agent design sketched above. The *reactive* agent mentioned in Sect. 1.1.3 is just one example of a minimalistic agent, whereas the *reflexive* agent would certainly belong to the complex agent category. However, there are also possible designs that range between these two extremes – such as the *Brownian agent* approach featured in this book (see Sect. 1.2). The Brownian agent is meant to be a subunit with the "intermediate" complexity. This means, on the one hand, that the agent is not assumed as a "physical" particle reacting only to external forces, but, on the other hand, should not already have the same complex capabilities as the whole system. The Brownian agent should be characterized by all of the criteria mentioned in Sect. 1.1.3 regarding its internal structure and its activities. However, a Brownian agent is minimalistic in the sense that it acts on the possible *simplest set of rules, without* deliberative actions, developing internal world models or strategies, calculating utilities, etc. Instead of specialized agents, the Brownian agent approach is based on a large number of "identical" agents, and the focus is mainly on *cooperative interaction* instead of autonomous action.

Such restrictions also set *limits* for applying the Brownian agent approach. Consequently, the concept would be less appropriate when agents display "intelligent" features, such as logical deduction, complex behavioral adaptation to the environment, development of individual strategies to pursue a certain goal, or development of an individual world view. The question, however, is to what degree these specifications need to be taken into account to explain or to reproduce an observed emergent behavior. In the Brownian agent approach, some features which might be considered important for a specific system are dropped to investigate a particular kind of interaction more carefully. Instead of describing a whole system most realistically, the Brownian agent approach focuses only on particular dynamic effects within the system dynamics but, as an advantage, also provides numerous quantitative methods to investigate the influence, e.g., of certain parameters or quantities. For example, bifurcations of the dynamics, the structure of attractors, conditions for stable nonequilibrium states, etc. can be investigated by advanced methods borrowed from statistical physics; this provides a clear-cut idea about the role of particular interactive features.

Like any other agent-based approach, Brownian agent models also are based on a specific kind of reductionism which should also be addressed here from a philosophy of science viewpoint. Compared to the reductionistic approaches especially in natural sciences, self-organization theory is often interpreted as a holistic approach that conquers classical reductionism. However, self-organization itself is a phenomenon which is realized only from a certain perspective, i.e., observing it depends on the specific level of description or on the focus of the "observer," respectively [365]. In this particular sense, self-organization theory is an *aithetical* theory (see the Greek meaning of "aisthetos", perceptible). The particular level of perception for self-organization has been denoted *mesoscopy* [446]. It is different from a real microscopic perception level that focuses primarily on the smallest entities or elements, as well as from the macroscopic perception level that rather focuses on the system as a whole. Instead, mesoscopy focuses on elements complex enough to allow those interactions that eventually result in emergent properties or complexity on the macroscopic scale. These elements are the "agents" in the sense denoted above; they provide an "intermediate complexity" and are capable of a certain level of activity, i.e., they do not just passively respond to external forces, but are actively involved, e.g., in nonlinear feedback processes.

Thus, for the further discussion we have to bear in mind the degree to which agent-based models are based on certain reductions regarding the system elements and their interactions. On the way toward a generalized self-organization theory, we have to understand carefully the nature of these reductions, especially when turning to the social and life sciences [206]. Self-organization in social systems is confronted with the mental reflections and purposeful actions of their elements, creating their own reality. While we are, on one hand, convinced that the basic dynamics of self-organization originates analogies between structure formation processes in very different fields, regardless of the elements involved, we should, on the other hand, not forget about the differences between these elements, especially between humans and physical particles. Thus, deeper understanding of self-organization, complex dynamics, and emergence in socioeconomic systems has to include also better insight into these reductions.

1.1.5 Agent Ecology

Agent models have originally been developed in the *artificial life community* [70, 295, 315, 336, 357, 482, 519, 547], but they recently turned out to be a suitable tool in various scientific fields, ranging from ecology to engineering [98, 290], and especially in economics and social sciences [9, 16, 206, 449, 461, 513]. However, agent-based models are not restricted to social and life sciences. They are also useful in *natural sciences* such as physics where continuous approximations are less appropriate and only a *small number* of agents (particles) govern the further evolution, for example, in dielectrical breakdown or filamentary pattern formation. Here deterministic approaches or mean-field equations are not sufficient to describe the behavior of a complex system. Instead, the influence of history, i.e., irreversibility, path dependence, the occurrence of random events/stochastic fluctuations play a considerable role.

Whereas in physical systems the basic entities of the same kind (e.g., the ions in an electrolytic solution) are almost "identical," this is certainly not true for the entities in other areas, such as population biology, economy, or sociology. Nevertheless, many agent concepts developed in those fields are still based on the assumption of the *representative agent* [273] – that means an "individual" equipped with the *averaged* properties of its kind, for example, the representative buyer or the representative voter. Such a reduction is certainly useful for obtaining theoretical results. A more realistic approach, called the *heterogeneous agent* [275], is based on two extensions: (i) the assumption of a (probability) distribution $\varrho(q)$ of a certain property q and (ii) the insight that individuals do not always follow certain specific rules in the same manner (as physical particles). Hence, if individuals who may be different from each other interact and produce collective behavior, this behavior may not correspond to that of an "average" agent.

In particular, in economic or social systems, there is a *heterogeneous world* of individual "microbehavior" – and it is not quite obvious how this generates the macroscopic regularities observed in a "society." In addition to *heterogeneity* and *bounded rationality* (see Sect. 1.1.4), a third issue makes agent models in the life sciences rather complex. For example, economic agents do not just interact *anonymously* and through the market mechanism as assumed in neoclassical economic theory; there are also *direct* interactions via *networks*. These can be (nontechnical) communication networks, such as "personal relations" in administration or management of companies, innovation networks used for the targeted spread of new products or technology [144], or social networks, for example, support networks in heterogeneous human societies [205].

Finally, we mention that some types of interactions in life science are *asymmetrical* interactions. Different from physical systems, where agents of two different kinds may attract each other in a symmetrical manner (such as electrons and protons), we find situations quite often where one agent A is attracted by another agent B (and thus tries to approach it as closely as possible), but B is repelled from A (and thus tries to avoid its neighborhood) [210]. This holds, for instance, in ecosystems for the interactions between two species, predator (fox) and prey (rabbit) but also in the economy where imitators try to follow closely the roots of innovators (or pioneers) who have invented new ideas or discovered new markets [435].

Rather complex relations in social and economic systems also involve another problem, i.e., the *evolution of interactions* in the course of time, which is rarely found in simpler systems. Interaction in physical systems, for instance, may depend on distance or on specific properties such as charge, but they usually do not change in time. The emergence of new structures in a MAS, such as the formation of coalitions – i.e., new types of "super"agents that act as a new entity with a new identity – on the other hand, may change the interaction structure of the whole system. This denotes a rather complex level of *adaptation*. Sometimes adaptation refers to the adjustment of a system to, e.g., changing boundary conditions, which may involve reconfiguration of relations, but does not necessarily mean an *evolution* of the basic interactions. In *complex adaptive systems*, for example immune systems, nervous systems, multicellular organisms, or ecologies, we may find, however, different levels of adaptation that may arise either from nonlinear spatiotemporal interaction among a large number of components, from the evolution of the different elements (or agents), or from the evolution of their interaction structure itself. Thus, *coevolution*, i.e., the common evolution of the system elements and their interactions, becomes an important feature of complex systems.

Though such a degree of complexity in natural systems can hardly be matched by an analytical approach, agent-based approaches allow us for the first time to model and to simulate it in close detail. For socioeconomic systems, so-called *artificial societies* can be used to investigate adaptation on different levels during the evolution. A prominent example is given by Epstein and Axtell in their book *Growing Artificial Societies: Social Science from the Bottom Up* [135], where discrete, spatially distributed agents interact and due to several feedback mechanisms, evolve "cultural identities." Other examples show the evolution of support networks among social agents [205]. Such computer simulations can be used to test hypotheses concerning the relationship of individual behavior to macroscopic regularities which can hardly be tested in social reality.

Many of the features mentioned above – such as heterogeneity or coevolution – are not just characteristic of socioeconomic systems, but in particular also of ecological systems that inhabit a large number of different species or subpopulations that each again consist of a large number of (not identical) individuals involved in complex direct and indirect interactions – therefore we have used the term *agent ecology* here. In fact, *natural* agent systems where collectives of large numbers of animals have evolved toward division of labor, task specialization etc., seem to be *the* most inspiring source of artificial multiagent systems. Therefore, to develop a sound concept of agent ecologies, it is useful to have a look also at suitable natural paradigms.

Insect societies, in particular, colonies of ants or termites, or swarms of bees or wasps, are among the most intriguing examples of natural "agent" systems and will also play their role in this book (see Sects. 3.2.6, 5.4). Collective behavior in social insects is primarily based on *chemical communication* using different kinds of chemical substances, called *pheromones*. This paradigm for (indirect) communication is also widely used in this book. The (positive or negative) response to different chemicals is denoted as *chemotaxis* and also involves in many cases the response to *chemical gradients*. This kind of behavior, as already noticed, can be adapted by rather simple "reactive" agents.

Trail formation in ants based on cooperative *chemotactic interaction* is one paradigmatic example of self-organized "meaningful" behavior. It has proven to be *robust* against disturbances (i.e., after moderate destruction, the trail is recovered in a rather short time) and *adaptive* to cope with a changing environment (i.e., exhausted food sources are left out and new ones are included); further *obstacles* are avoided (i.e., a deadlock almost never occurs) (see also Sect. 5.4). Such features are adopted in different multiagent models based on rather simple agents, for instance, for the development of exploratory strategies, for optimization routines (ACO – "ant colony optimization"), and for routing in networks [62, 63, 78, 107, 108, 456]. Chemotactic interaction is also involved in different aggregative phenomena in insects [103] (see also Sect. 3.2.6).

Another important feedback mechanism in social insects is named *stigmergy*, i.e., the coordination of activities through the environment. It plays an important role for sorting broods or corpses of dead nest mates [102] and in particular for nest building, e.g., of termites, where it was first described [182], and of wasps [548]. The mechanism is now used in multiagent models for sorting [102] and self-assembling three-dimensional structures [62, 106] – a development that gives further input to the vast field of *robotics* [10, 171, 286, 351, 367, 480]. We note that other insect capabilities such as navigation in a complex environment, are also used, for example, to simulate the behavior of *robots* and to build them in hardware [293, 349].[10]

Not only insect societies, but also other groups of interacting animals can be used as paradigms for designing multiagent systems. We just mention here *bacterial colonies* [47, 48], myxobacteria [112, 487], and slime molds [234] (see also Sect. 5.3). *Swarming* of animals [375, 381] that occurs on different scales ranging from the "microscopic level" of bacteria to the "macroscopic level" of hoofed animals is another example of collective interaction that gave input to the modeling of multiagent systems and also gave the name to the Swarm platform for simulating MAS (see also Sect. 1.1.6). It has been revealed that intriguing natural examples of swarming, such as the highly ordered motion of schools of fish or flocks of bird, can be explained by local or short-range interactions of the animals (see also Sect. 2.4). Last but not least, sophisticated hunting strategies for capturing animals, for instance, in wolves surrounding prey, that already involve task specification, have also been applied to multiagent systems, where they are investigated as *pursuit games* [88, 148, 233].

The principles underlying the behavior of natural multiagent systems have been, in a concise manner, included in the metaphor *swarm intelligence* [62]. This describes new computational approaches based on direct or indirect interaction among relatively simple agents for *solving* distributed problems, with emphasis on flexibility and robustness [382]. But the swarm metaphor, as already mentioned, is also used to describe a specific kind of computer architecture for *simulating* MAS, which will be discussed in the following.

---

[10] See also the international conference series on "Simulation of Adaptive Behaviour: *From Animals to Animats*", `http://www.isab.org.uk/conts/`

1.1.6 Simulation Approaches

Many of the (multi)agent features outlined above can hardly be matched to an analytical model. Although advanced methods separate timescales or spatial scales or handle time-delayed feedback processes, in many cases computer simulations seem to be the only way to get some insight into global dynamics. This raises the question of how to simulate MAS adequately.

Again, physics has provided some methodological approaches for tackling this problem. Spatial mobility, for example, is also a dynamic property of *physical* particles (such as ions) that may further interact via short-range or long-range *interaction potentials* (such as Coulomb interaction). Therefore, *molecular dynamics* may serve as a paradigm for microscopic or particle-based simulations.

However, there are notable differences in agent-based simulations. First, molecular dynamics is more or less interested in obtaining averaged results of some physical quantities rather than focusing on particular particles and their actions.[11] Physical particles are also hardly seen as autonomous instances with internal degrees of freedom, their behavior is more or less determined by the interaction potential. Although physical particles may be regarded as "reactive agents" in the sense described above, they lack, of course, other agent features such as proactivity. Consequently, molecular dynamics as a simulation technique does not provide concepts to deal with these features. Thus, while keeping the valuable idea of a *microscopic* dynamical simulation, we need to add some more advances to achieve a simulation system for MAS.

Another "microscopic" approach that has been successfully applied to physics and life and social sciences, is known as the *cellular automaton*, or CA for short [563, 564]. The main idea dates back to the 1940s when J. v. Neumann and St. Ulam developed a "(v. Neumann) machine" as a collection of cells on a grid (see Fig. 1.4). In this *discretized space*, each cell is characterized by a *discrete* state variable. The dynamics occurs in *discrete time steps* and may depend on the states of the neighboring cells, where different definitions exist for the neighborhood. Moreover, certain boundary conditions have to be considered that may correspond to conservation laws in physical systems.

Different variants of the CA approach, such as *lattice gas* models or Ising-like simulation systems, are used today [83, 170, 325, 508, 522]. Among the famous applications of CA in *artificial life* is Conway's *Game of Life*, where each cell is in one of two states: dead or alive, i.e., $\theta_i \in \{0, 1\}$. The "evolution" of each cell further depends on the states of its eight *nearest neighbors*; it will stay alive in the next time step if there are either two or three neighbors alive; in any other case, it will die of either overcrowding or exposure. On the other hand, a cell can be also reborn in a future time

---

[11] This restriction may, of course, not hold for molecular devices or quantum dots, etc., on the atomic level.

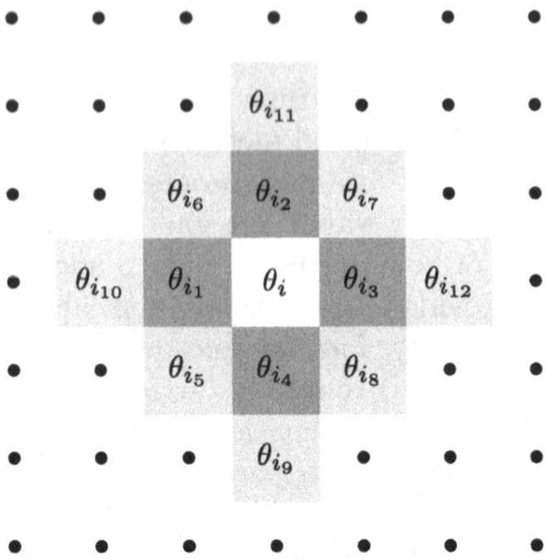

Fig. 1.4. Sketch of the two-dimensional cellular automaton. It shows a neighborhood of size n surrounding cell i where the neighbors are labeled by a second index $j = 1, ..., n-1$. The nearest neighbors are shown in darker gray, the second nearest neighbors in lighter gray. The so-called *von Neumann* neighborhood includes the cells $j = 1, ..., 4$; the so-called 3×3 *Moore* neighborhood includes the cells $j = 1, ..., 8$. The value θ_{i_j} characterizes the state of the cell i_j

step if three of its eight neighbors are alive. This kind of game is also very typical for many CA applications in population biology [111, 204, 328, 329]. In frequency-dependent invasion processes, for example, the occupation of a spatial patch depends on the frequency of the different species in the neighborhood [352].

But also in the social and economic sciences, CA has found broad application. In 1949, Sakoda already studied spatial segregation processes in two subpopulations based on a CA, where all individuals interact in a weighted manner [420]. In another CA by Schelling, the segregation process based on local interation was investigated [425]. Other CA-based examples describe the evolution of support networks [205], group decisions in Ising-like models [166, 277], and also influences of social impact [300, 353] and collective opinion formation [261, 262] (see also Chap. 10). Moreover, many game-theoretical models have been simulated by CA [1, 246, 302, 371, 372].

Although, on one hand, CAs provide the possibility of a "microscopic" approach, they have, on the other hand, several drawbacks in rather complex applications: (i) the discretization in space, time, and state variables makes it rather difficult to derive macroscopic dynamics from microscopic interactions, whereas a continuous description may ease the procedures a lot; (ii) interactions between distant cells are hard to define – in most applicable cases, either (nearest or second nearest) neighbor interactions or interactions via a *mean field* are taken into account; and (iii) properties of "microscopic" elements are assigned to cells that are fixed in "space," not to autonomous "objects." Taking, for instance, the rather simple example of a moving element, in a CA approach, this is not described as a motion of the element itself, but *indirectly* as a consecutive number of cells that become occupied and are assigned specific state variables for a short time before transiting back to their empty state. So, in conclusion, an *object-oriented* description *continuous in time and space* would be much more feasible in many cases.

Such a demand is fulfilled by the agent-based approach which is an object-oriented description by definition. "In mainstream computer science, the notion of an agent as a self-contained, concurrently executing software process, that encapsulates some state and is able to communicate with other agents via message passing, is seen as a natural development of the object-based concurrent programming paradigm" [565].

Here, any element properties are directly assigned to agents, which, of course, depend on the problem under consideration. Such a paradigmatic shift has an enormous influence on model building and simulation. On the computational level, it also results in novel design patterns and architectural structures that in turn open up new areas of application.

Classical software engineering often suggests a *functional decomposition* of the system, where specific functions are assigned to individual agents. This may be well suited to centralized systems, but as the discussion in Sects. 1.1.3 and 1.1.4 makes clear, it is more suitable for multiagent systems to let functionality emerge from the agent interaction. This also allows the agent architecture to adapt dynamically to changing conditions without top-down control from the system operator.

One of the available computer platforms is named Swarm[12], a *software package for multiagent simulation* that is written in Objective-C and in Java, more recently. It offers build-in capabilities, such as event management for swarms of concurrently interacting agents, an activity library, tools for information management and memory management, as well as GUI for graphical input and output. Recent applications include mainly economic simulations [312]. In fact, microeconomics creates a large demand for investigating complex interactions of economic agents. However the temptation to setup rather complex microsimulations in Swarm with parameters freely adjustable by graphical panels often prevents the users from investigating their dynamics systematically and in-depth.[13] This in turn underlines the need to link computer simulations of MAS to statistical theory and to analytical investigations.

It is *not* the aim of this book to review the many architectures for simulating multiagent systems of various complexity [135, 143, 175, 345, 547]. Only two specific architectures that can be applied for a range of different problems,

---

[12] `http://www.swarm.org`

[13] To quote from a book review of [312]: "The authors present the results of one or two runs, but in some cases it would be better to offer the results of hundreds or thousands of simulations for each parameter setting. That would allow readers to ascertain the variety of behaviours that are observed." `http://www.soc.surrey.ac.uk/JASSS/4/2/reviews/johnson.html`

including the examples in this book, shall be mentioned in the following. One rather general scheme to design agent-based models is called *Environment Rules Agents* (ERA) [175] and aims to separate the environment, with its rules, from the agents, which communicate through the environment. The agents, on one hand, receive information from the environment (for instance, about their neighbors) and, on the other hand, instructions from rather abstract entities called *Rule Master* and *Rule Maker* about the rules and metarules (i.e., the rules used to modify rules). The specific realization of this scheme must then be based on an object-oriented platform, such as Swarm or the FTA architecture described in the following.

In many applications, not *all* agents interact directly or need *complete* information about the system. Consider, for example, city traffic simulations where agents represent different kinds of cars, buses, trucks, pedestrians, bicyclists, etc. The action and the decisions of a particular agent in a given traffic situation is then affected mainly by those agents in a rather narrow vicinity than by others far away. Therefore, an efficient simulation system mainly has to process (only) the information needed to simulate the agent's (local and short-time) behavior appropriately.

A possible solution for this is proposed by the *blackboard architecture* [134] – one of the fundamental programming paradigms of artificial intelligence (AI) in the 1970s and 1980s. It resembles a blackboard in the real world, a data repository where agents can post and retrieve information. Early versions of blackboard concepts were focused on disseminating information via a *single* large blackboard. With the invention of distributed databases, the main idea was extended toward different distributed blackboards that can be accessed independently.

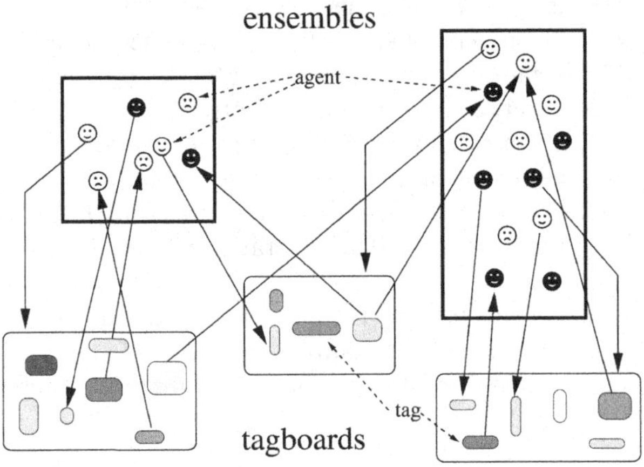

Fig. 1.5. A user's view of the Flip-Tick Architecture (FTA). Different ensembles of agents communicate via tags that are displayed for a certain lifetime on tagboards

The *Flip-Tick Architecture*, or FTA for short, is one example of a software platform based on distributed blackboards [409, 410, 520]. It is based on four types of objects (see Fig. 1.5): Agents (originally called *"actors"*) are functional units for processing data in a periodic operation, called actor cycles. They can be grouped into *ensembles*, i.e., functional units for managing a group of actors that proceed in a fully synchronized way. After all agents of an ensemble have performed their cycles, the cycle counter of the ensemble representing its local time is incremented by one – as if a "tick" of a clock had occurred. After the tick, a new ensemble cycle occurs.

Several agents may interact over many cycles by *reading* tags from tagboards, *processing* them, and *writing tags* to tagboards. A *tag* is a data object (later also called a piece of "structural information" in Sect. 1.2.2) produced by an agent to communicate with other agents. The tag is then displayed on a *tagboard* where it is present for a certain *lifetime* during which it can be read by any agent that has access to the tagboard. After a specific period of time called the tagboard cycle, the tagboard is "flipped." This can be envisioned as the tagboard having two faces, the face with the current display (read-only) and the face with the pending display (write-only). Agents can read only the current display and send tags for inclusion in the pending display. After the flip operation, the pending display along with continuing tags of the previous display is visible, and the display counter of the tagboard representing its local time is incremented by one; then a new tagboard cylce begins. Because each tagboard is associated with at least one ensemble of agents, the flip rate of the tagboard is related to the tick rate of its owning ensembles, i.e., the tagboard is flipped if the owners have completed at least one cycle.

The communication of agents via tagboards provides a very flexible mechanism for simulating interactions that evolve in time and/or space, as well as for *parallel* interactions on different *spatial* and *temporal* scales. Explicit point-to-point communication – widely used in message passing systems [333] such as PVM [496, 497] or MPI[14] – is useful for static communication structures like tree- or grid-like connections between known partners but will require rather global changes as the information flow changes during the evolution of the system. Using a broadcast system, on the other hand, would prevent this but could easily overload existing computer networks. Thus, distributed computer architectures which are based on cooperative/competitive ensembles of small or medium-grained agents, such as FTA, may be much more suitable for coping with time-varying interaction tasks.

[14] `http://www.mcs.anl.gov/mpi/`

1.2 Brownian Agents

1.2.1 Outline of the Concept

As already mentioned in Sect. 1.1.4, a *Brownian agent* is a particular type of agent that combines features of reactive and reflexive agent concepts. It is described by a set of state variables $u_i^{(k)}$, where the index $i = 1,...,N$ refers to the individual agent i and k indicates the different variables. These could be either *external* variables that can be observed from the outside or *internal degrees of freedom* that can be indirectly concluded only from observable actions. Important external variables are $u_i^{(1)} = \boldsymbol{r}_i$, which denotes the *space coordinate* (mostly a vector in two-dimensional physical space), or $u_i^{(2)} = \boldsymbol{v}_i$, which is the individual *velocity* of the moving agent. Both are assumed *continuous* variables.

The internal degrees of freedom, on the other hand, that cannot be directly observed, could be continuous or discrete variables. For instance, the state variable $u_i^{(3)} = \theta_i \in \{-1, 0, +1\}$ may describe three different responses to certain environmental conditions or to incoming information. For example, agents where $\theta = -1$ may not be affected by a particular signal, whereas agents where $\theta = +1$ may respond to it.

An important continuous state variable in the context of Brownian agents is the *internal energy depot* $u_i^{(4)} = e_i$, which determines whether or not agent i may perform a certain action. This includes the assumption that all actions – be it active motion, communication, or environmental changes – need to use "energy." In general, this term describes not just the physical free energy that is dissipated, e.g., during active motion; it intends also to cover other resources needed to perform a certain action – for instance, if an economic agent wants to buy stock at the market, this will certainly depend on its "liquidity." With an empty internal energy depot, the agent will not be able to "act", i.e., to interact actively with its environment or other agents, but may respond only *passively* to external forces that might be imposed on it, like a "physical" particle. We note that Brownian agents with just an internal energy depot in this book are mainly called *active particles* – if other, agent features will be neglected. This means a rather "simple" type of agent that does not deny its origin from the physical world, and therefore has also been denoted as a *particle agent* [124].

Noteworthy, the different (external or internal) state variables can change in the course of time, either due to impacts from the surroundings or due to internal dynamics. Thus, in a most general way, we may express the dynamics of the different state variables as follows:

$$\frac{d\,u_i^{(k)}}{dt} = f_i^{(k)} + \mathcal{F}_i^{\text{stoch}} \, . \tag{1.1}$$

As for other physical equations, this formulation is based on the *principle of causality*: any *effect* such as the temporal change of a variable u has some *causes* that are listed on the right-hand side. In the concept of Brownian agents, it is assumed that these causes may be described as a *superposition* of *deterministic* and *stochastic* influences imposed on agent i.

This picks up the ingenious idea first used by Langevin (see Sect. 1.3.2) to describe the motion of *Brownian particles*, and is basically the reason why this agent concept is denoted as a *Brownian* agent. Brownian particles (see Sect. 1.3.1 for details) move due to the impacts of surrounding molecules whose motion, however, can be observed only on much smaller time and length scales compared to the motion of a Brownian particle. Thus, Langevin invented the idea of summing up all of these impacts in a stochastic force with certain statistical properties.

For a Brownian agent, we will exploit Langevin's idea in a similar manner, i.e., we will sum up influences that may exist on a microscopic level but are not observable on the time and length scales of a Brownian agent, in a stochastic term $\mathcal{F}_i^{\mathrm{stoch}}$, whereas all of those influences that can be directly specified on these time and length scales are summed up in a *deterministic* term $f_i^{(k)}$.

Such a distinction basically defines the level of coarse-grained description for the multiagent system. The "cut" may prevent us from considering too much "microscopic" detail of the MAS, while focusing on particular levels of description. The summed stochastic influences might result from a more fine-grained deterministic description, but instead of taking this into detailed account, just some specific statistical (gross) properties are considered on the coarse-grained level. This implies, of course, that the "stochastic" part does *not* impose any *directed* influence on the dynamics (which would have counted as deterministic), but on the other hand, it does not necessarily mean a white-noise type of stochasticity. Instead, other types such as colored noise or multiplicative noise are feasible. Noteworthy, the strength of the stochastic influences may also vary for different agents and may thus depend on local parameters or internal degrees of freedom, as we will show in different examples.

The *deterministic* part $f_i^{(k)}$ contains all specified influences that cause changes in the state variable $u_i^{(k)}$. This could be

- nonlinear interactions with other agents $j \in N$; thus $f_i^{(k)}$ can be, in principle, a function of all state variables

$$\underline{u} = \left\{ u_1^{(1)}, u_1^{(2)}, ..., u_2^{(1)}, ..., u_j^{(k)}, ..., u_N^{(k)} \right\} \tag{1.2}$$

  describing any agent (including agent i).
- external conditions, such as forces resulting from external potentials, or the in/outflux of resources, etc. These circumstances shall be expressed as a set of (time-dependent) *control parameters*

$$\underline{\sigma} = \{\sigma_1, \sigma_2, ...\} . \tag{1.3}$$

- an *eigendynamics* of the system that does not depend on the action of the agents. In the example of an ecosystem, this eigendynamics may describe day/night or seasonal cycles or the agent-independent diffusion of resources within the system. This is expressed in an *explicit time-dependence*[15] of $f_i^{(k)}$.

Hence, in general, $f_i^{(k)} = f_i^{(k)}(\underline{u}, \underline{\sigma}, t)$. To set up a Brownian multiagent system we need to specify (i) the relevant state variables $u_i^{(k)}$; (ii) the dynamics for changing them, i.e., $\dot{u}_i^{(k)}$; and (iii) the external conditions, i.e. $\sigma_1, ..., \sigma_n$, or a possible eigendynamics of the system. Thus, basically the dynamics of the MAS is specified on the level of the individual agent, not on a macroscopic level.

In the following, we will briefly mention several possibilities for getting analytical expressions for the nonlinear function $f_i^{(k)}$. In the simplest case, where the Brownian agent can be treated as a *reactive* Brownian particle responding only to external forces, the dynamics is simply given by the Langevin equation for the two important state variables, the space coordinate $\boldsymbol{r}_i$ and the velocity $\boldsymbol{v}_i$. It reads (see Sect. 1.3.2 for more details)

$$\begin{aligned} \frac{d\,u_i^{(1)}}{dt} &= \frac{d\,\boldsymbol{r}_i}{dt} = f_i^{(1)} = \boldsymbol{v}_i\,, \\ \frac{d\,u_i^{(2)}}{dt} &= \frac{d\,\boldsymbol{v}_i}{dt} = f_i^{(2)} + \mathcal{F}_i^{\mathrm{stoch}}\,. \end{aligned} \tag{1.4}$$

Here, the change of the space coordinate results from the velocity of the particle without any additional stochastic influences, which are summed up in the dynamic equation for the velocity itself. The nonlinear deterministic term $f_i^{(2)}$ for simple Brownian particles may read:

$$f_i^{(2)} = \mathcal{F}_i^{\mathrm{diss}} + \mathcal{F}_i^{\mathrm{ext}}\,. \tag{1.5}$$

$\mathcal{F}_i^{\mathrm{diss}}$ denotes the dissipative forces that may result, for instance, from the friction of a moving particle (some appropriate expressions will be discussed in Sect. 1.3.2), and $\mathcal{F}_i^{\mathrm{ext}}$ describes the influence of external forces that may result, for instance, from an external potential $\mathcal{F}_i^{\mathrm{ext}} = -\boldsymbol{\nabla} U|_{\boldsymbol{r}_i}$.

If we extend the concept to *interacting* Brownian agents, we also have to consider for $f_i^{(2)}$, (1.5) an additional force $\mathcal{F}_i^{\mathrm{int}}$ that describes influences resulting from interactions with other agents, either via an interaction potential (see Sect. 2.4) or via an adaptive landscape (see Sect. 3.2). Further, if we want to model the *active motion* of agents, the nonconservative force $\mathcal{F}_i^{\mathrm{diss}}$ has to consider two contributions: (i) the friction γ_0 that *decelerates* the agent's motion and (ii) power from the internal energy depot e_i that is used for the *acceleration* of motion [126]:

$$\mathcal{F}_i^{\mathrm{diss}} = -\gamma_0\, \boldsymbol{v}_i + d_i\, e_i(t)\, \boldsymbol{v}_i\,. \tag{1.6}$$

---

[15] Physical systems with an explicit time dependence are called *nonautonomous*, but this does not mean here that the autonomy of the agents is violated.

d_i is the agent-specific rate of converting depot energy into kinetic energy. All terms will be specified later with respect to particular applications.

For the dynamics of the internal energy depot as another state variable of the Brownian agent, we may assume the following equation:

$$\frac{d\,u_i^{(4)}}{dt} = \frac{d\,e_i(t)}{dt} = q_i(\boldsymbol{r}_i, t) - p_i(\boldsymbol{r}_i, t) - c_i\, e_i(t)\,. \tag{1.7}$$

Here, the term $q_i(\boldsymbol{r}_i, t)$ describes the influx of resources into the agent's depot, for example, the take-up of energy that will later be used for active motion, the "liquidity" to act on the stock market, or the "power" to perform environmental changes. Since these resources may not be homogeneously distributed in the system or may not be available at all times, the influx may depend on the actual position of the agent, $\boldsymbol{r}_i$, and on time.

The term $p_i(\boldsymbol{r}_i, t)$, on the other hand, describes different kind of "outfluxes" from the agent's depot, i.e., what the depot energy is used for. Specifically, we will consider two different processes later:

$$p_i(\boldsymbol{r}_i, t) = s_i(\boldsymbol{r}_i, t) + d_i\, e_i(t)\,. \tag{1.8}$$

$s_i(\boldsymbol{r}_i, t)$ describes, for example, *environmental changes* performed by the agent at its current position. In particular, we will assume that the agent can change an "adaptive landscape" or establish a "communication field" to interact with other agents. The second term $d_i\, e_i(t)$ is used in Chap. 2 to describe the active (accelerated) motion of an agent, as described above.

The third term $c_i e_i(t)$ in (1.7) describes the internal dissipation of an energy depot at a specific loss rate c_i. In a biological context, internal dissipation is analogous to metabolism by the organism, but technical storage media such as batteries are also subject to internal dissipation. In the example of a financial depot, internal dissipation may result from processes such as inflation which may reduce the value of the depot in the course of time. According to (1.1), it is possible to consider also explicit stochastic influences on the internal energy depot. This is neglected here because (1.7) for the depot is implicitly coupled to the equations of motion, where stochastic forces are taken into account.

Eventually, we also have to specify the dynamics of the other state variables, e.g., of the internal degrees of freedom θ_i. This will be left to the specific examples discussed in this book. At this point, we just want to mention – in addition to the force ansatz of (1.5) – other suitable ways for getting *analytical expressions* for the key term $f_i^{(k)}$ from interaction dynamics. One possibility is to map interaction dynamics to so-called *symbolic reactions*, in close analogy to chemical kinetics. To take one example (in good fun), let the state variable u_i of agent i be a measure of its "scientific reputation." This reputation can be increased by publishing a new paper A in a particular scientific field n, but it can also fade out in the course of time.

This dynamics can be expressed by the symbolic reactions,

$$u_i + A_n \xrightarrow{k_{in}} 2u_i \ , \quad u_i \xrightarrow{k_i'} F \ , \tag{1.9}$$

and by using the rules of formal kinetics, we arrive at

$$f_i(\underline{u}, \underline{\sigma}, t) = \frac{du_i}{dt} = \sum_{n=1}^{m} k_{in}\, u_i\, A_n - k_i'\, u_i \ . \tag{1.10}$$

This expression for $f_i(\underline{u}, \underline{\sigma}, t)$ does not depend directly on the u_j of other agents or on an external dynamics, but there will be an indirect interaction among agents via the feasible publications, for instance,

$$\frac{dA_n}{dt} = \sigma_n(t) - \sum_{i=1}^{N} k_{in}\, u_i\, A_n \ . \tag{1.11}$$

Here, the number of new publications in a certain field can be increased by specific scientific progress in this field, expressed in terms of $\sigma_n(t)$ – an "influx" that plays the role of a (time-dependent) control parameter. On the other hand, the number of possible publications is also decreased by all of the papers already published by other agents (including agent i).[16] Consequently, indirect coupling between agents leads to competition, which for the assumed dynamics is known as *hyperselection*, a phenomenon well studied in molecular biology, population dynamics, and economics [72, 131]. Thus, this example shall not be discussed in more detail here; it shall simply show a feasible way to derive some analytical expression for agent dynamics.

In many cases, the interactions between agents are specified by *rules* that may describe the change of a state variable u_i, given that some preconditions are fulfilled (i.e., states of other agents $\underline{u}^\star$ or external conditions $\underline{\sigma}$). Let us take an example from *evolutionary game theory* where agents interact using different strategies that shall be indicated by (discrete) numbers. So, e.g., $u_i = 2$ means that agent i is currently using strategy number *2*. The change of a strategy $u_i \to u_i'$ is often given as an *operational* condition: `if ... then`, that may also depend on the reward (or utility) $g_i(u_i, \underline{u}^\star)$ received by agent i playing strategy u_i with (a subset of) other agents, denoted by $\underline{u}^\star$. In a formal approach, this condition has to be mapped to a transition rate $w(u_i'|u_i, \underline{u}^\star, \underline{\sigma})$ that gives either the number of transitions per time unit from u_i to any other possible u_i' or the probability of change. As will be outlined in more detail in Sect. 5.2, dynamics for the state variable u_i can then be derived in a first-order approximation in the form,

$$f_i(\underline{u}, \underline{\sigma}, t) = \sum_{u_i' \in \mathcal{L}} (u_i' - u_i)\, w(u_i'|u_i, \underline{u}^\star, \underline{\sigma}) \ . \tag{1.12}$$

---

[16] This example in good fun, of course, assumes that the content of each scientific paper is published only once, which may be true but not realistic.

Here, the summation goes over all possible values u_i' into which the strategy u_i can be changed.[17]

Another approach to deriving the deterministic part $f_i(\underline{u}, \underline{\sigma}, t)$ is inspired by the idea of a "generalized" *potential landscape* $V(\underline{u}, \underline{\sigma}, t)$ that is created by all possible state variables (see Fig. 1.6). Similar to classical mechanics or the theory of gradient systems in nonlinear dynamics, the force acting on agent i to change its state variable u_i may then be determined by deriving the potential in state space:

$$f_i(\underline{u}, \underline{\sigma}, t) = -\frac{\partial V(\underline{u}, \underline{\sigma}, t)}{\partial u_i} . \tag{1.13}$$

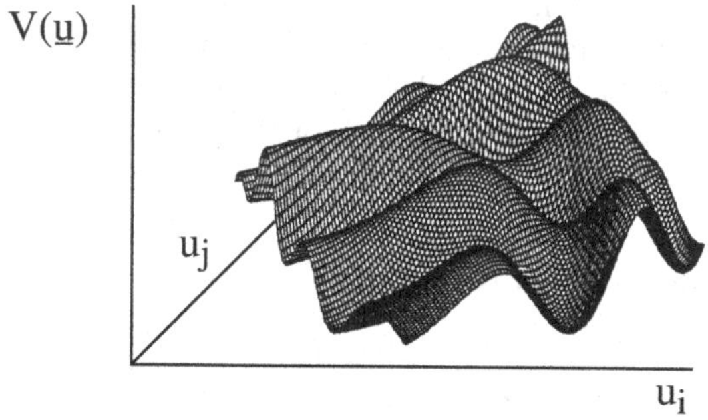

Fig. 1.6. Sketch of a high-dimensional landscape $V(\underline{u})$ that is shown for only two variables u_i, u_j

This idea, however, involves certain major restrictions, e.g., regarding the explicit time dependence of V. In particular, (1.13) is valid only for *gradient systems* where $\partial f_i/\partial u_j = \partial f_j/\partial u_i$ holds – a condition that is rarely fulfilled in high-dimensional systems.

To keep the graphic idea of a "landscape" but to ease the strict conditions on the existence of a gradient system, in this book we will adopt the idea of an *adaptive landscape*. Every action of each agent is assumed to change the state of the adaptive landscape – either *locally* or *globally*, depending on the model under consideration. On the other hand, changes in the landscape may affect the actions of other agents near or far distant. In this way, a *nonlinear feedback* occurs (see Fig. 1.7) that is important for all processes of structure formation and self-organization.

One of the major advances of the adaptive landscape concept is that it may allow local or global *couplings* between agents. Also *gradual, time delayed interactions* between agents can be realized, i.e., instead of simply

[17] This derivation is also based on the assumption that the possible change per time step, $u_i' - u_i$, is sufficiently small.

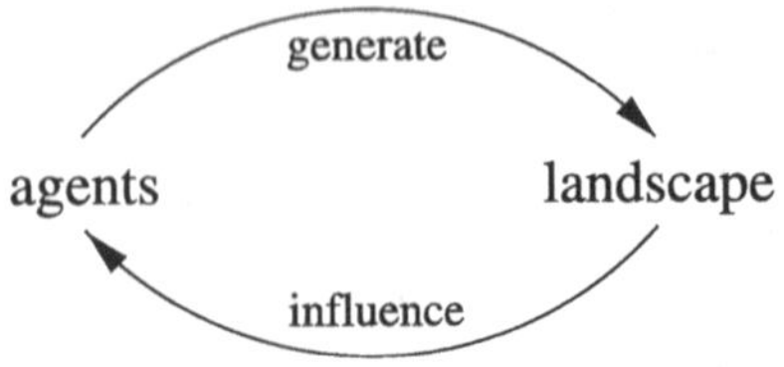

Fig. 1.7. Nonlinear feedback between agents and the adaptive landscape that is created by them but in turn also influences their further "behavior"

assuming nearest neighbor interactions (as in most CA) or interactions via a *mean field* equal for all agents (as in many Ising-like systems), the coupling of different local changes on a larger spatial or temporal scale can be taken into account, in this way modeling, e.g., hierarchical interactions. Further, an *eigendynamics* of the landscape may be taken into account that may result, for example, from cyclic changes (day and night cycles in biological systems or seasonal cycles in population dynamics), or time-dependent in- and outfluxes of resources, etc.

1.2.2 Interaction as Communication

From a very general perspective, the adaptive landscape basically serves as an *interaction medium* for the agents. This also reminds us of certain situations in physics, where entities interact with each other through a field they share. For example, electrons interact through the electromagnetic field created by them, or mass bodies interact through a gravitational field. Loosely speaking, these entities use the physical field to "communicate" to other entities. Similarly, one can imagine the adaptive landscape as a *medium for indirect communication* among the agents. This means that each action of the agents may generate certain *information* (in terms of changes of the adaptive landscape) that may propagate through the system via the medium, influencing the actions of other agents. We note that such an idea has certain similarities to the *communication via tagboards* discussed in Sect. 1.1.5. Therefore, the FTA design paradigm (see Fig. 1.5) may serve as an appropriate computer platform for simulating agents interacting via an adaptive landscape.

Using the analogy between interaction and communication, we will in this book mostly identify the adaptive landscape with a *spatiotemporal communication field* that is created by information resulting from the *local* actions of each agent. Because agents with different internal states θ may create different kinds of information, this field is assumed to be a scalar *multicomponent spatiotemporal field* $h_\theta(\boldsymbol{r}, t)$. To model the *dissemination* of information in the system, we may in a simple approach assume a diffusion-like process. Further, to consider a *finite lifetime* for the information generated, we may assume exponential decay of the information available. Then, the dynamics of the spatiotemporal communication field may obey the following *reaction–*

diffusion equation:

$$\frac{\partial}{\partial t} h_\theta(\boldsymbol{r},t) = \sum_{i=1}^{N} s_i \, \delta_{\theta,\theta_i} \, \delta(\boldsymbol{r}-\boldsymbol{r}_i) \; - \; k_\theta h_\theta(\boldsymbol{r},t) \; + \; D_\theta \Delta h_\theta(\boldsymbol{r},t). \qquad (1.14)$$

In (1.14), every agent contributes permanently to the communication field with its personal "strength" of information s_i. Here, δ_{θ,θ_i} is the Kronecker delta indicating that the agents contribute only to the field component that matches their internal parameter θ_i. $\delta(\boldsymbol{r}-\boldsymbol{r}_i)$ means Dirac's delta function used for continuous variables, which indicates that agents contribute to the field only at their current position $\boldsymbol{r}_i$. *Information* generated this way has a certain lifetime $1/k_\theta$; further, it can spread throughout the system where D_θ represents the diffusion constant for information exchange. Note that the parameters s_i, k_θ, D_θ describing the communication field, do not necessarily have to be the same for different agents i or internal parameters θ.

Further, we wish to emphasize the fact that the *effort* of an agent to establish the communication field, denoted by s_i, is also related to the internal energy depot e_i, (1.7), (1.8) of the agent. This shall reflect that every change in the agent's environment, such as generation of information or changes of the adaptive landscape in general, has some *costs* and therefore needs to be considered in the "energetic" balance equation. Despite this, in some of the applications discussed later, the energy for establishing the field $h_\theta(\boldsymbol{r},t)$ is simply taken for granted, in this way neglecting any influence on the agent's energy depot. But we wish to point out that this is simply an approximation.

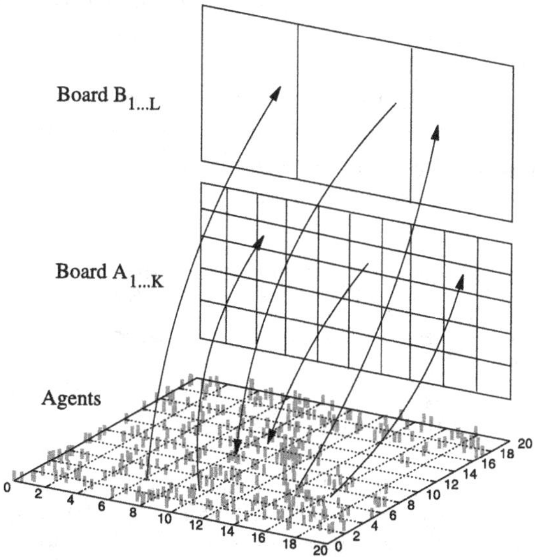

Fig. 1.8. Indirect interaction of spatially distributed agents via two different communication fields that can be imagined as arrays of blackboards

The assumption of reaction–diffusion dynamics in (1.14) purposefully stretches analogies to physicochemical dynamics; however, we note that communication in a variety of systems can be successfully described in this way – for instance, chemical communication among ants (Sect. 5.4), myxobacteria (Sect. 5.3), or larvae (Sect. 3.3). The dissemination of information in social systems can also be reasonably modeled by a communication field (Sect. 10.2).

For the computer simulations discussed in the following chapters, the spatiotemporal communication field of (1.14) needs to be *discretized* in space to describe the diffusion process. This will be discussed in more detail in Sect. 3.1.4, but we would like to point to the fact that it then can be regarded as a *spatial array of cells* each of which acts like a *blackboard* (or a tagboard) in the sense described for the FTA architecture in Sect. 1.1.5 (see Fig. 1.8). Agents locally generate information that is stored in the arrays of blackboards that can thus be regarded as a kind of *external memory*; further they are able to read and to process the information stored in the arrays, provided that they have access to it. In the following, we want to characterize this interaction process in terms of an information-theoretic description. To this end, we need to distinguish among three different kinds of information: functional, structural, and pragmatic information [121, 445] (see also Fig. 1.9).

Functional information denotes the capability of the agent to process external information (data) received. It can be regarded as an *algorithm* specific to the agent which may also depend on the agent's internal degrees of freedom. This algorithm is applied to "data" in a very general sense, which can also be denoted as *structural information* because it is closely related to the (physical) structure of the system. DNA is an example of structural information in a biological context. As a complex *structure*, it contains a mass of (structural) information in coded form that can be selectively activated depending on different circumstances. Another example of structural information is a book or a user manual, written in letters of a particular alphabet. Most important, structural information may be *stored* in external storage media (from computer discs, genome databases, to public libraries), and it may be *disseminated* in the system via exchange processes. Thanks to the

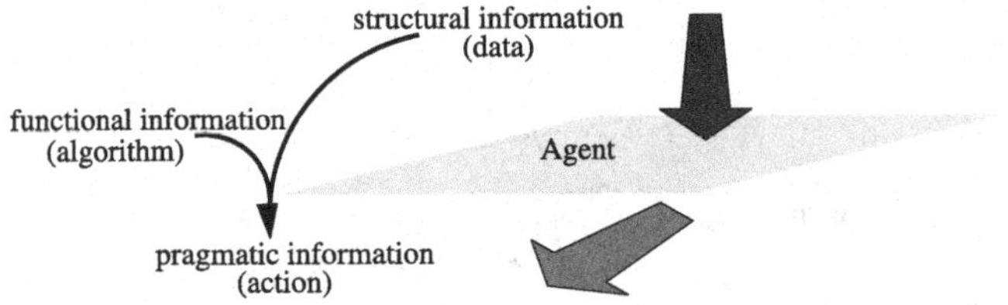

Fig. 1.9. Pragmatic information allows an agent to operate in a specific context. It emerges from an interplay of structural information (the data received) and functional information (the algorithm to process the data internally) and is thus specific to each agent

second law of thermodynamics, structural information may have only a *finite lifetime* depending on the storage medium.

Structural information is "meaningless" in the sense that it does not contain *semantic* aspects but only *syntactic* aspects. Hence, the content of structural information can be analyzed, for instance, by different physical measures (e.g., conditional or dynamic entropies, transformation, etc.) [121]. Functional information, on the other hand, is related to the semantic aspects of information; it reflects the contextual relations of the agent. It is the purpose of functional information to *activate* and to *interpret* the existing structural information with respect to the agent, in this way creating a new form of information denoted as *pragmatic information.* This means a type of operation-relevant information which allows the agent to *act.* In the examples above, cells can, by specific "functional" equipment, extract different (pragmatic) information from the genetic code, which then allows them to evolve differently, e.g., in morphogenesis. In the example of the user manual, the reader (agent) can, by specific functional information, i.e., an algorithm to process the letters, extract useful, "pragmatic" information from the text, which only allows him/her to act accordingly. If the functional information (algorithm) does not match the structural information (data), i.e., if the manual is written in Chinese and the reader can process only Latin letters, then pragmatic information will not emerge from this process, even though the structural and the functional information are still there. With respect to the term pragmatic information, we can express this relation as follows: *functional information transforms structural into pragmatic information* [445].

To characterize the Brownian agent model in terms of a generalized communication approach based on the exchange of information, we have to identify the kind of functional and structural information used in the system and then have to investigate which kind of pragmatic information may emerge from this interplay. As denoted above, functional information shall be a (simple) algorithm which can be steadily repeated by each agent. For example, we may assume that during every time step the agent is able (i) to read data, (ii) to write data, and (iii) to process the data currently read (e.g., to compare their value).

The data read and written are *structural* information which is stored on *blackboards* external to the agent. Because of this, the communication among the agents may be regarded as *indirect communication*; however, a medium always seems to be involved, even in "direct" oral communication.

The emergence of *pragmatic* information for a specific agent will, of course, depend (i) on the functional information, i.e., the "algorithm" to process specific structural information, and (ii) on the availability of this structural information, i.e., the access to the respective blackboard at a particular time or place. For different applications, we may consider various possibilities to restrict access to the blackboard both in space and time. In this way, communication between agents can be modeled as *local* or *global.* On the other hand, we may also assume that there are different *spatially distributed* blackboards in the system, modeling a spatially heterogeneous distribution of (structural) information (see Fig. 1.8).

Additionally, we may consider *exchange processes* between different blackboards, for instance, for the reconciliation of data. The data originally stored on a particular blackboard may then also propagate to other blackboards in the course of time. In this way, we observe a rather complex interaction dynamics determined by two quite different space- and time-dependent processes: (i) changes of blackboards caused by the *agents* and (ii) changes of blackboards caused by *eigendynamics* of the structural information. The latter may also involve dynamic elements such as (i) finite lifetime of the data stored which models the existence of a *memory*, (ii) an *exchange of data* in the system at a *finite* velocity, and (iii) the spatial heterogeneity of structural information available.

The concept of the spatiotemporal communication field introduced in (1.14) is just one specific realization of the general communication features outlined above. It models the eigendynamics of an array of spatially distributed blackboards. This communication field is external to the agents but is created by the various types of data (structural information) consecutively produced by them. The dissemination of the different kinds of structural information is spatially heterogeneous and time-dependent. However, it may affect other agents in different regions of the system, provided that they notice this information and can extract some meaning from it, which then may influence their own decisions, output, etc.

As this book will reflect, this concept has proved its applicability in a variety of models. We want to mention only two examples here. In a model of economic agglomeration (see Chap. 9), the communication field has been regarded as a spatially heterogeneous, time-dependent *wage field*, from which agents can extract meaningful information for their migration decisions. The model then describes the emergence of economic centers from a homogeneous distribution of productivity [447]. Another example deals with spatial self-organization in urban growth (see Sect. 8.3). To find suitable places for aggregation, "growth units" (specific urban agents) may use the information of an urban attraction field, which has been created by the existing urban aggregation, and this provides indirect communication between different types of urban agents [463].

1.2.3 A Short Survey of the Book

The Brownian agent approach in Sect. 1.2.1 is based on two key elements that play a major conceptual role in the outline of this book: (i) the existence of internal degrees of freedom, in particular, an internal energy depot, that may influence or change the behavior of the agent and (ii) the indirect interaction of agents via an adaptive landscape that is established and changed by them and then influences their further behavior. These two features each play a different role in the various applications discussed in the following chapters; therefore, at this point, it will be useful to sketch the outline of the book first.

It is the aim of this book to formalize the concept of Brownian agents by methods from statistical physics. Therefore, we first need to provide some basic knowledge about the main physical equations and relations used in this book. As already pointed out, the Brownian agent concept is inspired by Langevin's description of the motion of Brownian particles. This approach is outlined in detail in Sect. 1.3. On the level of the "individual" Brownian particle, we have the Langevin equation that considers the influence of different forces on the particle, namely, dissipative force resulting from friction, a force resulting from the existence of an (external) potential, and eventually a stochastic force resulting from random impacts of the surrounding molecules. In terms of a probabilistic description, the motion of Brownian particles can be described by a master equation for probability density, from which mean value equations for the macroscopic densities can be derived. In particular, we introduce the Fokker–Planck equation for probability density and discuss its stationary solution, which will later be needed as a kind of "reference state" to compare the macroscopic equations for Brownian agents with their physical counterparts.

Starting with Chap. 2, we focus on different features of Brownian agents. In general, Brownian agents can take up "energy" (or resources) from the environment, store it, and use it for different activities, such as accelerated motion, changes in their environment, or signal-response behavior. As pointed out in Sect. 1.1.2, in the book we follow a strategy to add more and more complexity gradually to the agents. So, the purely physical model of *passive* Brownian particles introduced in Sect. 1.3 will serve as a starting point. As a first step toward more complex phenomena, we consider particles with an *internal energy depot*, which is an additional *internal degree of freedom* with continuous values. The energy depot may allow a particle to perform *active motion* based on the consumption of energy; thus we may call Brownian particles with an internal energy depot *active particles*, if other agent features, such as further internal degrees of freedom or response to an adaptive landscape can be neglected. The *active particle* is a rather "simple" type of agent that does not deny its origin from the physical world, and therefore has been also denoted as *particle agent* [124].

The investigation of the *energetic aspects* of active motion is the major focus in Chap. 2. Provided a supercritical supply of energy, active particles, in addition to the passive (low velocity) mode of motion, can move in a *high velocity mode.* This active mode of motion is characterized by a *non-Maxwellian velocity distribution* and the possibility of limit cycle motion. Further, we could find new and interesting forms of complex motion for pumped Brownian particles in different types of external potentials. In one-dimensional systems, we could show the possibility of uphill motion of particles that provided supercritical energetic conditions. For an ensemble of active particles in a piecewise linear potential (ratchet potential), this results in a *directed net current* which results from undirected Brownian motion. The different critical parameters, which determine this occurrence, are investigated analytically and by computer simulations. In two-dimensional systems, we find, for instance, *deterministic chaotic motion* of active particles in the presence of obstacles or an *intermittent type* of motion in the presence of *localized energy sources.* Further, in Sect. 2.4. we also investigate *swarming* of active particles, i.e., coherent motion of an ensemble of particles in the presence of an external or interaction potential. We show how ordered collective motion emerges from a compromise of two counteracting influences, *spatial dispersion* resulting from the active motion of individual particles (driven by the energy pumping) and *spatial concentration* driven by the mutual interaction of active particles.

From Chap. 3 on, we are mainly interested in internal degrees of freedom that can be used to determine various types of "behavior" of Brownian agents. They are introduced as *discrete internal parameters* that may also change in the course of time, in this way alternating, e.g., the types of *response* of Brownian agents to external fields. Additionally, the internal parameters may also determine specific actions of the agents, namely, the generation of an *adaptive landscape* or a *self-consistent potential field*, respectively. Nonlinear feedback between the landscape which is, on one hand, generated by agents, and on the other hand, influences their further action, results in an *indirect interaction* of Brownian agents, as already mentioned in Sect. 1.2.2.

The general outline of interacting Brownian agents presented in Sect. 1.2 is specified for a number of different applications in Chaps. 3–6, 8–10. In particular, we assume that Brownian agents, depending on their internal parameter, can generate different *components* of the field, which then compose an *effective field*, which in turn influences the agents in a specific way. The dynamics of each field component is described by a reaction–diffusion equation that considers the creation of the field by agents and further diffusion and decay of the field. Moreover, we also consider dynamics for the internal parameter of agents that can change due to external or internal influences. In this way, we achieve a rather complex interaction between agents by the different field components, which serves as the base for a quite general model of *interactive structure formation.*

In Sect. 3.2, we investigate the simple case that the influence of internal parameters can be neglected, and only a one-component (chemical) field is generated, which locally attracts the agents. We find that nonlinear feedback between the particles and the field describes an *aggregation process* that displays significant analogies to biological aggregation phenomena, presented in Sect. 3.2.6. We can describe this aggregation process by an effective diffusion coefficient which can be locally negative, as a result of the generated field. On the other hand, we can also derive a selection equation similar to the Fisher–Eigen equation, which describes the competition between different aggregates.

We find that the possible range of parameters for the aggregation process is bound by a critical temperature, similar to first-order phase transitions in thermodynamics.

In Sect. 3.3, the model of Brownian agents is mainly used as a most efficient *simulation method* for structure formation processes in physicochemical systems. The basic idea of the simulations is the computation of a large number of coupled Langevin equations instead of the integration of the related macroscopic reaction–diffusion equations. In this way, we can simulate a variety of patterns that occur in physicochemical systems, for instance, periodic precipitation, spiral waves, traveling spots, or traveling waves in a temperature field. The results demonstrate the advantage of our particle-based approach, which provides a fast and stable simulation algorithm, even in the presence of large gradients.

In Chap. 4, we use the internal parameter by considering two different internal states of Brownian agents, which result in generating two different field components. Each of the components, which are not assumed to diffuse now, affects only those particles, which currently do not contribute to it. Further, we consider that agents can switch between the two internal states when they hit some nodes distributed on a surface. We find that nonlinear feedback between agents and the two-component field results in the self-organization of a *network* between the different nodes. The agents can connect almost all nodes, and the *self-assembled* network is characterized both by stability and *adaptivity*. We investigate different features of the network, such as connectivity or critical parameters for establishing links. Eventually, we suggest the construction of a *dynamic switch*, which switches between ON and OFF by connecting or disconnecting two nodes dependent on external influences. We show that the switch responds both reliably and fast in a certain range of parameters.

In Chap. 5, we use basic features of the above model to describe the formation of tracks and trail systems in biological systems. Different from other chapters, we prefer here a discrete description in terms of *active walkers*. The formation of tracks can be described as a *reinforced biased random walk*, where the reinforcement is due to nonlinear feedback between the walkers and a nondiffusing chemical field, again. The bias of the walk can be interpreted as a certain kind of short-term memory, which can also be described by the internal parameter. We further include that some biological species respond to chemical gradients within a spatial angle of perception, determined by their current moving direction, which has an effect on the resulting track patterns. The model outlined is applied to simulate the movement and aggregation of *Myxobacteria*, as presented in Sect. 5.3. In Sect. 5.4, we extend the model again to consider different chemical fields and different types of response of active walkers, dependent on their internal parameter. This allows us to simulate specific patterns of *trunk trail formation* by ants, which display remarkable similarity to biological observations, as described in Sect. 5.4.1.

Noteworthy, these foraging patterns are created by reactive agents, which, different from biological creatures, do not have capabilities of information storage or navigation.

Chapter 6, which is based on a continuous description of Brownian agents again, presents an application of the model to pedestrians. We explain that pedestrians can also be described by an overdamped Langevin equation, where the deterministic influences now result from a specific *social force*. This force considers features such as desired speed, desired destinations, and distance from borders and other pedestrians. The simulated trajectory patterns of pedestrian motion display striking analogies to "real" pedestrians, even in the presence of obstacles. In Sect. 6.2, we extend the model to consider also the formation of *human trails*, which are created by pedestrians and have a specific lifetime. The moving pedestrians have to find a compromise between existing trails and the direction of their destination, which is expressed by a specific *orientation relation*. The simulations show the occurrence of *minimized detour trail systems* commonly used by an ensemble of pedestrians, which can also be described by macroscopic trail equations.

Chap. 8 investigates the formation of urban aggregates. An analysis of existing urban aggregates which is restricted to *structural features* such as the spatial distribution of the built-up area, shows that the mass distribution of urban clusters displays *fractal properties*. Further, it is shown that the *rank–size distribution* of separated clusters that form large-scale urban agglomerations, approaches a Pareto–Zipf distribution in the course of historic evolution. This empirical result could be reproduced by computer simulations based on a master equation for urban growth, which is then used for a simple forecast model of spatial urban growth. We further show that the shift of growth zones, which is a characteristic feature of urban evolution, can be modeled in the framework of Brownian agents. Here, we consider agents in two different internal states, which create and respond to an attraction field characterizing the urban aggregate. Additionally, the depletion of free space is considered.

In Chap. 9, the model of Brownian agents is used to model economic aggregation. Here, the agents represent either employed workers who are immobile but generate a wage field as the result of their work, or unemployed workers who are mobile and respond to local gradients in the wage field to reach productive locations. Further, employed agents can become unemployed and vice versa, depending on the local economic situation. The economic assumptions are summarized in a *production function* that determines the transition rates and the wage field. The analytical investigations and, in particular, the computer simulations show a local concentration of employed and unemployed agents in different regions from a homogeneous distribution of agents. The agglomeration is due to a competition process in which *major economic centers* appear at the cost of numerous smaller centers that disappear. Noteworthy, there is a stable coexistence of the major economic regions which have a certain critical distance from each other. This is in accordance with predictions of the central-place theory. On the other hand, these centers still have stochastic eigendynamics and exist in a quasistationary nonequilibrium.

In Chap. 10, we describe another type of aggregation in social systems, namely, the formation of collective opinions. Here, the internal parameter represents different opinions of individuals that can be changed depending on external or internal influcences. Further, we consider migration of the individuals, which is described by a Langevin equation. The indirect interaction between individuals with the same or different opinions is described by a communication field, which has the same features as the basic model of Brownian agents. We show that under certain *critical conditions* expressed in terms of a critical population size or a critical "social" temperature, the community necessarily separates into minorities and majorities, which is again in close analogy to first-order phase transitions in thermodynamics. The influence of external support may change the coexistence between majorities and minorities. We also obtain certain critical conditions for a spatial separation of subpopulations with different opinions. In addition to indirect interaction mediated by the communication field, we have also considered a model of *direct* interactions. Depending on external conditions and the type of interaction, we find *multiple steady states* for the dynamic equations describing the social system and the occurrence of specific spatiotemporal patterns in the distribution of individuals.

Chapters 3–6 and 8–10 employ the same type of indirect interaction between Brownian agents by a (multicomponent) self-consistent field that plays the role of an *order parameter* in the system. In Chap. 7, on the other hand, we consider another type of interaction between agents. Here, it is assumed that agents move in a potential, which they *cannot* change. This potential should represent the search space of an optimization problem, and consequently Brownian agents are considered searchers that move through the state space, following local gradients in the potential. The Langevin dynamics ensures that the probability that the searchers accumulate in the minima of the potential increases in the course of time. If we further adjust the influence of the stochastic force properly by tuning the individual "temperature" of the searchers due to global or local rules, this results in an effective search algorithm.

Another way to describe the diffusion in the state space is *mutation*, which means the transfer of the searcher to a different point in the state space. An *additional coupling* can be introduced by considering *reproduction* of the searchers depending on their *fitness*, which is assumed to be the inverse of the potential minimum found. A combination of both reproduction and mutation leads to the mixed Boltzmann–Darwinian strategy used in evolutionary optimization, which is applied here to optimize a road network among a set of nodes. This is an example of a frustrated optimization problem because the optimization function considers two requirements which cannot be satisfied at the same time. Computer simulations show that evolutionary search can find optimized solutions, which represent a compromise between contradicting requirements. An investigation of the asymptotic results further allows us to characterize the search problem by the density of states.

The different examples in this book indicate that the model of Brownian agents is a versatile and tractable tool for describing different aspects of active motion and self-organization in complex systems. We would like to mention that these examples give just a selection of possible applications of the model. Noteworthy, some potential fields of interest had to be neglected in this book, namely, the description of *granular matter* by active Brownian particles [411], applications in *traffic models* [211], *geophysical problems* such as *erosion* [500, 501] and *river formation*, the formation of *sediments* and mineral deposits [280], and the modeling of *global systems*, where entities indirectly interact via climate impacts [60, 426].

Let us finally point out some common features in the seemingly different examples and give some generalizations. The central topic in particular with respect to the model of interacting Brownian agents is the occurrence of some kind of collective "behavior" among agents. The dominant phenomenon described in this book is *aggregation* that has been investigated in relation to biological aggregation (Sects. 3.2, 5.3), precipitation (Sect. 3.3), urban aggregation (Sect. 8.3), economic agglomeration (Sect. 9.2), and the formation of social groups with uniform opinions (Sect. 10.2). The ordered motion of agents either as swarms (Sect. 2.4) or along definite tracks or trails can also be considered a certain kind of *delocalized* aggregation, as discussed for network formation (Sect. 4.1) and the formation of tracks (Sect. 5.2) and trails by biological species (Sect. 5.4) and pedestrians (Sect. 6.2). A specific form of aggregation is also the agglomeration of searchers in the minima of a potential during the search for optima in high-dimensional state space (Sect. 7.1).

Common to all examples is the *basic dynamics* that includes the movement of Brownian agents with respect to deterministic influences and random forces, as described by the Langevin equation. This dynamics, in many cases, also includes the generation of an adaptive landscape that has an eigendynamics but also influences the further "behavior" of agents, either regarding their motion or regarding their transition into a different internal state. With these common elements in the dynamics, Brownian agents are very flexible "actors" that in this book represent chemical molecules or biological species such as ants, myxobacteria, cells, pedestrians, individuals with a certain opinion, employed or unemployed workers in economic systems, and also rather abstract entities, such as growth units in urban growth or just searchers in optimization space.

With respect to the modeling of complex systems, the Brownian agent approach provides a level of description, which, on one hand, considers specific features of the system under consideration, but, on the other hand, is not flooded with microscopic details. However, we should also bear in mind that this approach is based on certain reductions regarding the system elements and their interactions that need to be carefully understood. With these reservations in mind, the Brownian agent approach may be of value in the broader context of the agent-based models used today.

1.3 Brownian Motion

1.3.1 Observations

The aim of this section is to provide some basic knowledge of the main physical equations of Brownian motion for further use and reference within this book. Brownian motion denotes the erratic motion of a small, but larger than molecular, particle in a surrounding medium, e.g., a gas or a liquid. This erratic motion results from random impacts between the atoms or molecules of the medium and the (Brownian) particle, which cause changes in its direction and velocity $\boldsymbol{v}$.

The motion of the particle is named after the Britisch botanist Robert Brown (1773–1858), who in 1827 discovered the erratic motion of small pollen grains immersed in a liquid. Brown was not the first one who observed such a motion with a microscope. The Dutch Anton van Leeuwenhoek (1632–1723), who first discovered microorganisms with a simple microscope, already knew about the typical erratic motion; however, he considered it a feature of living entities. In 1785, the Dutch physician Jan Ingenhousz (1730–1799) also reported the erratic motion of inorganic material dispersed in a liquid, i.e., powdered charcoal floating on alcohol surfaces, but this was not known to the non-Dutch-speaking world.

Brownian motion itself is not a molecular motion, but it allows indirect conclusions about the molecular motion that drives only the motion of a Brownian particle. In this respect, the investigation of Brownian motion became of interest to physicists as an indirect method for determining molecular constants, such as the Avogadro number.

There are different experimental methods. One of them is based on measuring the displacement of a Brownian particle [46]. If $\boldsymbol{x}(t)$ denotes the $\boldsymbol{x}$ coordinate of a Brownian particle at time t, and the position of the particle is measured after time intervals 0, t_0, $2t_0$, ..., $\nu\, t_0$, ..., $N\, t_0$, then the mean squared displacement of Brownian particle in the x direction during the time interval t_0 is simply given by

$$\langle \Delta x^2(t_0) \rangle = \frac{1}{N} \sum_{\nu=1}^{N} \Big(\boldsymbol{x}_\nu - \boldsymbol{x}_{\nu-1} \Big)^2 . \tag{1.15}$$

For experimental observation, a suspension (Brownian particles in a liquid) is placed between two narrow glass plates, so that the system can be treated as two-dimensional in space. Under a microscope, the position of a particular

Brownian particle can be observed and recorded in constant time intervals t_0. Two examples are shown in Fig. 1.10. The graphs allow us to calculate the mean displacement in the x direction, (1.15), provided that N is large enough.

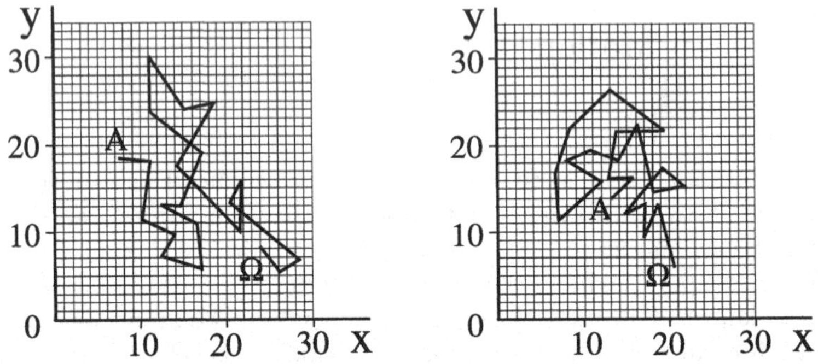

Fig. 1.10. Two-dimensional trajectories of a Brownian particle obtained from an experiment done by the author during his undergraduate courses (12/16/1980). The position of the Brownian particle (radius 0.4 μm) was documented on a millimeter grid in time intervals $t_0 = 30$ seconds. The diffusion constant was determined as 6×10^{-7} mm^2/s

Theoretical investigations of Brownian motion by Einstein [132, 133] in 1905 showed that the mean squared displacement of a Brownian particle is given by

$$\left\langle \left[\boldsymbol{x}(t) - \boldsymbol{x}(0)\right]^2 \right\rangle = \langle \Delta x^2(t) \rangle = 2D\,t\,. \tag{1.16}$$

Here, D is the diffusion constant, which also appears as a macroscopic quantity in the transport equation (in the x direction):

$$\boldsymbol{j} = -D\,\frac{\partial n}{\partial \boldsymbol{x}}\,, \tag{1.17}$$

where $\boldsymbol{j}$ is the particle flux resulting from the gradient of a spatial inhomogeneous particle density $n(\boldsymbol{x}, t)$. Conservation of the particle number yields the continuity equation, div$\boldsymbol{j} + \partial n/\partial t = 0$ or, for the one-dimensional case:

$$\frac{\partial n(\boldsymbol{x}, t)}{\partial t} = D\,\frac{\partial^2 n}{\partial x^2}\,. \tag{1.18}$$

So, the diffusion coefficient D has an ambiguous meaning: on one hand, it is related to a *one-particle* property, the mean displacement of one Brownian particle, and describes the strength of the stochastic force, as explained below. On the other hand, D is related to the *many-particle* or macroscopic properties of transport. Hence, (1.16) has the important consequence that the measurement of a one-particle property allows determining a *many-particle* property, such as the diffusion constant. From the perspective of statistical physics, this implies that the averaged statistical properties of a single particle are equivalent to the statistical properties of an ensemble of particles. In this particular case, the probability density $p(\boldsymbol{x}, t)$ to find a Brownian particle at location $\boldsymbol{x}$ at time t follows the same dynamic equation as the distribution function for an ensemble of particles, i.e., density $n(\boldsymbol{x}, t)$, (1.18).

Einstein proved that the above equivalence holds for the following relation between the diffusion constant D and the friction coefficient γ of a Brownian particle:

$$D = \frac{k_{\mathrm{B}}T}{\gamma m} = \frac{k_{\mathrm{B}}T}{\gamma_0}\,, \tag{1.19}$$

where m is the mass of the particle, T is the temperature, and k_{B} is the Boltzmann constant. For our convenience, we will mainly use the abbreviation $\gamma_0 = \gamma m$ (kg/s) for the friction coefficient.

We will come back to the relations between one-particle properties and macroscopic properties later in Sect. 3.1.3, when we discuss our setup for computer simulations. Here, (1.16) is used to determine the Avogadro number $N_{\mathrm{A}} = R_0/k_{\mathrm{B}}$; R_0 is the universal gas constant. Due to Stokes, the friction of a small sphere moving in a liquid can be expressed as

$$\gamma_0 = 6\pi\eta r\,, \tag{1.20}$$

where η is the dynamic viscosity of the liquid and r is the radius of a small sphere. Applying these considerations to the motion of a Brownian particle, we find from (1.16), (1.19), and (1.20) the relation between the mean squared displacement of one Brownian particle and the Avogadro number, N_{A} [183]:

$$\langle \Delta x^2(t)\rangle = \frac{R_0 T}{3\pi\eta r\, N_{\mathrm{A}}}\, t\,. \tag{1.21}$$

Equation (1.21) is only one example for exploiting Brownian motion to determine molecular constants. Another method is based on the barometric equation (1.22), which describes the equilibrium density of Brownian particles in a suspension as a function of the height z:

$$n(z) = n(0)\, \exp\left(-\frac{\mu g z}{k_{\mathrm{B}}T}\right), \tag{1.22}$$

where g is Earth's gravitation and $\mu = m(1-\varrho_l/\varrho_p)$ is the relative mass of the Brownian particles, which already considers buoyancy. Here, ϱ_l is the mass density of the liquid (e.g., water), and ϱ_p is the mass density of the immersed particles. Following the explanation by Einstein [132, 133], the distribution $n(z)$ in (1.22) is the result of a compromise between the diffusion of Brownian particles, on one hand, and gravitation, on the other hand. By counting the number of Brownian particles in a thin, horizontal plane of area A at height z, one can also determine the Avogadro number $N_{\rm A} = R_0/k_{\rm B}$. In 1908, the later Nobel laureate Jean Baptiste Perrin (1870–1942) was very successful in applying this method, while investigating Brownian motion in water [387].

According to the equipartition law, the average kinetic energy of a particle of mass m in thermodynamic equilibrium is given by

$$\frac{1}{2}\, m\, v_x^2 = \frac{1}{2}\, k_{\rm B}\, T\,, \tag{1.23}$$

where $\boldsymbol{v}_x$ is the velocity of the particle in the x direction. Equation (1.23) applies to the molecules of the liquid and also to the Brownian particles. The smaller the mass of the particle, the larger the velocity and vice versa. Hence, the observation technique basically determines whether the motion of a "Brownian" particle can be observed. For experiments with common microscopes, particles with a radius of about 0.5 μm are useful.

But Brownian motion is not only observable from particles; it also has consequences for constructing very sensitive measuring instruments. A small mirror of a size of about 2 mm^2, which is, for instance, part of a galvanometer, also displays random torsions caused by molecular impacts. In 1931, Kappler [266] was able to measure the torsional angle resulting from *thermal noise* and to determine the Avogadro number.

So, we have, on one hand, the benefits of Brownian motion for indirect measurement of molecular constants, and on the other hand, the random influences of Brownian motion set limits to the accuracy of experimental setups on the microscale. This ambiguous picture described the interest in Brownian particles and Brownian motion for quite a long time. But recently, the situation has changed a lot. This is in part due to more refined experimental methods which allow us to suppress the influence of thermal noise, e.g., in experiments at low temperatures.

On the other hand, progress in statistical physics has opened the door to the field of stochastic physics, where random events, fluctuation, and dissipation have an interest of their own. They are no longer considered undesired *perturbations* of equilibrium systems but play a *constructive* role in ordering processes, in structure formation, and in nonequilibrium motion, to mention a few areas [342, 430]. This book will use some of the basic concepts developed for the stochastic description of Brownian particles, and it will propose some novel ideas to extend these concepts to a flexible tool for active motion and interactive structure formation.

1.3.2 Langevin Equation of Brownian Motion

In the previous section, we already mentioned the important contribution of Albert Einstein (1875–1955) to understanding Brownian motion. His early papers, together with the pioneering work of Marian v. Smoluchowski (1872–1917) and Paul Langevin (1872–1946), laid the foundations for stochastic approaches in physics at the turn of the twentieth century.

Here, we are particularly interested in Langevin's ingenious explanation of Brownian motion because the resulting Langevin equation will play an important role thoughout this book. Due to the Newtonian ansatz, the equation of motion of a particle is given by

$$m\dot{\boldsymbol{v}} = \boldsymbol{F}(t), \tag{1.24}$$

where $\boldsymbol{F}(t)$ is the sum of all forces acting on a particle. Langevin considers two different forces:

$$\boldsymbol{F}(t) = -\gamma_0\,\boldsymbol{v} + \boldsymbol{F}^{\rm stoch}(t) = m\Big[-\gamma\,\boldsymbol{v} + \mathcal{K}(t)\Big]. \tag{1.25}$$

The first term of the r.h.s. of (1.25) describes a velocity-dependent *dissipative* force resulting from the friction γ_0, and the second term, $\boldsymbol{F}^{\rm stoch}(t)$, describes a time-dependent *stochastic* force that results from random interactions between a Brownian particle and the molecules of the medium. Since these random events are not correlated and compensate, on average, the time average over the stochastic influences should vanish, i.e., $\mathcal{K}(t)$ is a stochastic source with strength S and a δ-correlated time dependence:

$$\langle\mathcal{K}(t)\rangle = 0\,;\ \langle\mathcal{K}(t)\mathcal{K}(t')\rangle = 2S\,\delta(t-t'). \tag{1.26}$$

Due to the *fluctuation-dissipation theorem*, the loss of energy resulting from friction and the gain of energy resulting from the stochastic force are compensated for, on average, and S can be expressed as

$$S = \gamma\frac{k_{\rm B}T}{m}, \tag{1.27}$$

where T is the temperature and $k_{\rm B}$ is the Boltzmann constant. With respect to (1.24)–(1.27), the motion of a Brownian particle i with space coordinate $\boldsymbol{x}_i$ and velcocity $\boldsymbol{v}_i$ can be described by the *stochastic differential equations*:

$$\frac{d\boldsymbol{x}_i}{dt} = \boldsymbol{v}_i\ , \qquad m\frac{d\boldsymbol{v}_i}{dt} = -\gamma_0\boldsymbol{v}_i + \sqrt{2\,k_{\rm B}T\gamma_0}\;\boldsymbol{\xi}_i(t), \tag{1.28}$$

or, after division by mass m,

$$\frac{d\boldsymbol{x}_i}{dt} = \boldsymbol{v}_i\ , \qquad \frac{d\boldsymbol{v}_i}{dt} = -\gamma\boldsymbol{v}_i + \sqrt{2\,S}\;\boldsymbol{\xi}_i(t), \tag{1.29}$$

where the random function $\boldsymbol{\xi}_i(t)$ is assumed to be Gaussian white noise:

$$\langle\boldsymbol{\xi}_i(t)\rangle = 0\ , \quad \langle\boldsymbol{\xi}_i(t)\,\boldsymbol{\xi}_j(t')\rangle = \delta_{ij}\,\delta(t-t'). \tag{1.30}$$

Multiplying the Langevin equation by $\boldsymbol{x}$, we find with $d(\boldsymbol{x}\dot{\boldsymbol{x}})/dt = \dot{x}^2 + \boldsymbol{x}\ddot{\boldsymbol{x}}$,

$$m\left[\frac{d}{dt}(\boldsymbol{x}\dot{\boldsymbol{x}}) - \dot{x}^2\right] = -\gamma_0\,\boldsymbol{x}\dot{\boldsymbol{x}} + \boldsymbol{x}\boldsymbol{F}^{\text{stoch}}(t)\,. \tag{1.31}$$

Calculating the mean values from (1.31), the last term of the r.h.s. vanishes in accordance with (1.26), and for $\langle\dot{x}^2\rangle$, the equipartition law, (1.23), applies:

$$\langle\dot{x}^2\rangle = \langle v_x^2\rangle = \frac{k_{\text{B}}T}{m}\,. \tag{1.32}$$

Hence, for the mean values, the following ordinary differential equation results:

$$m\,\frac{d}{dt}\,\langle\boldsymbol{x}\dot{\boldsymbol{x}}\rangle + \gamma_0\,\langle\boldsymbol{x}\dot{\boldsymbol{x}}\rangle = k_{\text{B}}T\,. \tag{1.33}$$

Integration of (1.33) yields

$$\langle\boldsymbol{x}\dot{\boldsymbol{x}}\rangle = \frac{1}{2}\frac{d}{dt}\,\langle x^2\rangle = \frac{k_{\text{B}}T}{\gamma_0} + C\exp\left(-\frac{\gamma_0}{m}\,t\right). \tag{1.34}$$

Assuming that all particles have the initial condition $\boldsymbol{x}(0) = 0$, we find the solution:

$$\left\langle\left[\boldsymbol{x}(t) - \boldsymbol{x}(0)\right]^2\right\rangle = \langle\Delta x^2(t)\rangle = \frac{2k_{\text{B}}T}{\gamma_0}\left\{t - \frac{m}{\gamma_0}\left[1 - \exp\left(-\frac{\gamma_0}{m}t\right)\right]\right\}. \tag{1.35}$$

A comparison of (1.35) with (1.16) for the mean squared displacement leads to the following conclusions (see also Fig. 1.11):

1. For the prefactor, the relation $D = k_{\text{B}}T/\gamma_0$ for the diffusion coefficient, (1.19), is confirmed.
2. For times $t \ll (m/\gamma_0) = 1/\gamma$, (1.35) describes the *free* (uniform and frictionless) motion of a Brownian particle with mean velocity $\langle\boldsymbol{v}_x\rangle$, (1.32):

$$\langle\Delta x^2(t)\rangle = \langle v_x^2\rangle\, t^2\,, \quad \text{if } t \ll 1/\gamma\,. \tag{1.36}$$

3. For times $t \gg 1/\gamma$, (1.35) describes the mean displacement of a Brownian particle, which is subject to friction and to stochastic forces. As expected, the mean displacement increases with $t^{1/2}$ in the course of time:

$$\langle\Delta x^2(t)\rangle = 2D\,t - \frac{2D}{\gamma}\,, \quad \text{if } t \gg 1/\gamma\,. \tag{1.37}$$

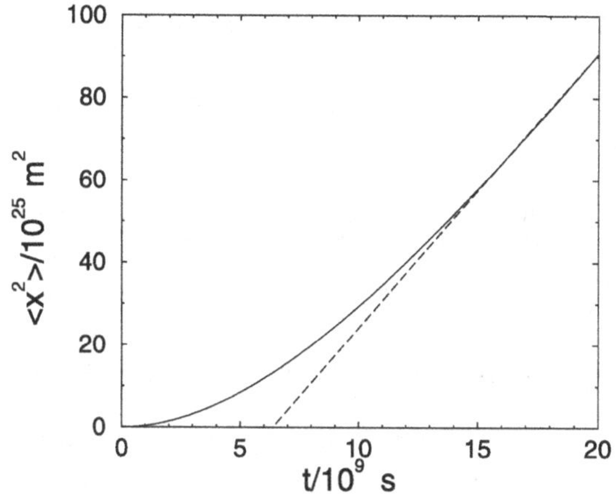

Fig. 1.11. Mean squared displacement, (1.35), of a Brownian particle vs. time (parameters: $\gamma_0 = 10^{-7}$ kg/s; $D = 4 \times 10^{-14}$ m^2/s, $m = 10^{-15}$ kg)

Hence, the diffusion approximation for Brownian particles is valid only for times which are large compared to the break time $t_b = 1/\gamma$, which is 10^{-8} s for a Brownian particle with mass $m = 10^{-15}$ kg and friction coeffficient $\gamma_0 = 10^{-7}$ kg/s. So, except on the nanotimescale, the Langevin approach to Brownian motion agrees with the result for the mean squared displacement, (1.16), because the constant in (1.37) can be neglected.

The above discussion already indicates that the value of the friction coefficient γ_0 plays a considerable role in determining the dynamic behavior of a Brownian particle. In the *overdamped limit*, i.e., for large values of γ_0, the time-dependent change of $\boldsymbol{v}_i$ in (1.28) can be approximately neglected. More precisely, we can assume that – compared to the timescale of changes in $\boldsymbol{x}_i$ – the velocity $\boldsymbol{v}_i$ can be treated as (quasi)stationary; $\dot{\boldsymbol{v}}_i \approx 0$. With the so-called adiabatic approximation, the Langevin equation, (1.28), can be transformed into the equation for the overdamped limit:

$$\boldsymbol{v}_i = \sqrt{\frac{2\,k_{\mathrm{B}}T}{\gamma_0}}\,\boldsymbol{\xi}_i(t) = \sqrt{2\,D}\,\boldsymbol{\xi}_i(t)\,, \tag{1.38}$$

where the velocity remains constant, but the direction can be changed by the noise.

If a Brownian particle moves in an external potential, $U(\boldsymbol{x})$, we have to consider the additional force, $\boldsymbol{F} = -\boldsymbol{\nabla} U$, in (1.25). This modifies the Langevin equation, (1.28), to

$$\frac{d\boldsymbol{x}_i}{dt} = \boldsymbol{v}_i\;, \quad m\,\frac{d\boldsymbol{v}_i}{dt} = -\gamma_0\boldsymbol{v}_i - \left.\frac{\partial U(\boldsymbol{x})}{\partial \boldsymbol{x}}\right|_{\boldsymbol{x}_i} + \sqrt{2\,k_{\mathrm{B}}T\gamma_0}\,\boldsymbol{\xi}_i(t)\,. \tag{1.39}$$

Consequently, the equation for the overdamped limit changes to

$$\boldsymbol{v}_i = -\frac{1}{\gamma_0} \frac{\partial U(\boldsymbol{x})}{\partial \boldsymbol{x}}\bigg|_{\boldsymbol{x}_i} + \sqrt{\frac{2\,k_\mathrm{B}T}{\gamma_0}}\, \boldsymbol{\xi}_i(t)\,. \tag{1.40}$$

Equations (1.39), (1.40) will serve as the starting point for the dynamics of active Brownian particles, which will be introduced below. But, first we want to discuss the distribution functions.

1.3.3 Probability Density and the Fokker–Planck Equation

In the Langevin equations (1.28), (1.39), the dynamics of a Brownian particle i is described by the space coordinate $\boldsymbol{x}_i$ and the velocity $\boldsymbol{v}_i$. In a d-dimensional space, these variables define a $2\,d$-dimensional phase space that contains all possible realizations of the dynamics. From the perspective of statistical physics, one is interested in the *probability* $p_i(\boldsymbol{x}, \boldsymbol{v}, t)$ of finding the particle i with a specific realization of the variables $\boldsymbol{x}$ and $\boldsymbol{v}$ at a particular time t. More precisely, this is a probability density with the normalization,

$$\int_{-\infty}^{\infty}\int_{-\infty}^{\infty} p_i(\boldsymbol{x}, \boldsymbol{v}, t)\; dx\; dv = 1\,. \tag{1.41}$$

Because we do not discuss any interaction between Brownian particles so far, we restrict ourselves to the case of only one particle and omit the index i. Further, only one variable, y, is considered, and the generalization is straightforward.

In general, the probability density of a particular realization $p(y,t)$ may depend on the previous history of the dynamics. However, usually the Markov approximation can be applied, which means that the dynamic process does not have a "memory", i.e., it is independent of the previous history. Considering further infinitely small time intervals, $t - t' = \Delta t \to 0$, the dynamics for the probability density can be described by the so-called *master equation*, which is the differential form of the Chapman–Kolmogorov equation known from probability theory:

$$\frac{\partial}{\partial t} p(y,t) = \sum_{y' \neq y} \Big[w(y|y')\, p(y',t) - w(y'|y)\, p(y,t) \Big]\,. \tag{1.42}$$

Equation (1.42) can be considered a rate equation for the probability density, where $w(y'|y)$ are the transition rates between two states $y \to y'$. Consequently, $\sum_{y'} w(y|y')\, p(y',t)$ describes the probability flux (influx) from all states $y' \neq y$ to state y, whereas $-\sum_{y'} w(y'|y)\, p(y,t)$ describes the probability flux (outflux) from state y to all possible states y'. Which of the states y' can be reached from a given state y (and vice versa) depends on the transition rates $w(y'|y)$ and has to be specified for each problem considered.

The master equation (1.42), holds for continuous time t and discrete states y. If y is also considered a continuous variable, the sums are replaced by integrals:

$$\frac{\partial}{\partial t} p(y,t) = \int dy' \left[w(y|y')\, p(y',t) - w(y'|y)\, p(y,t) \right] . \tag{1.43}$$

Equation (1.43) implies that the time-dependent change in the probability density is influenced by all possible y' in the state space. Alternatively, one can assume that the immediate change of $p(y,t)$ is influenced only by those values of $p(y',t)$, where y' is in the immediate neighborhood of y. In this case, a Taylor expansion of $p(y,t)$ can be carried out, which leads to the Kramers–Moyal form of (1.43):

$$\frac{\partial}{\partial t} p(y,t) = \sum_{n=1}^{\infty} \frac{(-1)^n}{n!} \left(\frac{\partial}{\partial y} \right)^n \left[\alpha_n(y,t)\, p(y,t) \right] . \tag{1.44}$$

Here, the expression $\alpha_n(y,t)$ is the nth moment of the transition rate $w(y'|y)$:

$$\alpha_n(y,t) = \int_{-\infty}^{\infty} dy' \; (y' - y) \; w(y'|y) \, . \tag{1.45}$$

If we consider terms up to $n = 2$, the Kramers–Moyal equation, (1.44), results in the Fokker–Planck equation:

$$\frac{\partial}{\partial t} p(y,t) = -\frac{\partial}{\partial y} \left[\alpha_1(y,t)\, p(y,t) \right] + \frac{1}{2} \frac{\partial^2}{\partial y^2} \left[\alpha_2(y,t)\, p(y,t) \right] . \tag{1.46}$$

The first term of the r.h.s. of (1.46) is the *drift term*; the second term is the *diffusion term* (fluctuation term) of the probability density. Consequently, with $\alpha_1(y,t) = 0$, $\alpha_2(y,t) \neq 0$, (1.46) simply describes a diffusion process in state space, whereas for $\alpha_2(y,t) \neq 0$, $\alpha_2(y,t) = 0$, a *deterministic* equation, or Liouville equation in classical mechanics, results.

The above formalism can now be used to derive the Fokker–Planck equation for the motion of a Brownian particle due to the Langevin equation, (1.28) [414]. Here, we haven't yet specified the transition rates, $w(y'|y)$. To calculate the moments $\alpha_n(y,t)$, we therefore rewrite (1.45) in the form,

$$\alpha_n(y) = \lim_{\Delta t \to 0} \frac{1}{\Delta t} \int_{-\infty}^{\infty} dy' \; (y' - y_0) \; p(y', t_0 + \Delta t | y_0, t_0) \, . \tag{1.47}$$

With respect to Brownian motion, $p(y', t_0 + \Delta t | y_0, t_0)$ denotes the conditional probability of finding a particle, e.g., with velocity $\boldsymbol{v}'$ at time $t_0 + \Delta t$, provided it had the velocity $\boldsymbol{v}_0$ at time t_0. The integral in (1.47) defines the *mean value* (or expectation value in probability theory) of a function $f(y)$:

$$\Big\langle f(y) \Big\rangle_{y_0} = \int dy' \; f(y') \; p(y', t_0 + \Delta t | y_0, t_0) \, . \tag{1.48}$$

Then, the nth moment of the velocity distribution for $y \to \boldsymbol{v}$ reads:

$$\alpha_n(\boldsymbol{v}) = \lim_{\Delta t \to 0} \frac{1}{\Delta t} \left\langle \Big[\boldsymbol{v}(t_0 + \Delta t) - \boldsymbol{v}_0\Big]^n \right\rangle_{v_0} . \tag{1.49}$$

The velocity $\boldsymbol{v}(t_0 + \Delta t)$ of a Brownian particle at time $t_0 + \Delta t$ has to be calculated from the Langevin equation, (1.29), which can be formally integrated:

$$\begin{aligned} \boldsymbol{v}(t) &= \boldsymbol{v}_0 \exp\Big[-\gamma(t - t_0)\Big] \\ &\quad + \sqrt{2\,S}\, \exp\Big[-\gamma(t - t_0)\Big] \int_{t_0}^{t} dt' \exp\Big[\gamma(t' - t_0)\Big] \boldsymbol{\xi}(t') . \end{aligned} \tag{1.50}$$

Because the mean value of the stochastic force vanishes, we find for the first moment, $\alpha_1(\boldsymbol{v})$, (1.49):

$$\alpha_1(\boldsymbol{v}) = \lim_{\Delta t \to 0} \frac{1}{\Delta t} \Big[\boldsymbol{v}_0 \exp(-\gamma \Delta t) - \boldsymbol{v}_0\Big] = -\gamma \boldsymbol{v}_0 = -\gamma \boldsymbol{v} \tag{1.51}$$

because of $\boldsymbol{v}_0 = \boldsymbol{v}(t_0) = \boldsymbol{v}(t_0 + \Delta t)$ in the limit $\Delta t \to 0$. For the second moment $\alpha_2(\boldsymbol{v})$, we have to consider the contributions from the stochastic force, which yield [414]:

$$\begin{aligned} &\left(\frac{2 k_B T \gamma}{m}\right) \exp\Big[-2\gamma(t - t_0)\Big] \\ &\qquad \times \int_{t_0}^{t} dt' \int_{t_0}^{t} dt'' \exp\Big[\gamma(t' + t'')\Big] \Big\langle \boldsymbol{\xi}(t')\, \boldsymbol{\xi}(t'') \Big\rangle_{v_0} \\ &= \frac{k_B T}{m} \Big\{1 - \exp\Big[-2\gamma(t - t_0)\Big]\Big\} . \end{aligned} \tag{1.52}$$

Hence, for the second moment, we find [414]

$$\begin{aligned} \alpha_2(\boldsymbol{v}) &= \lim_{\Delta t \to 0} \frac{1}{\Delta t} \left[\frac{k_B T}{m} + \left(v_0^2 - \frac{k_B T}{m}\right) \exp(-2\gamma \Delta t) \right. \\ &\qquad\qquad \left. - 2 v_0^2 \exp(-\gamma \Delta t) + v_0^2 \right] \\ &= \frac{2 \gamma k_B T}{m} = 2\,S . \end{aligned} \tag{1.53}$$

Consequently, the Fokker–Planck equation, (1.46), for Brownian motion with respect to the velocity distribution, reads

$$\frac{\partial p(\boldsymbol{v}, t)}{\partial t} = \gamma \frac{\partial}{\partial \boldsymbol{v}} \Big[\boldsymbol{v}\, p(\boldsymbol{v}, t)\Big] + S \frac{\partial^2}{\partial v^2} p(\boldsymbol{v}, t) . \tag{1.54}$$

It can be further proved [414] that for Brownian motion, all moments $\alpha_n(\boldsymbol{v})$ for $n \geq 3$ vanish.

If we consider again the motion of a Brownian particle in an external potential $U(\boldsymbol{x})$, as in the Langevin equation, (1.39), and further include the space coordinate $\boldsymbol{x}$, then the Fokker–Planck equation for the distribution function $p(\boldsymbol{x}, \boldsymbol{v}, t)$ reads [46]

$$\begin{aligned}\frac{\partial p(\boldsymbol{x},\boldsymbol{v},t)}{\partial t} &= \frac{\partial}{\partial \boldsymbol{v}}\left[\gamma\,\boldsymbol{v}\,p(\boldsymbol{x},\boldsymbol{v},t) + S\,\frac{\partial p(\boldsymbol{x},\boldsymbol{v},t)}{\partial \boldsymbol{v}}\right] \\ &\quad -\boldsymbol{v}\,\frac{\partial p(\boldsymbol{x},\boldsymbol{v},t)}{\partial \boldsymbol{x}} + \frac{1}{m}\,\nabla U(\boldsymbol{x})\,\frac{\partial p(\boldsymbol{x},\boldsymbol{v},t)}{\partial \boldsymbol{v}}\,. \end{aligned} \tag{1.55}$$

The stationary solution of the Fokker–Planck equations, (1.54), (1.55), can be found from the condition $\partial p(\boldsymbol{v}, t)/\partial t = 0$, or $\partial p(\boldsymbol{x}, \boldsymbol{v}, t)/\partial t = 0$, respectively. In the general case, (1.46), the stationary solution formally reads [414]

$$\lim_{t\to\infty} p(y,t) = p^0(y) = \frac{\text{const}}{\alpha_2(y)}\,\exp\left[-2\int\limits_0^y dy'\,\frac{\alpha_1(y')}{\alpha_2(y')}\right]. \tag{1.56}$$

From (1.54), as the stationary solution, we find the well-known *Maxwellian velocity distribution*:

$$p^0(\boldsymbol{v}) = C\,\exp\left(-\frac{m}{2\,k_\mathrm{B}T}\,v^2\right)\,,\quad C = \left(\frac{m}{2k_\mathrm{B}T\,\pi}\right)^{3/2}. \tag{1.57}$$

If we also consider the potential $U(\boldsymbol{x})$, from (1.55), we find the stationary solution in the form,

$$\begin{aligned} p^0(\boldsymbol{x},\boldsymbol{v}) &= C'\,\exp\left\{-\frac{1}{k_\mathrm{B}T}\left[\frac{m}{2}\,v^2 + U(\boldsymbol{x})\right]\right\} \\ &= C'\,\exp\left[-\frac{U(\boldsymbol{x})}{k_\mathrm{B}T}\right]\,\exp\left(-\frac{\gamma}{2\,S}\,v^2\right), \end{aligned} \tag{1.58}$$

where S is the strength of the stochastic source, (1.27), and C' results from the normalization condition, (1.41).

The master equations and Fokker–Planck equations introduced in this section for the probability density, will play an important role thoughout this book. Together with the Langevin equations introduced in the previous section, they allow a description of Brownian motion on different levels, the level of the individual particle and the level of the probability density, or distribution function, respectively.

2. Active Particles

2.1 Active Motion and Energy Consumption

2.1.1 Storage of Energy in an Internal Depot

Active motion is a phenomenon found in a wide range of systems. In physicochemical systems, *self-driven motion* of particles can already be observed [340]. On the biological level, *active*, self-driven motion can be found on different scales, ranging from cells [186, 511] or simple microorganisms up to higher organisms such as birds and fish [8, 375]. Last, but not least, human movement, as well as the motion of cars [211, 212], can also be described as active motion [222]. All of these types of active motion occur with energy consumption and energy conversion and may also involve processes of energy storage.

To describe both the *deterministic* and *random* aspects and the *energetic* aspects of active motion, we have introduced *active particles* [126, 453] as a first step to extend the concept of Brownian particles to Brownian agents. The term "active" means that a Brownian particle can take up energy from the environment, store in an internal depot, and convert internal energy into kinetic energy. As pointed out in Sect. 1.3.1, the internal energy depot $e_i(t)$ can be regarded as an additional internal degree of freedom for each particle i. Other internal degrees of freedom and hence other activities are not considered here; in particular, it is not assumed that particles create or respond to a self-consistent field. Instead, the main focus for investigating active particles is on *active motion*, which may become possible due to the external input of energy. In this respect, an active particle is a rather "simple" type of agent that does not deny its origin from the physical world and therefore has been also denoted as a *particle agent* [124].

In the following, we need to specify (i) the dynamics of the internal energy depot and (ii) its impact on the motion of a particle. As pointed out in Sect. 1.3.2, the motion of a simple Brownian particle in an external potential $U(\boldsymbol{r})$ is due to the deterministic force $-\nabla U(\boldsymbol{r})$ resulting from potential and stochastic influences, expressed by the noise term in the Langevin equation:

$$\dot{\boldsymbol{r}} = \boldsymbol{v} \ , \quad m\dot{\boldsymbol{v}} = -\gamma_0 \boldsymbol{v} - \boldsymbol{\nabla} U(\boldsymbol{r}) + m\sqrt{2S}\,\boldsymbol{\xi}(t) \, . \tag{2.1}$$

The fluctuation–dissipation relation, (1.27), tells us that even in the absence of an external potential, a Brownian particle keeps moving because its friction γ_0 is compensated for by stochastic force. We are now interested in how this known picture could be extended by considering an internal energy depot for a Brownian particle.

According to the explanation in Sect. 1.3.1, the internal energy depot $e(t)$ may be altered by three different processes:

1. Take-up of energy from the environment, where $q(\boldsymbol{r})$ is the (space-dependent) flux of energy into the depot.
2. Internal dissipation, which is assumed to be proportional to the internal energy. Here the rate of energy loss c is assumed constant.
3. Conversion of internal energy into motion, where $d(\boldsymbol{v})$ is the rate of conversion of internal into kinetic energy, which should be a function of the actual velocity of a particle.

Note that environmental changes, such as the production of a self-consistent field, are neglected in the following. Further, we add that other assumptions for the conversion term may be considered, for example, dependence on acceleration $d(\dot{\boldsymbol{v}})$, which holds, e.g., for cars in a (macroscopic) traffic application.

With the above assumptions, the resulting balance equation for the energy depot is given by

$$\frac{d}{dt}e(t) = q(\boldsymbol{r}) - c\, e(t) - d(\boldsymbol{v})\, e(t)\,. \tag{2.2}$$

If we think of *biological entities*, the take-up of energy is similar to feeding, which may occur at specific places or food sources. The internal dissipation of stored energy, on the other hand, is then analogous to metabolic processes inside an organism. Further, it is considered that the biological objects perform active motion, which also needs a supply of energy provided by the internal energy depot.

A simple ansatz for $d(\boldsymbol{v})$ reads,

$$d(\boldsymbol{v}) = d_2 v^2 \quad , \quad d_2 > 0\,. \tag{2.3}$$

For the special case $d_2 \equiv 0$, i.e., only internal dissipation of energy and take-up of energy are considered, the explicit solution of (2.2) is given by

$$e(t) = e(0) + \int_0^t d\tau \exp(-c\tau)\, q\Big[\boldsymbol{r}(\tau)\Big]\,. \tag{2.4}$$

Equation (2.4) shows that, in general, the depot is filled with a time lag and that the content of the energy depot is a function of the historical path $\boldsymbol{r}(t)$ of a particle.

The total energy of an active particle at time t is given by

$$E(t) = E_0(t) + e(t)\,,$$
$$E_0(t) = \frac{m}{2}v^2 + U(\boldsymbol{r})\,. \tag{2.5}$$

E_0 is the mechanical energy of a particle moving in potential $U(\boldsymbol{r})$. E_0 can be (i) increased by the conversion of depot energy into kinetic energy and (ii) decreased by the friction of a moving particle resulting in disspation of energy. Hence, the balance equation for the mechanical energy reads

$$\frac{d}{dt}E_0(t) = \Big[d_2 e(t) - \gamma_0\Big]\, v^2\,. \tag{2.6}$$

With respect to (2.2), (2.6), the resulting change in total energy is given by the balance equation

$$\frac{d}{dt}E(t) = q(\boldsymbol{r}) - ce(t) - \gamma_0 v^2\,. \tag{2.7}$$

Combining (2.5), (2.6), we can rewrite (2.6) in a more explicit form:

$$m\boldsymbol{v}\dot{\boldsymbol{v}} + \boldsymbol{\nabla} U(\boldsymbol{r})\,\dot{\boldsymbol{r}} = \Big[d_2 e(t) - \gamma_0\Big]\, v^2\,, \tag{2.8}$$

or

$$m\dot{\boldsymbol{r}}\ddot{\boldsymbol{r}} + \gamma_0\,\dot{r}^2 + \dot{\boldsymbol{r}}\,\boldsymbol{\nabla} U(\boldsymbol{r}) = d_2 e(t)\,\dot{r}^2\,. \tag{2.9}$$

Equation (2.9) indicates that for $\dot{\boldsymbol{r}} = 0$, $d(\boldsymbol{v} = 0) = 0$ holds for arbitrary values of $e(t)$, which is satisfied by the ansatz (2.3).

Based on (2.9), we postulate now a stochastic equation of motion for active Brownian particles [126, 453] which is consistent with the Langevin equation, (2.1):

$$m\dot{\boldsymbol{v}} + \gamma_0\,\boldsymbol{v} + \boldsymbol{\nabla} U(\boldsymbol{r}) = d_2 e(t)\boldsymbol{v} + m\sqrt{2S}\,\boldsymbol{\xi}(t)\,. \tag{2.10}$$

Compared to previous investigations [486], the first term of the right-hand side of (2.10) is the essential new element, reflecting the influence of the internal energy depot on the motion of Brownian particles. It describes the acceleration in the direction of movement, $\boldsymbol{e}_v = \boldsymbol{v}/v$. The right-hand side of the Langevin equation now contains *two driving forces* for motion: (i) the acceleration of motion due to the conversion of internal into kinetic energy and (ii) stochastic force with the strength $m\sqrt{2S}$. Assuming again that the loss of energy resulting from friction and the gain of energy resulting from the stochastic force, are compensated for, on average, (1.27), the balance equation for mechanical energy, (2.6), is modified for the *stochastic case* to

$$\frac{d}{dt}\left[\frac{1}{2}m\dot{r}^2 + U(\boldsymbol{r})\right] = d_2 e(t)\dot{r}^2\,. \tag{2.11}$$

Due to the additional driving force in (2.10), the system can be driven into nonequilibrium, and stochastic effects can be amplified. Thus, we expect qualitatively new behavior in the particle's motion, which will be discussed in detail in the following sections.

2.1.2 Velocity-Dependent Friction

The Langevin equation, (2.10), can be rewritten in the known form, (2.1), by introducing a *velocity-dependent friction function*, $\gamma(\boldsymbol{v})$:

$$m\dot{\boldsymbol{v}} + \gamma(\boldsymbol{v})\,\boldsymbol{v} + \boldsymbol{\nabla} U(\boldsymbol{r}) = m\sqrt{2S}\,\boldsymbol{\xi}(t)\,,$$
$$\gamma(\boldsymbol{v}) = \gamma_0 - d_2\,e(t)\,. \tag{2.12}$$

Here, the value of $\gamma(\boldsymbol{v})$ changes, depending on the value of the internal energy depot, which itself is a function of velocity. If the term $d_2\,e(t)$ exceeds the "normal" friction γ_0, the velocity-dependent friction function can be negative, which means that the active particle's motion is pumped with energy. To estimate the range of energy pumping, we assume a constant influx of energy into the internal depot:

$$q(\boldsymbol{r}) = q_0\,, \tag{2.13}$$

and introduce a formal parameter μ to describe the timescale of relaxation of the internal energy depot. With (2.3), (2.2) then reads

$$\mu\,\dot{e}(t) = q_0 - c\,e(t) - d_2\,v^2\,e(t)\,. \tag{2.14}$$

Using the initial condition $e(t=0)=0$, the depot is filled with a time lag in the general case, $\mu = 1$. The limit $\mu \to 0$, however, describes very fast adaptation of the depot, and, as an adiabatic approximation, we get

$$e_0 = \frac{q_0}{c + d_2 v^2}\,. \tag{2.15}$$

The quasi-stationary value e_0 can be used to approximate the velocity-dependent friction function $\gamma(\boldsymbol{v})$, (2.12):

$$\gamma(v) = \gamma_0 - \frac{q_0\,d_2}{c + d_2 v^2}\,, \tag{2.16}$$

which is plotted in Fig. 2.1.

Depending on the parameters, the friction function, (2.16), may have a zero at a definite velocity $\boldsymbol{v}_0$:

$$\boldsymbol{v}_0^2 = \frac{q_0}{\gamma_0} - \frac{c}{d_2}\,, \tag{2.17}$$

which allows us to express the friction function, (2.16), as

$$\gamma\,(\boldsymbol{v}) = \gamma_0\,\frac{(\boldsymbol{v}^2 - \boldsymbol{v}_0^2)}{(q_0/\gamma_0) + (\boldsymbol{v}^2 - \boldsymbol{v}_0^2)}\,. \tag{2.18}$$

We see that for $\boldsymbol{v} < \boldsymbol{v}_0$, i.e., in the range of low velocities, pumping due to *negative friction* occurs, as an additional source of energy for a Brownian par-

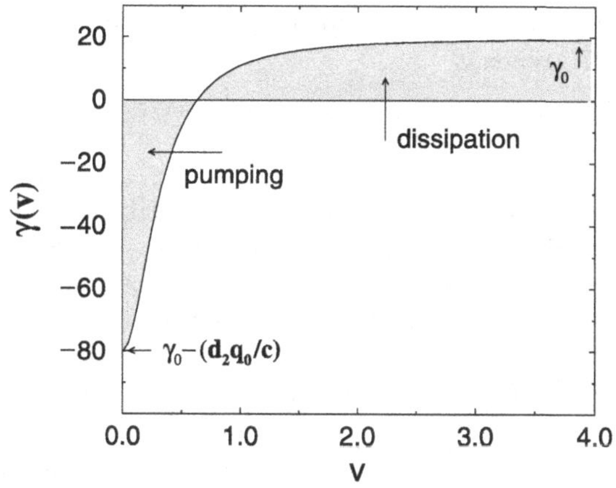

Fig. 2.1. Velocity-dependent friction function $\gamma(v)$, (2.16), vs. velocity v. The velocity ranges for "pumping" $[\gamma(v) < 0]$ and "dissipation" $[\gamma(v) > 0]$ are indicated. Parameters: $q_0 = 10$; $c = 1.0$; $\gamma_0 = 20$; $d_2 = 10$ [137]

ticle (cf. Fig. 2.1). Hence, slow particles are accelerated, whereas the motion of fast particles is damped. Due to the pumping mechanism introduced in our model, the conservation of energy clearly does not hold for active particles, i.e., we now have a nonequilibrium, canonical-dissipative system [142, 454]. instead of an equilibrium canonical system, which will be discussed in more detail in Sect. 2.4.1.

Negative friction is known, i.e., from technical constructions, where moving parts cause a loss of energy, which is from time to time compensated for by the pumping of mechanical energy. For example, in grandfather's mechanical clock, the dissipation of energy by moving parts is compensated for by a heavy weight (or a spring). The weight stores potential energy that is gradually transferred to the clock via friction, i.e., the strip with the weight is pulled down. Another example is violin strings, where friction-pumped oscillations appear if the violin bow transfers energy to the string via friction. Provided with a supercritical influx of energy, a self-sustained periodic motion can be obtained; the clock keeps moving or the violin string emits acoustic waves.

The latter example was already discussed by Lord Rayleigh at the end of the nineteenth century, when he developed his Theory of Sound [401]. Here, the velocity-dependent friction function can be expressed as

$$\gamma(\boldsymbol{v}) = -\gamma_1 + \gamma_2\, \boldsymbol{v}^2 = \gamma_1 \left(\frac{\boldsymbol{v}^2}{\boldsymbol{v}_0^2} - 1 \right) . \tag{2.19}$$

The Rayleigh-type model is a standard model for self-sustained oscillations also studied in the context of Brownian motion [276]. We note that $\boldsymbol{v}_0^2 = \gamma_1/\gamma_2$ defines here the special value where the friction function, (2.19), is zero (cf. Fig. 2.2).

Another example of a velocity-dependent friction function with a zero $\boldsymbol{v}_0$ introduced in [427], reads

$$\gamma(\boldsymbol{v}) = \gamma_0 \left(1 - \frac{v_0}{v}\right) . \tag{2.20}$$

It has been shown that (2.20) allows us to describe the active motion of different cell types [186, 427]. Here, the speed $\boldsymbol{v}_0$ expresses the fact that the motion of cells is not driven only by stochastic forces, instead cells are also capable of self-driven motion. This example will be discussed in more detail in the next section. First, we want to mention that the velocity-dependent friction function in (2.16) avoids some drawbacks of the two ansatzes of (2.19), (2.20), that are shown in Fig. 2.1. $\gamma(\boldsymbol{v})$ in Fig. 2.1 is bound to a maximum value γ_0 reached for $\boldsymbol{v} \to \infty$ and avoids the singularity for $\boldsymbol{v} \to 0$, on the other hand.

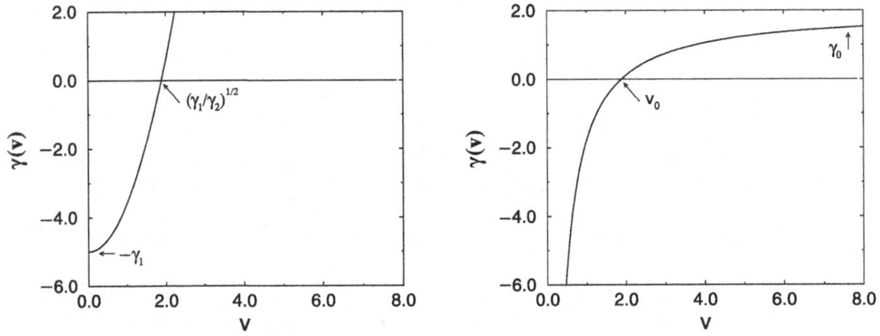

Fig. 2.2. Two different types of a velocity-dependent friction function $\gamma(\boldsymbol{v})$ with a zero $\boldsymbol{v}_0$: (*left*) (2.19), (*right*) (2.20). Parameters: $\gamma_1 = 5$, $\gamma_2 = 1.4$, $\gamma_0 = 2$, $v_0 = 1.889$. For $\gamma(\boldsymbol{v}) < 0$, "pumping" dominates, whereas for $\gamma(\boldsymbol{v}) > 0$, "dissipation" dominates [137]

2.1.3 Active Motion of Cells

Looking through a microscope, the motion of Brownian particles resembles that of small living creatures. Therefore, Leeuwenhoek, Brown, and others who discovered the erratic motion of these particles early, considered them living entities. However, the previous sections made clear that the motion of Brownian particles is due to fluctuations in the surrounding medium. This type of motion would be rather considered *passive motion*, simply because the Brownian particle does not play an active part in this motion. In fact, even in the organic world, passive motion is a widespread phenomenon. A simple example is microorganisms driven by a current or by convection. If the organism is microscopically small, passive motion can also occur via random impacts from the surrounding medium, in which case the distance moved in the course of time is given by the mean squared displacement, $\langle \Delta x \rangle^2$, (1.16).

However, a mean displacement $\langle \Delta x \rangle \sim t^{1/2}$ does not always indicate passive motion via diffusion. It could be also an *active*, but *undirected* motion, i.e., the direction of motion is changed randomly. In this respect, the motion of a broad variety of microorganisms could be described within the framework of Brownian motion, even if this is an active motion. For example, in 1920, Fürth [163] already applied the Brownian motion approach to the motion of infusoria, which are randomly moving small animals. In recent years, stochastic differential equations have been widely used to describe the motion of different biological entities [4, 5] such as cells [68, 105, 110, 368, 369, 427, 511], ants [77, 222, 341], bacteria [54, 310], amoebas [200], and insects [267].

In this section, we discuss the undirected migration of *cells*, which gives an example of an active biological motion described by Langevin-type equations. For instance, Gruler et al. [149, 186, 427] experimentally investigated migrating human *granulocytes* and compared the results with theoretical approaches. The movement of a cell in a two-dimensional plane is described by its space vector $\boldsymbol{v}(t)$ that reads, in *polar coordinates*,

$$\boldsymbol{v}(t) = [v(t) \cos\varphi(t),\, v(t) \sin\varphi(t)] \,. \tag{2.21}$$

As experimental results indicate, the temporal variations of $v(t)$ and $\varphi(t)$ are statistically independent of each other; further, the mean velocity is independent of the direction of migration. This behavior is not only specific for granulocytes [149, 187, 398]; it also holds for other cells like monocytes [61], fibroblasts, and neural crest cells [188]. Hence, the migration of cells is described by a set of two stochastic equations, one for speed v, the other for the turning angle φ:

$$\dot{v}(t) = -\gamma \left(1 - \frac{v_\mathrm{s}}{v}\right) v + \sqrt{q_v}\, \xi_v(t) \,, \tag{2.22}$$

$$\dot{\varphi}(t) = -k_p E \sin\varphi(t) + \sqrt{q_\varphi}\, \xi_\varphi(t) \,. \tag{2.23}$$

Here, q_v and q_φ are the strengths of the stochastic speed change and the stochastic torque, respectively. The speed v_s is a minimum value, indicating that the motion of cells is not driven only by stochastic forces. Its consideration results in the velocity-dependent friction function, (2.20), Fig. 2.2(*right*), already mentioned in the previous section. The equation for $\dot{\varphi}$ further considers the influence of an external electric field E that may influence the direction of the move. It is known that human granulocytes can perform a directed movement in the presence of electric fields. However, for the discussion here, $E = 0$ is considered. The mean squared displacement in the x direction can be calulated from (2.22) via [6]

$$\langle x^2(t) \rangle = \left\langle \left[\int_0^t v(t') \cos\varphi(t')\, dt' \right]^2 \right\rangle \,. \tag{2.24}$$

The result for two-dimensional motion has been found as [427]

$$\langle x^2 \rangle + \langle y^2 \rangle \simeq 2\, v_s^2\, \tau_\varphi \left\{ t - \tau_\varphi \left[1 - \exp\left(-\frac{t}{\tau_\varphi} \right) \right] \right\} . \tag{2.25}$$

The characteristic time, $\tau_\varphi = 2/q_\varphi$, incorporates the effect of *persistence*, i.e., the time the cell moves before it changes its direction. Consequently, τ_φ is the inverse of the strength of the stochastic torque q_φ. Figure 2.3 shows experimental results for the mean squared displacement of granulocytes, which are compared with (2.25).

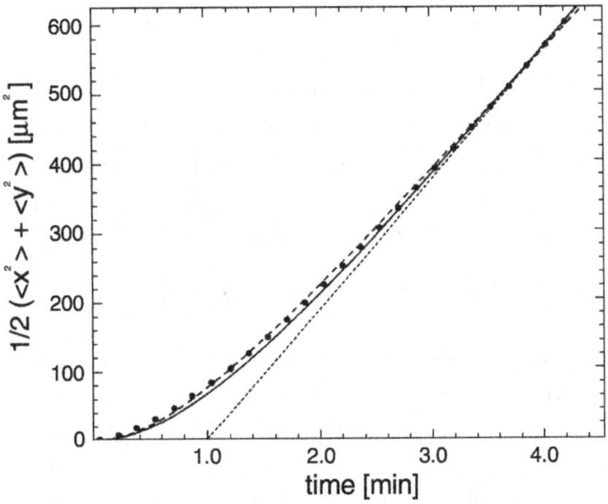

Fig. 2.3. Mean squared displacement as a function of time. The *dots* were obtained from migrating granulocytes ($\Delta t = 10\,\mathrm{s}$). The *dotted straight line* is fitted to the data points at large times. The fitting parameters are the interception ($\tau_\varphi = 58\,\mathrm{s}$) and the slope ($187.5\,\mu\mathrm{m}^2/\mathrm{min}$). The approximate expression, (2.25), using the above values, is shown by the *solid line.* The *dashed line* is obtained by fitting (2.25) directly to the data where $\tau_\varphi \approx 42\,\mathrm{s}$ and $v_s = 15.6\,\mu\mathrm{m/min}$ are the fitting parameters [427]

Equation (2.25), which holds for two dimensions, can be compared with the mean squared displacement, (1.35), for one dimension, obtained from the Langevin equation, (1.28), in Sect. 1.3.2. Applying the equipartition law, (1.23), and the definition of the diffusion coefficient D, (1.19), we find that both equations agree completely provided that the characteristic variables have the following relations:

$$\begin{aligned} \tau_\varphi &= \frac{m}{\gamma_0} = \frac{1}{\gamma} , \\ v_s^2\, \tau_\varphi &= \frac{m\, v_s^2}{\gamma_0} = \frac{2k_B T}{\gamma_0} = 2\, D . \end{aligned} \tag{2.26}$$

The factor of 2 results from the fact that for two-dimensional motion $v_s^2 = (v_x^2 + v_y^2)$ holds, where each degree of freedom contributes $k_B T$ according to (1.23).

The calculation of the mean squared displacement of human granulocytes indicates that the active random motion of cells can be well described within the Langevin approach. The differences between the simple "passive" motion of Brownian particles and the "active" motion of living entities, which result from the velocity-dependent friction function $\gamma(\boldsymbol{v})$ can be seen in the *velocity distribution function* of cells, $W(v,t)$. It can be obtained from a Fokker–Planck equation equivalent to (2.22) [427]:

$$\frac{\partial W(v,t)}{\partial t} = \left[\gamma_v \frac{\partial}{\partial v}(v - v_s) + \frac{1}{2} q_v \frac{\partial^2}{\partial v^2}\right] W(v,t) \,. \tag{2.27}$$

The stationary speed distribution of granulocytes, $W_{st}(v)$, results from (2.27):

$$W_{st}(v) \simeq \sqrt{\frac{\gamma_v}{q_v \pi}} \exp\left[-\frac{\gamma_v}{q_v}(v - v_s)^2\right] . \tag{2.28}$$

The theoretical result, a Gaussian distribution centered around the mean value $v_s \neq 0$, is in good agreement with the experimental observations as Fig. 2.4 shows. The difference in "simple" Brownian particles which move only passively and therefore have only an absolute velocity $|v| \sim k_B T$, (2.26), results clearly from the existence of the velocity v_s that describes the active motion of cells. To understand the origin of the velocity v_s from the bottom up, we would need a detailed microbiological description, which is far beyond the intentions of this book [61, 105, 110, 184, 185].

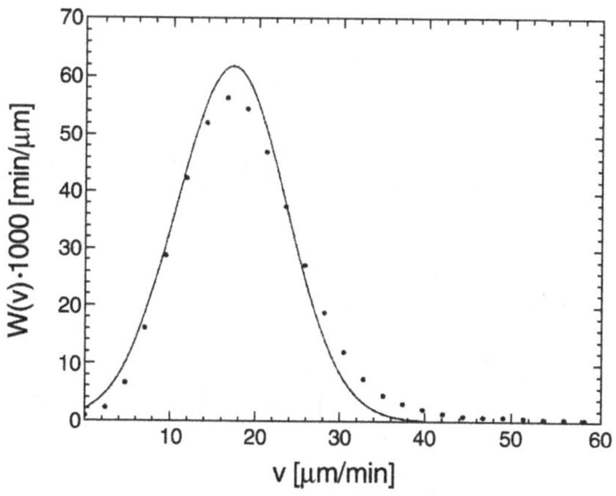

Fig. 2.4. Stationary speed distribution of granulocytes. The *dots* represent experimental results (original data of Franke & Gruler [149]; $\Delta t = 10\,\text{s}$). The *solid line* is the theoretical function, (2.28), obtained by a least squares fit ($v_s = 17\,\mu\text{m/min}$, $\sqrt{q_v/\gamma_v} = 9\,\mu\text{m/min}$) [427]

2.1.4 Pumping by Space-Dependent Friction

In the previous section, we discussed the influence of a *velocity-dependent* friction function that can become negative under certain conditions, in this way describing an influx of energy needed for active motion. Another, but quite simple mechanism to take up the additional energy required is the pumping of energy by a *space-dependent* friction function $\gamma(\boldsymbol{r})$ [486].

If we consider the one-dimensional motion of an active particle in a parabolic potential,

$$U(x) = \frac{a}{2}\,x^2 \;, \qquad \nabla_x U(x) = a\,x \,, \tag{2.29}$$

and restrict the discussion to the deterministic case, the motion of the particle is described by two coupled first-order differential equations:

$$\dot{x} = v \;, \quad m\ddot{x} = -\gamma(x)\,\dot{x} - ax \,, \tag{2.30}$$

where the vectorial character is omitted for further investigations. Here, $\gamma(x)$ is a space-dependent friction function that can also be negative within certain regions. For example (cf. Fig. 2.5),

$$\gamma(x) = \begin{cases} \gamma_- < 0 & \text{if } b_1 \le x \le b_2 \\ \gamma_0 > 0 & \text{else}\,. \end{cases} \tag{2.31}$$

That means that in a distance b_1 from the origin of the parabolic potential, $x = 0$, there is a patch of width $\Delta b = b_2 - b_1$ within which the friction coefficient is negative, γ_-. Outside of this region, γ_0 holds, which is the "normal" friction coefficient.

Equation (2.30) with respect to (2.31) can be rewritten as

$$\ddot{x} + 2\beta_k\,\dot{x} + \omega_0^2\,x = 0 \;, \quad k = 1, 2 \,, \tag{2.32}$$

with the abbreviations:

$$\omega_0^2 = \frac{a}{m} \;, \qquad \beta_k = \begin{cases} \beta_1 = \gamma_0/2m > 0 & \text{if } b_1 \le x \le b_2 \\ \beta_2 = \gamma_-/2m < 0 & \text{else}\,. \end{cases} \tag{2.33}$$

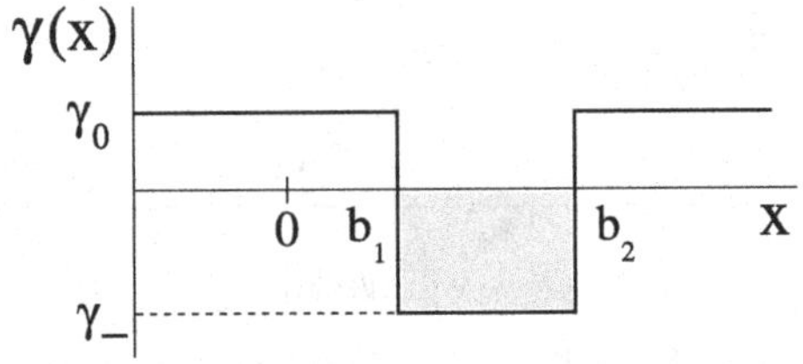

Fig. 2.5. Sketch of the space-dependent friction function $\gamma(x)$, (2.31). Note that for $0 < b_1 < b_2$, the region of negative friction does not contain the minimum, $x = 0$, of the potential $U(x)$, (2.29)

The dynamic behavior of (2.32) depends on three parameters, ω_0 and the ratios γ_-/γ_0 and b_2/b_1. With the ansatz $x(t) = x_0 e^{\lambda t}$, we find from (2.32) the characteristic equation,

$$\lambda_{1,2} = -\beta_k \pm \sqrt{\beta_k^2 - \omega_0^2}\,. \tag{2.34}$$

If we consider the case of low (positive or negative) friction, i.e., $\beta_k^2 < \omega_0^2$, then the solution for $x(t)$ is described by growing or shrinking oscillations:

$$x(t) = e^{\beta_k t}[A_1 \cos(\omega t) + A_2 \sin(\omega t)]\ , \quad \omega = \sqrt{\omega_0^2 - \beta_k^2}\,. \tag{2.35}$$

Note that β_k varies for the two different regions with positive and negative friction; hence the constants A_1, A_2 have to be specified by both the initial and the boundary conditions between the two regions.

Without any additional pumping, the particle's motion in the parabolic potential, (2.29), has the only stationary solution $x = 0$ because of friction. If, on the other hand, the initial conditions are chosen so that the particle can reach the region of negative friction, it could gain additional energy. Provided this amount is large enough, i.e., the particle gains as much energy at each cycle as it loses because of friction during the same time period, we can expect the particle to move on a deterministic limit cycle in space which corresponds to elliptic trajectories in the $\{x, v\}$ phase space.

The crossover from damped motion to motion along a stable periodic orbit occurs if the loss and gain of energy during one cycle are just balanced. With $F = -\gamma_0\, v$ as the dissipative force, the critical condition for the existence of stable periodic orbits can be obtained from the energy balance equation for one cycle:

$$\frac{d}{dt}E = -\oint \gamma(x)\, v(x)\, dx = -\int\limits_{x \notin [b_1, b_2]} \gamma_0\, v(x)\, dx \;-\int\limits_{x \in [b_1, b_2]} \gamma_-\, v(x)\, dx\,. \tag{2.36}$$

The critical condition is given by just $dE/dt = 0$, which results in the following relation for critical friction [486]:

$$\left(\frac{\gamma_-}{\gamma_0}\right)_{\text{crit}} = \frac{\int\limits_{x \notin [b_1, b_2]} v(x)\, dx}{\int\limits_{x \in [b_1, b_2]} v(x)\, dx}\,. \tag{2.37}$$

Depending on the value of the width $\Delta b = b_2 - b_1$, (2.37) determines the minimum ratio between γ_- and γ_0, which is necessary to compensate for the dissipation of energy caused by "normal" friction. The solution, of course, depends also on the initial condition $x(t=0) = x_0$ of the particle. To ensure that the particle in the parabolic potential reaches the region of negative friction for any value of γ_0, $x_0 > b_2$ should be chosen.

To evaluate (2.37), we need the space-dependent velocity of the particle $v(x)$, which results from the damped equation (2.32). In [486], a simplified version of the problem is discussed that is restricted to the limit of (very) low friction. In this case, $v(t)$ has been simply approximated by the *nondamped* (ideal) solution of (2.32) that reads

$$v(x) = \omega\sqrt{x_0^2 - x^2}\,. \tag{2.38}$$

For the nondamped case, $|x_0|$ also gives the maximum distance. i.e., the turning point for motion in the parabolic potential. Using the ideal solution for $v(x)$, (2.38), (2.36) can be written explicitly:

$$\begin{aligned} \frac{d}{dt}E &= -2\int_{-x_0}^{x_0} \gamma(x)\, v(x)\, dx \\ &= -2\gamma_0\omega_0 \left[\int_{-x_0}^{x_0} \sqrt{x_0^2 - x^2}dx + \left(\frac{\gamma_-}{\gamma_0} - 1\right)\int_{b_1}^{b_2} \sqrt{x_0^2 - x^2}dx\right]. \end{aligned} \tag{2.39}$$

Calculating the integrals and applying the critical condition, $dE/dt = 0$, we find from (2.39):

$$\left(\frac{\gamma_-}{\gamma_0}\right)_{\text{crit}} = 1 - \frac{\pi x_0^2}{x_0^2\left(\arcsin\frac{b_2}{x_0} - \arcsin\frac{b_1}{x_0}\right) + b_2\sqrt{x_0^2 - b_2^2} - b_1\sqrt{x_0^2 - b_1^2}}\,. \tag{2.40}$$

Here, the critical ratio still depends on the initial condition x_0. It can be proved that $|\gamma_-/\gamma_0|$ has a *minimum* for

$$x_0^{\text{crit}} = \sqrt{b_1^2 + b_2^2}\,. \tag{2.41}$$

In this case, x_0 in (2.40) can be eliminated, and the critical ratio reads

$$\left(\frac{\gamma_-}{\gamma_0}\right)_{\text{min}} = 1 - \frac{\pi}{\arcsin\frac{b_2}{\sqrt{b_1^2 + b_2^2}} - \arcsin\frac{b_1}{\sqrt{b_1^2 + b_2^2}}}\,, \tag{2.42}$$

which agrees with the result in [486]. For a given width of the patch with negative friction, $\Delta b = b_2 - b_1$, Fig. 2.6 shows the negative friction γ_- which is at least necessary to *compensate* for the dissipation resulting from normal friction γ_0. The region of parameters, γ_-/γ_0 and Δb, where continuous motion is possible, is bound, on one hand, by the value $\Delta b = 0$ because the patch should have a finite width, and, on the other hand, by the value $\gamma_-/\gamma_0 = -1$ because the negative friction should at least compensate for normal friction.

We finally note that the *deterministic* motion of an active particle in the case of negative friction has some analogies to mechanical models, such as

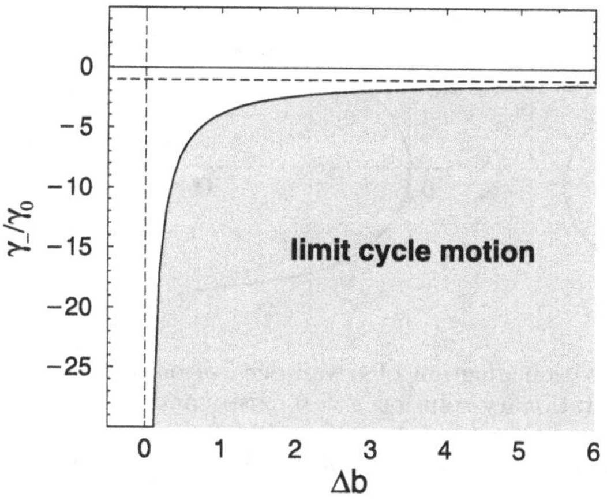

Fig. 2.6. Minimum ratio of negative and normal friction, γ_-/γ_0, (2.42), vs. patch width of negative friction, $\Delta b = b_2 - b_1$, to allow cyclic motion in the deterministic case. The initial condition of the particle is x_0, (2.41). See also Fig. 2.5

the kicked rotator or the Van-der-Pol oscillator [136]. The dynamic equation for a driven mechanical system can be written in the general form [117]:

$$m\ddot{x} - \left(a - bx^2 - c\dot{x}^2\right)\dot{x} + \nabla U = 0\,. \tag{2.43}$$

Equation (2.43) covers different special cases, i.e., for $b = 0$, the dynamic equation for the Rayleigh oscillator is obtained, see (2.19), and for $c = 0$, the dynamic equation for the Van-der-Pol oscillator results. In the case of a parabolic potential, $U(x) = x^2/2$, we have, e.g., the Van-der-Pol equation:

$$v = \dot{x} \quad , \quad \dot{v} = (\alpha - x^2)\,v - x\,. \tag{2.44}$$

Equation (2.44) formally agrees with the equation for an active particle, (2.30), provided that the space-dependent friction function $\gamma(x)$ in (2.30) is identified as $\gamma(x) = x^2 - \alpha$.

For small amplitudes, $x \ll \alpha^{1/2}$, (2.44) has the solution [142],

$$x(t) = x(0)\,\exp\left(\frac{\alpha}{2}\,t\right)\,\cos\left[t\left(1 - \frac{\alpha^2}{4}\right)^{1/2} + \delta\right]. \tag{2.45}$$

The dynamic behavior of (2.44) depends significantly on the sign of the parameter α. For $\alpha < 0$, the dynamic system has a stable fixed point where $x = 0$, $v = 0$, and for $\alpha > 0$, two nontrivial solutions $v_{1,2} \neq 0$ exist, which depend on the initial conditions $x(0)$, $v(0)$. As (2.45) further indicates, the phase space trajectories correspond either to shrinking spirals in the case of $\alpha < 0$, or spirals which eventually evolve into a closed trajectory (*limit cycle*). This behavior is qualitatively shown in Fig. 2.7.

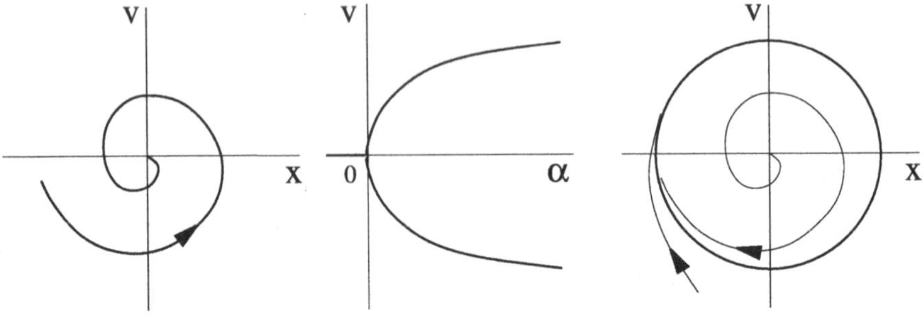

Fig. 2.7. Bifurcation diagram of a van-der-Pol oscillator, (2.44) (*middle*). For $\alpha < 0$, only the stationary solution $v = 0$ exists, and for $\alpha > 0$, two solutions (of equal amount, but different sign) are found (*left*, *right*). Phase space trajectories for $\alpha < 0$ (*left*) and $\alpha > 0$ (*right*). In the latter case, a stable limit cycle for motion is obtained

In the case of the Van-der-Pol oscillator, new stationary solutions for velocity appear for a supercritical value of parameter α, and limit cycle behavior is obtained. As shown above, the same situation is also obtained for active particles that are locally pumped by negative friction. For subcritical conditions, the stable fixed point is also given by $x = 0$, $v = 0$. Provided with a supercritical influx of energy, determined by γ_-/γ_0, (2.40), we find cyclic motion of a particle in the parabolic potential because the dissipation is compensated for by the gain of energy via negative friction. The bifurcation value, where new solutions become possible, is determined by (2.42), which describes the minimum value for limit cycle behavior.

At the end of this section, we may draw a comparison between the space-dependent friction function $\gamma(x)$, discussed above, and the velocity-dependent friction function $\gamma(v)$, (2.16). In both cases, a negative value of γ indicates pumping of energy into the motion of a particle. For space-dependent friction, we have demonstrated that, provided some critical conditions given in (2.40) are fulfilled, pumped motion may lead to a periodic or *limit cycle motion*. Therefore, similar behavior can also be expected for the more complicated case of a velocity-dependent friction function. This will be discussed in more detail in the following sections.

Despite this, some conceptual differences will also be noticed. In fact, a space-dependent friction function can be used to model the *spatially heterogeneous supply* of energy. However, it has the drawback of not considering processes of storage and conversion of energy. With only space-dependent friction, an active particle is *instantaneously* accelerated or slowed down like a physical particle, whereas, for instance, biological systems can *stretch* their supply of energy over a certain time interval. To allow a more general description of active motion, in the following we will use the concept of the internal energy depot for a particle that also features storage and conversion of energy and considers internal dissipation. The spatially heterogeneous supply of energy can be still considered by the space-dependent take-up rate of energy $q(\boldsymbol{r})$, as shown, e.g., in Sect. 2.3.6.

2.2 Active Motion in One-Dimensional Systems

2.2.1 Adiabatic Approximations and Stationary Solutions

In the following, we want to investigate the dynamics of active particles with an internal energy depot in more detail. To this end, we first restrict ourselves to the one-dimensional motion that already provides some characteristic insights.

In the one-dimensional case with the space coordinate $\boldsymbol{x}$ and with $m = 1$ for the mass, the dynamics for the active particle, (2.10), (2.2), is given by the following set of equations:

$$\begin{aligned} \dot{\boldsymbol{x}} &= \boldsymbol{v}\,, \\ \dot{\boldsymbol{v}} &= -\Big[\gamma_0 - d_2 e(t)\Big]\boldsymbol{v} - \frac{\partial U(\boldsymbol{x})}{\partial \boldsymbol{x}} + \sqrt{2\,S}\,\boldsymbol{\xi}(t)\,, \\ \dot{e} &= q(\boldsymbol{x}) - ce - d_2 v^2 e\,. \end{aligned} \tag{2.46}$$

To find a solution for the coupled equations (2.46), let us now consider relaxation of the dynamics on *different timescales.* If we assume that the velocity $\boldsymbol{v}(t)$ is changing much faster than the space coordinate $\boldsymbol{x}(t)$, for $\boldsymbol{v}(t)$, (2.46), a formal solution can be given [169, 469]:

$$\begin{aligned} \boldsymbol{v}(t) &= \boldsymbol{v}(0)\exp\left[-\gamma_0 t + d_2\int_0^t e(t')dt'\right] \\ &\quad + \exp\left[-\gamma_0 t + d_2\int_0^t e(t')dt'\right]\int_0^t \exp\left[\gamma_0 t' - d_2\int_0^{t'} e(t'')dt''\right] \\ &\quad \times \left[-\boldsymbol{\nabla} U + \sqrt{2k_{\mathrm{B}}T\gamma_0}\,\boldsymbol{\xi}(t')\right] dt'\,. \end{aligned} \tag{2.47}$$

This solution, however, depends on the integrals over $e(t)$, reflecting the influence of the energy depot on velocity. If we further assume fast relaxation of the depot $e(t)$, compared to the relaxation of the velocity $\boldsymbol{v}(t)$, the corresponding equation of (2.46) can be solved and we find with the initial condition $e(0)$, that

$$e(\boldsymbol{x},t) = e(0) + \frac{q(\boldsymbol{x})}{c + d_2 v^2}\left\{1 - \exp\Big[-(c + d_2 v^2)\,t\Big]\right\}. \tag{2.48}$$

With the intitial condition $e(0) = 0$, we get in the asymptotic limit, $t \to \infty$,

$$e_0(\boldsymbol{x}) = \frac{q(\boldsymbol{x})}{c + d_2 v^2}\,, \tag{2.49}$$

which yields a possible *maximum value* of $e_0^{\max} = q(\boldsymbol{x})/c$. If we use $e_0(\boldsymbol{x})$, (2.49), to rewrite (2.47),

$$\begin{aligned} \boldsymbol{v}(t) &= \boldsymbol{v}(0) \exp\Big\{ - [\gamma_0 - d_2 e_0(\boldsymbol{x})]\, t \Big\} \\ &+ \int_0^t \exp\Big\{ - [\gamma_0 - d_2 e_0(\boldsymbol{x})]\,(t-t')\Big\} \Big[-\boldsymbol{\nabla} U + \sqrt{2k_\mathrm{B}T\gamma_0}\,\boldsymbol{\xi}(t') \Big]\, dt' , \end{aligned} \tag{2.50}$$

the following solution for the adiabatic approximation results:

$$\begin{aligned} \boldsymbol{v}(t) &= \boldsymbol{v}(0) \exp\Big\{ - [\gamma_0 - d_2 e_0(\boldsymbol{x})]\, t \Big\} \\ &- \frac{\boldsymbol{\nabla} U}{\gamma_0 - d_2 e_0(\boldsymbol{x})} \Big(1 - \exp\Big\{ - [\gamma_0 - d_2 e_0(\boldsymbol{x})]\, t \Big\} \Big) \\ &+ \sqrt{2k_\mathrm{B}T\gamma_0} \int_0^t \exp\Big\{ - [\gamma - d_2 e_0(\boldsymbol{x})]\,(t-t')\Big\}\, \boldsymbol{\xi}(t')\, dt' . \end{aligned} \tag{2.51}$$

The overdamped limit is obtained by considering fast relaxation of the velocities, in which case the stochastic equation of motion, (2.51), can be further reduced to [469]

$$\boldsymbol{v}(t) = - \frac{1}{\gamma_0 - d_2 e_0(\boldsymbol{x})} \frac{\partial U}{\partial \boldsymbol{x}} + \frac{\sqrt{2k_\mathrm{B}T\gamma_0}}{\gamma_0 - d_2 e_0(\boldsymbol{x})}\, \boldsymbol{\xi}(t) . \tag{2.52}$$

We note that, due to the dependence of $e_0(\boldsymbol{x})$ on $v^2 = \dot{x}^2$, (2.52) is coupled to (2.49). Thus, the overdamped equation (2.52) could also be written in the form,

$$\left[\gamma_0 - d_2 \frac{q(\boldsymbol{x})}{c + d_2\, \dot{x}^2} \right] \dot{\boldsymbol{x}} = - \frac{\partial U}{\partial \boldsymbol{x}} + \sqrt{2k_\mathrm{B}T\gamma_0}\, \boldsymbol{\xi}(t) . \tag{2.53}$$

Equation (2.53) indicates a cubic equation for the velocities in the overdamped limit, i.e., the possible existence of nontrivial solutions for the stationary velocity. For further discussion, we neglect the stochastic term in (2.53) and denote the stationary values of $\boldsymbol{v}(t)$ by $\boldsymbol{v}_0(\boldsymbol{x})$. Further, the force resulting from the gradient of the potential, $\boldsymbol{F}(\boldsymbol{x}) = -\boldsymbol{\nabla} U$, is introduced. Then, (2.53) can be rewritten as

$$\Big\{ d_2 \gamma_0\, v_0^2 - d_2 \boldsymbol{F}\, \boldsymbol{v}_0 - \Big[q(\boldsymbol{x}) d_2 - c\gamma_0 \Big] \Big\}\, \boldsymbol{v}_0 = c\boldsymbol{F} . \tag{2.54}$$

Depending on the value of $\boldsymbol{F}$ and in particular on the sign of the term $[q(\boldsymbol{x})d_2 - c\gamma]$, (2.54) has either one or three real solutions for the stationary velocity $\boldsymbol{v}_0$. The always existing solution,

$$\boldsymbol{v}_0^{(p)}(\boldsymbol{x}) \sim \boldsymbol{F}(\boldsymbol{x}) , \tag{2.55}$$

expresses a direct response to the force, i.e., it results from the analytic continuation of Stokes' law, $\boldsymbol{v}_0 = \boldsymbol{F}/\gamma_0$, which is valid for $d_2 = 0$. We denote this solution as the "normal" mode of motion because the velocity $\boldsymbol{v}$ has the same direction as the force $\boldsymbol{F}$ resulting from the external potential $U(x)$.

As long as the supply from the energy depot is small, we will also name the normal mode as the *passive mode* because the particle is simply driven by the external force. More interesting is the case of three stationary velocities, $\boldsymbol{v}_0$, which significantly depends on the (supercritical) influence of the energy depot. We will denote the two additional solutions, $\boldsymbol{v}_0^{(a)}(\boldsymbol{x})$, as the "high velocity" or *active mode* of motion [469, 506]. For one-dimensional motion, in the active mode only two different directions are possible, but already in the two-dimensional case there are infinitely many different possibilities. This conclusion is of importance when discussing stochastic influences.

In Sects. 2.2.2, 2.2.4, and 2.3, we will discuss the passive and active modes of motion for three different potentials, a constant, a piecewise linear, and a two-dimensional parabolic potential.

2.2.2 Stationary Velocities and Critical Parameters for $U = \mathrm{const}$

In the following, we restrict the discussion to the *deterministic* motion of a particle, corresponding to $S = 0$ in (2.53), and a constant potential that results in $\boldsymbol{F} = 0$. That means that the particle is not driven by stochastic or deterministic forces, but it is initially nonstationary. In our investigations, the particle is assumed to have a certain initial velocity $\boldsymbol{v}(t=0) \neq 0$. Initially, the internal energy depot has no energy: $e(0) = 0$. During its motion, the particle takes up energy (q), but it also loses energy because of internal dissipation (c) and because of the friction (γ_0), which is not compensated for, now. We may assume a supply of energy continuous in space, i.e., $q(\boldsymbol{x}) = q_0$, (2.13). Then, (2.54) for stationary velocities is reduced to

$$\left[d_2\gamma_0\, v_0^2 - (q_0 d_2 - c\gamma_0)\right] \boldsymbol{v}_0 = 0\,. \tag{2.56}$$

The "normal" or *passive mode* for stationary motion is simply given by the solution, $\boldsymbol{v}_0^{(p)} = 0$. For the passive mode, the equation for $e(t)$ can be easily integrated, and we find with $e(0) = 0$ that

$$e(t) = \frac{q_0}{c}\left[1 - \exp(-ct)\right]. \tag{2.57}$$

That means that the value of the internal energy depot reaches, asymptotically, a constant saturation value, $e_0^{(p)} = q_0/c$, while the particle is at rest.

The "high-velocity" or *active mode* for stationary motion is given by the solution,

$$(v_0^{(a)})^2 = \frac{q_0}{\gamma_0} - \frac{c}{d_2}\,. \tag{2.58}$$

These solutions are real only if the condition

$$d_2 q_0 \geq \gamma_0 \, c \tag{2.59}$$

is satisfied. Otherwise, the dissipation of energy exceeds the take-up of energy, and the particle comes to rest. Thus, we find $q_0 d_2 = c\gamma_0$ as the *critical condition* for the existence of the active mode when U = const. Inserting the solution for the active mode, $(v_0^a)^2$, (2.58), in the equation for the energy depot, we find the stationary solution,

$$e_0^{(a)} = \frac{\gamma_0}{d_2} \, . \tag{2.60}$$

The three possible stationary solutions can be shown in a bifurcation diagram, Fig. 2.8, where the bifuraction point is determined by

$$d_2^{\mathrm{crit}} = \frac{\gamma_0 \, c}{q_0} \, . \tag{2.61}$$

This relation defines a critical conversion rate d_2^{crit}, dependent on the internal dissipation c, the energy influx q_0, and the friction coefficient, γ_0. Alternatively, a critical friction coefficient γ_0^{crit} can also be defined by the same relation, as used in Fig. 2.9.

Below d_2^{crit}, the passive mode of motion, $v_0^{(1)} = 0$, is the only stable solution. Above the critical conversion rate, however, Fig. 2.8 shows the occurrence of two new solutions corresponding to the active modes of motion. For $\boldsymbol{F} = 0$, $v_0^{(2)}$ and $v_0^{(3)}$ both have the same amount but different directions, (2.58). Which of the possible directions is realized depends on the initial conditions in the deterministic case.

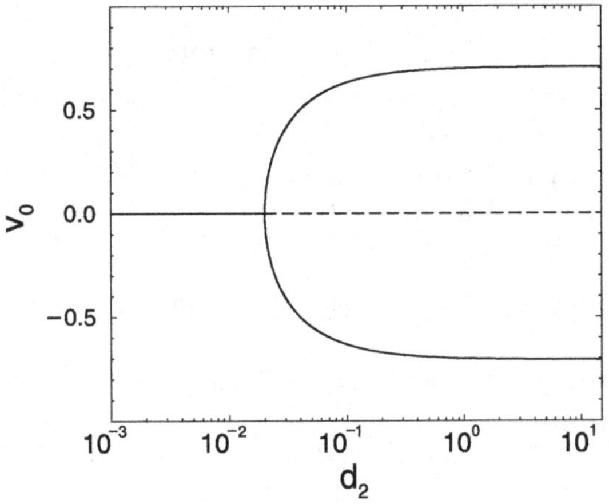

Fig. 2.8. Bifurcation diagram for stationary velocity v_0, dependent on the conversion rate of internal into kinetic energy d_2 for $\boldsymbol{F} = 0$. Parameters: $q_0 = 10$; $c = 0.01$; $\gamma_0 = 20$ [469]

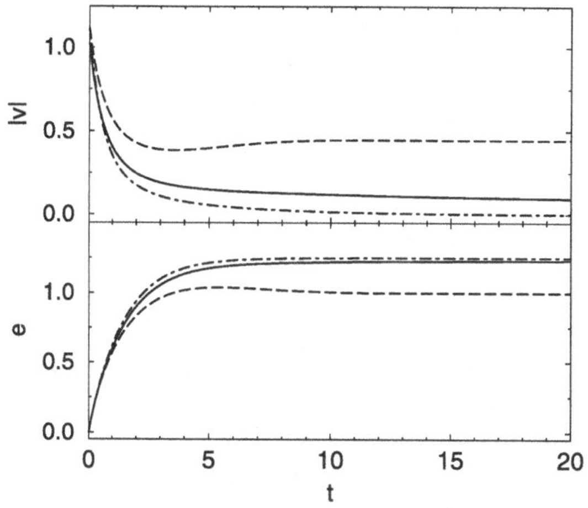

Fig. 2.9. Absolute value of velocity $|\boldsymbol{v}|$ and internal depot energy e vs. time for three different values of the friction coefficient γ_0: 1.0 (*dashed line*), which is below the critical value; 1.25 (*solid line*), which is the critical value; 1.5 (*dot-dashed line*), above the critical value. Parameters: $q_0 = 1.0$, $d_2 = 1.0$, $c = 0.8$; initial conditions: $\boldsymbol{x} = 0.0$, $\boldsymbol{v} = 1.0$ [126]

To demonstrate how the stationary values for the internal energy depot and the velocity are reached, the set of equations (2.46) has been integrated with $\boldsymbol{\nabla} U = 0$, $S = 0$, $q(\boldsymbol{x}) = q_0$ and different values of γ_0 (see Fig. 2.9).

The asymptotic values in Fig. 2.9 agree with the stationary solutions in (2.57) and (2.60), (2.58), respectively. For γ_0^{crit}, the critical slowing down for relaxation into the stationary state is also shown in Fig. 2.9.

With respect to energy input, conversion of energy, loss of energy due to internal and external dissipation, and energy output (motion in a potential), we may consider the active particle as a *micromotor* [126, 453]. Molecular motors based on Brownian motion have been recently introduced [22, 260, 317]. Taking into account the energy balance, we can discuss the *efficiency ratio* σ, which is defined as the ratio between the input of energy per time interval, dE_{in}/dt, and the output of energy per time interval, dE_{out}/dt. The input is simply given by $q(\boldsymbol{x})$, which describes the take-up of external energy per time interval. Other energy sources, like active friction or stochastic forces, are neglected here. The output is defined as the amount of mechanical energy available from the micromotor, $d(\boldsymbol{v})e(t)$, (2.11). With the ansatz for $q(\boldsymbol{x})$ (2.13) and $d(\boldsymbol{v})$ (2.3), the efficiency ratio can be expressed as follows:

$$\sigma = \frac{dE_{\text{out}}/dt}{dE_{\text{in}}/dt} = \frac{d(\boldsymbol{v})\, e(t)}{q(\boldsymbol{x})} = \frac{d_2\, e\, v^2}{q_0}\,. \tag{2.62}$$

Assuming again very fast relaxation of the internal energy depot, which then can be described by (2.126), (2.62) is modified to

$$\sigma = \frac{d_2\, v^2}{c + d_2\, v^2}\,. \tag{2.63}$$

Inserting the stationary velocity in the active mode, (2.58), we find for the efficiency ratio in the stationary limit,

$$\sigma = 1 - \frac{c\,\gamma_0}{d_2\, q_0}\,. \tag{2.64}$$

Equation (2.64) concludes the above discussions. The efficiency ratio, which is 1 only in the ideal case, is decreased by dissipative processes, like (passive) friction (γ_0) and internal dissipation (c). Moreover, σ is larger than zero only if the conversion rate is above the critical value $d_2^{\rm crit}$ (2.61).

We would like to mention that these considerations are based on a deterministic description of the moving particle. However, for microscopically small objects, the influence of noise, i.e., the stochastic force S as another source of energy, cannot be neglected, which may result in a modification of (2.62)–(2.64).

2.2.3 Stationary Solutions for a Linear Potential $U = ax$

We restrict the discussion again to the one-dimensional deterministic motion of a particle, corresponding to $S = 0$ in (2.46). The flux of energy into the internal depot of the particle is constant: $q(\boldsymbol{x}) = q_0$. Further, we may assume that the force resulting from the gradient of the potential is piecewise constant or very slowly varying in time, respectively: $\boldsymbol{F} = -\boldsymbol{\nabla} U = \text{const}$ in a fixed region of $\boldsymbol{x}$.

Then, we have two coupled equations for $\boldsymbol{v}(t)$ and $e(t)$:

$$\begin{aligned} \dot{\boldsymbol{v}} &= -\Big[\gamma_0 - d_2 e(t)\Big]\boldsymbol{v} + \boldsymbol{F}\,, \\ \dot{e} &= q_0 - ce - d_2 v^2 e\,. \end{aligned} \tag{2.65}$$

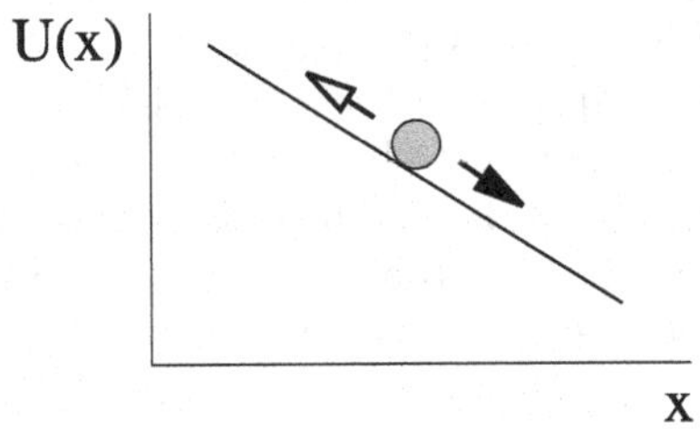

Fig. 2.10. Sketch of the one-dimensional deterministic motion of a particle in the presence of a constant force $\boldsymbol{F} = -\boldsymbol{\nabla} U(x) = \text{const}$. Provided with a supercritical amount of energy from the depot, the particle might be able to move "uphill", i.e., against the direction of the force

The stationary solutions of (2.65) are obtained from $\dot{\boldsymbol{v}} = 0$ and $\dot{e} = 0$:

$$\boldsymbol{v}_0 = \frac{\boldsymbol{F}}{\gamma_0 - d_2 e_0} \quad , \quad e_0 = \frac{q_0}{c + d_2 v_0^2} \, , \tag{2.66}$$

which lead to the known cubic polynomial for the amount of constant velocity v_0, (2.54), which is reprinted here:

$$d_2\gamma_0 \boldsymbol{v}_0^3 - d_2 \boldsymbol{F} v_0^2 - (q_0 d_2 - c\gamma_0)\boldsymbol{v}_0 - c\boldsymbol{F} = 0 \, . \tag{2.67}$$

$\boldsymbol{v}_0^{n+1}$ is defined as a vector $|v_0|^n \, \boldsymbol{v}_0$.

For $\boldsymbol{F} = \text{const} \neq 0$, the bifurcation diagram resulting from (2.67) is shown in Fig. 2.11 that should be compared with Fig. 2.8 for $\boldsymbol{F} = 0$.

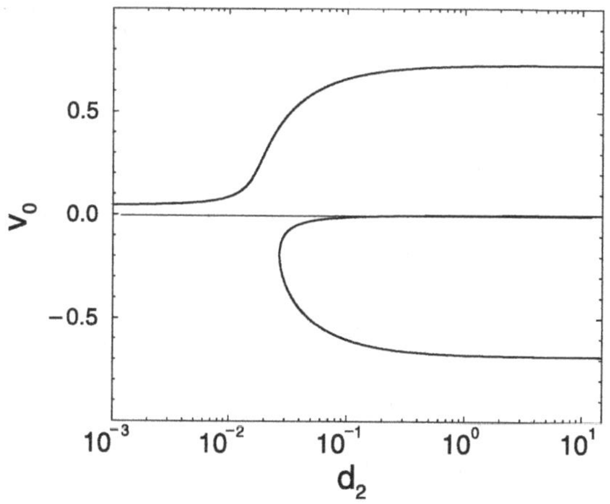

Fig. 2.11. Stationary velocities v_0, (2.67), vs. conversion rate d_2 for $\boldsymbol{F} = +7/8$, which should be compared to Fig. 2.8. Above a critical value of d_2, a negative stationary velocity indicates the posssibility of moving against the direction of the force. For parameters, see Fig. 2.8 [469]

As discussed in Sect. 2.2.1, (2.67) has either one or three real solutions for the stationary velocity $\boldsymbol{v}_0$. The passive mode $\boldsymbol{v} \sim \boldsymbol{F}$, (2.55), where only one solution exists, corresponds to a situation where the driving force $\boldsymbol{F}$ is small, but the friction of the particle is strong. If the influence of the energy depot is neglected, we find simply that $\boldsymbol{v}_0^{id} = \boldsymbol{F}/\gamma_0$ is the (small) stationary velocity in the direction of the driving force.

Above a critical supply of energy, two high velocity or active modes of motion appear. One of these active modes has the same direction as the driving force; thus it can be understood as the continuation of the normal solution. As Fig. 2.11 shows, the former passive normal mode, which holds for subcritical energetic conditions, is transformed into an active normal mode, where the particle moves in the same direction, but at a much higher velocity.

Additionally, in the active mode, a new high-velocity motion *against* the direction of the force $\boldsymbol{F}$ becomes possible. Although the first active mode would be considered rather a *normal response* to the force $\boldsymbol{F}$, the second active mode appears as an unnormal (or nontrivial) response, which corresponds to an "uphill" motion (see Fig. 2.10).

It is obvious that the particle's motion "downhill" is stable, but the same does not necessarily apply to the possible solution of an "uphill" motion. Thus, in addition to (2.67) that provides the *values* of the stationary solutions, we need a second condition that guarantees the *stability* of these solutions.

For the stability analysis, we consider small fluctuations around the stationary values, v_0 and e_0:

$$v = v_0 + \delta v \ , \quad e = e_0 + \delta e \ , \quad \left|\frac{\delta v}{v_0}\right| \sim \left|\frac{\delta e}{e_0}\right| \ll 1 \, . \tag{2.68}$$

Inserting (2.68) in (2.65), we find after linearization:

$$\begin{aligned} \dot{\delta v} &= \delta v \left(-\gamma_0 + d_2 e_0 \right) + \delta e \left(d_2 v_0 \right) , \\ \dot{\delta e} &= \delta v \left(-2 d_2 e_0 v_0 \right) \ \ + \delta e \left(-c - d_2 v_0^2 \right) . \end{aligned} \tag{2.69}$$

With the ansatz,

$$\delta v \sim \delta e \sim \exp(\lambda t) \, , \tag{2.70}$$

we find from (2.69) the following relation for λ [469]:

$$\begin{aligned} \lambda^{(1,2)} &= -\frac{1}{2}\left(\gamma_0 + c + d_2 v_0^2 - d_2 e_0\right) \\ &\pm \sqrt{\frac{1}{4}\left(\gamma_0 + c + d_2 v_0^2 - d_2 e_0\right)^2 - c(\gamma_0 - d_2 e_0) - d_2 v_0^2(\gamma_0 + d_2 e_0)} \, . \end{aligned} \tag{2.71}$$

In general, we need to discuss (2.71) for the three possible solutions of v_0, which result from (2.67). Depending on whether the λ for each solution has real or complex positive or negative values, we can classify the types of the possible stationary solutions in this case. The results are summarized in Table 2.1. The phase plots shown in Fig. 2.12a,b present more details. Further, Fig. 2.13 shows the real part $\Re(\lambda)$ of (2.71) for the active mode corresponding to the "uphill" motion of the particle, which is the most interesting.

We find that below the bifurcation point which is $d_2^{\mathrm{bif}} = 0.027$ for the given set of parameters, only one stable node exists in the $\{v, e\}$ phase space, which corresponds to the passive normal mode. Then, a subcritical bifurcation occurs that leads to three stationary solutions: a stable and an unstable node and a saddle point because all of the λ are real. At $d_2 = 0.046$, however, the nodes turn into focal points. With respect to the "uphill motion", we find in Fig. 2.13 the occurrence of an unstable node at $d_2 = 0.027$, which then becomes an unstable focus for $0.046 < d_2 < 2.720$. The respective real parts

Table 2.1. Results of the stability analysis, (2.71) [469]

0	$< d_2 <$	0.027	:	1 stable node
0.027	$< d_2 <$	0.046	:	1 stable node
				1 unstable node
				1 saddle point
0.046	$< d_2 <$	2.720	:	1 stable focal point
				1 unstable focal point
				1 saddle point
2.720	$< d_2$		:	2 stable focal points
				1 saddle point

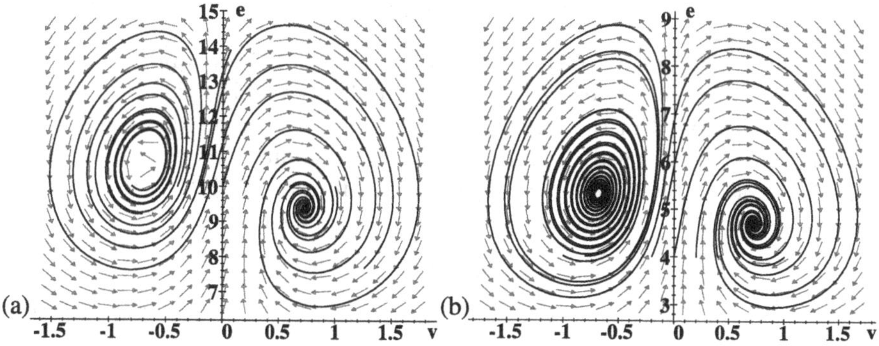

Fig. 2.12. Phase trajectories in the $\{v, e\}$ phase space for the motion in or against the direction of the driving force, which correspond either to positive or negative velocities. (**a**) $d_2 = 2.0$ corresponding to an unstable "uphill" motion; (**b**) $d_2 = 4.0$ corresponding to a stable "uphill" motion, respectively. For other parameters, see Fig. 2.8 [469]

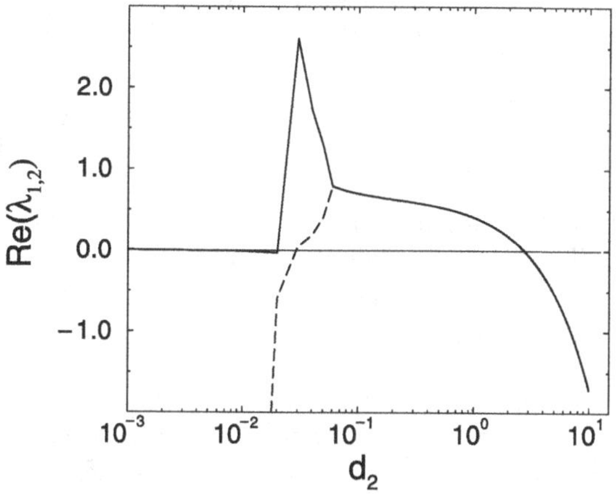

Fig. 2.13. Real part of $\lambda^{(1,2)}$, (2.71), vs. conversion parameter d_2 for the stationary motion against the force $\boldsymbol{F}$. For parameters, see Fig. 2.8 [469]

of λ are equal in this range, i.e., the $\lambda^{1,2}$ are complex. The stability condition is satisfied only if $\Re(\lambda) \leq 0$, which is above a second critical value $d_2^{\mathrm{crit}} = 2.72$ for the given set of parameters. That means, for $d_2 > 2.72$, the unstable focal point becomes a stable focus, which is also clearly shown in the phase plots of Fig. 2.12a,b. In both figures, we see a stable focal point for positive values of velocity v, which correspond to the stable motion "downhill", i.e., in the direction of the driving force. For $d_2 = 2.0 < d_2^{\mathrm{crit}}$, the phase plot for negative values of v shows an unstable focal point, which turns into a stable focal point for $d_2 = 4.0 > d_2^{\mathrm{crit}}$.

Thus, we can conclude that for $d_2 > d_2^{\mathrm{bif}}$, an active mode of motion becomes possible, which also implies the possibility of an "uphill" motion of a particle. However, only for values $d_2 > d_2^{\mathrm{crit}}$, can we expect a *stable* motion against the direction of the force.

For our further investigations, it will be useful to have a handy expression for the critical supply of energy, d_2^{crit}, which allows a stable "uphill" motion. This will be derived in the following, with only a few approximations [469]. For the parameters used during the computer simulations discussed later, Fig. 2.13 and Table 2.1 indicate that the square root in (2.71) is imaginary, thus the stability of the solutions depends on the condition,

$$\gamma_0 + c + d_2(v_0^2 - e_0) \geq 0 \,. \tag{2.72}$$

If we insert the stationary value e_0, (2.66), (2.72) leads to a fourth-order inequality for v_0 to obtain stability:

$$(\gamma_0 c - d_2 q_0) \leq v_0^4 \, d_2^2 + v_0^2 \, (\gamma_0 d_2 + 2cd_2) + c^2 . \tag{2.73}$$

For stable stationary motion of a particle, both (2.67) and (2.73) have to be satisfied.

The *critical condition* for stability results just from the equality in (2.73), which then provides a replacement for the prefactor $(\gamma_0 c - d_2 q_0)$ in (2.67). If we insert the critical condition in (2.67), we arrive at a fifth-order equation for v_0:

$$\boldsymbol{v}_0^5 + \boldsymbol{v}_0^3 \left(\frac{2c}{d_2}\right) + v_0^2 \left(\frac{\boldsymbol{F}}{d_2}\right) + \boldsymbol{v}_0 \left(\frac{c}{d_2}\right)^2 + \frac{c\boldsymbol{F}}{d_2^2} = 0 \,. \tag{2.74}$$

To simplify the further discussion, we assume that internal dissipation is negligible, $c = 0$. Then, (2.74) gives the simple nontrivial solution,

$$\boldsymbol{v}_0^3 = -\frac{\boldsymbol{F}}{d_2} \,, \qquad \text{if} \quad c = 0 \,. \tag{2.75}$$

This expression can be used to eliminate the stationary velocity $\boldsymbol{v}_0$ in (2.73). Assuming that $c = 0$, we obtain now from the critical condition, i.e., from the equality in (2.73), a relation between the force $\boldsymbol{F}$ and the conversion parameter d_2. Combining (2.73) and (2.75) results in

$$(-F)^{4/3} d_2 \, + \, \gamma_0 (-F)^{2/3} d_2^{2/3} \, - \, q_0 d_2^{4/3} = 0 \,. \tag{2.76}$$

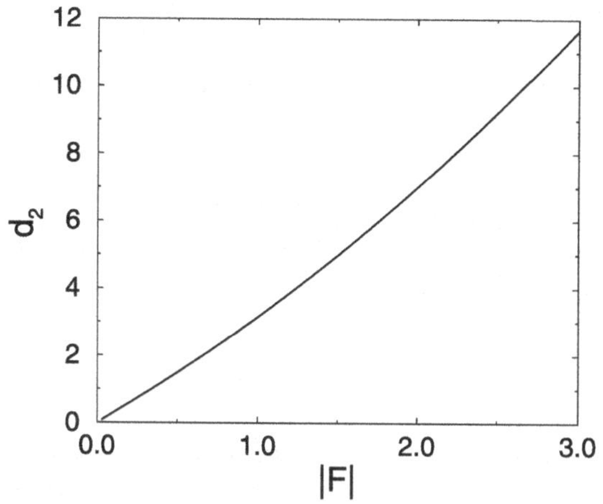

Fig. 2.14. Critical conversion rate $d_2^{\rm crit}$, (2.77), vs. amount of driving force $|F|$ to allow stable motion of a particle both "downhill" and "uphill" (see Fig. 2.10). For $d_2 > d_2^{\rm crit}$, the particle can also move against the direction of the force. Parameters: $\gamma_0 = 20$, $q_0 = 10$ [469]

Because $d_2 > 0$, the trivial and the negative solution of (2.76) can be neglected, and we finally arrive at the following *critical relation* for $d_2(F)$ [469]:

$$d_2^{\rm crit} = \frac{F^4}{8q_0^3}\left(1 + \sqrt{1 + \frac{4\gamma_0 q_0}{F^2}}\right)^3 . \tag{2.77}$$

In the limit of negligible internal dissipation, this relation describes how much power has to be supplied by the internal energy depot to allow stable motion of a particle in *both* directions, in particular, *stable uphill motion* of the particle. Figure 2.14 shows the function $d_2^{\rm crit}(F)$ for a strongly damped motion of a particle.

The interesting result of a possible stationary "uphill motion" of particles with an internal energy depot will be elucidated in the following section, where we turn to a more sophisticated, piecewise linear potential.

2.2.4 Deterministic Motion in a Ratchet Potential

For further investigation of the motion of active particles, we specify the potential $U(\boldsymbol{x})$ as a piecewise linear, asymmetrical potential (see Fig. 2.15), which is known as a *ratchet potential*:

$$U(x) = \begin{cases} \frac{U_0}{b}(x - nL) & \text{if } nL \le x \le nL + b \quad (n = 0, 1, 2, \dots) \\ \frac{U_0}{L-b}[(n+1)L - x] & \text{if } nL + b \le x \le (n+1)L \; (n = 0, 1, 2, \dots) . \end{cases} \tag{2.78}$$

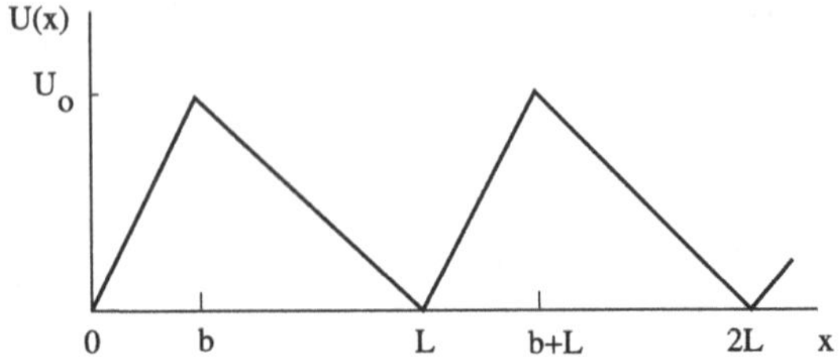

Fig. 2.15. Sketch of the asymmetrical potential $U(x)$, (2.78). For the computer simulations, the following values are used: $b = 4$, $L = 12$, $U_0 = 7$ in arbitrary units

The actual position $\hat{x}$ of a particle moving in the periodic ratchet potential can then also be expressed in cyclic coordinates, $x \equiv \hat{x} \bmod L$ with $\hat{x} = x + kL$, $k = 0, \pm 1, \pm 2, ...$ and $0 \leq x \leq L$. The motion of active particles is still described by the set of equations, (2.46).

Further, we will use the following abbreviations with respect to the potential $U(x)$, (2.78). The index $i = \{1, 2\}$ refers to the two pieces of the potential, $l_1 = b$, $l_2 = L - b$. The asymmetry parameter a should describe the ratio of the two pieces, and $\boldsymbol{F} = -\boldsymbol{\nabla} U = \text{const}$ is the force resulting from the gradient of the potential, which is assumed constant. Hence, for the potential $U(x)$, (2.78), the following relations hold:

$$
\begin{aligned}
F_1 &= -\frac{U_0}{b}, \quad F_2 = \frac{U_0}{L-b}, \quad a = \frac{l_2}{l_1} = \frac{L-b}{b} = -\frac{F_1}{F_2}, \\
F_1 &= -\frac{U_0}{L}(1+a), \quad F_2 = \frac{U_0}{L}\frac{1+a}{a}.
\end{aligned} \tag{2.79}
$$

Whether or not a particle can leave one of the potential wells described by (2.78) depends in a first approximation on the height of the potential barrier U_0 and on the kinetic energy of the particle. For particles with an internal energy depot, the actual velocity also depends on the conversion of internal into kinetic energy, (2.46). To elucidate the class of possible solutions for the dynamics specified, let us first discuss the phase-space trajectories for the *deterministic* motion, i.e, $S = 0$. Due to friction, a particle moving in the ratchet potential will eventually come to rest in one of the potential wells because the dissipation is not compensated for by the energy provided from the internal energy depot. The series of Fig. 2.16 shows the corresponding attractor structures for the particle's motion depending on the supply of energy expressed in terms of the conversion rate d_2. In Fig. 2.16a, we see that for a subcritical supply of energy expressed in terms of the conversion rate d_2, only *localized* states of the particles exist. The formation of limit cycles inside each minimum corresponds to stable oscillations in the potential well, i.e., the particles cannot escape from the potential well.

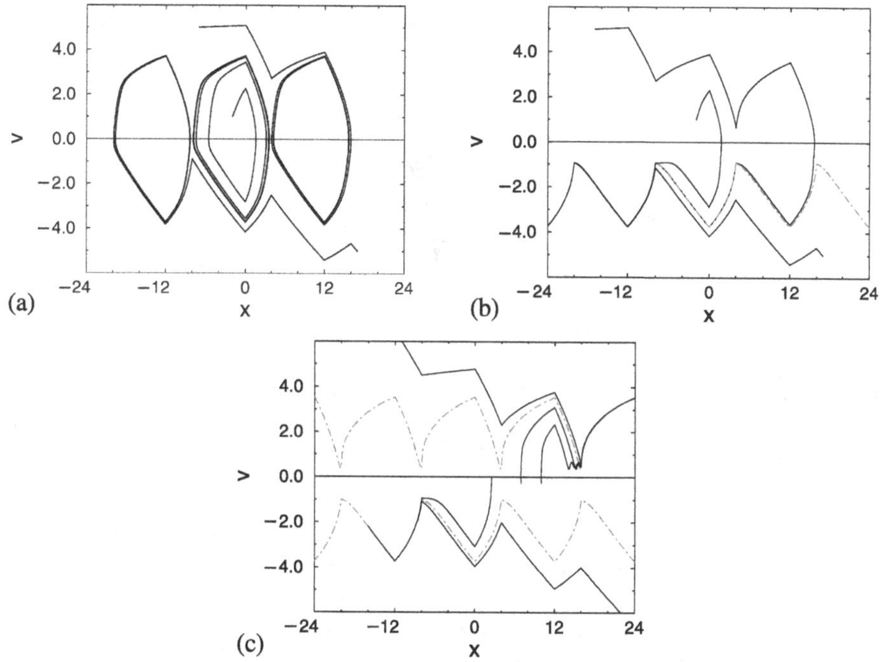

Fig. 2.16. Phase-space trajectories of particles starting with different initial conditions for three different values of the conversion parameter d_2: (**a**) $d_2 = 1$, (**b**) $d_2 = 4$, (**c**) $d_2 = 14$. Other parameters: $q_0 = 1$, $c = 0.1$, $\gamma_0 = 0.2$. The *dashed-dotted lines* in the middle and bottom parts show the unbounded attractor of the delocalized motion which is obtained in the long time limit [506]

With increasing d_2, the particles can climb up the potential flank with the lower slope, and in this way escape from the potential well in the negative direction. As Fig. 2.16b shows, this also holds for particles that initially start in the positive direction. Thus, we find an unbounded attractor corresponding to *delocalized motion* for negative values of v. Only if the conversion rate d_2 is large enough to allow uphill motion along the flank with the steeper slope, can particles escape from the potential well in *both* directions, and we find two unbounded attractors corresponding to *delocalized* motion in both positive and negative directions.

To conclude these investigations, we find that the structure of the phase space of an active particle in an unsymmetrical ratchet potential may be rather complex. Although bounded attractors (limit cycles) in each potential well are observed at low pumping rates, at increasing pumping rates, one/two new unbounded attractors are formed, which correspond to a directed stationary transport of particles in negative/positive directions. Both types of motion have analogies to the localized and delocalized states of electrons in solid-state physics.

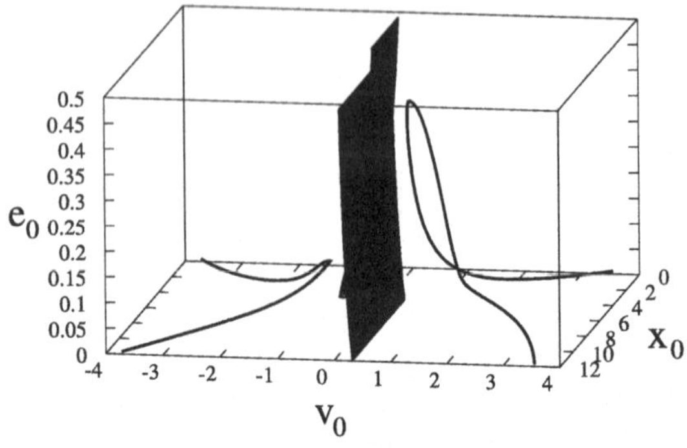

Fig. 2.17. Separatrix and asymptotic trajectories in the $\{x, v, e\}$ phase space. For parameters, see Fig. 2.16c. To obtain the separatrix, we have computed the velocity v_0 for given values of x_0 and e_0 (in steps of 0.01) using the Newtonian iteration method with a accuracy of 0.0001 [506]

Considering the three-dimensional phase space $\{x, v, e\}$, the two stationary solutions for the unbounded motion are separated by a two-dimensional separatrix plane. This plane has to be periodic in space because of the periodicity of the ratchet potential. To get an idea of the shape of the separatrix, we performed computer simulations which determined the direction of motion of one particle for various initial conditions. Figure 2.17 shows the respective trajectories for movement in both directions and the separatrix plane. We can conclude that, if a particle moves in the positive direction, most of the time the trajectory is very close to the separatrix. That means it will be rather susceptible to small perturbations, i.e., even small fluctuations might destabilize the motion in the positive direction. The motion in the negative direction, on the other hand, is not susceptible in the same manner because the respective trajectory remains at a considerable distance from the separatrix or comes close to the separatrix only for a very short time. This conclusion will be of importance when discussing current reversal in a ratchet potential in Sect. 4.

First, we discuss some more computer simulations of the *deterministic* motion of *one* active particle in a ratchet potential. The particle (mass $m = 1$) starts its motion outside the potential minimum; hence, there is an initial force on the particle. The results for a single particle are shown in Fig. 2.18, where two different sets of parameters are used:

1. a small internal dissipation c, which means a nearly ideal energy depot, and a large friction coefficient γ_0, resulting in a *strongly overdamped motion*;
2. an internal dissipation c that is 10 times larger, an energy influx q_0 ten times smaller, and a friction coefficient γ_0 100 times smaller than in (1), resulting in *weakly damped motion*.

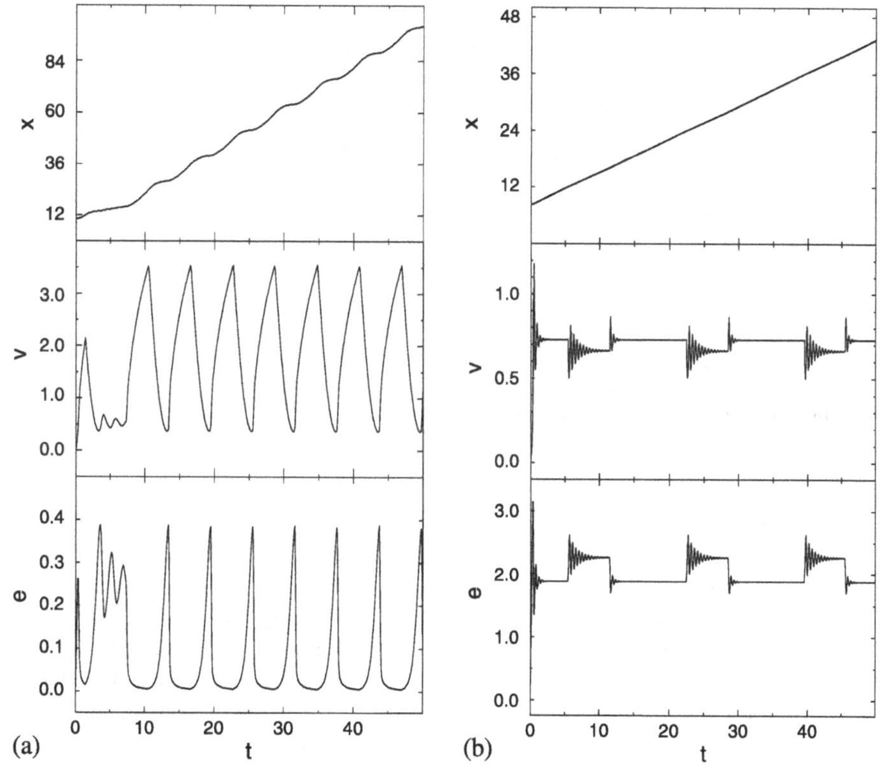

Fig. 2.18. Trajectory $x(t)$, velocity $v(t)$, and energy depot $e(t)$ for a single particle moving in a ratchet potential (see Fig. 2.15). Parameters: (**a**) $q_0 = 1.0$, $\gamma_0 = 0.2$, $c = 0.1$, $d_2 = 14.0$; (**b**) $q_0 = 10$, $\gamma_0 = 20$, $c = 0.01$, $d_2 = 10$, initial conditions: $x(0) \in [4, 12]$, $v(0) = 0$, $e(0) = 0$ [469]

We note that in the computer simulations the complete set of (2.46) for particles is always solved, regardless of these approximations. The trajectories $\boldsymbol{x}(t)$ in Fig. 2.18 indicate nearly uniform motion of a particle in one direction, which means that $\dot{\boldsymbol{x}}$ is almost constant. Figure 2.19a, however, reveals small periodic modulations in the trajectory for the overdamped case, which map the shape of the ratchet potential. These modulations occur even stronger if the motion of the particle is less damped.

As shown in Fig. 2.18a, the less damped motion of a single particle may result in steady oscillations in the velocity and energy depot. Only if the damping is large enough, may the velocity and the energy depot reach constant values (see Fig. 2.18b). These values are, of course, different for each piece of the potential; hence the periodic movement through every maximum or minimum of the potential results in jumps in both the velocity and the energy depot, which are followed by oscillations. In the phase space shown

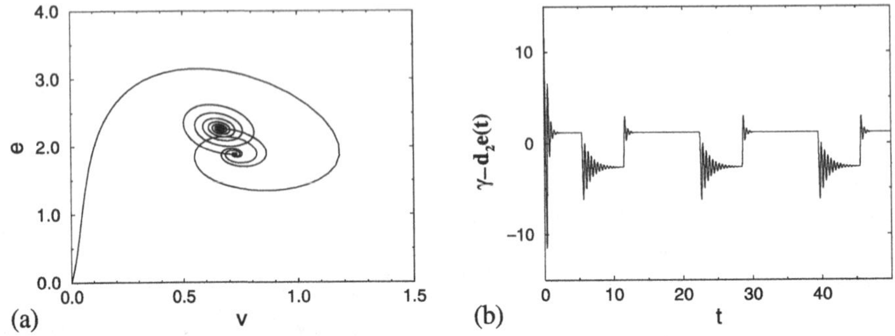

Fig. 2.19. (**a**) Energy depot $e(t)$ vs. velocity $v(t)$; (**b**) prefactor: $\gamma_0 - d_2 e(t)$ for the overdamped motion of a single particle (data from Fig. 2.18b) [469]

in Fig. 2.19a, the motion of the particle appears as a transition between two stable fixed points, each of which describes the stable motion on one flank.

In the following, we restrict the discussion to the *strongly damped case.* The oscillations that occur in v and e are damped out on a characteristic timescale $\tau = 1/\gamma_0$. If we assume that a particle moves on the different pieces of the potential $\{b,\ L-b\}$ (see Fig. 2.15) during the two characteristic time intervals, $T_b = b/v_b$ and $T_L = (L-b)/v_{(L-b)}$, then the particle is subject to a constant force only as long as $\tau \ll T_b$ or $\tau \ll T_L$, respectively. For times larger than the characteristic time τ, the motion of the particle can be described by the equation of the overdamped limit, (2.52). If we neglect again stochastic influences, (2.52) can be rewritten in the form,

$$0 = -[\gamma_0 - d_2 e_0]\, \boldsymbol{v}_0 + \boldsymbol{F}\,, \tag{2.80}$$

where $|\dot{\boldsymbol{x}}| = |\boldsymbol{v}_0| = \text{const}$ is the velocity in the overdamped limit and $\boldsymbol{F} = \{F_1, F_2\}$ is defined by (2.79). The stationary value for the internal energy depot, e_0, is given by (2.66), which, in the limit of a nearly ideal energy depot, reads

$$e_0 = \frac{q_0}{d_2 v_0^2}\,, \qquad \text{if } c \ll d_2 v_0^2\,. \tag{2.81}$$

The constant velocity can be calculated from the stationary condition, (2.67), with $\boldsymbol{F}$ specified as F_1 or F_2, respectively. If we assume a nearly ideal internal energy depot, (2.67) can be simplified with the assumption that $c = 0$, and we find for the three stationary velocities,

$$v_0^{(1)} = 0\,, \quad v_{0\,i}^{(2,3)} = \frac{F_i}{2\gamma_0} \pm \sqrt{\frac{F_i^2}{4\gamma_0^2} + \frac{q_0}{\gamma_0}} \qquad (i = 1, 2)\,. \tag{2.82}$$

For any constant force F, there are two possible nontrivial solutions of (2.82): positive and negative velocities with different, but constant amounts, which depend on the gradient of the potential. The nontrivial values $v_0 \neq 0$ of (2.82) can be compared with the constant values obtained in the simulations, and we find that

	Fig. 2.18b	**(2.82)**	
Lower value of v_0	0.665	0.664	(2.83)
Upper value of v_0	0.728	0.729 .	

The slight differences result from the assumption that $c = 0$ used for deriving (2.82).

Equations (2.52), (2.80) for the overdamped limit indicate that the dynamics depends remarkably on the sign of the prefactor $\gamma_0 - d_2 e_0$, which governs the influence of the potential and the stochastic force. Therefore, the prefactor should be discussed in more detail now. Figure 2.19b shows that the prefactor $\gamma_0 - d_2 e(t)$ displays behavior similar to that of the velocity, Fig. 2.18b. The prefactor jumps between a positive and a negative constant value, which can be approximated by the constant, e_0, (2.66), reached after the oscillations damp out. It is shown that the jump occurs at the same time when the gradient of the potential changes its sign. This can also be proved analytically. Using (2.81), (2.82), the prefactor $\gamma_0 - d_2 e_0$ in (2.52) and (2.80), respectively, can be rewritten, and we find after a short calculation,

$$\frac{1}{\gamma_0 - d_2 e_0} = \frac{1}{2\gamma_0 F_i}\left(F_i \pm \sqrt{F_i^2 + 4q_0\gamma_0}\right) . \tag{2.84}$$

This means that the product of the prefactor and the potential gradient always has the same positive (or negative) sign, and the direction of motion for the particle is determined only by the initial condition.

The prefactor $\gamma_0 - d_2 e_0$ describes the balance between dissipation and the energy supply from the internal depot of the particle. Therefore, it is expected that the time average of the prefactor,

$$\langle \gamma_0 - d_2 e_0 \rangle = \frac{1}{T}\int_0^T (\gamma - d_2 e_0)\, dt , \tag{2.85}$$

should be zero, when averaged over one time period T. Considering the ratchet potential, Fig. 2.15, with the abbreviations, (2.79), we assume that $T = t_1 + t_2$, where t_1 is the time the particle moves on piece $l_1 = b$ of the potential U and t_2 is the time the particle moves on the remaining piece, $l_2 = L - b$. v_{01}, v_{02} and e_{01}, e_{02} should be the related velocity and energy depot along these pieces. Let us consider forward motion, i.e., $v_{0i} > 0$; then the velocities are given by the positive solutions of (2.82). Because $t_i = v_{0i} l_i$ holds for the motion along the piecewise linear potential, the time average, (2.85), can be calculated for this case as

$$\langle \gamma_0 - d_2 e_0 \rangle \; T = \gamma_0 \left(v_{01} l_1 + v_{02} l_2\right) - d_2 \left(e_{01} v_1 l_1 + e_{02} v_2 l_2\right) . \tag{2.86}$$

Due to (2.80), $d_2 e_{0i} v_{0i} = \gamma_0 v_{0i} - F_i$ holds. Using again the assumption of a nearly ideal energy depot, we find with (2.82) for v_{0i},

$$\gamma_0 v_{0i} l_i = \frac{l_i}{2}\left(F_i + \sqrt{F_i^2 + 4\gamma_0 q_0}\right),$$

$$d_2 e_i v_{0i} l_i = l_i \left(\frac{F_i}{2\gamma_0} + \sqrt{\frac{F_i^2}{4\gamma_0} + \frac{q_0}{\gamma_0}} - F_i\right), \tag{2.87}$$

which results in a time average, (2.85), equal to zero:

$$\langle \gamma_0 - d_2 e_0 \rangle \, T = 0 \,. \tag{2.88}$$

2.2.5 Investigation of the Net Current

Let us now discuss the *deterministic* motion of an *ensemble* of N active particles in a ratchet potential [469]. For the computer simulations, we have assumed that the starting locations of the particles are equally distributed over the first period of the potential, $\{0, L\}$. In the deterministic case, the *direction of motion* and the *velocity* at any time t are determined mainly by the initial conditions. Hence, particles with an initial position between $\{0, b\}$, which initially feel a force in the negative direction, most likely move at a negative velocity, whereas particles with an initial position between $\{b, L\}$ most likely move in the positive direction. This is also shown in Fig. 2.20a, where the velocity v is plotted versus the initial positions of the particles. Oscillations occur only at the minima and maxima of the related potential, indicating a strong sensitivity to the initial condition in these regions. The distribution of the final velocity is shown in Fig. 2.20b.

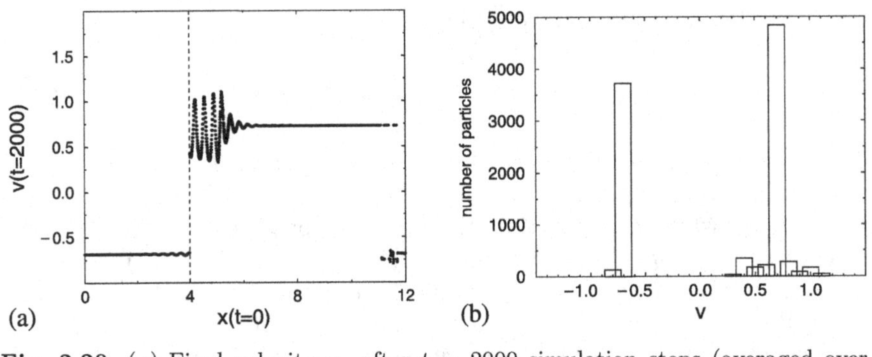

Fig. 2.20. (**a**) Final velocity v_e after $t = 2000$ simulation steps (averaged over 10,000 particles) vs. initial location x_0 of the particles. (**b**) Distribution of the final velocity v_e. For parameters, see Fig. 2.18b. The initial locations of the particles are equally distributed over the first period of the ratchet potential $\{0, L\}$ [469]

From Fig. 2.20b, we see two main currents of particles occurring, one with a positive and one with a negative velocity, which can be approximated by (2.82). The net current, however, has a positive direction because most of the particles start with the matching initial condition. The time dependence of the averages is shown in Fig. 2.21. The long-term oscillations in the average velocity and the average energy depot result from the superposition of velocities, which are sharply peaked around the two dominating values (see Fig. 2.20b).

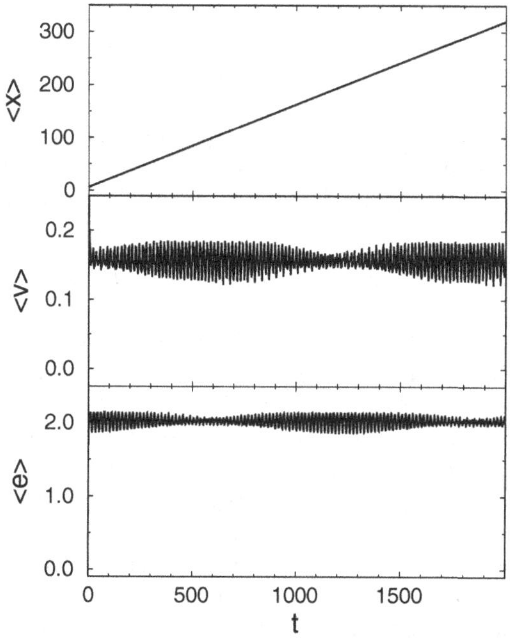

Fig. 2.21. Average location $\langle x \rangle$, velocity $\langle v \rangle$ and energy depot $\langle e \rangle$ of 10,000 particles vs. time t. For parameters, see Fig. 2.18b [469]

The existence of periodic stationary solutions, $\boldsymbol{v}_0(x) = \boldsymbol{v}_0(x \pm L)$, requires that the particles are able to escape from the initial period of the potential; hence, they must be able to move "uphill" on one or both flanks of the ratchet potential. In Sect. 2.2.3, we already investigated the necessary conditions for such a motion for a single flank and found a critical condition for the conversion rate d_2, (2.77). To demonstrate the applicability of (2.77) to the ratchet potential, we investigated the dependence of the *net current*, expressed by the mean velocity $\langle v \rangle$, on the conversion rate d_2 for the overdamped case. The results of computer simulations are shown in Fig. 2.22.

In Fig. 2.22, we see the existence of *two different critical values* for parameter d_2, which correspond to the onset of a *negative net current* at $d_2^{\rm crit1}$

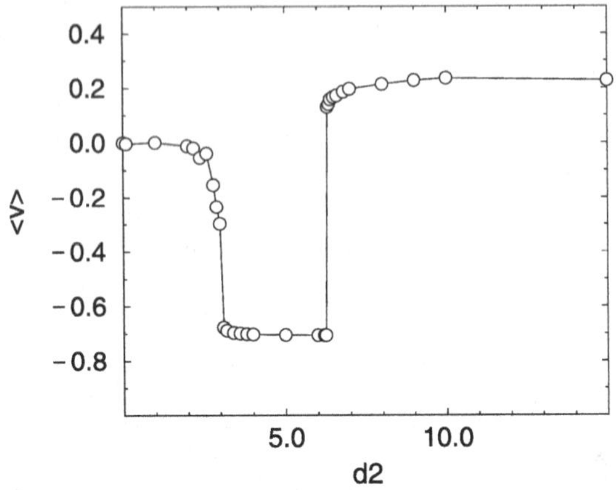

Fig. 2.22. Average velocity $\langle v \rangle$ vs. conversion parameter d_2. The data points are obtained from simulations of 10,000 particles with arbitrary initial positions in the first period of the ratchet potential. For other parameters, see Fig. 2.18b [469]

and a *positive net current* at d_2^{crit2}. For values of d_2 near zero and less than d_2^{crit1}, there is no net current at all. This is due to the subcritical supply of energy from the internal depot, which does not allow an uphill motion on any flank of the potential. Consequently, after the initial downhill motion, all particles come to rest in the minima of the ratchet potential, with $v_0 = 0$ as the only stationary solution for the velocity. With an increasing value of d_2, we see the occurrence of a negative net current at d_2^{crit1}. That means that the energy depot provides enough energy for the uphill motion along the flank with the lower slope, which, in our example, is that where $F = 7/8$ (see Fig. 2.15). If we insert this value for F in the critical condition, (2.77), a value $d_2^{\mathrm{crit1}} = 2.715$ is obtained, which agrees with the onset of the negative current in the computer simulations, Fig. 2.22.

For $d_2^{\mathrm{crit1}} \leq d_2 \leq d_2^{\mathrm{crit2}}$, stable motion of particles up and down the flank with the lower slope is possible, but the same does not necessarily apply to the steeper slope. Hence, particles that start on the lower slope at a positive velocity cannot continue their motion in the positive direction because they cannot climb up the steeper slope. Consequently, they turn their direction on the steeper slope, then move downhill driven by the force in the negative direction and continue to move in the negative direction while climbing up the lower slope. Therefore, for values of the conversion rate between d_2^{crit1} and d_2^{crit2}, we have only a *unimodal* distribution of the velocity centered around the negative value:

$$v_1 = \frac{F}{2\gamma_0} - \sqrt{\frac{F^2}{4\gamma_0^2} + \frac{q_0}{\gamma_0}} = -0.6855 \,, \quad \text{for } F = \frac{7}{8} \,. \tag{2.89}$$

For $d_2 > d_2^{\text{crit2}}$, the energy depot also supplies enough energy for the particles to climb up the steeper slope; consequently, periodic motion of particles in the positive direction becomes possible now. In our example, the steeper slope corresponds to the force $F = -7/4$ (see Fig. 2.15) which yields a critical value $d_2^{\text{crit1}} = 5.985$, obtained by means of (2.77). This result agrees with the onset of the positive current in the computer simulations, Fig. 2.22.

For $d_2 > d_2^{\text{crit2}}$, we have a *bimodal* velocity distribution, also as shown in Fig. 2.20b. The net current, which results from the average of the two main currents, has a positive direction in the deterministic case because most of the particles start in a positive direction, as discussed above. We may simply assume that the number of particles in each direction is roughly proportional to the length of the flank from which they started, which is also indicated by the velocity distribution, Fig. 2.20b. Then the mean velocity in the overdamped case can be approximated by

$$\langle v \rangle = \frac{1}{N} \sum_{i=1}^{N} v_i = \frac{1}{3} v_1 + \frac{2}{3} v_2 , \tag{2.90}$$

where v_1 and v_2 are the stationary velocities on each flank, which, in the limit of an nearly ideal energy depot, can be determined from (2.82). With the negative velocity, v_1, (2.89), and the positive velocity,

$$v_2 = \frac{F}{2\gamma_0} + \sqrt{\frac{F^2}{4\gamma_0^2} + \frac{q_0}{\gamma_0}} = 0.664 \quad \text{for} \quad F = -\frac{7}{4} , \tag{2.91}$$

we find from (2.90) for $d_2 > d_2^{\text{crit2}}$ an average velocity, $\langle v \rangle = 0.216$, which also agrees with the computer simulations, Fig. 2.22.

The results of the computer simulations have demonstrated that in the deterministic case, the direction of the net current can be adjusted by choosing the appropriate values of the conversion rate d_2. The critical values for d_2, on the other hand, depend on the slope of the two flanks of the potential, expressed by the force F. Lower slopes also correspond to lower values of the conversion rate because less power is needed for the uphill motion.

We conclude our results by investigating the influence of the slope on the establishment of a positive or negative net current. With a fixed height of the potential barrier U_0 and a fixed length L, the ratio of the two different slopes is described by the asymmetry parameter $a = l_2/l_1 = -F_1/F_2$, (2.79). The occurrence of a current in the ratchet potential requires the possibility of uphill motion, which depends on the critical supply of energy, described by (2.77). To obtain the critical value for the asymmetry of the potential, we replace the force F in (2.77) by the parameter a, (2.79). In our example, the flank l_1 of the potential has the steeper slope, so the critical condition is determined by $F_1 = U_0/L\,(1 + a)$. As a result, we find [469]

$$a^{\text{crit}} = \frac{L}{U_0} \left(-\frac{\gamma_0}{2} d_2^{-1/3} + \sqrt{\frac{\gamma^2}{4} d_2^{-2/3} + q_0 d_2^{1/3}} \right)^{3/2} - 1 . \tag{2.92}$$

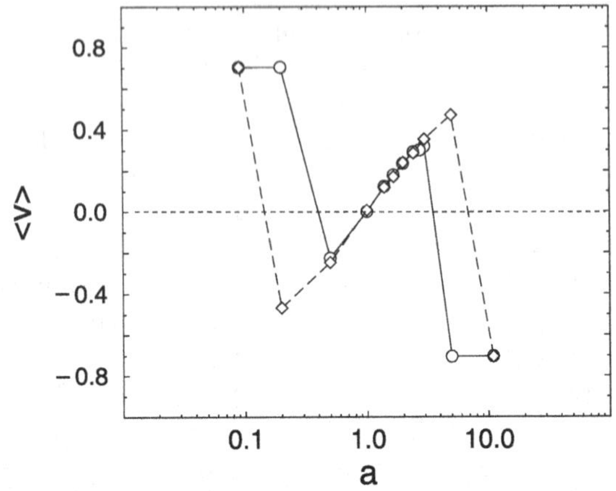

Fig. 2.23. Average velocity $\langle v \rangle$ vs. asymmetry parameter a, (2.79). The data points are obtained from simulations of 10,000 particles with arbitrary initial positions in the first period of the ratchet potential. (○): $d_2 = 10$ (—), (◇): $d_2 = 20$ (– –). For other parameters, see Fig. 2.18b [469]

$a^{\text{crit}} \geq 1$ gives the critical value for the asymmetry, which may result in a reversal of the net current. For $a > a^{\text{crit}}$, the flank l_1 is too steep for the particles, therefore only a negative current can occur which corresponds to the unimodal velocity distribution discussed above. For $1 < a < a^{\text{crit}}$, however, the particles can move uphill either flank. Hence, also a positive current can be established, and the velocity distribution becomes bimodal, which results in a positive net current.

The current reversal from a negative to a positive net current is shown in Fig. 2.23. Depending on the value of the conversion rate d_2, we see the switch from the negative to the positive value of the net current at a critical value of the asymmetry parameter a. Because of the definition of a, the results for $a < 1$, are the inverse of the results for $a > 1$. Obviously, for a symmetrical ratchet, $a = 1$, no net current occurs because the two main currents compensate. From (2.92), we obtain $a^{\text{crit}} = 3.5$ for $d_2 = 10$ and $a^{\text{crit}} = 6.5$ for $d_2 = 20$, both of which agree with the results of the computer simulations, Fig. 2.23. Further, the results of Fig. 2.23 show that the stationary velocities are independent of d_2 in the limit of a nearly ideal energy depot, which is also indicated by (2.82).

2.2.6 Stochastic Influences on the Net Current

To demonstrate the influence of fluctuations on the mean values $\langle x \rangle$, $\langle v \rangle$, $\langle e \rangle$, we add a *stochastic force* where $S > 0$ to the previous simulation of the ensemble of active particles. The first part of Fig. 2.24a shows again the simulation of the deterministic motion in the *overdamped case*, Fig. 2.21, for $t \leq 2000$, whereas the second part demonstrates the changes after a stochastic

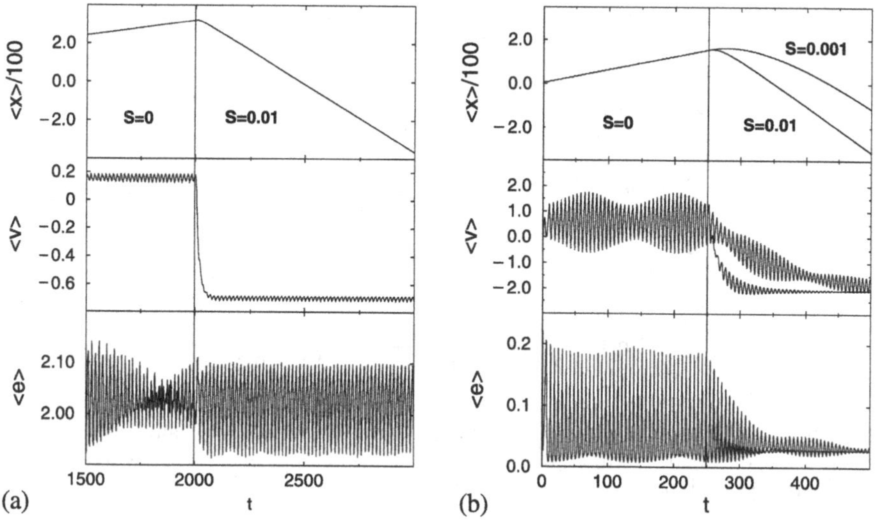

Fig. 2.24. Averaged location $\langle x \rangle$, velocity $\langle v \rangle$, and energy depot $\langle e \rangle$ versus time t for 10,000 particles. (**a**) Overdamped case: the stochastic force $S = 0.01$ is switched on at $t = 2000$ (for parameters, see Fig. 2.18b). (**b**) Less damped case: the stochastic force, either $S = 0.01$ or $S = 0.001$, is switched on at time $t = 250$ (for parameters, see Fig. 2.18a) [506]

force where $S = 0.01$ is switched on at $t = 2000$. The simulations show (i) that the current changes its direction when noise is present and (ii) that the average velocity, instead of oscillating, approaches a constant value. Different from the deterministic case, Fig. 2.20b, the related velocity distribution in the stochastic case now approaches a one-peak distribution (cf. Fig. 2.25). Hence, stochastic effects can stabilize the motion in the vicinity of the unbounded attractor which corresponds to the negative current, while they destabilize the motion in the vicinity of the unbounded attractor corresponding to the positive current, both shown in Fig. 2.16c.

A similar situation can also be observed when the motion of particles is *less damped* (see Fig. 2.24b). In this case also, we see now both the mean velocity and the mean energy depot of an ensemble of particles approaching constant values, instead of oscillating. Further, the timescale of the relaxation into stationary values decreases if the intensity of the stochastic force is increased. The related velocity distribution of the particles is plotted for different intensities of the stochastic force in Fig. 2.26. For $S = 0$, there are two sharp peaks at positive values for the speed and one peak for negative values; hence, the average current is positive. The three maxima correspond to the three existing solutions for stationary velocity in the deterministic case. For $S = 0.001$, the distribution of positive velocities is declining because most of the particles already move at a negative velocity, and for $S = 0.01$, after the same time, all particles have a negative vel-

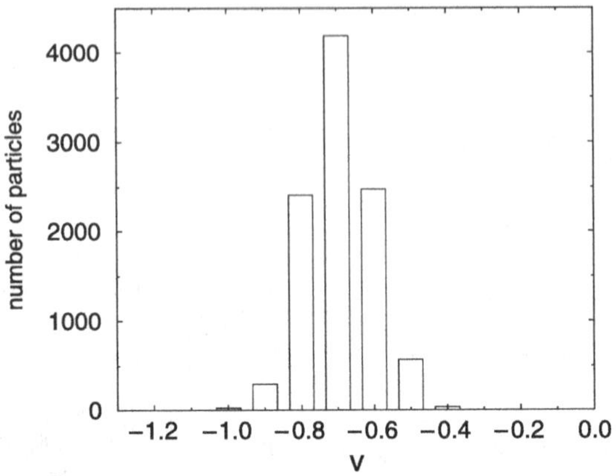

Fig. 2.25. Distribution of velocity v for the simulation of Fig. 2.24a (overdamped case) at time $t = 4000$, which means 2000 simulation steps after the stochastic force was switched on [506]

ocity. Further, the velocity distribution becomes broader in the presence of noise.

To conclude the above computer simulations for an ensemble of active particles moving in a ratchet potential, for supercritical conditions in the *deterministic case*, we find two currents in opposite directions related to a sharply peaked bimodal velocity distribution. The direction of the resulting net current is determined by the initial conditions of the majority of the particles. In the *stochastic case*, however, we find only a broad and symmetrical unimodal velocity distribution, resulting in a stronger net current. The direction of this net current is opposite that of the deterministic case and points in the negative direction; hence, stochastic forces can *change the direction* of motion of particles moving in the positive x direction.

With respect to the shape of the ratchet potential (see Fig. 2.15), this means that in the stochastic case, all particles move uphill along the flank with the *lower* slope. In the previous section, we have shown for the deterministic case, that above a critical value for the conversion parameter d_2^{crit2}, particle motion along the flank with the steeper slope would also be possible. This, however, depends on the initial conditions, which have to point in the positive direction. In the deterministic case, the particles will keep this direction provided that the energy supply allows them to move "uphill," which is the case for $d_2 > d_2^{\mathrm{crit2}}$. In the stochastic case, however, the initial conditions will be "forgotten" after a short time; hence, due to stochastic influences, the particle's "uphill" motion along the steeper flank will soon turn into a "downhill" motion. This motion in the negative direction will most likely be kept because less energy is needed. Thus, the stochastic fluctuations reveal the instability of an "uphill" motion along the steeper slope.

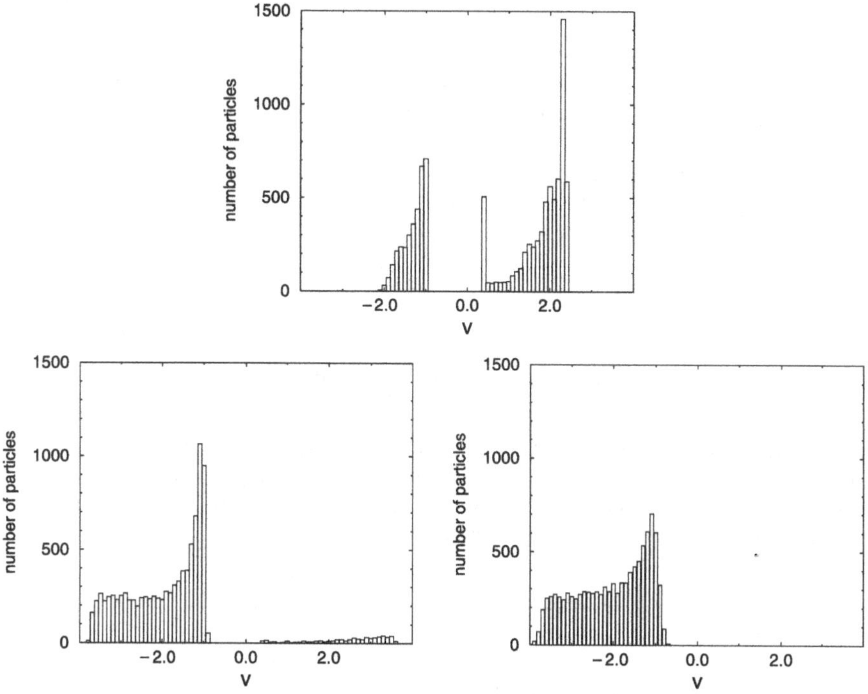

Fig. 2.26. Distribution of velocity v for the simulation of Fig. 2.24b (less damped case). (*above*) $S = 0.0$ ($t = 250$), (*left*) $S = 0.001$ ($t = 500$), (*right*) $S = 0.01$ ($t = 500$). The stochastic force is switched on at $t = 250$ [506]

Figure 2.27 shows, for less damped conditions, the average velocity for both the stochastic and the deterministic cases, the latter can be compared with Fig. 2.22 for the overdamped case. In the stochastic case, the net current is always negative in agreement with the explanation above. This holds even if the supercritical supply of energy, expressed by the conversion parameter $d_2 > d_2^{\mathrm{crit2}}$ would allow a *deterministic* motion in the positive direction (see the dashed line in Fig. 2.27). In addition, we find a *very small* positive net current in the range of small d_2 (see the insert in Fig. 2.27). Whereas in the deterministic case, for the same values of d_2, no net current at all is obtained; the fluctuations in the stochastic case allow some particles to escape the potential barriers.

To investigate how much the strength S of the stochastic force may influence the magnitude of the net current in the negative direction, we varied S for a fixed conversion parameter $d_2 = 1.0$ for the less damped case. As Fig. 2.27 indicates, for this setup, there will be only a negligible net current, $\langle v \rangle \approx 0$ in the deterministic case ($S = 0$), but a remarkable net current, $\langle v \rangle = -0.43$ in the stochastic case for $S = 0.1$. As Fig. 2.28 shows, there is a *critical strength* of the stochastic force, $S^{\mathrm{crit}}(d_2 = 1.0) \simeq 10^{-4}$,

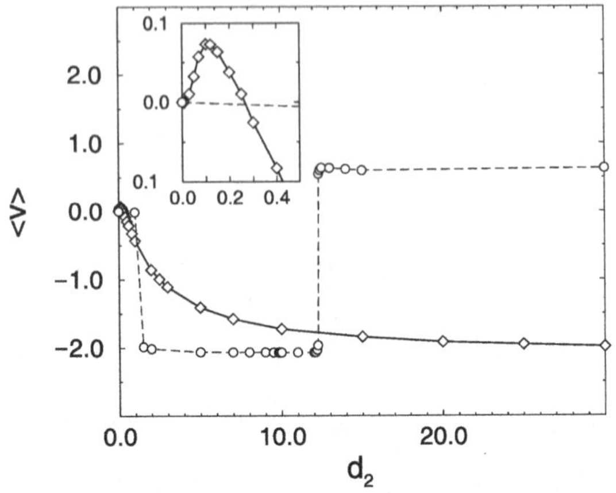

Fig. 2.27. Average velocity $\langle v \rangle$ vs. conversion parameter d_2. The data points are obtained from simulations of 10,000 particles with arbitrary initial positions in the first period of the ratchet potential. (◇) stochastic case ($S = 0.1$), (○) deterministic case ($S = 0$); for other parameters, see Fig. 2.18a [506]

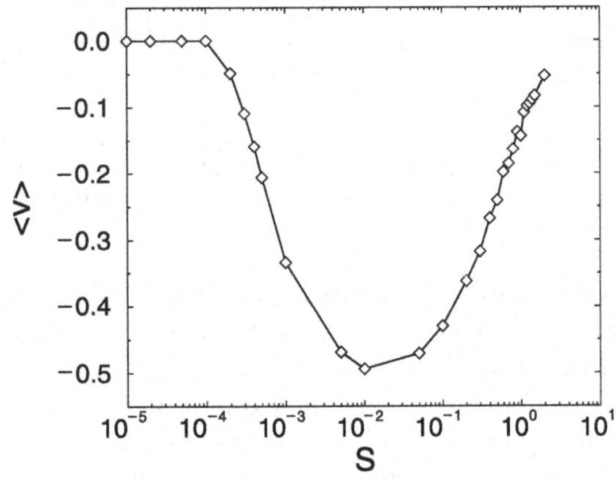

Fig. 2.28. Average velocity $\langle v \rangle$ vs. strength of stochastic force S. The data points are obtained from simulations of 10,000 particles with a fixed conversion parameter $d_2 = 1.0$; for the other parameters, see Figs. 2.18a, 2.27 [506]

where an onset of the net current can be observed, whereas for $S < S^{\mathrm{crit}}$, no net current occurs. On the other hand, there is an *optimal strength* of the stochastic force, S^{opt}, where the amount of the net current, $|\langle v \rangle|$, reaches a *maximum*. An increase in the stochastic force above S^{opt} will only increase the *randomness* of the particle's motion; hence, the net current decreases again. In conclusion, this sensitive dependence on the stochastic force may be used to adjust a *maximum net current* for particle movement in the ratchet potential.

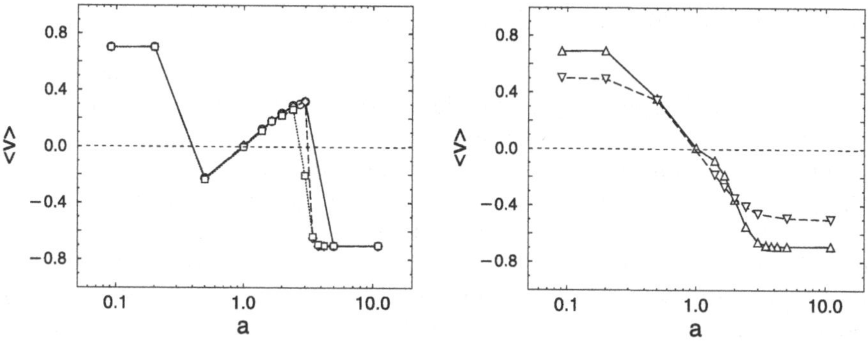

Fig. 2.29. Average velocity $\langle v \rangle$ vs. asymmetry parameter a, (2.79). The data points are obtained from simulations of 10,000 particles with arbitrary initial positions in the first period of the ratchet potential. (*left*) Subcritical stochastic force: (○) $S = 0$ (—), (◇) $S = 0.001$ (– –), (□) $S = 0.01$ (···); (*right*) supercritical stochastic force: (△) $S = 0.05$ (—), (▽) $S = 0.1$ (– –). $d_2 = 10$; for other parameters, see Fig. 2.18b [506]

Finally, we also investigated the influence of the slope on the establishment of a net current for the stochastic case. Figure 2.29 shows the average velocity $\langle v \rangle$ depending on the asymmetry parameter $a = l_2/l_1 = -F_1/F_2$, (2.79), for different values of the stochastic force in the *overdamped case* for a fixed value $d_2 = 10$. The deterministic curve for $S = 0$ is identical with that of Fig. 2.23. As long as the stochastic forces are below a critical value, $S < S^{\mathrm{crit}}$, which is about $S^{\mathrm{crit}}(d_2 = 10) \simeq 0.02$, the curves for the stochastic case are not very different from those of the deterministic one. Hence, the conclusions in Sect. 2.2.5 about the current reversal in the deterministic case, Fig. 2.23, apply. However, for $S > S^{\mathrm{crit}}$, we do not find a critical asymmetry a^{crit} for a current reversal. Instead, the net current keeps its positive (for $a < 1$) or negative (for $a > 1$) direction for any value of $a < 1$ or $a > 1$, respectively.

The right part of Fig. 2.29 further indicates the existence of an optimal strength of the stochastic force. Similar to the investigations in Fig. 2.28, we find that an increase in S does not necessarily result in a increase in the amount of the net current. In fact, the maximum value of $|\langle v \rangle|$ is smaller for $S = 0.1$ than for $S = 0.05$, which indicates an optimal strength of the stochastic force S^{opt} between 0.05 and 0.1 for the set of parameters considered.

We note that the values of both the critical strength S^{crit} and the optimal strength S^{opt} of the stochastic force depend on the value of the conversion parameter d_2, as a comparison of Figs. 2.28 and 2.29 shows. Further, these values may also be functions of the other parameters, such as q_0, γ_0, and therefore differ for the strongly overdamped and the less damped cases.

2.2.7 Directed Motion in a Ratchet Potential

In Sects. 2.2.5 and 2.2.6, we have shown by computer simulations that an ensemble of active particles moving in a ratchet potential can produce a *directed net current.* Hence, by an appropriate *asymmetrical potential* and an additional mechanism to drive the system into *nonequilibrium,* we can convert the genuinely *nondirected* Brownian motion into *directed motion.*

The directed motion in a ratchet potential recently attracted much interest with respect to transport phenomena in cell biology. Here, the directed movement of "particles" (e.g., kinesin or myosin molecules) occurs along periodic structures (e.g., microtubules or actin filaments) in the *absence of a macroscopic force,* which may have resulted from temperature or concentration gradients. To reveal the microscopic mechanisms resulting in directed movement, different physical ratchet models have been proposed [313, 314], such as *forced thermal ratchets* [316, 318], *stochastic ratchets* [311, 343], or *fluctuating ratchets* [417, 572]. To compare these ratchet models with those suggested in Sects. 2.2.5, 2.2.6, we want to summarize some basic principles.

Ratchet models that describe the directed transport of particles are usually based on *three ingredients*: (i) an *asymmetrical periodic potential,* known as ratchet potential, (2.78), Fig. 2.15, that lacks reflection symmetry; (ii) *stochastic forces* $\boldsymbol{\xi}(t)$, i.e., the influence of noise resulting from thermal fluctuations on the microscale; and (iii) *additional correlations* that push the system *out of thermodynamic equilibrium.* For the last, different assumptions can be made.

In the example of the *flashing ratchet,* correlations result from an fluctuating energy profile, i.e., the potential is periodically switched ON and OFF (see Fig. 2.30). In the overdamped limit, the dynamics of a Brownian particle i moving in a ratchet potential $U(x)$ can be described by the equation of motion:

$$\frac{dx_i}{dt} = -\zeta(t)\,\frac{\partial U(x)}{\partial x}\bigg|_{x_i} + \sqrt{2D}\,\xi_i(t)\,. \tag{2.93}$$

The *nonequilibrium forcing* $\zeta(t)$ that governs the time-dependent change of the potential can either be considered a periodic, deterministic modulation with period τ, $\zeta(t) \to F(t) = F(t+\tau)$, or a stochastic *nonwhite* process $\zeta(t)$. In the special case where $\zeta(t)$ or $F(t)$ have only the values $\{0, 1\}$, the periodic potential is switched ON and OFF. As the left part of Fig. 2.30 shows, a particle distribution that is initially located at the minimum of the potential will spread symmetrically by diffusion while the potential is switched OFF. When the potential is switched on again, a net part of the distribution will settle in the minimum located toward the left. Hence, on average, a particle current flows to the left, which means the direction of the *steeper* (but shorter) slope.

It has been pointed out [203] that for $\zeta(t)$ approaching white noise, the noise-induced current is vanishing. Thus, a nonuniform, but periodic, white noise of the form $-\zeta(t)\,U'(x) + \sqrt{2D}\xi(t)$ alone does not result in a finite

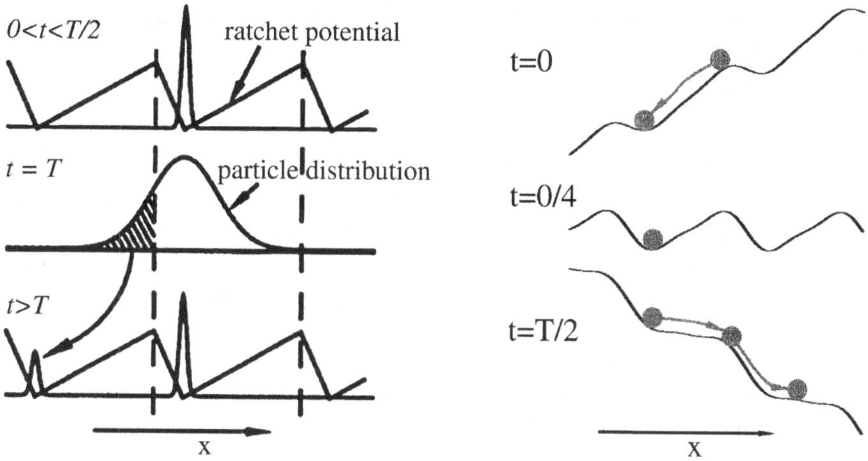

Fig. 2.30. Schematic representation of two different mechanisms inducing directed motion in a ratchet potential: (*left*) the "flashing ratchet"; (*right*) the "rocking ratchet." Here, T denotes the period τ. See text for details [32, 203]

current. Also, with $D = 0$, a nonuniform source of nonwhite diffusion, $-\zeta(t)\, U'(x)$, alone yields no net current.

In another class of ratchet models, the additional correlations result from *spatially uniform forces* of temporal or statistical *zero average*. In the example of the *rocking ratchet* (cf. Fig. 2.30), particles are subject to a spatially uniform, time-periodic deterministic force, $F(t) = F(t + \tau)$, for instance:

$$\frac{dx_i}{dt} = -\left.\frac{\partial U(x)}{\partial x}\right|_{x_i} - A\,\cos\left(\frac{2\pi\, t}{\tau}\right) + \sqrt{2D}\,\xi_i(t)\,. \tag{2.94}$$

Here, the potential $U_s(x,t) = U(x) + A\,x\,\cos\,(\Omega\, t)$ is periodically rocked, as shown in the right part of Fig. 2.30. If the value of A is adjusted properly, in the deterministic case, a net current of particles in the direction of the *lower slope* can be observed, whereas the movement in the direction of the steeper slope is still blocked. However, consideration of stochastic influences results in *current reversal* for a certain range of parameters A and D [33]. Then, the net current occurs again in the direction of the *steeper slope*. Because this phenomenon also depends on the mass of the particles, it can be used for mass separation [303].

Another example of the same class, the *correlation ratchet* [311, 316, 343], is also driven by a spatially uniform, but stochastic force, $\zeta(t)$. Here, the equation for the overdamped motion reads, for instance,

$$\frac{dx_i}{dt} = -\left.\frac{\partial U(x)}{\partial x}\right|_{x_i} + \zeta(t) + \sqrt{2D}\,\xi_i(t)\,, \tag{2.95}$$

where $\zeta(t)$ is a time correlated (colored) noise of zero average. As a third example, the *diffusion ratchet* [406] is driven by a spatially uniform, time-periodic diffusion coefficient $D(t) = D(t + \tau)$, which may result, e.g., from an oscillating temperature. In the overdamped limit, the equation of motion reads, for example,

$$\frac{dx_i}{dt} = -\left.\frac{\partial U(x)}{\partial x}\right|_{x_i} + \zeta(t) + \left[1 + A\,\sin(\Omega\, t)\right]\,\sqrt{2D}\,\xi_i(t)\,. \tag{2.96}$$

In spite of the different ways to introduce additional correlations, all of these models have in common transferring the undirected motion of Brownian particles into directed motion; hence, the term *Brownian rectifiers* [203] has been established.

Our own model can also serve this purpose. For a comparison with the models above, we rewrite (2.52) for the overdamped motion of active particles with an internal energy depot:

$$\frac{dx_i}{dt} = -\frac{1}{\gamma_0 - d_2 e_0}\left.\frac{\partial U(x)}{\partial x}\right|_{x_i} + \frac{\sqrt{2k_\mathrm{B}T\gamma_0}}{\gamma_0 - d_2 e_0}\,\xi_i(t)\,. \tag{2.97}$$

In a more general way,

$$\frac{dx}{dt} = -f(t)\,\frac{\partial U(x)}{\partial x} + C\,f(t)\,\sqrt{2D}\,\xi_i(t)\;,\quad f(t) = \frac{1}{\gamma_0 - d_2 e(t)}\,. \tag{2.98}$$

In (2.98), the prefactor $f(t)$ appears *twice*, up to a constant C. It changes *both* the influence of the ratchet potential and the magnitude of the diffusion coefficient. In *this* respect, our model is between a flashing ratchet model, (2.93), and a diffusion ratchet model, (2.96).

For $q(x) = q_0$, $f(t) = 1/[\gamma_0 - d_2 e(t)]$ is *in general* a time-dependent function, because of $e(t)$, (2.48). For *overdamped motion* in a ratchet potential, the time-dependent term $\gamma_0 - d_2 e(t)$ is shown in Fig. 2.19b. In the limit of a stationary approximation, $e(t) \to e_0$, (2.15), we found that $f(t)$ may switch between two constant values $f_1(x) > 0$, $f_2(x) < 0$, the values of which are given by (2.84). In the limit $c \to 0$, they read:

$$f_i(x) = \frac{1}{2\gamma_0 F_i}\left(F_i \pm \sqrt{F_i^2 + 4q_0\gamma_0}\right) \tag{2.99}$$

depending on the moving direction and the flank, on which the particle is moving. Hence, the function $f(x)$ does not represent a spatially uniform force. However, it has been shown in (2.88), that when the stationary approximation $\langle\gamma_0 - d_2 e_0\rangle\;\tau = 0$ holds, i.e., the force is of zero average with respect to one period, τ. Hence, we conclude that the mechanism of motion on which (2.97) is based, should be different from the previous mechanisms which originate directed motion in a ratchet potential. On the other hand, it still fulfills the definition of a *ratchet system* given in [203, p. 295]: "We define a ratchet system as a system that is able to transport particles in a periodic structure with nonzero macroscopic velocity although on *average* no macroscopic force is acting."

Finally, we note that the model of active particles with an internal energy depot is not restricted to ratchet systems. In a general sense, these particles can be described as Brownian machines or *molecular motors* that convert chemical energy into mechanical motion. Although different ideas for Brownian machines have been suggested [22, 260, 317], our model aims to add a new perspective to this problem. The ability of particles to take up energy from the environment, to store it in an internal depot, and to convert internal energy to perform different activities may open the door to a more refined description of microbiological processes based on physical principles.

2.3 Active Motion in Two-Dimensional Systems

2.3.1 Distribution Function for $U = \text{const}$

So far, we have investigated the motion of active particles in one-dimensional space. As pointed out before, this case already allows some insight into the dynamics of active motion; however, only two different directions of motion are possible. In two- or three-dimensional systems, new possibilities for spatial dynamics will arise that will be investigated in the following.

We start with the discussion of the velocity distribution function $p(\boldsymbol{r}, \boldsymbol{v}, t)$. In Sect. 1.3.3, we showed that the dynamics of $p(\boldsymbol{r}, \boldsymbol{v}, t)$ for "simple" Brownian particles can be described by a Fokker–Planck equation, (1.55). If we are particularly interested in the velocity distribution $p(\boldsymbol{v}, t)$, the Fokker–Planck equation reads (see Sect. 1.3.3),

$$\frac{\partial p(\boldsymbol{v},t)}{\partial t} = \gamma \frac{\partial}{\partial \boldsymbol{v}} \left[\boldsymbol{v}\, p(\boldsymbol{v},t) \right] + S \frac{\partial^2}{\partial v^2} p(\boldsymbol{v},t) \,, \tag{2.100}$$

and the stationary solution is given by the *Maxwellian velocity distribution*:

$$p^0(\boldsymbol{v}) = C \exp\left(-\frac{m}{2\, k_\mathrm{B} T} v^2 \right) \,, \quad C = \left(\frac{m}{2 k_\mathrm{B} T \, \gamma \pi} \right)^{3/2} \,. \tag{2.101}$$

We are now interested in how this known picture changes if we extend the description to *active* particles. The possible existence of an *active mode* for the stationary velocities discussed before may then result in deviations from the equilibrium Maxwell distribution, (2.101), which are discussed in the following.

The presence of an internal energy depot means an additional degree of freedom for a particle, which extends the phase space $\Gamma = \{\boldsymbol{r}, \boldsymbol{v}, e\}$. Let $p(\boldsymbol{r}, \boldsymbol{v}, e, t)$ denote the probability density of finding the particle at time t at location $\boldsymbol{r}$ with velocity $\boldsymbol{v}$ and internal depot energy e. The Fokker–Planck equation has to consider now both the Langevin equation, (2.10), for the motion of the particle and the equation for the energy depot, (2.2). Therefore, the Fokker–Planck equation for "simple" Brownian particles, (1.55), has to be extended to [137]

$$\begin{aligned}\frac{\partial p(\boldsymbol{r},\boldsymbol{v},e,t)}{\partial t} &= \frac{\partial}{\partial \boldsymbol{v}}\left[\frac{\gamma_0 - d_2 e}{m}\,\boldsymbol{v}\,p(\boldsymbol{r},\boldsymbol{v},e,t) + S\,\frac{\partial p(\boldsymbol{r},\boldsymbol{v},e,t)}{\partial \boldsymbol{v}}\right] \\ &\quad -\boldsymbol{v}\,\frac{\partial p(\boldsymbol{r},\boldsymbol{v},e,t)}{\partial \boldsymbol{r}} + \frac{1}{m}\,\boldsymbol{\nabla} U(\boldsymbol{r})\,\frac{\partial p(\boldsymbol{r},\boldsymbol{v},e,t)}{\partial \boldsymbol{v}} \\ &\quad -\frac{\partial}{\partial e}\left[q(\boldsymbol{r}) - c\,e - d_2 v^2 e\right] p(\boldsymbol{r},\boldsymbol{v},e,t)\,. \end{aligned} \tag{2.102}$$

Further, for the probability density, the following normalization condition holds:

$$\int_{-\infty}^{\infty} dr \int_{-\infty}^{\infty} dv \int_{0}^{\infty} de\; p(\boldsymbol{r},\boldsymbol{v},e,t) = 1\,. \tag{2.103}$$

For a comparison with the Maxwellian velocity distribution, (2.101), we are interested only in $p(\boldsymbol{v},t)$ now. Further, the discussion is restricted again to the case $U(\boldsymbol{r}) = \text{const}$ and the assumption of a quasi-stationary energy depot e_0, (2.15). Also, $m = 1$ is considered, which means $\gamma_0 = \gamma$ in this special case. With these approximations, the probability density for the velocity $p(\boldsymbol{v},t)$ is described by the following Fokker–Planck equation:

$$\frac{\partial p(\boldsymbol{v},t)}{\partial t} = \frac{\partial}{\partial \boldsymbol{v}}\left[\left(\gamma_0 - \frac{d_2 q_0}{c + d_2\, v^2}\right)\boldsymbol{v}\,p(\boldsymbol{v},t) + S\,\frac{\partial p(\boldsymbol{v},t)}{\partial \boldsymbol{v}}\right], \tag{2.104}$$

which with respect to the changed *drift term* can be compared with (2.100) for Brownian motion without an energy depot. We further note similarities of (2.104) to the well-known *laser equation* [198, 199, 412]; thus analogies can be expected for the discussion of the stationary solution of (2.104). Using the condition $\dot{p}(\boldsymbol{v},t) = 0$, we find that

$$p^0(\boldsymbol{v}) = C'\left(1 + \frac{d_2 v^2}{c}\right)^{q_0/2S} \exp\left(-\frac{\gamma_0}{2S}\,v^2\right). \tag{2.105}$$

The normalization constant C' can be obtained from $\int d\boldsymbol{v}\,p^0(\boldsymbol{v}) = 1$. Compared to (2.101), which describes the Maxwellian velocity distribution of "simple" Brownian particles, a new prefactor appears now in (2.105) which results from the internal energy depot. In the range of small values of v^2, the prefactor can be expressed by a power series truncated after the first order, and (2.105) reads then,

$$p^0(\boldsymbol{v}) \sim \exp\left[-\frac{\gamma_0}{2S}\left(1 - \frac{q_0 d_2}{c\gamma_0}\right) v^2 + \cdots\right]. \tag{2.106}$$

The sign of the expression in the exponent depends significantly on the parameters that describe the balance of the energy depot. For subcritical pumping of energy, $q_0 d_2 < c\gamma_0$, which corresponds to $\sigma < 0$, (2.64), the expression in the exponent is negative and only a *unimodal velocity distribution* results, centered around the maximum $v_0 = 0$. This is the case of the "low velocity" or *passive mode* for stationary motion, discussed in Sect. 2.2.2, which corresponds here to the *Maxwellian velocity distribution.*

However, for supercritical pumping, $q_0 d_2 > c\gamma_0$, which corresponds to $\sigma > 0$, (2.64), the expression in the exponent becomes positive, and a *crater-like velocity distribution* results. This is also shown in Fig. 2.31. The maxima of $p^0(\boldsymbol{v})$ correspond to the solutions for $(v_0^{(a)})^2$, (2.58). Hence, the "high velocity" or *active mode* for stationary motion is described by strong deviations from the Maxwell distribution. However, as Fig. 2.31 shows, the stationary velocity distribution is centered around $\boldsymbol{v} = 0$ for both subcritical and supercritical pumping.

The crossover from the unimodal to the crater-like velocity distribution occurs at the critical condition $q_0 d_2 = c\gamma_0$, which also describes the bifurcation point, (2.61), in Fig. 2.8. Here, we note again analogies to the *laser* [198, 199, 412]. If we assume that q_0, c, and γ_0 are fixed, then an increase in d_2 accounts for the transition from passive to active motion, as also shown

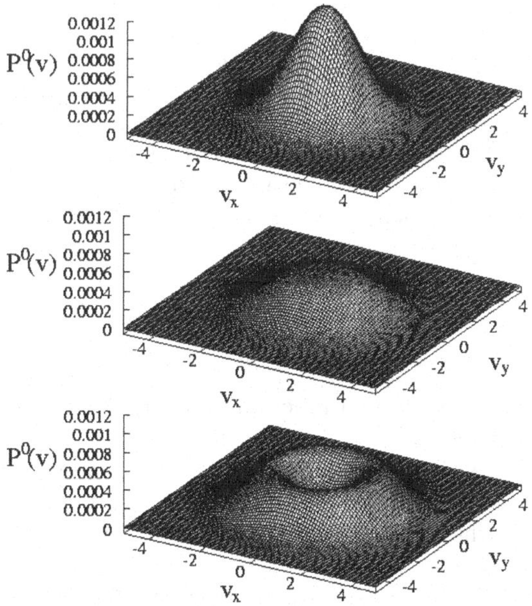

Fig. 2.31. Normalized stationary solution $P^0(\boldsymbol{v})$, (2.105), for $d_2 = 0.07$ (*top*), $d_2 = 0.2$ (*middle*) and $d_2 = 0.7$ (*bottom*). Other parameters: $\gamma_0 = 2$, $S = 2$, $c = 1$, $q_0 = 10$. Note that $d_2 = 0.2$ is the bifurcation point for the given set of parameters [137]

in Fig. 2.31. Because this parameter describes the conversion of internal energy into energy of motion, energy is pumped into fluctuations via the internal energy depot. Therefore, we may conclude that our model provides a simple mechanism to create *fluctuations far from equilibrium* and *strongly nonequilibrium Brownian motion* [137].

From the stationary distribution, (2.105), we recover the known Maxwellian distribution, (2.101), in the limit of strong noise $S \sim T \to \infty$, i.e., at high temperatures, using (1.27) further. Hence, in this limit, some characteristic quantities are explicitly known from Brownian motion in two dimensions, as, e.g., the dispersion of velocities $\boldsymbol{v} = v_1 + v_2$; see (1.33):

$$\langle \boldsymbol{v}^2 \rangle = 2k_\mathrm{B}T \tag{2.107}$$

and the Einstein relation for mean squared displacement, see (1.37):

$$\langle \Delta r^2(t) \rangle = \left\langle \Big[\boldsymbol{r}(t) - \boldsymbol{r}(0) \Big]^2 \right\rangle = 4D_r t\,, \tag{2.108}$$

where $D_r = k_\mathrm{B}T/\gamma_0 = S/\gamma_0^2$ is the spatial diffusion coefficient, (1.19).

In the other limiting case of strong activation, i.e., relatively weak noise $S \sim T \to 0$ and/or strong pumping, we find a δ distribution instead:

$$p^0(\boldsymbol{v}) = C\delta\left(\boldsymbol{v}_0^2 - \boldsymbol{v}^2\right)\,, \quad \langle \boldsymbol{v}^2 \rangle = \boldsymbol{v}_0^2\,, \tag{2.109}$$

where $\boldsymbol{v}_0$ is given by (2.17). This distribution is characteristic of a microcanonical ensemble. An interesting feature of the probability distribution, (2.109), can be observed when looking at the integrated probability density according to one definite velocity component of $\boldsymbol{v} = v_1 + v_2$, say, e.g., v_1, $p_1^0(v_1)$. This is a projection obtained by integrating (2.109) with respect to the other velocity components. In the most general d-dimensional case, the integrated distribution is given by [400]

$$p_1^0(v_1) = \frac{\Gamma\left(\frac{d}{2}\right)}{\sqrt{\pi}\,\Gamma\left(\frac{d-1}{2}\right) v_0^{d-2}} \left(v_0^2 - v_1^2\right)^{\frac{d-3}{2}}\,, \tag{2.110}$$

where $\Gamma(x)$ denotes the gamma function and d is the dimensionality of the system. The d dependence leads to the notable result [124] that for one-dimensional and two-dimensional systems, i.e., $d = 1$ and $d = 2$, the shape of the probability distribution is crater-like, as shown in Fig. 2.31 bottom. However, for $d = 3$, the distribution is constant for $v_1 < v_0$, i.e., the characteristic dip disappears at $v_1 = 0$.

That means that only for one-dimensional and two-dimensional systems and supercritical pumping, the velocity distribution of active particles shows some clear maxima at $v^2 \approx v_0^2$ and a clear minimum at $v^2 = 0$, indicating the tendency of the particles to move with non-Maxwellian velocity. In the following we restrict our consideration to $d = 2$ again.

To treat this case in the full phase space, we follow [340, 427] and introduce first an amplitude-phase representation in the velocity space:

$$v_1 = v_0 \cos(\varphi) , \quad v_2 = v_0 \sin(\varphi) . \tag{2.111}$$

This allows us to separate the variables, and we get a distribution function of the form,

$$\begin{aligned} p(\boldsymbol{r}, \boldsymbol{v}, t) &= p(x_1, x_2, v_1, v_2, t) \\ &= p(x_1, x_2, t) \cdot \delta\left(v_1^2 + v_2^2 - v_0^2\right) \cdot p(\varphi, t) . \end{aligned} \tag{2.112}$$

The distribution of the phase φ satisfies the Fokker–Planck equation:

$$\frac{\partial}{\partial t} p(\varphi, t) = D_\varphi \frac{\partial^2}{\partial \varphi^2} p(\varphi, t) . \tag{2.113}$$

By means of the known solution of (2.113), we get for the mean square,

$$\langle \varphi^2(t) \rangle = D_\varphi t , \quad D_\varphi = \frac{S}{v_0^2} , \tag{2.114}$$

where D_φ is the angular diffusion constant. Using (2.114), the mean squared spatial displacement $\langle r^2(t) \rangle$, (2.108), of the particles can be calculated according to [340] as

$$\langle \Delta r^2(t) \rangle = \frac{2 v_0^4 t}{S} + \frac{v_0^6}{S^2} \left[\exp\left(-\frac{2St}{v_0^2} \right) - 1 \right] . \tag{2.115}$$

For times $t \gg v_0^2/S$, we find from (2.115) that

$$\langle \Delta r^2(t) \rangle = \frac{2 v_0^4}{S} t . \tag{2.116}$$

Consequently, the diffusion coefficient D_r in (2.108) for supercritical pumping has to be replaced by an *effective* spatial diffusion coefficient [137]:

$$D_r^{\mathrm{eff}} = \frac{v_0^4}{2S} = \frac{1}{2\,S} \left(\frac{q_0}{\gamma_0} - \frac{c}{d_2} \right)^2 , \tag{2.117}$$

where v_0, (2.17), considers the additional pumping of energy resulting from the friction function $\gamma(\boldsymbol{v})$. Due to this additional pumping, we obtain high sensitivity with respect to noise expressed in scaling with $(1/S)$.

The analytical results of (2.108), (2.116) are valid only for the limiting cases mentioned and will therefore give a *lower* and an *upper* limit for the mean squared displacement. Figure 2.32 shows the mean squared displacement averaged over 2000 simulations for both sub- and supercritical pumping together with the theoretical results of (2.108), (2.116). We see that for long

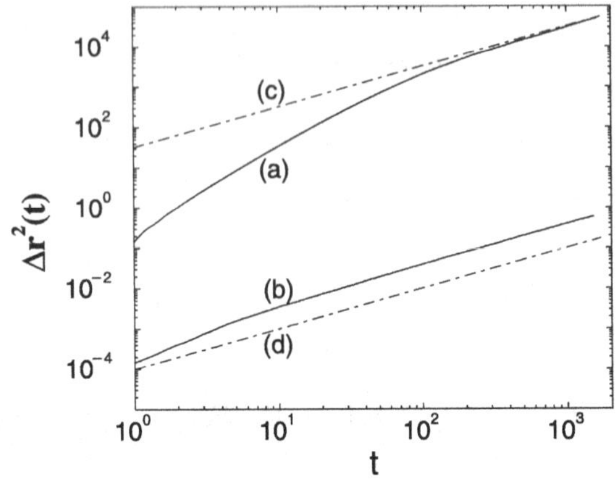

Fig. 2.32. Mean squared displacement $\Delta r^2(t)$, averaged over 2000 simulations as a function of time. (a) Supercritical pumping, $d_2 = 10.0$; (b) subcritical pumping, $d_2 = 1.0$. The additional curves give the theoretical results of (2.116) (*c: upper limit*) and (2.108) (*d: lower limit*). Parameters: $S = 10^{-2}$, $q_0 = 10.0$, $\gamma_0 = 20.0$, $c = 1.0$ [454]

times, the computer simulations for supercritical pumping agree very well with (2.116).

We wish to point to the fact that a non-Maxwellian velocity distribution can also be found for other negative friction functions, (2.19), (2.20). For the Rayleigh-type model, (2.19), the stationary solution of the corresponding Fokker–Planck equation in the isotropic case reads

$$p^0(\boldsymbol{v}) = C \exp\left(\frac{\gamma_1}{2D}\,\boldsymbol{v}^2 - \frac{\gamma_2}{4D}\,\boldsymbol{v}^4\right) . \tag{2.118}$$

The distribution $p^0(\boldsymbol{v})$, (2.118), has its maximum for $\boldsymbol{v} = 0$, if $\gamma_1 \leq 0$. For $\gamma_1 > 0$, the mean of the velocity distribution is still zero, but the distribution then has the form of a hat similar to Fig. 2.31, indicating that a particle most likely moves at a constant absolute velocity $\boldsymbol{v}^2 = \boldsymbol{v}_0^2 = \gamma_1/\gamma_2$.

To calculate the normalization constant C in (2.118), we use a method described in [493] and find with $\gamma_2 = 1$, the explicit expression [136],

$$C^{-1} = \pi\sqrt{\pi D}\,\exp\left(\frac{\gamma_1^2}{4D}\right)\left[1 + \mathrm{erf}\left(\frac{\gamma_1}{2\sqrt{D}}\right)\right] . \tag{2.119}$$

Using (2.118), (2.119), we can calculate the different moments of the stationary velocity distribution and find [137]

$$\left\langle v^{2n}\right\rangle = \frac{(2D)^n}{Z_{2D}(\gamma_1)}\,\frac{\partial^n}{\partial\gamma_1^n}\,Z_{2\mathrm{d}}(\gamma_1) \tag{2.120}$$

with the function

$$Z_{\mathrm{2d}}(\gamma_1) = \pi\sqrt{\pi D}\,\exp\left(\frac{\gamma_1^2}{4D}\right)\left[1+\mathrm{erf}\left(\frac{\gamma_1}{2\sqrt{D}}\right)\right]. \tag{2.121}$$

In particular, the second moment, which is proportional to the temperature, and the fourth moment, which is proportional to the fluctuations of the temperature, read

$$\begin{aligned}\langle \boldsymbol{v}^2\rangle &= \gamma_1 + \sqrt{\frac{D}{\pi}}\,\frac{2\exp\left(\frac{\mathrm{d}\gamma_1^2}{\mathrm{d}4D}\right)}{1+\mathrm{erf}\left(\frac{\mathrm{d}\gamma_1}{\mathrm{d}2\sqrt{D}}\right)},\\ \langle \boldsymbol{v}^4\rangle &= \gamma_1^2 + 2D + \gamma_1\langle \boldsymbol{v}^2\rangle .\end{aligned} \tag{2.122}$$

A similar discussion can be carried out also for the friction function, (2.20). In this case, the stationary solution of the Fokker–Planck equation is of particular simplicity [427]:

$$p^0(\boldsymbol{v}) = C\exp\left[\frac{\gamma_0}{2D}\,(v-v_0)^2\right], \tag{2.123}$$

as also discussed in Sect. 2.1.3. This has also been validated by experiments on *migrating granulocytes*, as Fig. 2.4 shows.

In the following, we will consider the nonlinear friction function, (2.16), that is based on the internal energy depot of active particles and provides a generalized description of the other velocity-dependent friction functions, as shown in Sect. 2.1.1.

2.3.2 Deterministic Motion in a Parabolic Potential

So far, we considered the case $U(\boldsymbol{r}) = \mathrm{const}$, which in the deterministic case and for supercritical conditions, (2.59), implies an unbounded motion of an active particle keeping its initial direction. To bound the particle's motion to a certain area of two-dimensional space (x_1, x_2), we have to specify the potential $U(x_1, x_2)$ with the condition $\lim_{x\to\infty} U(x) = \infty$. Let us start with the simple case of a parabolic potential:

$$U(x_1, x_2) = \frac{a}{2}\left(x_1^2 + x_2^2\right). \tag{2.124}$$

This potential originates a force directed to the minimum of the potential. In a biological context, it simply models a "home," and the moving object always feels a driving force pointing back to its "nest."

With (2.124) and $S = 0$, the Langevin dynamics, (2.10), for the movement of an active Brownian particle can be specified for two-dimensional space. We get five, coupled, first-order differential equations:

$$
\begin{aligned}
\dot{x_1} &= v_1\,, \\
\dot{x_2} &= v_2\,, \\
m\dot{v_1} &= d_2 e v_1 - \gamma_0 v_1 - a\; x_1\,, \\
m\dot{v_2} &= d_2 e v_2 - \gamma_0 v_2 - a\; x_2\,, \\
\mu\dot{e} &= q(x_1,x_2) - ce - d_2 e\left[v_1^2 + v_2^2\right].
\end{aligned}
\tag{2.125}
$$

The formal parameter μ may be used again for adiabatic switching of the depot variable $e(t)$. In the limit $\mu \to 0$, which describes very fast adaptation of the depot, we get as an adiabatic approximation,

$$
e(t) = \frac{q\Big[x_1(t), x_2(t)\Big]}{c + d_2\Big[v_1^2(t) + v_2^2(t)\Big]}\,. \tag{2.126}
$$

In this approximation, the depot energy is simply proportional to the gain of energy q from external sources at the given location and decreases with the increase in kinetic energy. In the following, we assume $q(\boldsymbol{r}) = q_0$, (2.13), again.

Figure 2.33 shows the movement of a particle in the potential U, (2.124), as the result of integrating (2.125). The figure clearly indicates that the deterministic motion, after an initial relaxation time with transient trajectories, tends toward stationary motion in a limit cycle. The corresponding change of the internal energy depot of the particle is presented in Fig. 2.34.

We see that the stationary motion of a particle in a limit cycle corresponds to a saturation value of the internal energy depot, obtained after a period of

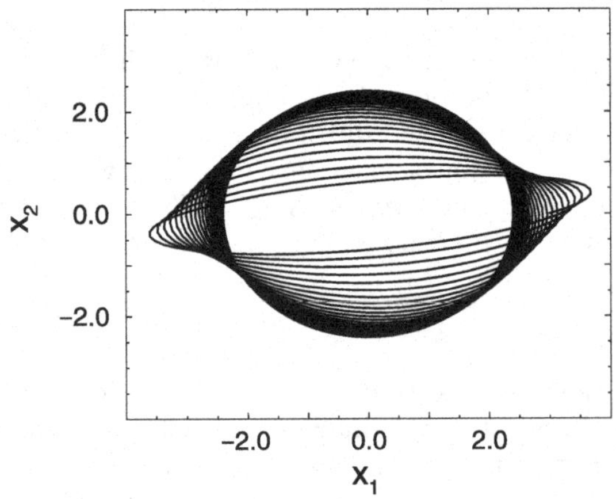

Fig. 2.33. Trajectories in the x_1, x_2 space for the deterministic motion of a particle. Initial conditions: $x_1(0) = 3$, $x_2(0) = 1$, $v_1(0) = 3$, $v_2(0) = 1$, $e(0) = 0$. Parameters: $q_0 = 1.0$, $d_2 = 1.0$, $c = 0.9$, $\gamma_0 = 0.1$ [126, 505]

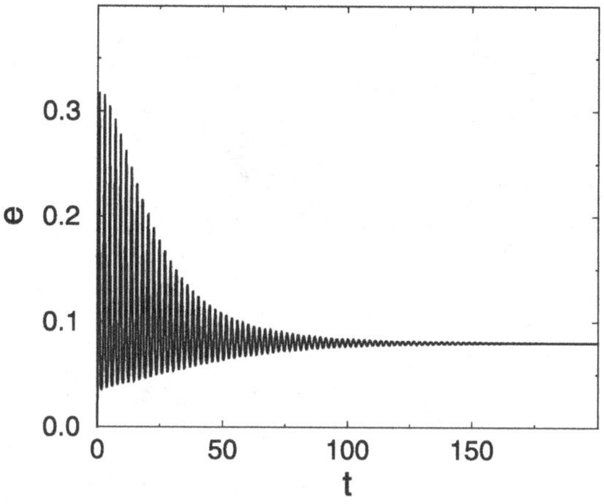

Fig. 2.34. Evolution of the internal energy depot for the motion shown in Fig. 2.33 [126, 505]

relaxation. An analytic approximation for the stationary level of the internal energy will be given in the next section.

2.3.3 Analytical Solutions for Deterministic Limit Cycle Motion

As shown in the previous section, the deterministic motion of an active particle depends on the internal energy depot e, which itself is determined by the take-up q_0, the loss c of energy, and the conversion rate (d_2). Near the origin of the potential, the energy of the internal depot, (2.126), can be approximated by the constant

$$e = \frac{q_0}{c}, \quad \text{if} \quad (v_1^2 + v_2^2) \ll c/d_2 \,. \tag{2.127}$$

Inserting (2.127), in the equation of motion, (2.125), we find

$$\begin{aligned} &\ddot{x}_i + 2\alpha\dot{x}_i + \omega_0^2\, x_i = 0 \\ \text{where} \quad &\alpha = \frac{\gamma_0}{2m} - \frac{d_2 q_0}{2m\; c} \;, \quad \omega_0^2 = \frac{a}{m} \quad (i = 1, 2)\,. \end{aligned} \tag{2.128}$$

For $\alpha^2 < \omega_0^2$, the phase-space trajectories resulting from (2.128) are described by the spiral solution

$$\{x_i, v_i\} = e^{\alpha t}[A_1 \cos(\omega t) + A_2 \sin(\omega t)] \;, \quad \omega = \sqrt{\omega_0^2 - \alpha^2} \quad (i = 1, 2)\,, \tag{2.129}$$

where A_1, A_2 are specified by the intial conditions. Equation (2.129) means that the projection of the motion of a particle on any pair of axes $\{x_1\; x_2\; v_1\; v_2\}$

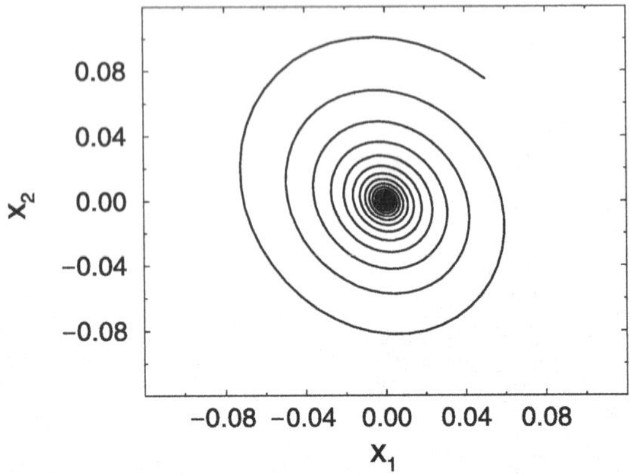

Fig. 2.35. Trajectories in the x_1, x_2 space for subcritical motion of a particle ($q_0 d_2 < \gamma_0 c$). Parameters: $q_0 = 0.1$, $d_2 = 0.1$, $c = 0.1$, $\gamma_0 = 0.2$ [126, 505]

corresponds to expanding or shrinking ellipses. Figure 2.35 presents the subcritical motion of a particle, i.e., the take-up of energy or the conversion of internal energy into kinetic energy is not large enough to continue moving, and the active particle finally rests in the minimum of the potential. Hence, the passive mode for particle motion in the parabolic potential is described by the stationary solution:

$$\boldsymbol{v}_0^{(p)} = 0\,, \tag{2.130}$$

which agrees with the solution for $U = \text{const}$.

For larger amplitudes, the condition, (2.127), is violated, and we have to solve the full equations of motion resulting from (2.125). With $\mu \equiv 0$, i.e., (2.126), we find that

$$m\ddot{x}_i = \frac{d_2 q_0 \dot{x}_i}{c + d_2(v_1^2 + v_2^2)} - a x_i - \gamma_0 \dot{x}_i \qquad (i = 1, 2)\,. \tag{2.131}$$

The solution of this equation is unknown but, with a little trick [142], we can find an approximate solution. Counting on the fact that, at least in the harmonic case, the average of the potential energy is equal or rather close to the average of the kinetic energy,

$$\left\langle \frac{m}{2}\left(v_1^2 + v_2^2\right)\right\rangle \cong \left\langle \frac{a}{2}\left(x_1^2 + x_2^2\right)\right\rangle\,, \tag{2.132}$$

we can modify the function $d(\boldsymbol{v})$ as follows:

$$d_2\left(v_1^2 + v_2^2\right) = d_2\left[\frac{1}{2}\left(v_1^2 + v_2^2\right) + \frac{a}{2m}\left(x_1^2 + x_2^2\right)\right]\,, \tag{2.133}$$

which can be reinserted in (2.131). Specifying further the mechanical energy of the particle, (2.5),

$$E_0 = \frac{m}{2}\left(v_1^2 + v_2^2\right) + \frac{a}{2}\left(x_1^2 + x_2^2\right), \tag{2.134}$$

we get from (2.131) for the change of the mechanical energy [126],

$$\frac{dE_0}{dt} = \left(\frac{d_2 q_0}{c + d_2 E_0/m} - \gamma_0\right)\left(\frac{\partial E_0}{\partial v_i}\right)^2 \frac{1}{m^2}. \tag{2.135}$$

Equation (2.135) indicates that stable orbits with constant mechanical energy exist in the stationary limit:

$$E_0 = m\frac{d_2 q_0 - \gamma_0 c}{d_2 \gamma_0}, \tag{2.136}$$

if the constraint $d_2 q_0 > \gamma_0 c$ is fulfilled. For the adiabatic approximation considered, (2.136) provides a relation between energy dissipation due to friction (γ_0) and the different parameters that determine the level of the internal energy depot: take-up of energy q_0, internal dissipation c, and conversion of internal to kinetic energy d_2.

Of course, the adiabatic appproximation does not apply during the initial period of motion; however, in the asymptotic limit, our approximation remains valid. This is also shown in Fig. 2.36, which is a numerical solution of the equation of motion (2.131) and thus allows estimating the *time lag* before the stationary regime of motion is reached.

We note that the critical condition obtained from (2.136) agrees with the result, (2.59), derived in the previous section for the potential U = const. It can be further shown that the stationary velocity for the active mode of motion in the potential $U(x_1, x_2)$, (2.124), also agrees with the previous result

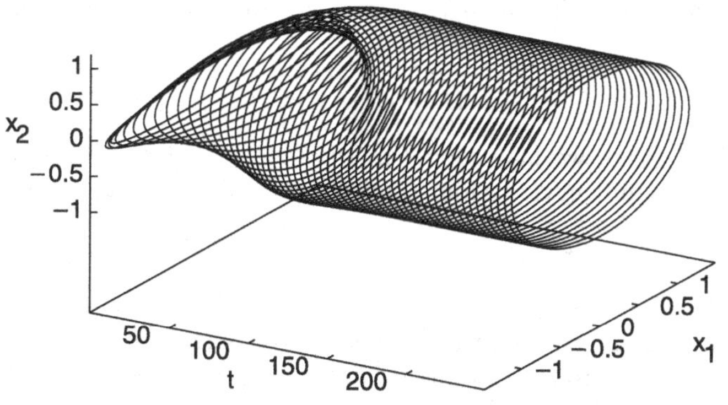

Fig. 2.36. Trajectories in the x_1, x_2 space for the deterministic motion of a particle. Parameters: $q_0 = 0.5$, $\gamma_0 = 0.2$, $c = 0.01$, $d_2 = 0.9$, $a^2 = 2$; initial conditions: $x_1(0) = 1$, $x_2(0) = 0$, $v_1(0) = 1$, $v_2(0) = 0.33$ [126]

of (2.58). By taking advantage of (2.132), (2.134), the stationary velocity can be derived from (2.136):

$$v_0^{(a)\,2} = v_1^2 + v_2^2 = \frac{d_2 q_0 - \gamma_0 c}{d_2 \gamma_0}\,. \tag{2.137}$$

Inserting this expression in (2.126), the stationary value of the internal energy depot can be estimated as

$$e_0 = \frac{\gamma_0}{d_2}, \quad \text{if } v_1^2 + v_2^2 = v_0^{(a)\,2}\,, \tag{2.138}$$

which agrees with (2.60) and with the result of the computer simulations shown in Fig. 2.34.

Accordingly, the stationary limit predicts that beyond a critical value of the energy take-up,

$$q_0 > q_0^{\text{crit}} = \frac{\gamma_0 c}{d_2}\,, \tag{2.139}$$

limit cycles for the motion of the object exist which are closed trajectories on the ellipsoid [126]

$$\frac{m}{2}\left(v_1^2 + v_2^2\right) + \frac{a}{2}\left(x_1^2 + x_2^2\right) = m\left(\frac{q_0}{\gamma_0} - \frac{c}{d_2}\right). \tag{2.140}$$

In four-dimensional phase space, the ellipsoid defines the *hyperplane of constant energy*, which contains all limit cycles. The projection of this limit cycle onto the $\{x_1, x_2\}$ plane corresponds to a circle

$$x_1^2 + x_2^2 = r_0^2 = \text{const.} \tag{2.141}$$

The energy of motion, (2.134), on the limit cycle reads

$$E_0 = \frac{m}{2} v_0^2 + \frac{a}{2} r_0^2\,, \tag{2.142}$$

and any initial value of the energy converges (at least in the limit of strong pumping) to

$$H \longrightarrow E_0 = v_0^2\,. \tag{2.143}$$

In explicit form, we may represent the deterministic motion of an active particle on the limit cycle in four-dimensional space by the four equations,

$$\begin{aligned} x_1 &= r_0 \cos(\omega t + \varphi_0)\,, & v_1 &= -r_0\,\omega \sin(\omega t + \varphi_0)\,, \\ x_2 &= r_0 \sin(\omega t + \varphi_0)\,, & v_2 &= r_0\,\omega \cos(\omega t + \varphi_0)\,. \end{aligned} \tag{2.144}$$

The frequency ω is given by the time an active particle needs for one period while moving on a circle of radius r_0 at constant speed v_0. This leads to the relation

$$\omega_0 = \frac{v_0}{r_0} = \sqrt{a} = \omega\,, \tag{2.145}$$

which means that even in the case of strong pumping, a particle oscillates at the frequency given by the linear oscillator frequency ω.

The trajectory defined by the four equations (2.144) is like a hoop in four-dimensional space (see also Fig. 2.39 in Sect. 2.3.5 for the stochastic case). Therefore, most projections to two-dimensional subspaces are circles or ellipses. However there are two subspaces, $\{x_1, v_2\}$ and $\{x_2, v_1\}$, where the projection is like a rod. A second limit cycle is obtained by time reversal:

$$t \to -t\,, \quad v_1 \to -v_1\,, \quad v_2 \to -v_2\,. \tag{2.146}$$

This limit cycle also forms a hula hoop which is different from the first one in that the projection onto the $\{x_1, x_2\}$ plane has a rotational direction opposite to the first one. However, both limit cycles have the same projections onto the $\{x_1, x_2\}$ and the $\{v_1, v_2\}$ planes. The separatrix between the two attractor regions is given by the following plane in four-dimensional space:

$$v_1 + v_2 = 0\,. \tag{2.147}$$

For the harmonic potential discussed above, equal distribution between potential and kinetic energy, $mv_0^2 = a\, r_0^2$, is valid, which leads to the relation: $\omega_0 = v_0/r_0 = \omega$. For other radially symmetrical potentials $U(\boldsymbol{r})$, this relation has to be replaced by the condition that on the limit cycle, attracting radial forces are in equilibrium with centrifugal forces. This condition leads to

$$\frac{v_0^2}{r_0} = |U'(r_0)|\,. \tag{2.148}$$

For a given v_0, this equation defines an implicit relation for the equilibrium radius r_0: $v_0^2 = r_0\, |U'(r_0)|$. Then, the frequency of the limit cycle oscillations is given by

$$\omega_0^2 = \frac{v_0^2}{r_0^2} = \frac{|U'(r_0)|}{r_0}\,. \tag{2.149}$$

For linear oscillators, this leads to $\omega_0 = \sqrt{a}$ as before, (2.145). But, e.g., for a quartic potential

$$U(r) = \frac{k}{4} r^4\,, \tag{2.150}$$

we get the limit cycle frequency

$$\omega_0 = \frac{k^{1/4}}{v_0^{1/2}}\,. \tag{2.151}$$

If (2.148) has several solutions for the equilibrium radius r_0, the dynamics might be much more complicated; for example, we could find Kepler-like orbits oscillating between the solutions for r_0. In other words, we then also find

– in addition to the driven rotations already mentioned – driven oscillations between the multiple solutions of (2.148).

In the general case where the potential $U(x_1, x_2)$ does not obey radial symmetry, the local curvature of potential levels $U(x_1, x_2) = \text{const}$ replaces the role of radius r_0. We thus define $\varrho_U(x_1, x_2)$ as the radius of local curvature. In general, global dynamics will be very complicated; however, local dynamics can be described as follows: Most particles will move with constant kinetic energy

$$\boldsymbol{v}^2 = v_1^2 + v_2^2 = v_0^2 = \text{const} \tag{2.152}$$

along the equipotential lines, $U(x_1, x_2) = \text{const}$. In dependence on the initial conditions, two nearly equally probable directions of the trajectories, a counterclockwise and a clockwise direction, are possible. Among the different possible trajectories, the most stable will be one that fulfills the condition that the potential forces and centrifugal forces are in equilibrium:

$$v_0^2 = \varrho_U(x_1, x_2) \left| \nabla U(x_1, x_2) \right| . \tag{2.153}$$

This equation is a generalization of (2.148).

2.3.4 Deterministic Chaotic Motion in the Presence of Obstacles

Real biological motion occurs in complicated landscapes which typically contain localized areas of energy supply and also obstacles. This may lead to rather complex active motion. In our simple model, the existence of an obstacle can be implemented in the potential U (2.124), for example,

$$\begin{aligned} U(x_1, x_2) = {} & \frac{a}{2}(x_1^2 + x_2^2) \\ & + U_0\, \Theta\left[R^2 - (x_1 - b_1)^2 + (x_2 - b_2)^2\right] , \end{aligned} \tag{2.154}$$

where $U_0 \to \infty$ and Θ is the Heaviside function,

$$\Theta\left[R^2 - r^2\right] = \begin{cases} 1 \text{ if } r^2 \leq R^2 \\ 0 \text{ else} . \end{cases} \tag{2.155}$$

Equation (2.155) models an obstacle centered around the point (b_1, b_2) which has the form of a hard core of radius R. If a particle hits the obstacle, it is simply reflected at the boundary R. We want to consider again a continuous supply of energy, $q(x_1, x_2) = q_0$ but only a deterministic motion of a particle, i.e., $S = 0$. Then, the initial conditions, $\boldsymbol{v} \neq 0$, determine whether or not the particle can start its motion and will hit the obstacle. As Fig. 2.37 shows, the existence of reflecting obstacles results in a complex motion of an active particle, even in the deterministic case.

In pure Hamiltonian mechanics, there is an analogy to the Sinai billiard, where Hamiltonian chaos has been found [405, 477]. In our model, the situation is different because we do not have energy conservation, but energy

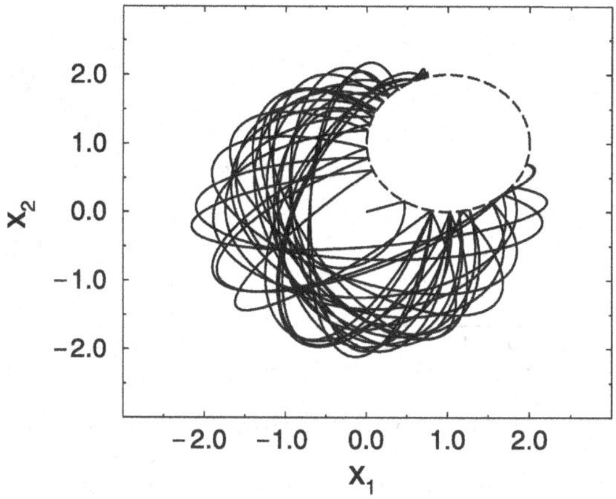

Fig. 2.37. Trajectories in the x_1, x_2 space for the *deterministic* motion of an active Brownian particle in a parabolic potential, (2.154), where the circle [cooordinates (1,1), radius $R = 1$] indicates the reflecting obstacle. Parameters: $q_0 = 1.0$, $\gamma_0 = 0.2$ $d_2 = 1.0$, $c = 0.1$; initial conditions: $(x_1, x_2) = (0, 0)$, $(v_1, v_2) = (1.0, 0.33)$, $e(0) = 0$ [453, 505]

dissipation, on one hand, and an influx of energy, on the other hand. To decide whether the motion shown in Fig. 2.37 is chaotic, we have calculated the Ljapunov exponent λ, which characterizes how a small perturbation of trajectory $\boldsymbol{r}$, $\Delta\boldsymbol{r}(t_0) = \boldsymbol{r}(t_0) - \boldsymbol{r}'(t_0)$, evolves in the course of time, $t > t_0$. The motion becomes asymptotically unstable, if $\lambda > 0$. Using standard methods described, e.g., in [299], we found, for the parameters used in Fig. 2.37, that the largest Ljapunov exponent is $\lambda = 0.17$. This indicates chaotic motion of an active particle in phase space $\Gamma = \{x_1, x_2, v_1, v_2, e\}$. Hence, we conclude that for the motion of active particles with energy depots, reflecting obstacles have an effect similar to stochastic influences (external noise); both produce interesting forms of complex motion.

2.3.5 Stochastic Motion in a Parabolic Potential

In the following, we discuss the *stochastic* motion of an active particle with an internal energy depot in the parabolic potential, $U(x_1, x_2)$, (2.124). The random force in (2.10) keeps the particle moving, even without the take-up of energy, as Fig. 2.38a shows. Moving from simple Brownian motion to active Brownian motion, Fig. 2.38b demonstrates that the constant in space take-up of energy, $q(x_1, x_2) = q_0$ and its conversion to kinetic energy allow a particle to reach out farther.

The active mode of motion, however, is only possible if the critical conditions already discussed in Sect. 2.2.2 are satisfied:

$$q_0 > q_0^{\text{crit}} = \frac{\gamma_0 c}{d_2} \,. \tag{2.156}$$

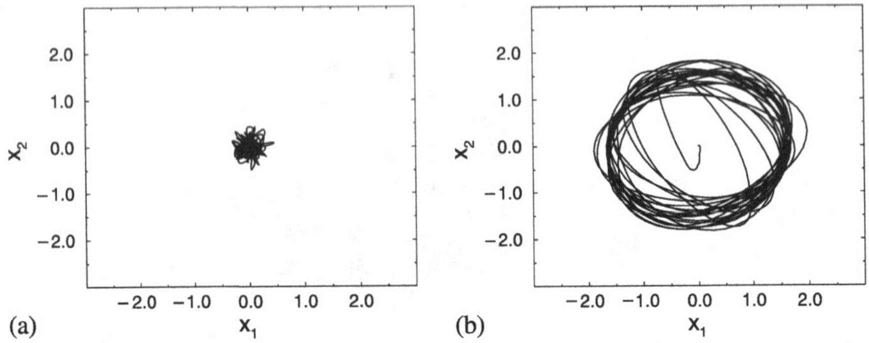

Fig. 2.38. Stochastic motion of an active Brownian particle in a parabolic potential, (2.124) ($a = 2$). (**a**) $q = 0$ (simple Brownian motion); (**b**) $q_0 = 1.0$ Other parameters: $\gamma_0 = 0.2$ $d_2 = 1.0$, $c = 0.1$, $S = 0.01$; initial conditions: $(x_1, x_2) = (0, 0)$, $(v_1, v_2) = (0, 0)$, $e(0) = 0$ [453]

Provided with a supercritical supply of energy, a particle periodically moves on the *stochastic limit cycle* shown in Fig. 2.38b, where the direction of motion may depend on stochastic influences.

Similar to the deterministic motion on the limit cycle discussed in Sect. 2.3.3, we may describe the stochastic motion of an active particle in four-dimensional space by

$$\begin{aligned} x_1 &= \varrho\,\cos(\omega t + \varphi)\,, \qquad & v_1 &= -\varrho\,\omega\,\sin(\omega t + \varphi)\,, \\ x_2 &= \varrho\,\sin(\omega t + \varphi)\,, \qquad & v_2 &= \varrho\,\omega\,\cos(\omega t + \varphi)\,. \end{aligned} \tag{2.157}$$

In this amplitude-phase representation, the radius ϱ is now a slow *stochastic variable*, and the phase φ is a fast *stochastic variable.* Again, the trajectory defined by (2.157) is like a hoop in four-dimensional space. The projections of the distribution onto the $\{x_1, x_2, v_2\}$ subspace and the $\{v_1, v_2, x_2\}$ subspace are two-dimensional rings, as shown in Fig. 2.39. We find in the *stochastic* case two embracing hoops with finite size, which for strong noise convert into two embracing tires in four-dimensional space [137].

Although either left- or right-handed rotations are found in the deterministic case, in the stochastic case, the system may switch randomly between the left- and right-hand rotations because the separatrix becomes permeable. This has been also confirmed by looking at the projections of the distribution onto the $\{x_1, v_2\}$ plane and the $\{x_2, v_1\}$ plane [137].

Equation (2.157) can be also used to get an explicit form of the probability distribution of the radii, $p(\varrho)$. By using the standard procedure of averaging with respect to the fast phases for the friction function (2.16) resulting from the depot model, we get the following stationary solution for the distribution of radii [137]:

$$p^0(\varrho) \simeq \left(1 + \frac{d_2}{c} a \varrho^2\right)^{\frac{q_0}{2D}} \exp\left(-\frac{\gamma_0}{D} a \varrho^2\right), \tag{2.158}$$

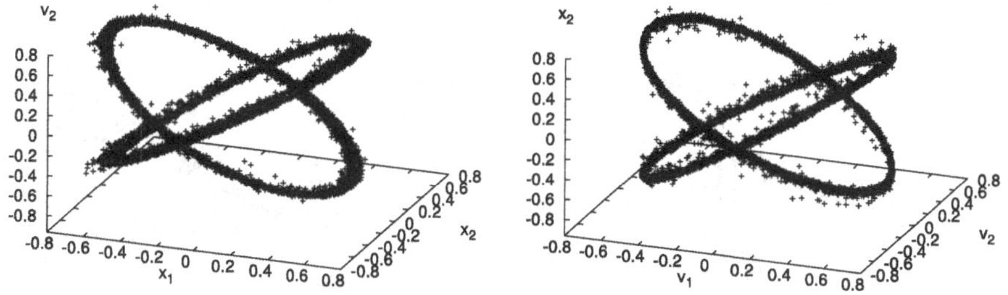

Fig. 2.39. Projections of two stochastic tracjectories on three-dimensional subspaces $\{x_1, x_2, v_2\}$ and $\{x_1, x_2, v_2\}$ for very low noise ($S = 10^{-5}$). One of the trajectories represents right-handed and the other left-handed rotations. Other parameters: $\gamma_0 = 20$, $c = 1$, $d_2 = 10$, $q_0 = 10.0$, $a = 2.0$

with the the maximum located at

$$\varrho^2 = r_0^2 = \frac{v_0^2}{\omega^2} = \left(\frac{q_0}{\gamma_0} - \frac{c}{d_2}\right)\frac{1}{\omega^2}\,. \tag{2.159}$$

Due to the special form of the attractor that consists, as pointed out above, of two hula hoops embracing each other, the distribution in the phase space cannot be constant at $\varrho = \text{const}$. An analytical expression for the distribution in four-dimensional phase space is not yet available. But it is expected that the probability density is concentrated around the two deterministic limit cycles.

2.3.6 Stochastic Motion with Localized Energy Sources

In the previous sections, we have always assumed a constant in space supply of energy, i.e., $q(\boldsymbol{r}) = q_0$. If, on the other hand, energy sources are localized in space, the internal depot of active Brownian particles can be refilled only in a restricted area. This is reflected in a space dependence of the energy influx $q(\boldsymbol{r})$, for example,

$$q(x_1, x_2) = \begin{cases} q_0 & \text{if } \left[(x_1 - b_{1'})^2 + (x_2 - b_{2'})^2\right] \le R^2 \\ 0 & \text{else}\,. \end{cases} \tag{2.160}$$

Here, the energy source is modeled as a circle, whose center is different from the minimum of the potential. Noteworthy, an active particle is *not* attracted by the energy source due to long-range attractive forces. In the beginning, the internal energy depot is empty, and active motion is not possible. So, a particle may hit the supply area because of the action of the stochastic force. But once the energy depot is filled, it increases the particle's motility, as presented in Fig. 2.40. Most likely, motion into the energy area becomes accelerated; therefore an oscillating movement between the energy source and the potential minimum occurs after an initial period of stabilization.

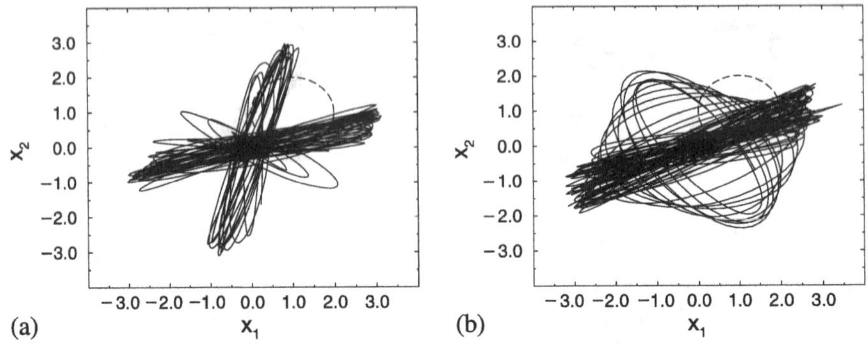

Fig. 2.40. Trajectories in the x_1, x_2 space for stochastic motion of an active Brownian particle in a parabolic potential, (2.124) ($a = 2$). The circle [cooordinates (1,1), radius $R = 1$] indicates the area of energy supply, (2.160). Parameters: $q_0 = 10$, $\gamma_0 = 0.2$, $c = 0.01$, $S = 0.01$; initial conditions: $(x_1, x_2) = (0, 0)$, $(v_1, v_2) = (0, 0)$, $e(0) = 0$, (**a**) $d_2 = 1$ [453], (**b**) $d_2 = 0.1$ [126]

Figure 2.41 presents more details of the motion of the active Brownian particle, shown in Fig. 2.40. Considering the time-dependent change of the internal energy depot, the velocities, and the space coordinates, we can distinguish between two different stages. In the first stage, the particle has not found the energy source; thus its energy depot is empty while the space coordinates fluctuate around the coordinates of the potential minimum. The second stage starts when the particle by chance, due to stochastic influences, reaches the localized energy source. Then the internal depot is soon filled, which in turn allows the particle to reach out farther, shown in the larger fluctuations of the space coordinates and the velocities. This accelerated movement, however, leads the active particle away from the energy source at the expense of the internal energy depot, which is decreased until the particle reaches the energy source again. Figure 2.41 shows the corresponding oscillations in the energy depot.

Interestingly, due to stochastic influences, the oscillating motion breaks down after a certain time, as shown in Fig. 2.41. Then the active particle, with an empty internal depot, again moves like a simple Brownian particle, until a new cycle starts. In this way, the particle motion is an intermittent type [439]. We found that the trajectories eventually cover the whole area inside certain boundaries; however, during an oscillation period, the direction is most likely kept.

Every new cycle starts with a *burst of energy* in the depot, indicated by the larger peak in $e(t)$, Fig. 2.41, which can be understood on the basis of (2.2). At the start of each cycle, e is small and de/dt is very large; thus a burst follows, which is used to accelerate the particle. However, an increase in $d_2 v^2 e$ makes the last term in (2.2) more negative, and the growth of e is more rapidly cut off. For the beginning of the first cycle in Fig. 2.41, this is shown in more detail in Fig. 2.42, which also clearly indicates the oscillations. The transition

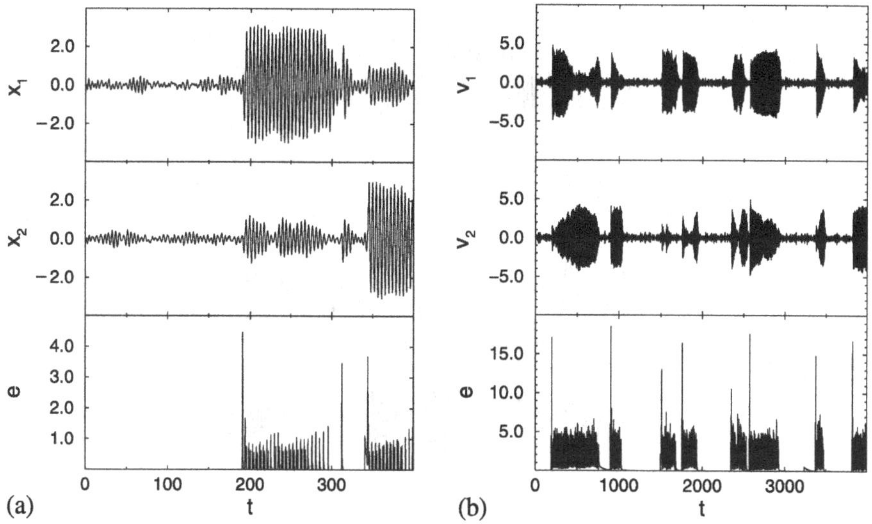

Fig. 2.41. (**a**) Space coordinates x_1, x_2 and internal energy depot e vs. time for the stochastic motion with $d_2 = 1$ shown in Fig. 2.40a [126]. (**b**) Velocities v_1, v_2 and internal energy depot e vs. time for stochastic motion with $d_2 = 0.1$ shown in Fig. 2.40b [453]

time into the oscillation regime, as well as the duration of the oscillatory cycle, depend remarkably on the conversion parameter d_2. The bottom part of Fig. 2.42, compared with the part above, indicates that an increase in d_2 reduces the amount of bursts and abridges the cycle. For a larger d_2, more depot energy is converted into kinetic energy. Hence, with less energy in stock, particle motion is more susceptible to becoming Brownian motion again, if stochastic influences prevent the particle from returning to the source in time.

The basic types of motion for active Brownian particles with an internal energy depot can be extended to describe more complex situations. We just want to mention the case of many separated potential minima of $U(\boldsymbol{r})$. On the other hand, we can also assume many separated energy sources, randomly distributed on the surface, which means a specification for $q(\boldsymbol{r})$. Further, it is possible to consider different obstacles, as discussed in Sect. 2.3.4, which may also allow us to draw analogies to the search for food in biological systems.

Finally, we can generalize the situation by considering that the availablility of energy is both space- and time-dependent, which means additional conditions for the take-up of energy $q(\boldsymbol{r}, t)$. Let us consider a situation, where the supplied external energy ("food") grows with a given flow density $\Phi(\boldsymbol{r}) = \eta q_f(\boldsymbol{r})$, where η is a dimensional constant. Then, we assume that the change of the energy take-up function may depend on both the increase and the decrease in energy supplied, i.e.,

$$\dot{q}(\boldsymbol{r}, t) = \Phi(\boldsymbol{r}) - \eta q(\boldsymbol{r}, t) = \eta [q_f(\boldsymbol{r}) - q(\boldsymbol{r}, t)] \,. \tag{2.161}$$

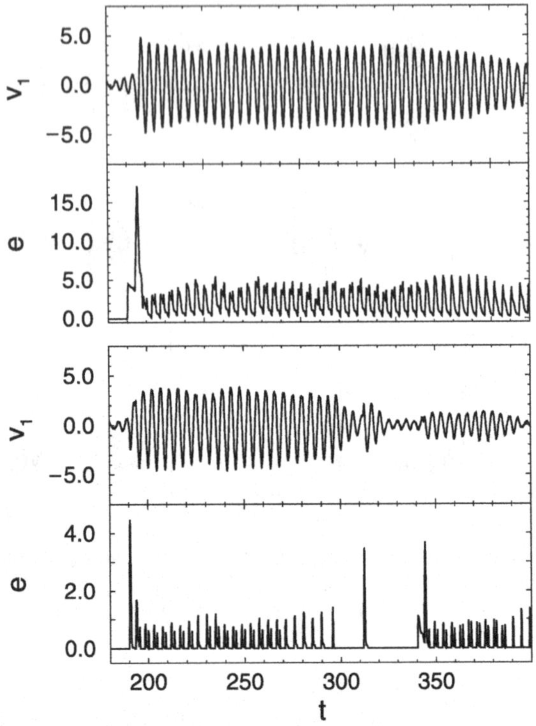

Fig. 2.42. Velocity v_1 and internal energy depot e vs. time for two different values of d_2: (*top*) $d_2 = 0.1$ (enlarged part of Fig. 2.41b), (*bottom*) $d_2 = 1.0$. For other parameters, see Fig. 2.40 [453]

The formal solution for the energy take-up yields

$$q(\boldsymbol{r}, t) = e^{-\eta t} q(\boldsymbol{r}, 0) - \eta \int_{-\infty}^{t} d\tau \, e^{-\eta(t-\tau)} q_f \Big[\boldsymbol{r}(\tau), \tau \Big] \, . \tag{2.162}$$

As we see, the actual value of the energy influx now depends on the whole prehistory of the motion of the particle and reflects a certain kind of memory. However, those memory effects will be neglected here, which means that $\eta \to \infty$.

2.4 Swarming of Active Particles

2.4.1 Canonical-Dissipative Dynamics of Swarms

So far, we have investigated the motion of either a *single* active particle or an *ensemble* of active particles *without* interactions. In this section, we want to apply the general description outlined above to swarm dynamics. Basically, at a certain level of abstraction, a swarm can be viewed as an active, many-particle system with some additional coupling that would account for the typical correlated motion of the entities. In addition, some energetic conditions must also be satisfied to keep the swarm moving. These conditions have been investigated in depth in the previous sections; thus in the following, we concentrate on the coupling of individual motion.

Let us consider a swarm of N active particles, each characterized by its spatial position $\boldsymbol{r}_i$, its individual velocity $\boldsymbol{v}_i$, and its internal energy depot e_i. The last is assumed to relax fast and can therefore be described by the quasi-stationary value $e_0(v_i^2)$, (2.15). This allows us to reduce the number of independent variables to $\boldsymbol{r}_i$ and $\boldsymbol{v}_i$. The N-particle distribution function that describes the dynamics of the swarm is given by

$$p(\underline{r}, \underline{v}) = p(\boldsymbol{r}_1, \boldsymbol{r}_2, ..., \boldsymbol{r}_N, \boldsymbol{v}_1, \boldsymbol{v}_2, ..., \boldsymbol{v}_N) . \tag{2.163}$$

In the following, $m_i \equiv m = 1$ is used, i.e., $v_i = p_i$, where p_i is the momentum of a particle. The center of mass of the swarm is defined as

$$\boldsymbol{R} = \frac{1}{N} \sum \boldsymbol{r}_i , \tag{2.164}$$

whereas the mean velocity $\boldsymbol{V}$ and the mean angular momentum $\boldsymbol{L}$ are defined as

$$\boldsymbol{V} = \frac{1}{N} \sum \boldsymbol{v}_i , \tag{2.165}$$

$$\boldsymbol{L} = \frac{1}{N} \sum \boldsymbol{L}_i = \frac{1}{N} \sum \boldsymbol{r}_i \times \boldsymbol{v}_i . \tag{2.166}$$

To be consistent with the previous discussion, the equations of motion for each active particle shall be given by a modified Langevin equation, (2.12), with a velocity-dependent friction function, (2.16). At this point, however, we would like to point out first that such a description fits into the rather general theoretical approach of *canonical-dissipative systems.*

This theory – which is not so well known even among experts – results from an extension of the statistical physics of Hamiltonian systems to a special type of dissipative system, the so-called canonical-dissipative system [113, 116, 118, 142, 181, 197, 242]. The term *dissipative* means here that the system is nonconservative, and the term *canonical* means that the dissipative as well as the conservative parts of the dynamics are both determined by a Hamiltonian function H (or a larger set of invariants of motion).

This special assumption in many cases allows exact solutions for the distribution functions of many-particle systems, even far from equilibrium. Therefore, it can be also useful for describing swarms. We have already pointed out that swarming is based on *active motion* of particles; hence we need to consider that a many-particle system is basically an *open* system that is driven into nonequilibrium.

The dynamics of a *mechanical* many-particle system with N degrees of freedom and with the Hamiltonian H is described by the equations of motion:

$$\frac{dr_i}{dt} = \frac{\partial H}{\partial v_i}, \quad \frac{dv_i}{dt} = -\frac{\partial H}{\partial r_i}. \tag{2.167}$$

Each solution of the system of equations (2.167), $v_i(t)$, $r_i(t)$, defines a trajectory on the plane $H = E = \text{const.}$ This trajectory is determined by the initial conditions, and also the energy $E = H(t = 0)$ is fixed due to the initial conditions. Now, we construct a *canonical-dissipative system* with the same Hamiltonian:

$$\frac{dv_i}{dt} = -\frac{\partial H}{\partial r_i} - g(H)\frac{\partial H}{\partial v_i}, \tag{2.168}$$

where $g(H)$ denotes the *dissipation function.* In general, we will only assume that $g(H)$ is a nondecreasing function of the Hamiltonian. The canonical-dissipative system, (2.168) [113, 142, 181], does not conserve energy because of the following relation:

$$\frac{dH}{dt} = -g(H) \sum_i \left(\frac{\partial H}{\partial v_i}\right)^2. \tag{2.169}$$

Whether the total energy increases or decreases, consequently, depends on the form of the dissipation function $g(H)$. For example, if we consider constant friction, $g(H) = \gamma_0 > 0$, energy always decays.

Instead of a driving function $g(H)$ which depends only on the Hamiltonian, we may include the dependence on a larger set of invariants of motion, $I_0, I_1, I_2, ..., I_s$, for example:

- Hamilton function of a many-particle system: $I_0 = H$
- total momentum of a many-particle system: $I_1 = P = mV$
- total angular momentum of a many-particle system: $I_2 = L$

The dependence on this larger sets of invariants may be considered by defining a *dissipative potential* $G(I_0, I_1, I_2, ...)$. The canonical-dissipative equation of motion, (2.168), is then generalized to

$$\frac{dv_i}{dt} = -\frac{\partial H}{\partial r_i} - \frac{\partial G(I_0, I_1, I_2, ...)}{\partial v_i}. \tag{2.170}$$

Using this generalized canonical-dissipative formalism, by an appropriate choice of the dissipative potential G, the system may be driven to particular subspaces of the energy surface, e.g., the total momentum or the angular momentum may be prescribed.

The Langevin equations are obtained by adding a white noise term $\xi_i(t)$ to the deterministic (2.168),

$$\frac{dv_i}{dt} = -\frac{\partial H}{\partial r_i} - g(H)\frac{\partial H}{\partial v_i} + \sqrt{2S(H)}\,\xi_i(t)\,, \tag{2.171}$$

or to (2.170) in the more general case:

$$\frac{dv_i}{dt} = -\frac{\partial H}{\partial r_i} - \frac{\partial G(I_0, I_1, I_2, ...)}{\partial v_i} + \sqrt{2S(H)}\,\xi_i(t)\,. \tag{2.172}$$

The essential assumption is that both the strength of the noise, expressed in terms of $D(H)$, and the dissipation, expressed in terms of the friction function $g(H)$, depend only on the Hamiltonian H. In the following, however, we will use the more restrictive assumption that the strength of the noise is a constant that can be expressed by the fluctuation-dissipation theorem, $S = \gamma_0 k_{\mathrm{B}} T$. This would also allow us to solve the Fokker–Planck equation for the distribution function $p(\underline{r}, \underline{v})$ which corresponds to (2.172):

$$\frac{\partial p(\underline{r},\underline{v})}{\partial t} + \sum \boldsymbol{v}_i \frac{\partial p(\underline{r},\underline{v})}{\partial \boldsymbol{r}_i} - \sum \frac{\partial H}{\partial \boldsymbol{v}_i}\frac{\partial p(\underline{r},\underline{v})}{\partial \boldsymbol{v}_i} =$$
$$\sum \frac{\partial}{\partial \boldsymbol{v}_i}\left[\frac{\partial G(I_0, I_1, I_2, ...)}{\partial \boldsymbol{v}_i} p(\underline{r},\underline{v}) + S\frac{\partial p(\underline{r},\underline{v})}{\partial \boldsymbol{v}_i}\right]. \tag{2.173}$$

Due to the invariant character of the I_k, the l.h.s. of (2.173) disappears for all functions of I_k. Therefore, we have to search only for a function $G(I_0, I_1, I_2, ...)$, for which the collision term also disappears. In this way, we find the stationary solution of (2.173):

$$p^0(\underline{r},\underline{v}) = Q^{-1}\ \exp\left[-\frac{G(I_0, I_1, I_2, ...)}{S}\right]. \tag{2.174}$$

The derivative of $p^0(\underline{r},\underline{v})_0$ vanishes if $G(I_0, I_1, I_2, ..) = \min$, which means that the probability is maximal for the attractor of the dissipative motion.

Let us first discuss the limit case of a *free swarm*, i.e., without interactions between particles. The Hamiltonian of the many-particle system then has the simple form,

$$H = \sum_{i=1}^{N} H_i = \sum_{i=1}^{N} \frac{v_i^2}{2}\,. \tag{2.175}$$

This allows us to reduce the description level from the N-particle distribution function, (2.163), to the one-particle distribution function:

$$p(\underline{r},\underline{v}) = \prod_{i=1}^{N} p(\boldsymbol{r_i}, \boldsymbol{p_i})\,. \tag{2.176}$$

With $g(H) = g(v_i^2)$, (2.175), and $S = \text{const}$, the Langevin equation, (2.171), for each particle i in the absence of an external potential reads

$$\dot{\boldsymbol{r}}_i = \boldsymbol{v}_i\,, \quad \dot{\boldsymbol{v}}_i = -g(v_i^2)\,\boldsymbol{v}_i + \sqrt{2S}\,\boldsymbol{\xi}_i(t)\,. \tag{2.177}$$

For the dissipative function $g(H) = g(v_i^2)$, we will use the velocity-dependent friction function $\gamma(v)$, (2.16), that has been derived based on the assumption of a quasi-stationary energy depot:

$$g(H) = g(v_i^2) = \gamma_0 - \frac{q_0\, d_2}{c + d_2 v_i^2}\,, \tag{2.178}$$

which is plotted in Fig. 2.1. We note that all noninteracting systems, where $g = g(v^2)$, are of the canonical-dissipative type. Hence, many of the previous results are still valid in the limit of a free swarm. For instance, the velocity distribution function $p(\boldsymbol{v}, t)$ is given by the Fokker–Planck equation, (2.104). The stationary solution is given by (2.174), which reads in the case considered, explicitly,

$$p^0(\boldsymbol{v}) = C_0 \exp\left[-\frac{G_0(v^2)}{S}\right], \tag{2.179}$$

where C_0 results from the normalization condition. $G_0(v^2)$ is the special form of the dissipation potential $G(I_0, I_1, I_2, ...)$ considering only $I_0 = H$ as the invariant of motion, and further H as given by (2.175). It reads explicitly:

$$G(I_0) = G_0(v^2) = \gamma_0 \frac{v^2}{2} - \frac{q_0}{2} \ln\left(1 + \frac{d_2 v^2}{c}\right). \tag{2.180}$$

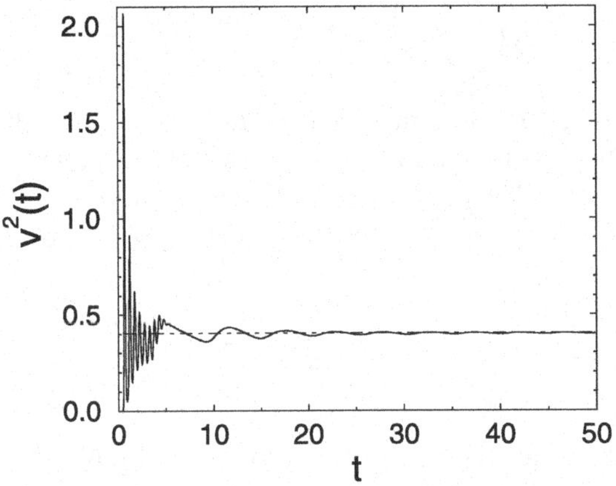

Fig. 2.43. Averaged squared velocity $v^2(t) = 1/N \sum v_i^2(t)$, (2.177), of a swarm of 2000 active particles as a function of time. Parameters: $S = 10^{-4}$, $d_2 = 10.0$, $q_0 = 10.0$, $\gamma_0 = 20.0$, $c = 1.0$; initial conditions: $\boldsymbol{r}_i(0) = \{0.0, 0.0\}$, $\boldsymbol{v}_i(0) = \{0.0, 0.0\}$ for all particles. The *dash-dotted line* gives the stationary velocity, (2.181) [454]

Depending on the parameters q_0, d_2, c, the velocity distribution may show strong deviations from the known Maxwellian distribution; this has been shown in Fig. 2.31. Also, the investigations of the mean squared displacement $\langle \Delta r^2(t) \rangle$, (2.116), shown in Fig. 2.32 remain valid for free swarms.

Another unchanged quantity is the stationary velocity of a free swarm for supercritical pumping, given by (2.17):

$$v_0^2 = \frac{q_0}{\gamma_0} - \frac{c}{d_2} . \tag{2.181}$$

Figure 2.43 shows the results of computer simulations for $v^2(t)$ for supercritical pumping. The convergence toward the theoretical result, (2.181), can be clearly observed.

2.4.2 Harmonic Swarms

So far we have neglected any coupling within an active particle ensemble. This leads to the effect that the swarm eventually disperses in the course of time, whereas a "real" swarm would maintain its coherent motion. A common way to introduce correlations between the moving particles in physical swarm models is the coupling to a mean value. For example, in [92, 93] the coupling of the particles' individual *orientation* (i.e., direction of motion) to the mean orientation of the swarm is discussed. Other versions assume the coupling of the particles' velocity to a *local average velocity*, which is calculated over a space interval around the particle [91, 93].

Here we are mainly interested in *global couplings* of the swarm, which fit it into the theory of canonical-dissipative systems outlined in Sect. 2.4.1. As the simplest case, we first discuss global coupling of the swarm to the *center of mass*, (2.164). That means that the particle's position $\boldsymbol{r}_i$ is related to the mean position of the swarm $\boldsymbol{R}$ via a potential $U(\boldsymbol{r}_i, \boldsymbol{R})$. For simplicity, we may assume a parabolic potential:

$$U(\boldsymbol{r}_i, \boldsymbol{R}) = \frac{a}{2} (\boldsymbol{r} - \boldsymbol{R})^2 . \tag{2.182}$$

This potential generates a force directed to the center of mass, which can be used to control the dispersion of the swarm. In the case considered, it reads

$$\boldsymbol{\nabla} U(\boldsymbol{r}) = a (\boldsymbol{r}_i - \boldsymbol{R}) = \frac{a}{N} \sum_{j=1}^{N} (\boldsymbol{r}_i - \boldsymbol{r}_j) . \tag{2.183}$$

Then, the Hamiltonian for the harmonic swarm reads

$$H = \sum_{i=1}^{N} \frac{v_i^2}{2} + \frac{a}{4} \sum_{i=1}^{N} \sum_{j=1}^{N} (\boldsymbol{r}_i - \boldsymbol{r}_j)^2 . \tag{2.184}$$

With (2.183), the corresponding Langevin equation, (2.177), for an active particle system reads explicitly:

$$\dot{\boldsymbol{v}}_i = -g(v_i^2)\,\boldsymbol{v}_i - \frac{a}{N}\sum_{j=1}^{N}(\boldsymbol{r}_i - \boldsymbol{r}_j) + \sqrt{2S}\,\boldsymbol{\xi}_i(t)\,. \tag{2.185}$$

Hence, in addition to the dissipative function, there is now an attractive force between each two particles i and j which depends linearly on the distance between them. With respect to the harmonic interaction potential, (2.182), we call such a swarm a *harmonic* swarm [124].

Strictly speaking, the dynamic system of (2.185) is not canonical-dissipative, but it may be reduced to this type by some approximations, which will be discussed below. We note that this kind of swarm model has been previously investigated in [337] for one dimension, however, with a different dissipation function $g(v^2)$, for which we use (2.178) again.

To get some insight into the dynamics of a swarm, Fig. 2.44 presents snapshots of a computer simulation of (2.185) for 2000 particle agents.[1] Here, we have assumed that the particles are initially at rest and at the same spatial position. Due to a supercritical take-up of energy, the agents can move actively; the interaction, however, prevents the swarm from simply dispersing in space. Thus, the collective motion of the swarm becomes rather complex, as a compromise between *spatial dispersion* (driven by energy pumping) and *spatial concentration* (driven by mutual interaction).

With the assumed coupling to the center of mass $\boldsymbol{R}$, the motion of the swarm can be considered a superposition of two motions: (i) motion of the center of mass itself and (ii) motion of the particles relative to the center of mass. For motion of the center of mass, we find for the assumed coupling,

$$\dot{\boldsymbol{R}} = \boldsymbol{v}\,, \quad \dot{\boldsymbol{v}} = -\frac{1}{N}\sum_{i=1}^{N} g(v_i^2)\,\boldsymbol{v}_i + \sqrt{2S}\,\boldsymbol{\xi}(t)\,. \tag{2.186}$$

Because of the nonlinearities in the dissipative function $g(v^2)$, both motions (i) and (ii) cannot be simply separated. The term $g(v^2)$ vanishes only for two cases: the trivial one that is free motion without dissipation/pumping and the case of supercritical pumping where $v_i^2 = v_0^2$, (2.181), for each particle. Then, the mean momentum becomes an invariant of motion, $\boldsymbol{v}(t) = \boldsymbol{v}_0 = \mathrm{const}$, and the center of mass moves according to $R(t) = R(0) + v_0(t)$, where the stochastic forces in (2.186) may result in changes of the direction of motion. This behavior may also critically depend on the initial conditions of the particles, $\boldsymbol{v}_i(0)$, and shall be investigated in more detail now.

In [337], an approximation of the mean velocity $\boldsymbol{v}(t)$ of a swarm in one dimension is discussed that shows the existence of two different asymptotic

---

[1] A movie of these computer simulations – with the same parameters, but a different random seed – can be found at `http://www.ais.fhg.de/~frank/swarm-tb.html`

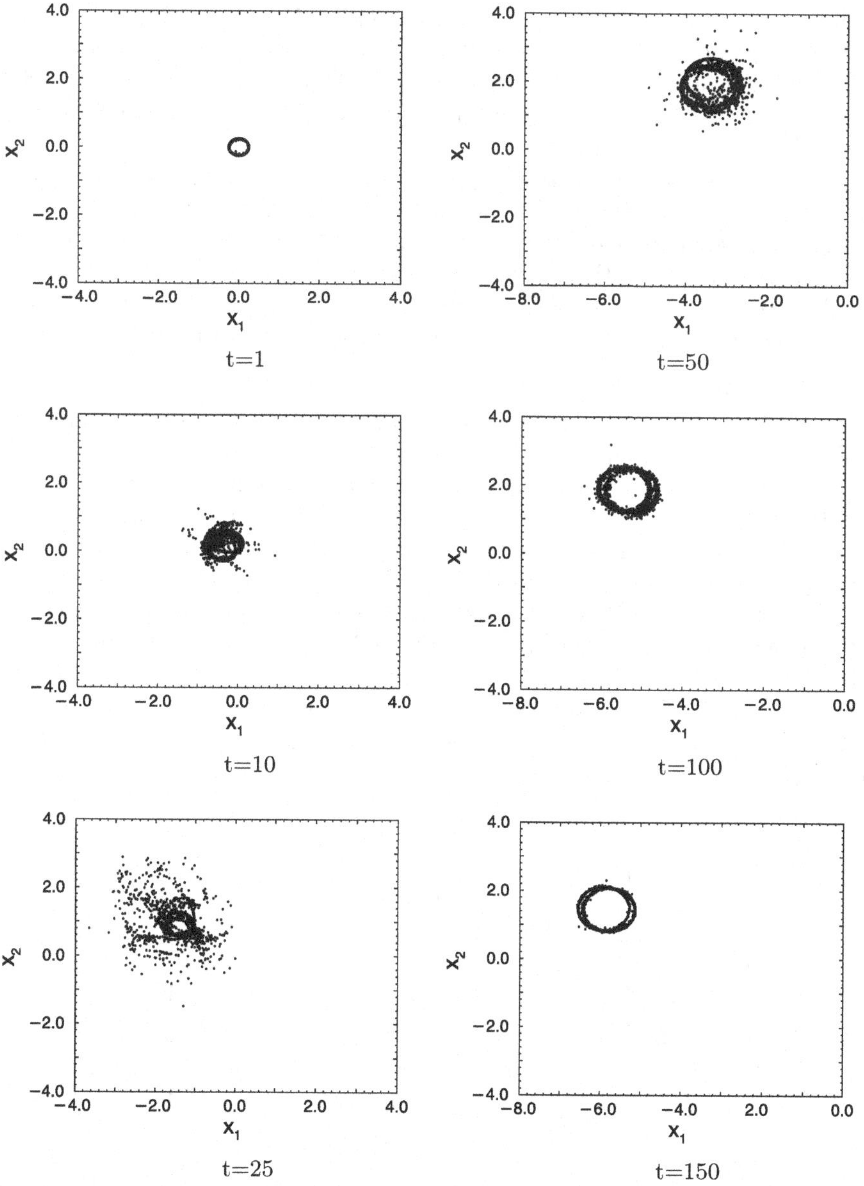

Fig. 2.44. Snapshots (spatial coordinates) of a swarm of 2000 particle agents moving according to (2.185). t gives the different times. Note that the pictures in the right column have a shifted x_1 axis compared to the left column. Parameters: $a = 1$, $D = 10^{-8}$; for the other parameters and initial conditions, see Fig. 2.43 [124]

solutions dependent on the noise intensity S and the initial momentum $\boldsymbol{v}_i(0)$ of the particles. Below a critical noise intensity S_c, the initial condition $v_i^2(0) > v_0^2$ leads to a swarm whose center travels at a constant nontrivial mean velocity, whereas for the initial condition $v_i^2(0) < v_0^2$, the center of the swarm is at rest.

We can confirm these findings by two-dimensional computer simulations presented in Figs. 2.45 and 2.46, which show the mean squared displacement, the average squared velocity of the swarm, and the squared mean velocity of the center of mass for the two different initial conditions.

For (a) $v_i^2(0) > v_0^2$, we find a continuous increase in $\Delta r^2(t)$ (Fig. 2.45a), whereas the velocity of the center of mass reaches a constant value: $V^2(t) = \left(N^{-1}\sum_i \boldsymbol{v}_i(t)\right)^2 \to v_0^2$, known as the stationary velocity of the force-free case, (2.181). The average squared velocity reaches a constant nontrivial value, too, that depends on the noise intensity and the initial conditions, $v_i^2(0) > v_0^2$, i.e., on the energy initially supplied (see Fig. 2.46 *top*).

For (b) $v_i^2(0) < v_0^2$, we find that the mean squared displacement after a transient time reaches a constant value, i.e., the center of mass comes to rest (Fig. 2.45b), which corresponds to $V^2(t) \to 0$ in Fig. 2.46 (*bottom*). In this case, however, the averaged squared velocity of the swarm reaches the known stationary velocity, $v^2(t) = 1/N \sum v_i^2(t) \to v_0^2$. Consequently, in this case, the energy provided by pumping goes into motion *relative to the center of mass*, and motion of the center of mass is damped out (see also [124]). Thus, in the following, we want to investigate the relative motion of the particles in more detail.

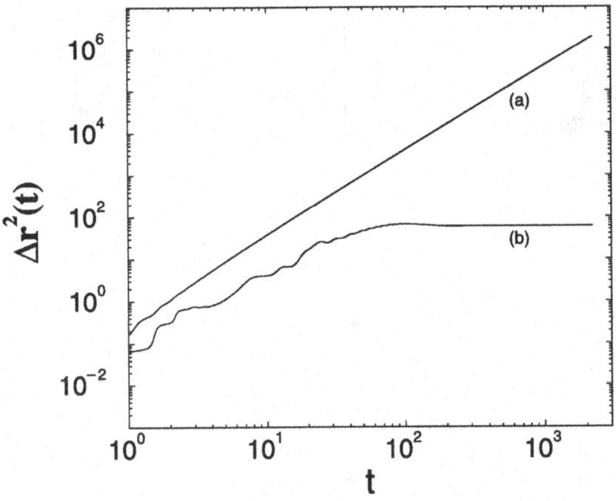

Fig. 2.45. Mean squared displacement $\Delta r^2(t)$, (2.108), of a swarm of 2000 particles coupled to the center of mass. Initial conditions: $\boldsymbol{r}_i(0) = \{0.0, 0.0\}$, (a) $\boldsymbol{v}_i(0) = \{1.0, 1.0\}$; (b) $\boldsymbol{v}_i(0) = \{0.0, 0.0\}$, for all particles. Parameters: $a = 1$, $D = 10^{-8}$, $d_2 = 10.0$, $q_0 = 10.0$, $\gamma_0 = 20.0$, $c = 1.0$ [454]

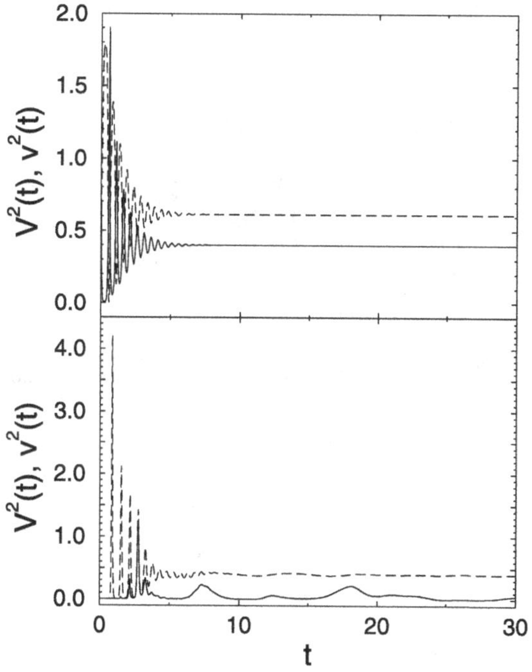

Fig. 2.46. Squared velocity of the center of mass, $V^2(t) = \left[N^{-1}\sum_i \boldsymbol{v}_i(t)\right]^2$ (*solid lines*) and averaged squared velocity $v^2(t) = N^{-1}\sum v_i^2(t)$ (*dashed lines*) for the simulations shown in Fig. 2.45: (*top*) initial conditions (a), (*bottom*) initial conditions (b) [454]

Using relative coordinates, $\{x_i, y_i\} \equiv \boldsymbol{r}_i - \boldsymbol{R}$, the dynamics of each particle in two-dimensional space is described by four, coupled, first-order differential equations:

$$\begin{aligned} \dot{x}_i &= v_{xi} - V_x \,, \quad \dot{v}_{xi} - \dot{V}_x = -g\left(v_i^2\right) v_{xi} - a\, x_i + \sqrt{2S}\,\xi_i(t) \,, \\ \dot{y}_i &= v_{yi} - V_y \,, \quad \dot{v}_{yi} - \dot{V}_y = -g\left(v_i^2\right) v_{yi} - a\, y_i + \sqrt{2S}\,\xi_i(t) \,. \end{aligned} \tag{2.187}$$

For $\boldsymbol{v} = 0$, i.e., for the initial conditions $v_i^2 < v_0^2$ and sufficiently long times, this dynamics is equivalent to the motion of free (or uncoupled) particles in a parabolic potential $U(x, y) = a(x^2 + y^2)/2$ with origin $\{0, 0\}$. Thus, within this approximation, the system becomes a canonical-dissipative system again, in the strict sense used in Sect. 2.4.1.

Figure 2.47 presents computer simulations of (2.187) for the relative motion of a particle swarm in a parabolic potential.[2] (Note that in this case, all particles started from the same position slightly *outside* the origin of

---

[2] The reader is invited to view a *movie* of the respective computer simulations ($t =$ 0–130), which can be found at `http://www.ais.fhg.de/~frank/swarm1.html`

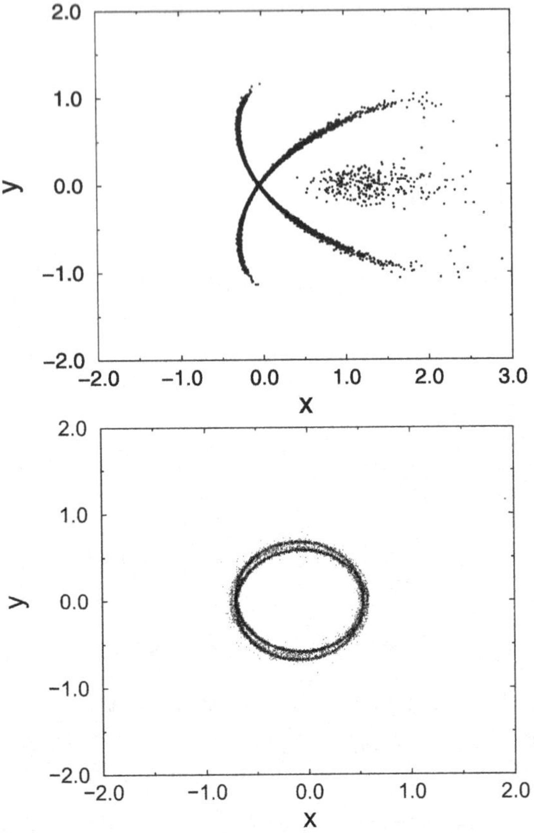

Fig. 2.47. Snapshots (relative coordinates) of a swarm of 10,000 particles moving according to (2.187) with $\boldsymbol{v} = 0$. (*top*) $t = 15$; (*bottom*) $t = 99$. Initial conditions: $\{x_i, y_i\} = \{0.5, 0.0\}$, $\{v_{xi}, v_{yi}\} = \{0.0, 0.0\}$ for all particles. Parameters: $a = 1$, $D = 10^{-5}$, $q_0 = 10.0$; $c = 1.0$; $\gamma_0 = 20$, $d_2 = 10$ [454]

the parabolic potential. This has been chosen to make the evolution of the different branches more visible.) As the snapshots of the spatial dispersion of the swarm show, we find after an inital stage the occurrence of two branches of the swarm which results from a *spontaneous symmetry break* (see Fig. 2.47 *top*). After a sufficiently long time, these two branches will move on two limit cycles (as already indicated in Fig. 2.47 *bottom*). One of these limit cycles refers to the left-handed, the other to the right-handed direction of motion in the 2-D space. This finding also agrees with the theoretical investigations of the deterministic case [126] (see Sect. 2.2.5) which showed the existence of a limit cycle of amplitude

$$r_0 = |v_0| \, a^{-1/2} , \tag{2.188}$$

provided that the relation $q_0 d_2 > \gamma_0 c$ is fulfilled. In the small noise limit, the radius of the limit cycles shown in Fig. 2.47 agrees with the value of r_0. Further, Fig. 2.46 has shown that the averaged squared velocity $v^2(t)$ of the swarm approaches the theoretical value of (2.181).

The existence of two opposite rotational directions of the swarm can also be clearly seen from the distribution of the angular momenta L_i of the particles. Figure 2.48 shows the existence of a bimodal distribution for $p(L)$. (The observant reader may notice that each of these peaks actually consists of two subpeaks resulting from the initial conditions, which are still not forgotten at $t = 99$). Each of the main peaks is centered around the theoretical value

$$|\boldsymbol{L}| = L_0 = r_0\, v_0\,, \tag{2.189}$$

where r_0 is given by (2.188) and v_0 is given by (2.181).

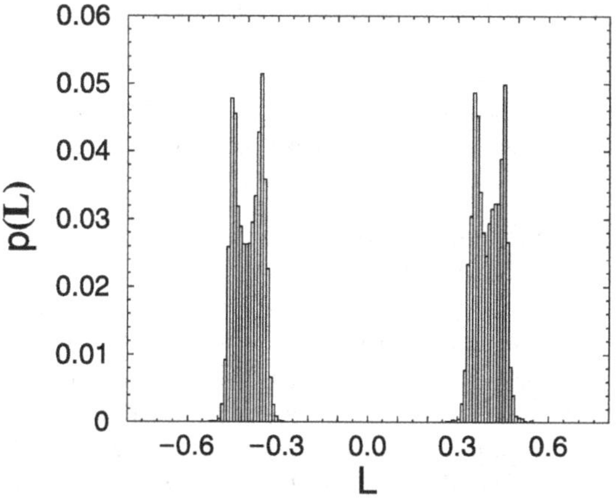

Fig. 2.48. Angular momentum distribution $p(L)$ for a swarm of 10,000 particles at $t = 99$. The figure refers to the spatial snapshot of the swarm shown in Fig. 2.47 (*bottom*) [454]

The emergence of the two limit cycles means that the dispersion of the swarm is kept within certain spatial boundaries. This occurs after a transient time used to establish the correlation between the individual particles. In the same manner as the motion of the particles becomes correlated, the motion of the center of mass is slowed down until it comes to rest, as already shown in Fig. 2.46.

This, however, is not the case if initial conditions $v_i^2(0) > v_0^2$ are chosen. Then, the energy provided by pumping does not go completely into the relative motion of the particles and the establishment of the limit cycles, as discussed above. Instead, the center of mass keeps moving, as shown in Fig. 2.45, whereas the swarm itself does not establish an internal order.

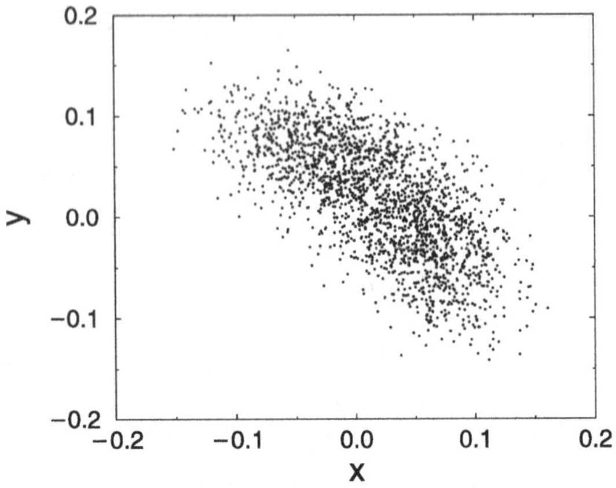

Fig. 2.49. Snapshot (relative coordinates) of a swarm of 10,000 particles moving according to (2.187) at $t = 99$. Initial conditions: $\{x_i, y_i\} = \{0.5, 0.0\}$, $\{v_{xi}, v_{yi}\} = \{1.0, 1.0\}$ for all particles. For parameters, see Fig. 2.47 [454]

Figure 2.49 displays a snapshot of the relative positions of the particles in this case (note the different scales of the axes compared to Fig. 2.47).

2.4.3 Coupling via Mean Momentum and Mean Angular Momentum

In the following, we want to discuss two other ways of *globally* coupling the swarm which fit into the general framework of canonical-dissipative systems outlined in Sect. 2.4.1. There, we introduced a dissipative potential $G(I_0, \boldsymbol{I}_1, \boldsymbol{I}_2, \dots)$ that depends on the different invariants of motion, $\boldsymbol{I}_i$. So far, we have only considered $I_0 = H$, (2.175), in the swarm model. If we additionally include the mean momentum $\boldsymbol{I}_1 = \boldsymbol{V}$, (2.165) as the first invariant of motion, the dissipative potential reads

$$G(I_0, \boldsymbol{I}_1) = \sum_{i=1}^{N} G_0(v_i^2) + G_1(\boldsymbol{V}), \tag{2.190}$$

$$G_1(\boldsymbol{V}) = \frac{C_V}{2} \left(\sum_{i=1}^{N} \boldsymbol{v}_i - N\, \boldsymbol{V}_1 \right)^2 . \tag{2.191}$$

Here, $G_0(v^2)$ is given by (2.180). The stationary solution of the probability distribution $p^0(\boldsymbol{v})$ is again given by (2.174). In the absence of an external potential $U(\boldsymbol{r})$, the Langevin equation that corresponds to the dissipative potential of (2.190) reads now

$$\dot{\boldsymbol{v}}_i = -g(v_i^2)\, \boldsymbol{v}_i - C_V \left(\sum \boldsymbol{v}_i - N\, \boldsymbol{V}_1 \right) + \sqrt{2S}\, \boldsymbol{\xi}_i(t) . \tag{2.192}$$

The term $G_1(\boldsymbol{V})$ is chosen so that it may drive the system toward the prescribed momentum $\boldsymbol{V}_1$, where the relaxation time is proportional to C_V^{-1}. If we would have a vanishing dissipation function, i.e., $g = 0$ for $v_i^2 = v_0^2$, it follows from (2.192) for the mean momentum that

$$\boldsymbol{V}(t) = \frac{1}{N}\sum_{i=1}^{N} \boldsymbol{v}_i(t) = \boldsymbol{V}_1 + \Big[\boldsymbol{V}(0) - \boldsymbol{V}_1\Big] e^{-C_V t} . \tag{2.193}$$

The existence of two terms, G_0 and G_1, however, could lead to competing influences of the resulting forces, and a more complex dynamics of the swarm results. As before, this may also depend on the initial conditions, i.e., $V_1^2 \geq v_0^2$ or $V_1^2 \leq v_0^2$.

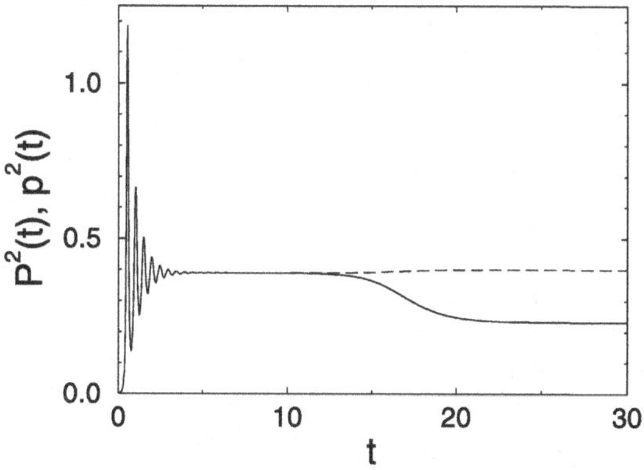

Fig. 2.50. Squared velocity of the center of mass, $V^2(t) = \Big[N^{-1}\sum_i \boldsymbol{v}_i(t)\Big]^2$ (*solid lines*) and averaged squared velocity $v^2(t) = N^{-1}\sum v_i^2(t)$ (*dashed lines*) for the simulations shown in Fig. 2.51 [454]

Figure 2.50 shows the squared velocity of the center of mass $V^2(t)$ and the averaged squared velocity of the swarm $v^2(t)$ for $V_1^2 \leq v_0^2$, We find an intermediate stage, where both velocities are equal, before the global coupling drives the mean momentum $\boldsymbol{V}$ toward the prescribed value $\boldsymbol{V}_1$, i.e., $V^2(t) \to (V_{1x}^2 + V_{1y}^2)$. On the other hand, $v^2(t) \to v_0^2$, as we have found before for the force-free case and for the linearly coupled case for similar initial conditions. The noticeable decrease of V^2 after the initial time lag can be best understood by looking at the spatial snapshots of the swarm provided in Fig. 2.51. For $t = 10$, we find a rather compact swarm where all particles move in the same (prescribed) direction. For $t = 50$, the correlations between the particles have already become effective, which means that the swarm begins to establish a circular front, which, however, does not become a full circle[3]. Eventually,

[3] The movie that shows the respective computer simulations for $t = 0$–100 can be found at `http://www.ais.fhg.de/~frank/swarm2.html`

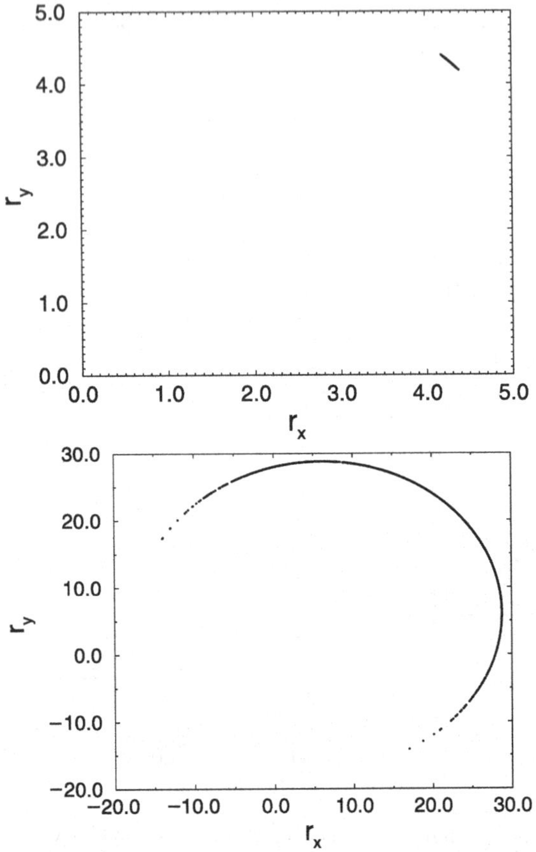

Fig. 2.51. Snapshots of a swarm of 2000 particles moving according to (2.192). (*top*) $t = 10$, (*bottom*) $t = 50$. Initial conditions: $\{r_{xi}, r_{yi}\} = \{0.0, 0.0\}$, $\{v_{xi}, v_{yi}\} = \{0.0, 0.0\}$ for all particles. Parameters: $\{V_{1x}, V_{1y}\} = \{0.344, 0.344\}$, $C_V = 10^{-3}$ $D = 10^{-8}$, $q_0 = 10.0$, $c = 1.0$, $\gamma_0 = 20$, $d_2 = 10$ [454]

we find again that the energy provided by pumping goes into the motion of the particles relative to the center of mass, whereas the motion of the center of mass itself is driven by the prescribed momentum.

For the initial condition $V_1^2 \geq v_0^2$, the situation is different again, as Fig. 2.52 shows. Apparently, both curves are the same for a rather small noise intensity, i.e., $V^2(t) = v^2(t)$ are both equal, but different from v_0^2, (2.181) and the prescribed momentum V_1^2. This can be realized only if all particles move in parallel in the same direction. Thus, a snapshot of the swarm would look much like the top part of Fig. 2.51.

Finally, we may also use the second invariant of motion, $\boldsymbol{I}_2 = \boldsymbol{L}$, (2.166), for global coupling of the swarm. In this case, the dissipative potential may be defined as follows:

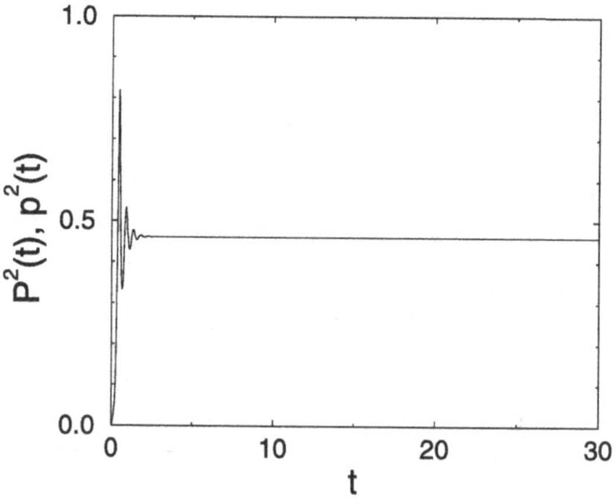

Fig. 2.52. Squared velocity of the center of mass, $V^2(t) = \left[N^{-1}\sum_i \boldsymbol{v}_i(t)\right]^2$ (*solid lines*) and averaged squared velocity $v^2(t) = N^{-1}\sum v_i^2(t)$ (*dashed lines*) for a swarm of 2000 particles moving according to (2.192). $\{V_{1x}, V_{1y}\} = \{1.0, 1.0\}$; for the other parameters and initial conditions, see Fig. 2.51 [454]

$$G(I_0, \boldsymbol{I}_2) = \sum_{i=1}^{N} G_0(v_i^2) + G_2(\boldsymbol{L})\,, \tag{2.194}$$

$$G_2(\boldsymbol{L}) = \frac{C_L}{2}\left(\sum \boldsymbol{r}_i\boldsymbol{v}_i - N\,\boldsymbol{L}_1\right)^2 . \tag{2.195}$$

$G_0(v^2)$ is again given by (2.180). The term $G_2(\boldsymbol{L})$ shall drive the system to a prescribed angular momentum $\boldsymbol{L}_1$ with a relaxation time proportional to C_L^{-1}.

$\boldsymbol{L}_1$ can be used to break the symmetry of the swarm toward a prescribed rotational direction. In Sect. 2.4, we observed the spontaneous occurrence of left-hand and right-hand rotations of a swarm of linearly coupled particles. Without additional coupling, both rotational directions are equally probable in the stationary limit. Considering both the parabolic potential $U(\boldsymbol{r}, \boldsymbol{R})$, (2.182), and the dissipative potential, (2.194), the corresponding Langevin equation may read now

$$\begin{aligned}\dot{\boldsymbol{v}}_i = &-g(v_i^2)\,\boldsymbol{v}_i - \frac{a}{N}\sum_{j=1}^{N}(\boldsymbol{r}_i - \boldsymbol{r}_j)\\ &+C_L\,\boldsymbol{r}_i\left(\sum \boldsymbol{r}_i\boldsymbol{v}_i - N\,\boldsymbol{L}_1\right) + \sqrt{2S}\,\boldsymbol{\xi}_i(t)\,.\end{aligned} \tag{2.196}$$

The computer simulations shown in Fig. 2.53 clearly display a unimodal distribution of the angular momenta L_i of the particles, which can be compared to Fig. 2.48 without coupling to the angular momentum. Consequently, we find in the long time limit, only one limit cycle corresponding to the movement

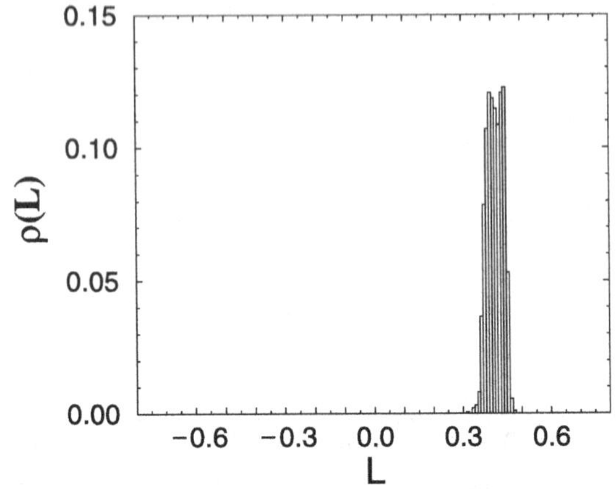

Fig. 2.53. Angular momentum distribution $p(L)$ for a swarm of 2000 particles at $t = 99$. For comparison with Fig. 2.48, the parameters, initial conditions, and snapshot times are the same as in Fig. 2.47. Additional coupling: $C_L = 0.05$, $\{L_{1z}\} = \{0.4\}$ [454]

of the swarm in the same rotational direction. The radius r_0 of the limit cycle is again given by (2.188).

We would like to add that in this case the dynamics also depends on the initial condition $\boldsymbol{L}_1$. For simplicity, we have assumed here $|\boldsymbol{L}_1| = L_0 = r_0 v_0$, (2.189), which is also reached by the mean angular momentum $\boldsymbol{L}$ in the course of time (cf. Fig. 2.53). For initial conditions $|\boldsymbol{L}_1| \ll L_0$, there is, of course, no need for the rotation of *all* particles in the same direction. Hence, we observe both left- and right-handed rotations of the particles in different proportions, so that the mean angular momentum is still $\boldsymbol{L} \to \boldsymbol{L}_1$. This results in a broader distribution of the angular momenta of particles instead of the clear unimodal distribution shown in Fig. 2.53. For initial conditions $|\boldsymbol{L}_1| \gg L_0$, on the other hand, the stable rotation of the swarm breaks down after some time because the driving force $\boldsymbol{L} \to \boldsymbol{L}_1$ tends to destabilize the attractor $\boldsymbol{L} \to \boldsymbol{L}_0$. This effect will be investigated in a forthcoming paper, together with some combined effects of different global couplings.

Finally, we can also combine the different global couplings discussed above by defining the dissipative potential as

$$\begin{aligned} G(I_0, \boldsymbol{I}_1, \boldsymbol{I}_2) &= G(v^2, \boldsymbol{V}, \boldsymbol{L}) \\ &= G_0(v^2) + G_1(\boldsymbol{V}) + G_2(\boldsymbol{L}) \,. \end{aligned} \tag{2.197}$$

$G_0(v^2)$ is given by (2.180), $G_1(\boldsymbol{V})$ by (2.191), and $G_2(\boldsymbol{L})$ by (2.195). Considering further an additional – external or interaction – potential, the corresponding Langevin equation can be written in the more general form:

$$
\begin{aligned}
\dot{\boldsymbol{r}}_i &= \boldsymbol{v}_i\,, \\
\dot{\boldsymbol{v}}_i &= -g(v_i^2)\boldsymbol{v}_i - \frac{\partial}{\partial \boldsymbol{r}_i}\left[\alpha_0\, U\left(\boldsymbol{r}, \boldsymbol{R}\right)\right] \\
&\quad - \frac{\partial}{\partial \boldsymbol{v}_i}\left[\alpha_1\left(\boldsymbol{V} - \boldsymbol{V}_1\right)^2 + \alpha_2\left(\boldsymbol{L} - \boldsymbol{L}_1\right)^2\right] + \sqrt{2S}\,\boldsymbol{\xi}_i(t)\,.
\end{aligned} \tag{2.198}
$$

The mean momentum $\boldsymbol{V}$ and mean angular momentum $\boldsymbol{L}$ are given by (2.165), (2.166), whereas the constant vectors $\boldsymbol{V}_1$ and $\boldsymbol{L}_1$ are used to break the spatial or rotational symmetry of the motion toward a preferred direction. The different constants α_i may decide whether or not the respective influence of the conservative or dissipative potential is effective; they further determine the timescale when global coupling becomes effective. The term $g(v^2)$, (2.178), on the other hand, considers the energetic conditions for the active motion of the swarm, i.e., it determines whether the particles of the swarm are able to "take off" at all.

The combination of the different types of coupling may lead to rather complex swarm dynamics, as already indicated in the examples discussed in this book. In particular, we note that the different terms may have competing influences on the swarm, which then would lead to "frustrated" dynamics with many possible attractors.

3. Aggregation and Physicochemical Structure Formation

3.1 Indirect Agent Interaction

3.1.1 Response to External Stimulation

This chapter marks a next step in our strategy to add complexity gradually to our concept of *Brownian agents*. So far, we have generalized the idea of Brownian particles (see Sect. 1.3) in Chap. 2 by considering an additional internal degree of freedom, i.e., an *internal energy depot* of the particles that can be used to take up, store, and transfer energy into motion. In this way, a simplified model of active biological motion can be derived – but we should note again that in this way only a rather simple type of "agent" has been introduced that has been therefore denoted as a *particle agent*, or *active particle* for simplicity.

So far, mainly the energetic aspects of active motion have been discussed. However, biological entities, such as cells, insects, bacteria, and other microorganims, are not only capable of active, i.e., self-driven motion, but also of more complex activities; in particular, they can respond to external signals. This can be considered in the *reactive agent* approach (see Sect. 1.1.3) as a next step toward a more complex agent design.

In Sect. 2.1.3, we already mentioned that human granulocytes in the presence of an external electric field can perform a *directed movement* either in or opposite the direction of the field. In Fig. 3.1, various modes of cell migration dependent on an external stimulating field are shown. If a cell preferentially moves toward (or away from) the direction of the stimulus gradient, this activity is known as *taxis*. In addition to the response to electric fields known as *galvanotaxis*, *phototaxis*, the response to light, *haptotaxis*, the response to an adhesion ligand, and *chemotaxis*, the response to chemicals, have also been observed. For instance, chemotaxis of white blood cells has been investigated both experimentally [512, 557] and theoretically [297, 511], and galvanotaxis of different cell types is discussed in [149, 188, 398]. A glossary of terms that describe oriented movement is given in [509].

To conclude the examples given, an extension of the model of Brownian particles to a model of active biological motion also has to consider different levels of activities of the microorganisms, each level corresponding to a particular response to a stimulus. Further, these active modes can be changed

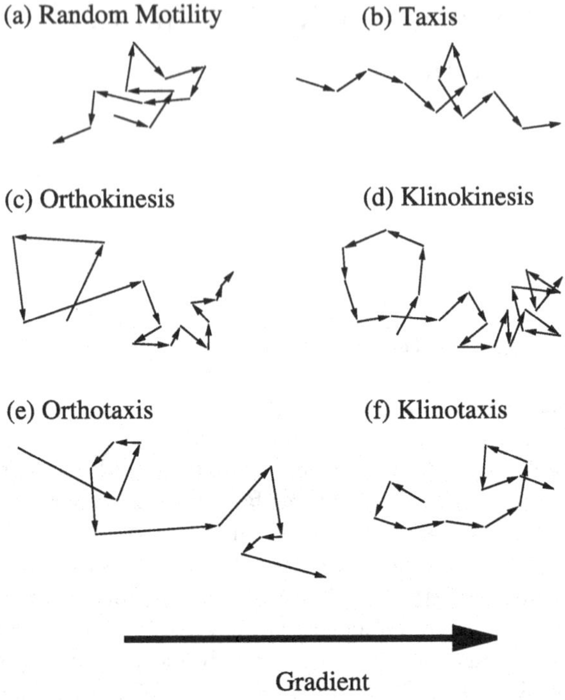

Fig. 3.1. Possible modes of cell migration. Each vector represents cell displacement over a constant time increment τ_0. The vector length indicates the magnitude of the cell speed. The modes of cell migration considered are (**a**) *random motility*, where there is no bias in turning or cell speed; (**b**) *taxis*, where a cell preferentially turns toward the gradient; (**c**) *orthokinesis*, where cell speed is dependent on the stimulus magnitude; (**d**) *klinokinesis*, where cell turning behavior is dependent on the stimulus magnitude; (**e**) *orthotaxis*, where cell speed is dependent on the direction of movement; and (**f**) *klinotaxis*, where cell turning behavior is dependent on the direction of movement. Dependences in (**c**)–(**f**) consistent with drift up the gradient are illustrated (after [105])

depending on the stimulus. To cope with these features within our approach to *Brownian agents*, we assume that each agent should be characterized by an additional *internal degree of freedom*, expressed by the parameter $\theta_i(t)$. This is assumed to be a *discrete value.* The internal energy depot as an internal degree of freedom will be not considered in this chapter together with other activities of the agent, i.e., questions of energy take-up and conversion are *neglected* in the following.

The internal parameter θ_i will be used to characterize different *modes of activity*. For instance, a Brownian agent with $\theta_i = 0$ should behave like a simple Brownian particle that is not affected at all by external stimuli of chemical gradients, etc. However, under certain circumstances that may depend on the stimulus, the agent's internal state could change to, e.g.,

$\theta_i = 1$, which describes an active mode, in which the agent may respond to an external stimulus in a particular way. Within a stochastic approach, the change of the internal parameter $\theta_i(t)$ in the course of time could be either described by a Langevin-type ansatz given in (1.1) or alternatively by a master equation (see Sect. 1.3.3), as will be shown in the different chapters of this book. In the following, our investigations will focus mainly on a specific activity, namely, *taxis*, where Brownian agents would change the direction of their movement according to the *gradient* of an external stimulus.

With respect to the Langevin approach to "simple" Brownian motion, Sect. 1.3.2, the response of a Brownian particle to the gradient of an external potential $U(\boldsymbol{r})$ is already implemented in (1.39) and (1.40) for the overdamped case. This potential is assumed fixed in space and time, it is thought, in analogy to a mechanical potential. However, if we consider a chemical stimulus, we would rather think in terms of a chemical concentration, which may change in the course of time, e.g., due to external dynamics which includes diffusion or absorption of the chemical substance. Hence, the response to the gradient of a (static) potential should be replaced by the response to the gradient of a more general "potential," which is denoted as an *effective potential field*, $h^e(\boldsymbol{r}, t)$. The term "field" might be misleading, if one draws simply analogies, e.g., to the electric (vector) field $\boldsymbol{E}$, which is already the gradient of the potential $\boldsymbol{E}(\boldsymbol{r}) = -\mathrm{grad}\, U(\boldsymbol{r})$. Therefore, we note explicitly that the effective field $h^e(\boldsymbol{r}, t)$ is meant to be a *scalar potential field*, measured in units of a *density* of the potential energy: $U(\boldsymbol{r}) \to \Omega\, h^e(\boldsymbol{r}, t)$, where Ω means the size of the volume element d^3r. Another, better known synonym for naming the effective field $h^e(\boldsymbol{r}, t)$ is the term *"adaptive landscape"* that has been already used in Sect. 1.2.

With this generalization, the Langevin equation, (1.39), in the case of *reactive Brownian agents*, reads

$$\frac{d\boldsymbol{r}_i}{dt} = \boldsymbol{v}_i \quad , \quad m\,\frac{d\boldsymbol{v}_i}{dt} = -\gamma_0 \boldsymbol{v}_i + \alpha_i \left.\frac{\partial h^e(\boldsymbol{r}, t)}{\partial \boldsymbol{r}}\right|_{\boldsymbol{r}_i} + \sqrt{2\,\varepsilon_i\,\gamma_0}\,\boldsymbol{\xi}_i(t)\,, \qquad (3.1)$$

which, in the overdamped limit, can be reduced to

$$\frac{d\boldsymbol{r}_i}{dt} = \frac{\alpha_i}{\gamma_0}\left.\frac{\partial h^e(\boldsymbol{r}, t)}{\partial \boldsymbol{r}}\right|_{\boldsymbol{r}_i} + \sqrt{\frac{2\varepsilon_i}{\gamma_0}}\,\boldsymbol{\xi}_i(t)\,. \qquad (3.2)$$

In (3.1), (3.2), two "individual" parameters appear, which, in general, can be different for each agent and may additionally depend on the internal parameter θ_i:

1. the individual response to the gradient of the field, α_i, which weights the *deterministic* influences in (3.1), (3.2),
2. the individual intensity of the noise ε_i, which is related to the temperature [see (1.39), (1.40)] and weights the *stochastic* influences. ε_i can be regarded as a measure of the individual *sensitivity* $1/\varepsilon_i$ of the agent.

The parameter α_i can be used to describe different responses of a reactive agent to the field, for instance,

1. attraction to the field, $\alpha_i > 0$, or repulsion, $\alpha_i < 0$.
2. response only if the local value of the field is above a certain threshold $h_{\rm thr}$: $\alpha_i = \Theta[h^{\rm e}(\boldsymbol{r}, t) - h_{\rm thr}]$, where $\Theta[y]$ is the Heaviside function: $\Theta = 1$, if $y > 0$, otherwise $\Theta = 0$.
3. response only if the agent is in a specific internal state θ: $\alpha_i = \delta_{\theta,\theta_i}$, where δ is the Kronecker delta used for discrete variables: $\delta_{\theta,\theta_i} = 1$ if $\theta_i = \theta$ and zero otherwise.

The relation between α_i and ε_i has some impact on the behavior of an agent. If both the response to the field and the sensitivity are low, the agent nearly behaves as a normal Brownian particle. On the other hand, a strong response or high sensitivity may result in a decrease of stochastic influences, and the agent is more or less guided by the gradient of the potential field. We want to mention that the idea of adjustable sensitivity, which, at first hand seems a rather artificial assumption in a physical context, will be elucidated and successfully applied in the context of search problems that can be solved by means of Brownian agents; see Sects. 5.4 and 7.1.1.

3.1.2 Generation of an Effective Potential Field

So far, the model describes the stochastic motion of agents and their response to the gradient of a potential field $h^{\rm e}(\boldsymbol{r}, t)$. Even with some additional assumptions about the variables α_i and ε_i, the dynamics given by (3.1), (3.2) is not significantly different from that of usual Brownian particles. As an important extension of the Brownian agent model, we now introduce *nonlinear feedback* between the potential field and the Brownian agents, i.e., the agents should have the ability to change the potential field $h^{\rm e}(\boldsymbol{r}, t)$. On the other hand, the potential field should also have an eigendynamics similar to that of a chemical field, which does not depend on agents. With these two elements, the effective potential field really becomes an *adaptive landscape*.

To be specific, we consider $i = 1, \dots, N$ Brownian agents and assume that agents with the internal parameter θ generate a field $h_\theta(\boldsymbol{r}, t)$ that obeys the following reaction–diffusion equation [see also (1.14)]:

$$\frac{\partial h_\theta(\boldsymbol{r}, t)}{\partial t} = -k_\theta\, h_\theta(\boldsymbol{r}, t) + D_\theta\, \Delta h_\theta(\boldsymbol{r}, t) + \sum_{i=1}^{N} s_i(\theta_i, t)\, \delta_{\theta,\theta_i(t)}\, \delta\Big[\boldsymbol{r} - \boldsymbol{r}_i(t)\Big]\,. \tag{3.3}$$

The spatiotemporal evolution of the field, $h_\theta(\boldsymbol{r}, t)$, is governed by three processes: (i) decay at a rate k_θ; (ii) diffusion coefficient D_θ; and (iii) production at an individual production rate, $s_i(\theta_i, t)$; which in general may depend on the internal parameter θ and can change in time. Here, $\delta_{\theta,\theta_i(t)}$ is the Kronecker delta indicating that agents contribute only to the field component, which matches their internal parameter θ, and $\delta[\boldsymbol{r} - \boldsymbol{r_i}(t)]$ means Dirac's delta function used for continuous variables, which indicates that agents contribute to the field component only at their current position $\boldsymbol{r}_i$.

The gradient of the effective field $h^e(\boldsymbol{r},t)$ which eventually influences agent motion due to (3.1), (3.2), is in general a specific function of different field components $h_\theta(\boldsymbol{r},t)$:

$$\boldsymbol{\nabla} h^e(\boldsymbol{r},t) = \boldsymbol{\nabla} h^e\Big(\dots, h_\theta(\boldsymbol{r},t), h_{\theta'}(\boldsymbol{r},t), \dots\Big). \tag{3.4}$$

This rather general outline of the model will be specified for various applications in the following chapters. Here, we want to conclude with some general remarks. As Fig. 3.2 shows, an important feature of the Brownian agent approach is *nonlinear feedback* between agents of different kinds θ and different components $h_\theta(\boldsymbol{r},t)$ of the effective field. Whereas "simple" Brownian particles usually do not interact with each other (except via some random collisions which change their impulses), Brownian agents can change their *environment* by generating/changing certain components of the self-consistent field $h^e(\boldsymbol{r},t)$ that in this way plays the role of an *adaptive landscape*. Changes in the field in turn influence the further movement and response behavior of the Brownian agents, in this way establishing *circular causation*.

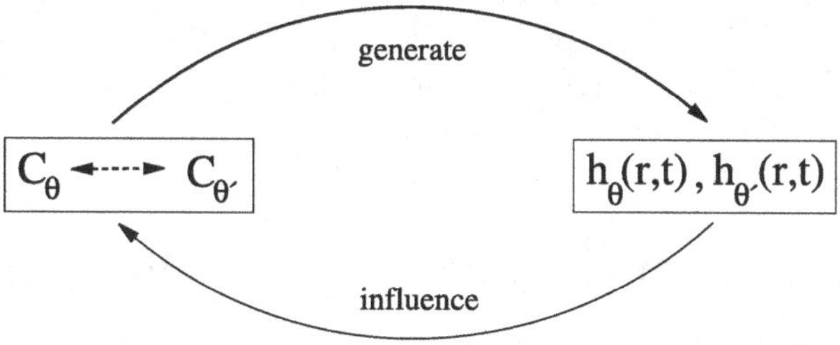

Fig. 3.2. Circular causation between Brownian agents, (C_θ, $C_{\theta'}$) with different internal parameters θ, θ' and different components $h_\theta(\boldsymbol{r},t)$ of the effective field. Specific realizations of this general scheme are shown in Figs. 3.3, 4.3, 8.26, 9.3, and 10.1

The effective potential field acts as a medium for *indirect interaction* – or as a medium for *communication* (see Sect. 1.2.2) – between Brownian agents. Because the components of the field are assumed to diffuse, local actions of the agents, i.e., a local increase in the field via s_i, can propagate through the whole system; hence the field realizes *global coupling* between agents. On the other hand, these changes only have a certain "lifetime" given by the decay constant k_θ. In this way, the effective field represents a certain kind of *external memory* for the actions of agents. Further, the possibility that agents in different active modes can generate separate field components that contribute to the effective field in a nonlinear manner opens the door to simulating complex types of indirect interaction between agents.

As we will show in the following chapters, indirect interaction between agents will result in specific *spatiotemporal structures* on the macroscopic level, such as aggregation, swarming, or directed motion. This structure formation is accompanied by evolution of the effective field, or, more precisely, there is a *co-evolution* of agents' spatial structure and the potential field. This can be described by moving from the agent's perspective to the density perspective in the following section.

3.1.3 Master Equations and Density Equations

Let us now consider an ensemble of $i = 1, \ldots, N$ Brownian agents with internal parameters $\theta_1, \ldots, \theta_N$, at positions $\boldsymbol{r}_1, \ldots, \boldsymbol{r}_N$. The equation of motion for each agent i is given by the Langevin equations, (3.1), (3.2). If we assume that the *total number* of particles is constant, the complete dynamics of the ensemble can be formulated in terms of the canonical N-particle distribution function:

$$P(\underline{r}, \underline{\theta}, t) = P(\boldsymbol{r}_1, \theta_1, \ldots, \boldsymbol{r}_N, \theta_N, t)\,. \tag{3.5}$$

$P(\underline{r}, \underline{\theta}, t)$ describes the *probability density* of finding N Brownian agents with the distribution of internal parameters $\underline{\theta}$ and positions $\underline{r}$ considered. Considering both "reactions" (i.e., changes of the internal parameters θ_i) and movement of the agents, this probability density changes in the course of time due to a *multivariate master equation*:

$$\begin{aligned}\frac{\partial}{\partial t} P(\underline{r}, \underline{\theta}, t) = & \sum_{\underline{r}' \neq \underline{r}} \Big\{ w(\underline{r}|\underline{r}') P(\underline{r}', \underline{\theta}, t) - w(\underline{r}'|\underline{r}) P(\underline{r}, \underline{\theta}, t) \Big\} \\ & + \sum_{\underline{\theta}' \neq \underline{\theta}} \Big\{ w(\underline{\theta}|\underline{\theta}') P(\underline{r}, \underline{\theta}', t) - w(\underline{\theta}'|\underline{\theta}) P(\underline{r}, \underline{\theta}, t) \Big\}\,.\end{aligned} \tag{3.6}$$

The first line of the right-hand side of (3.6) describes the change in the probability density due to the motion of the agents; the second line describes the "loss" and "gain" of agents with respect to changes in their internal parameter. The transition rates $w(\underline{r}|\underline{r}')$ and $w(\underline{\theta}|\underline{\theta}')$ refer to any possible transition within the distributions $\underline{r}'$ and $\underline{\theta}'$, respectively, which leads to the assumed distribution $\underline{r}$ and $\underline{\theta}$.

If we consider a *continuous space*, where the motion of agents is described by the *overdamped* Langevin equation, (3.2), and further assume $\alpha_i = \alpha$, $\varepsilon_i = \varepsilon$, the multivariate master equation reads

$$
\begin{aligned}
\frac{\partial}{\partial t} P(\underline{r},\underline{\theta},t) = & -\sum_{i=1}^{N} \Big\{ \boldsymbol{\nabla}_i \Big[(\alpha/\gamma_0)\, \boldsymbol{\nabla}_i h^e(\boldsymbol{r},t)\, P(\underline{r},\underline{\theta},t) \Big] - D_n\, \boldsymbol{\Delta}_i P(\underline{r},\underline{\theta},t) \Big\} \\
& + \sum_{i=1}^{N} \sum_{\theta_i' \neq \theta_i} \Big[w(\theta_i|\theta_i') P(\theta_i',\underline{\theta}^\star,\underline{r},t) - w(\theta_i'|\theta_i) P(\theta_i,\underline{\theta}^\star,\underline{r},t) \Big] .
\end{aligned} \tag{3.7}
$$

Here, $w(\theta_i'|\theta_i)$ denotes the transition rate to change the internal state θ_i into one of the possible θ_i' during the next time step, and $\underline{\theta}^\star$ denotes all elements of the configuration, which are not explicitly written. The above description assumes "reactions" $\theta \to \theta'$, while the total number of agents is constant. In Sect. 3.3.1, we will further discuss reactions that change the total number of agents via an *influx* and an *outflux* of agents. Then, the dynamics of the ensemble will be described in terms of the grand-canonical distribution function.

Equation (3.7) for the probability density $P(\underline{r},\underline{\theta},t)$ is coupled to the equation for the effective field, $h^e(\boldsymbol{r},t)$, via the reaction–diffusion equations for the different components, $h_\theta(\boldsymbol{r},t)$, (3.3). We note that (3.3) is a stochastic partial differential equation where

$$
n_\theta^{\mathrm{micr}}(\boldsymbol{r},t) = \sum_{i=1}^{N} \delta_{\theta,\theta_i(t)}\, \delta\Big[\boldsymbol{r} - \boldsymbol{r}_i(t)\Big] \tag{3.8}
$$

is the *microscopic density* [431] of agents with internal parameter θ, changing their position due to (3.1). For the canonical ensemble, the macroscopic density of the agents can be introduced by means of $P(\underline{r},\underline{\theta},t)$ as follows:

$$
n_\theta(\boldsymbol{r},t) = \sum_{i=1}^{N} \delta_{\theta,\theta_i} \int_A \delta(\boldsymbol{r}-\boldsymbol{r}_i)\, P(\theta_1,\boldsymbol{r}_1 \ldots, \theta_N, \boldsymbol{r}_N, t)\, d\boldsymbol{r}_1 \ldots d\boldsymbol{r}_N \tag{3.9}
$$

with the boundary condition:

$$
N = \sum_\theta \int_A n_\theta(\boldsymbol{r},t)\, dr = \mathrm{const}\,. \tag{3.10}
$$

Here, A denotes the system size. If we integrate (3.9) with respect to (3.7) and neglect higher correlations, we obtain the following *reaction–diffusion equation* for $n_\theta(\boldsymbol{r},t)$:

$$
\begin{aligned}
\frac{\partial}{\partial t} n_\theta(\boldsymbol{r},t) = & -\boldsymbol{\nabla} \Big[n_\theta(\boldsymbol{r},t)\, (\alpha/\gamma_0)\, \boldsymbol{\nabla} h^e(\boldsymbol{r},t) \Big] + D_n\, \Delta n_\theta(\boldsymbol{r},t) \\
& - \sum_{\theta' \neq \theta} \Big[w(\theta'|\theta)\, n_\theta(\boldsymbol{r},t) + w(\theta|\theta')\, n_{\theta'}(\boldsymbol{r},t) \Big] .
\end{aligned} \tag{3.11}
$$

We cannot easily use $n_\theta(\boldsymbol{r},t)$ to replace the microscopic density in the reaction–diffusion equation for the field components, $h_\theta(\boldsymbol{r},t)$, (3.3), because the production rates, $s_i(\theta_i,t)$, are, in general, different for every agent. However, if we assume a constant rate, $s_i(\theta_i,t) = s_\theta$, equal for all agents with the internal parameter θ, $n_\theta(\boldsymbol{r},t)$ can be inserted in (3.3), which then becomes a *linear deterministic equation*:

$$\frac{\partial}{\partial t} h_\theta(\boldsymbol{r},t) = -k_\theta h_\theta(\boldsymbol{r},t) \; + \; D_\theta \Delta h_\theta(\boldsymbol{r},t) \; + \; s_\theta \, n_\theta(\boldsymbol{r},t) \,. \tag{3.12}$$

Finally, we have a set of coupled differential equations for the densities of Brownian agents, $n_\theta(\boldsymbol{r},t)$, and the components of the effective field, $h_\theta(\boldsymbol{r},t)$. In general, field components can also determine transition rates, $w(\theta'|\theta)$, which then become implicit functions of time and space coordinates, as discussed, for example, in Sect. 10.2.

So far, we have introduced *two different levels* of description for Brownian agents: the "individual" description by Langevin equations, (3.1), (3.2), and the density description by Fokker–Planck equations, or reaction–diffusion equations, (3.11), (3.12), respectively. In the limit of a *large number of agents*, $N \to \infty$, both descriptions will, of course, lead to equivalent results. The *advantage* of the agent-based or particle-based approach is that it is also applicable where only *small numbers* of agents govern the structure formation. In physicochemical structure formation, this happens, for instance, in gas discharges [558], on catalytic surfaces [416], and in cell membranes [161]. Here, the continuous limit becomes questionable, and partial differential equations are not sufficient to describe the behavior of the system. The final pattern is *path-dependent*, which means it is intrinsically determined by the history of its creation, and irreversibility and early symmetry breaks play a considerable role, as will be shown, e.g., in the examples of Chaps. 4 and 5.

Although we find a continuous description in terms of coupled reaction–diffusion equations useful, e.g., for deriving *critical parameters*, we will use the agent-based approach for *computer simulations* of the spatiotemporal patterns because it provides quite a *stable* and *fast* numerical algorithm. This holds especially for *large density gradients*, which may considerably decrease the time step Δt allowed to integrate the related dynamics. If we solve the Langevin equation, (3.2), the gradient $\boldsymbol{\nabla} h^e(\boldsymbol{r},t)$ appears only in a *linear* manner. Hence, for the time step, the restriction results in

$$\Delta t < \left[\frac{\alpha}{\gamma_0} \left| \frac{\partial h^e(\boldsymbol{r},t)}{\partial \boldsymbol{r}} \right| \right]^{-1} . \tag{3.13}$$

In the corresponding reaction–diffusion equations, however, the time step required for the integration is determined mainly by the nonlinearities of the equations. If we suppose that $n_\theta(\boldsymbol{r},t)$ and $h_\theta(\boldsymbol{r},t)$ are of the same order, the allowed time step Δt should be less than

$$\Delta t < \left[\frac{\alpha}{\gamma_0} \left| \frac{\partial n_\theta(\boldsymbol{r},t)}{\partial \boldsymbol{r}} \frac{\partial h^e(\boldsymbol{r},t)}{\partial \boldsymbol{r}} \right| \right]^{-1} \sim \left[\frac{\alpha}{\gamma_0} \left[\frac{\partial h^e(\boldsymbol{r},t)}{\partial \boldsymbol{r}} \right]^2 \right]^{-1} , \tag{3.14}$$

i.e., it should decrease according to the *square* of $\boldsymbol{\nabla} h$.

One could argue that for an appropriate agent-based or particle-based simulation of reaction–diffusion equations, a large number of coupled Langevin equations need to be solved. This might be considered a disadvantage that compensates for the advantage of a fast algorithm. But, as our experience has shown [429, 431] (see Sect. 3.3), consideration of about 10^4 particles already results in sufficently smooth patterns. Hence, the main idea of our approach is to solve the Langevin equations for an *ensemble of Brownian agents* instead of integrating the related nonlinear partial differential equations.

3.1.4 Stochastic Simulation Technique

Based on the explanations above, we want to introduce our simulation algorithm for Brownian agents. The computer program has to deal with three different processes, which have to be discretized in time for the simulation: (i) the movement of Brownian agents, (ii) transitions of the internal states due to rates $w(\theta'|\theta)$, and (iii) the generation of the field components $h_\theta(\boldsymbol{r},t)$. For simplicity, we will assume, for example, agents with two internal states $\theta_i \in \{0,+1\}$. The changes of the internal states should not explicitly depend on i, which means only two possible transitions described by rates $k^+ = w(0|+1)$, $k^- = w(+1|0)$. Further, only a one-component field, $h^e(\boldsymbol{r},t) = h_0(\boldsymbol{r},t)$, is considered.

The three processes have to be discretized in time now, and the time steps Δt could be either constant or nonconstant. Considering a two-dimensional surface $\{x,y\}$ for the motion of agents with respect to the Langevin equation, (3.1), the new x position and new velocity v_x of agent i at time $t+\Delta t$ are given by

$$
\begin{aligned}
x_i(t+\Delta t) &= x_i(t) + v_x^i(t)\,\Delta t\,, \qquad (3.15)\\
v_x^i(t+\Delta t) &= v_x^i(t)\Big[1-\gamma_0\,\Delta t\Big] + \Delta t\,\alpha_i\,\left.\frac{\partial h_0(x,y,t)}{\partial x}\right|_{x_i} + \sqrt{2\varepsilon_i\gamma_0\,\Delta t}\;\mathrm{GRND}\,.
\end{aligned}
$$

The equation for the y position reads accordingly. For the overdamped limit, a discretization of (3.2) would result in

$$
x_i(t+\Delta t) = x_i(t) + \Delta t\,\frac{\alpha_i}{\gamma_0}\,\left.\frac{\partial h_0(x,y,t)}{\partial x}\right|_{x_i} + \sqrt{2D_n\Delta t}\;\mathrm{GRND}\,. \qquad (3.16)
$$

$D_n = \varepsilon_i/\gamma_0$ is the diffusion coefficient, and GRND is a Gaussian random number whose mean equals zero and standard deviation equals unity.

To calculate the spatial gradient of the field $h_0(x,y,t)$, we have to consider that $h_0(x,y,t)$ is a density defined over a volume element of a certain size Ω. Thus, it would be convenient to divide the the surface into boxes with unit length Δl, which is a sufficiently small, but constant value. The spatial gradient is then defined as

$$\frac{\partial h_0(x,y,t)}{\partial x} = \frac{h_0(x+\Delta l,y,t) - h_0(x-\Delta l,y,t)}{2\Delta l} ,$$
$$\frac{\partial h_0(x,y,t)}{\partial y} = \frac{h_0(x,y+\Delta l,t) - h_0(x,y-\Delta l,t)}{2\Delta l} . \tag{3.17}$$

We further note that *periodic boundary conditions* are used for the simulations; therefore, the neighboring box, $x \pm \Delta l$, $y \pm \Delta l$, is always specified.

The actual values of $h_0(x,y,t)$ are determined by a reaction–diffusion equation, (3.3), which has to be discretized in space and time, too. If we assume again a constant production rate, s_0, the discrete dynamics of the field can be approximated by

$$\begin{aligned} h_0(x,y,t+\Delta t) &= (1 - k_0\,\Delta t)\,h_0(x,y,t) + \Delta t\; s_0\,N_0(x,y,t) \\ &\quad - D_0 \frac{\Delta t}{(\Delta l)^2}\Big[4\,h_0(x,y,t) - h_0(x{+}\Delta l,y,t) - h_0(x{-}\Delta l,y,t) \\ &\quad - h_0(x,y+\Delta l,t) - h_0(x,y-\Delta l,t)\Big] . \end{aligned} \tag{3.18}$$

$N_0(x,y,t)$ denotes the number of agents at position $\{x,y\}$ which generate the field, $h_0(x,y,t)$. To solve (3.18) numerically, we have to respect two conditions for choosing the parameters D_0 and k_0 together with the time step Δt and the unit length Δl:

$$k_0\,\Delta t \ll 1\,, \quad D_0 \frac{\Delta t}{(\Delta l)^2} < \frac{1}{4}\,; \tag{3.19}$$

otherwise no stable numerical results would be obtained.

Finally, we have to consider the transition rates k^+, k^-, which determine the *average number* of transitions of an agent in the internal state θ_i during the next time step Δt. In a stochastic simulation, however, the actual number of transitions is a *stochastic* variable, and we have to ensure that the number of transitions during the simulation (i) does not exceed the actual number of agents available during Δt and (ii) is equal to the average number of reactions in the limit $t \to \infty$.

This problem can be solved by using the *stochastic simulation technique* for reactions which defines the appropriate time step Δt as a *random variable* [119]. Let us assume that we have exactly N_1 agents with internal state $\theta_i = 1$ in the system at time $t = t_0$. Then the probability $P(N_1, t_0)$ equals one, and the probability for any other number N_1' is zero. With this intial condition, the master equation to change N_1 reads

$$\frac{\partial P(N_1,t)}{\partial t} = \sum_{i=1}^{N} (k_i^+ + k_i^-)\,P(N_1,t)\;, \quad P(N_1,t_0) = 1\,, \tag{3.20}$$

where N is the total number of agents and k_i^+, k_i^- are the transition rates for each agent. The solution of this equation yields

$$P(N_1, t - t_0) \sim \exp\left(-\frac{t - t_0}{t_\mathrm{m}}\right),$$

$$t_\mathrm{m} = \frac{1}{\sum_{i=1}^{N} k_i}, \quad \sum_{i=1}^{N} k_i = \sum_{i=1}^{N_1} k_i^+ + \sum_{i=N_1+1}^{N} k_i^- . \tag{3.21}$$

Here, t_m is the mean lifetime of the state N_1. For $t - t_0 \ll t_\mathrm{m}$, the probability $P(N_1, t)$ of still finding N_1 is almost one, but for $t - t_0 \gg t_\mathrm{m}$, this probability goes to zero. The time when the change of N_1 occurs is most likely about the mean lifetime $t - t_0 \approx t_\mathrm{m}$. In a stochastic process, however, this time varies; hence, the *real lifetime* $t - t_0 = \tau$ is a randomly distributed variable. Because we know that $P(N_1, t)$ has values between $[0, 1]$, we find from (3.21) that

$$\tau = t - t_0 = -\,t_\mathrm{m} \ln\left\{\mathrm{RND}[0,1]\right\}. \tag{3.22}$$

RND is a random number drawn from the interval $[0, 1]$ (zero actually has to be excluded). It yields $\langle \ln\{\mathrm{RND}[0,1]\}\rangle = 1$. That means, after real lifetime τ, one of the possible processes that changes N_1 occurs. Each of these processes has the probability,

$$p(\theta_i = 0, t \to \theta_i = 1, t + \tau) = k_i^- / \sum_{i=1}^{N} k_i ,$$

$$p(\theta_i = 1, t \to \theta_i = 0, t + \tau) = k_i^+ / \sum_{i=1}^{N} k_i . \tag{3.23}$$

Thus, a second random number $\mathrm{RND}[0, \sum k_i]$ drawn from the interval $[0, \sum k_i]$ will determine which of these possible processes occurs. It will be the process b which satisfies the condition:

$$\sum_{i=1}^{b-1} k_i < \mathrm{RND} , \quad \sum_{i=1}^{b} k_i > \mathrm{RND} . \tag{3.24}$$

For the transition probabilities to change N_1, we find, in particular,

$$\mathrm{Prob}[N_1, t \to N_1 + 1, t + \tau] = \sum_{i=1}^{N_0} k_i^- t_\mathrm{m} ,$$

$$\mathrm{Prob}[N_1, t \to N_1 - 1, t + \tau] = \sum_{i=1}^{N_1} k_i^+ t_\mathrm{m} . \tag{3.25}$$

Obviously, the sum over these probabilities is one, which means that during the time interval τ, one of these processes occurs with certainty. Further, using the definition of τ, (3.22), we see that

$$t = \sum_{l=1}^{z} t_\mathrm{m}^l = \sum_{l=1}^{z} \tau^l \tag{3.26}$$

yields asymptotically, i.e., after a large number of simulation steps, $z \gg 1$ because of $\langle \ln\{\mathrm{RND}[0,1]\}\rangle = 1$.

In conclusion, determining the time step as $\Delta t = \tau$, (3.22), ensures that only one transition occurs during one time step and the number of transitions does not get out of control. Further, in the asymptotic limit, the actual (stochastic) number of transitions is equal to the average number. The (numerical) disadvantage is that time step Δt is not constant, so it has to be recalculated after each cycle before moving the agents with respect to (3.16). This might slow down the speed of the simulations considerably. Another point to be mentioned is that, depending on the random number drawn, $\Delta t = \tau$ could be also rather large, thus violating the conditions of (3.19). In this case, the calculations of (3.15)–(3.18) have to be rescaled appropriately.

The above procedure seems to be very painstaking, but we have had rather satisfying experiences with it. It is appropriate especially when the transition rates differ for each agent i and may further depend on other variables, including the local value of the field, $h_0(\boldsymbol{r}, t)$, as discussed, for example, in Sect. 9.2. On the other hand, in some applications, it will also be sufficient simply to choose a constant time step. In Sect. 3.3.1, we will discuss an example, where the transition rates, $w(+1 \mid 0)$, $w(0 \mid +1)$, describe *adsorption* and *desorption* of particles on a surface of area A. For the adsorption process, a homogeneous influx of particles with a rate Φ is assumed, and desorption is described as an inhomogeneous outflux of agents at a rate k_n, *constant* for all particles. In this case, the mean lifetime t_m, (3.21), is a function only of the stochastic variable N_1 of particles on the surface:

$$t_\mathrm{m} = \frac{1}{\Phi\, A + k_n\, N_1}\,, \tag{3.27}$$

which can be approximately treated as constant, after the adsorption and desorption processes are balanced. We may then use a *fixed* time step, $\Delta t \simeq t_\mathrm{m}$, and transition probabilities to change N_1, (3.25), can be simplified to

$$\begin{aligned} \mathrm{Prob}[N_1, t \to N_1 + 1, t + \Delta t] &= \Phi\, A\, \Delta t\,, \\ \mathrm{Prob}[N_1, t \to N_1 - 1, t + \Delta t] &= k_n\, N_1\, \Delta t\,. \end{aligned} \tag{3.28}$$

In the absence of transitions, $w(\theta'|\theta) = 0$, the time step can be chosen constant in any case. We have, however, to ensure that Δt is still sufficiently small, (i) because of (3.19) and (ii) because we want to simulate agent motion approximately as a "continuous" process rather than as a *hopping process* between boxes of size $(\Delta l)^2$. This means that the spatial move of an agent in the x direction during time step Δt, $\Delta x = x(t + \Delta t) - x(t)$, has to be sufficiently small, i.e., $\Delta x \ll \Delta l$.

3.2 Aggregation of Brownian Agents

3.2.1 Chemotactic Response

To elucidate the interactive structure formation that results from nonlinear feedback between Brownian agents and an effective potential field, we start with a simple model: we assume that all agents are identical, which means that the "individual" parameters introduced in Sect. 3.1.1 are the same for all agents: $\alpha_i = \alpha > 0$, $\varepsilon_i = \varepsilon$, $\theta_i = 0$. Here, α describes the kind of response by agents to the gradient of an effective field; ε is the strength of the stochastic force, and thus weights the "sensitivity" of the particles. Because of $\theta_i = 0$, we have only a *one-component field*, $h_0(\boldsymbol{r}, t)$, that shall be assumed as a *chemical field*. Further, no transitions between different internal states θ' have to be considered in the master equation, (3.7). Figure 3.3 illustrates simple feedback between agents and the field, which results from this simple model.

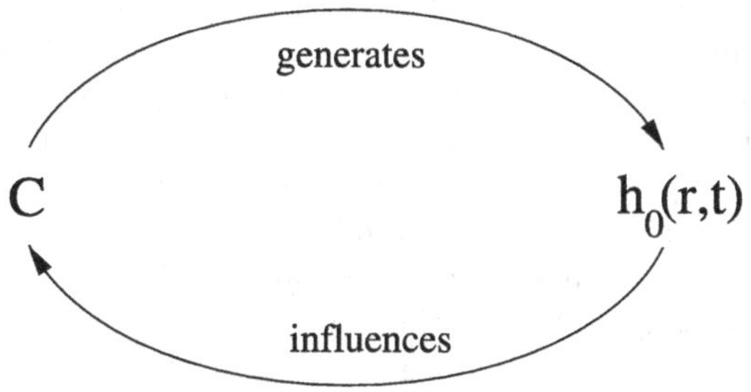

Fig. 3.3. Circular causation between Brownian agents (C) and the effective field, $h_0(\boldsymbol{r}, t)$ generated by them

For simplicity, we can further assume that the production rate of the field, $s_i(\theta_i, t)$, (3.3), does not explicitly change in time: $s_i(\theta_i, t) = s_0 = \text{const}$. Then, the equations that describe the effective potential field, (3.3), (3.4), simply read

$$\boldsymbol{\nabla_i} h^e(\boldsymbol{r}, t) = \boldsymbol{\nabla_i} h_0(\boldsymbol{r}, t)\,, \tag{3.29}$$

$$\frac{\partial h_0(\boldsymbol{r}, t)}{\partial t} = -k_0\, h_0(\boldsymbol{r}, t) + D_0\, \boldsymbol{\Delta} h_0(\boldsymbol{r}, t) + s_0 \sum_{i=1}^{N} \delta[\boldsymbol{r} - \boldsymbol{r}_i(t)]\,.$$

The equation for the potential field is coupled to the Langevin equation that describes the motion (position) of the agents. Now, (3.1) reads explicitly

$$\frac{d\boldsymbol{r}_i}{dt} = \boldsymbol{v}_i \quad , \quad \frac{d\boldsymbol{v}_i}{dt} = -\gamma_0 \boldsymbol{v}_i + \alpha \left.\frac{\partial h_0(\boldsymbol{r}, t)}{\partial \boldsymbol{r}}\right|_{\boldsymbol{r}_i} + \sqrt{2\,\varepsilon\,\gamma_0}\,\xi_i(t) \tag{3.30}$$

and can, in the overdamped limit, be reduced to

$$\frac{d\boldsymbol{r}_i}{dt} = \frac{\alpha}{\gamma_0}\,\frac{\partial h_0(\boldsymbol{r},t)}{\partial \boldsymbol{r}}\bigg|_{\boldsymbol{r}_i} + \sqrt{\frac{2\varepsilon}{\gamma_0}}\,\xi_i(t)\,. \tag{3.31}$$

The first term of (3.31) describes the fact that Brownian agent i follows the concentration gradient, whereas the second term represents the noise that keeps it moving away. As long as the gradient of $h_0(\boldsymbol{r},t)$ is small and the noise is large, the agent behaves nearly as a Brownian particle. But for temperatures below a critical temperature T_c, introduced in the next section, the action of the agent, i.e., the local production of the chemical field, can turn this Brownian motion into a locally restricted motion in the course of time. Hence, with the assumed nonlinear feedback, our model can describe the aggregation of agents due to a *positive response* ($\alpha > 0$) to the chemical field that is generated by them. This shall be elucidated by some computer simulations in Fig. 3.4.

Initially, N Brownian agents are randomly distributed on a surface with periodic boundary conditions, as described in detail in Sect. 3.1.4. Figure 3.4 shows some snapshots of the actual positions of agents after different time intervals. The pictures clearly indicate that agents "gather" in certain regions and, in the course of time, tend to form aggregates. These *aggregates* are different from *compact clusters* where every agent sticks to its neighbors. Instead, these agents still have a certain kinetic energy due to stochastic force, so their position is fluctuating while they are trapped in certain locations.

Figure 3.4 already indicates that agents first form a number of small aggregates containing only a few agents, but in the course of time, some of these aggregates grow, and their total number is decreasing. This process has some analogies to the phenomenon of Ostwald ripening in classical nucleation theory [516], which describes a competitive process among existing clusters. As the result of this dynamics, eventually, all agents are contained in only one cluster. To discuss this analogy in more detail, we would need to know about the critical conditions for such an aggregation process. In the following section, these critical parameters will be derived from a mean-field analysis of the process described above.

3.2.2 Stability Analysis for Homogeneous Distributions

In Sect. 3.1.3, we already described the formal derivation of the density equations from the probability distribution $P(\boldsymbol{r}_1,\theta_1,\ldots,\boldsymbol{r}_N,\theta_N,t)$, (3.5), of Brownian agents by means of (3.9) for $N = \text{const.}$ In the case of identical agents, $\theta_i = 0$, the macroscopic density of Brownian agents, $n_\theta(\boldsymbol{r},t)$, (3.11), obeys the following Fokker–Planck equation:

$$\frac{\partial}{\partial t}n(\boldsymbol{r},t) = \frac{\partial}{\partial \boldsymbol{r}}\left[-\frac{\alpha}{\gamma_0}\,\frac{\partial h_0(\boldsymbol{r},t)}{\partial \boldsymbol{r}}n(\boldsymbol{r},t) + D\frac{\partial n(\boldsymbol{r},t)}{\partial \boldsymbol{r}}\right]\,. \tag{3.32}$$

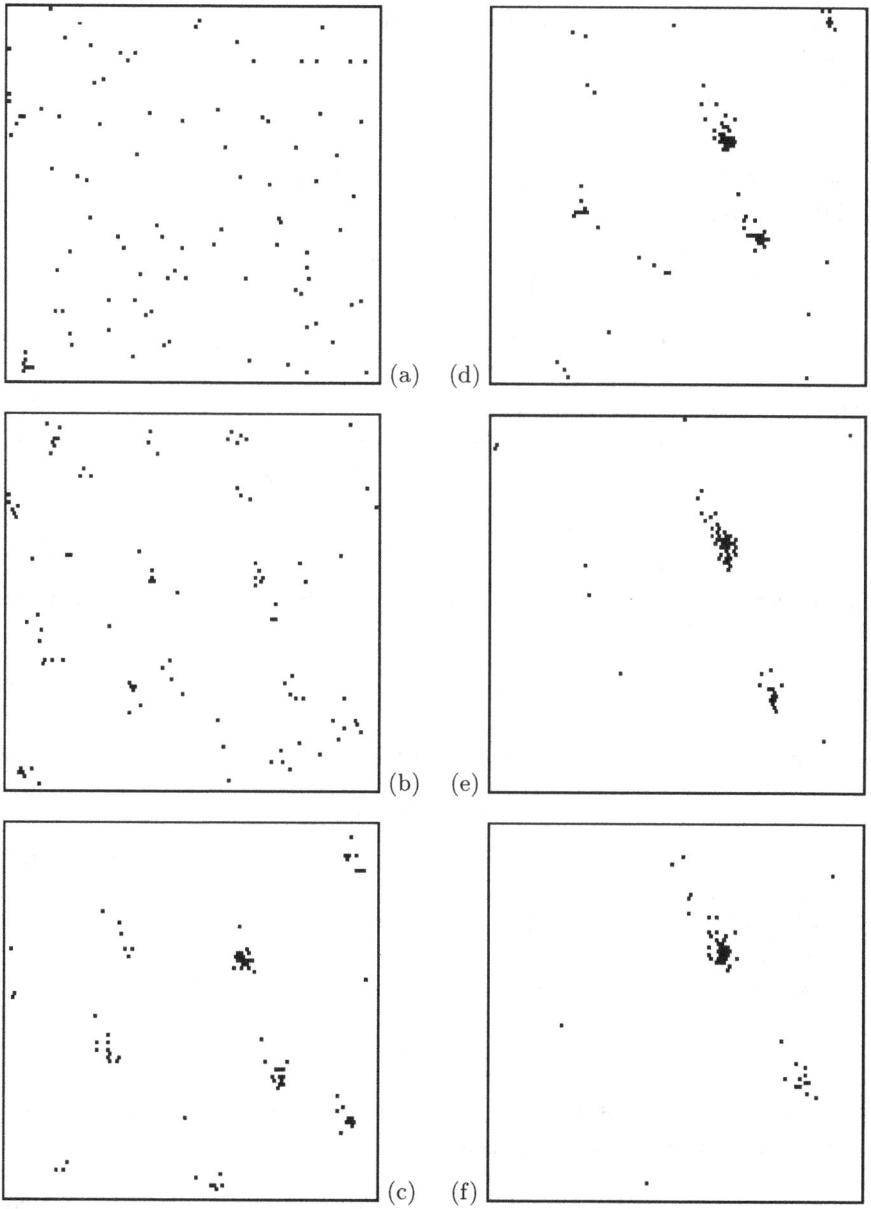

Fig. 3.4. Snapshots of the position of 100 Brownian agents moving on a triangular lattice with periodic boundary conditions (size: $A = 100 \times 100$). Time in simulation steps: (**a**) $t = 100$; (**b**) $t = 1000$; (**c**) $t = 5000$; (**d**) $t = 10,000$; (**e**) $t = 25,000$; and (**f**) $t = 50,000$. Parameters: $T = 0.4\,T_c$ ($\eta = 0.4$), $s_0 = 80$, $k_0 = 0.001$, $D_n = 0.5$, $D_0 = 0.01$ [457]

Using the density $n(\boldsymbol{r},t)$, (3.29) for the chemical field becomes a linear deterministic equation

$$\frac{\partial}{\partial t}\,h_0(\boldsymbol{r},t) = s_0\,n(\boldsymbol{r},t) - k_0\,h_0(\boldsymbol{r},t) + D_0\,\frac{\partial^2 h_0(\boldsymbol{r},t)}{\partial \boldsymbol{r}^2}\,. \tag{3.33}$$

The *coupled differential equations* (3.32) and (3.33) have a homogeneous solution, given by the average densities:

$$\begin{aligned} \bar{n} &= \langle n(\boldsymbol{r},t)\rangle = \frac{N}{A}\,, \qquad N = \int_A n(\boldsymbol{r},t)\,d\boldsymbol{r}\,, \\ \bar{h}_0 &= \langle h_0(\boldsymbol{r},t)\rangle = \frac{H_{\mathrm{tot}}(t)}{A}\,, \qquad H_{\mathrm{tot}}(t) = \int_A h_0(\boldsymbol{r},t)\,d\boldsymbol{r}\,. \end{aligned} \tag{3.34}$$

The surface with area A on which agents move is treated as a torus; that means it acts like a closed system. Therefore, diffusion cannot change the total amount H_{tot} of chemical, which obeys the equation,

$$\frac{dH_{\mathrm{tot}}}{dt} = -k_0 H_{\mathrm{tot}} + s_0 N\,. \tag{3.35}$$

Assuming the initial condition $H_{\mathrm{tot}}(t=0) = 0$, the solution of (3.35) is given by

$$H_{\mathrm{tot}}(t) = \frac{s_0}{k_0}\,N\left[1 - \exp\left(-k_0 t\right)\right] \xrightarrow{t\to\infty} \frac{s_0}{k_0}N\,. \tag{3.36}$$

Equation (3.36) means that after an initial period, given by the time

$$t = \tau \geq 5/k_0\,, \tag{3.37}$$

the total amount of chemical H_{tot} in the system has reached more than 99% of a constant that depends on the ratio between the production and the decomposition rates, s_0/k_0, and on the total number N of agents that produce the chemical field.

Assuming $H_{\mathrm{tot}} = \mathrm{const}$, the homogeneous solution $\bar{h}_0$ is then given by

$$\bar{h}_0 = \frac{s_0}{k_0}\,\frac{N}{A} = \frac{s_0}{k_0}\,\bar{n}\,. \tag{3.38}$$

To investigate the stability of the homogeneous state, (3.34), (3.38), for times $t \geq \tau$, we allow small fluctuations around $\bar{n}$ and $\bar{h}$:

$$\begin{aligned} n(\boldsymbol{r},t) &= \bar{n} + \delta n \\ h_0(\boldsymbol{r},t) &= \bar{h} + \delta h \end{aligned} \qquad \left|\frac{\delta n}{\bar{n}}\right| \sim \left|\frac{\delta h}{\bar{h}}\right| \ll 1\,. \tag{3.39}$$

Inserting (3.39) in (3.32), (3.33), we find after linearization:

$$\begin{aligned} \frac{\partial \delta n}{\partial t} &= -\frac{\alpha}{\gamma_0}\bar{n}\,\Delta\delta h \,+\, D\,\Delta\delta n\,, \\ \frac{\partial \delta h}{\partial t} &= -k_0\,\delta h + s_0\,\delta n \,+\, D_0\,\Delta\delta h\,. \end{aligned} \tag{3.40}$$

With the ansatz

$$\delta n \sim \delta h \sim \exp\{\lambda t + i\boldsymbol{\kappa}\boldsymbol{r}\}\,, \tag{3.41}$$

we find from (3.40) the following dispersion relation for small inhomogeneous fluctuations with wave numbers $\boldsymbol{\kappa}$:

$$\begin{aligned}\lambda_{1,2} = &-\frac{1}{2}\left[k_0 + \kappa^2(D + D_0)\right] \\ &\pm\sqrt{\frac{1}{4}\left[k_0 + \kappa^2(D+D_0)\right]^2 - \kappa^2\left(D\,k_0 + \kappa^2 D\,D_0 - \frac{\alpha}{\gamma_0}\,s_0\,\bar{n}\right)}\,.\end{aligned} \tag{3.42}$$

For homogeneous fluctuations ($\boldsymbol{\kappa} = 0$), we obtain

$$\lambda_1 = -k_0\;,\quad \lambda_2 = 0\,,\qquad \text{for}\quad \boldsymbol{\kappa} = 0\,, \tag{3.43}$$

expressing the conservation of the total number of agents and the stability of the field $h_0(\boldsymbol{r}, t)$ for this case. On the other hand, the system is stable to inhomogeneous fluctuations ($\boldsymbol{\kappa} \neq 0$) as long as the relation holds: $\lambda_1(\boldsymbol{\kappa}) < 0$, $\lambda_2(\boldsymbol{\kappa}) < 0$. This implies for (3.42),

$$\kappa^2\left(D\,k_0 + \kappa^2 D\,D_0 - \frac{\alpha}{\gamma_0}\,s_0\,\bar{n}\right) \geq 0\,, \tag{3.44}$$

which results in the *stability condition*:

$$D > D_{\text{crit}} \tag{3.45}$$

with

$$D = \frac{k_{\text{B}}T}{\gamma_0}$$
$$\text{and}\quad D_{\text{crit}} = \frac{1}{\gamma_0}\left(\frac{\alpha\,s_0\,\bar{n}}{k_0\,+\,D_0\kappa^2}\right)\,.$$

Equation (3.45) determines a *critical diffusion coefficient* for agents, below which inhomogeneous fluctuations of a certain size may result in self-collapsing behavior of the system at a given temperature. From the limit $\boldsymbol{\kappa} \to 0$, we can obtain a critical temperature T_{c} from (3.45), where for $T > T_{\text{c}}$, the system remains stable even to large fluctuations [457]:

$$T_{\text{c}} = \frac{\alpha}{k_{\text{B}}}\,\frac{s_0}{k_0}\,\frac{N}{A}\,. \tag{3.46}$$

Introducing the parameter $\eta = T/T_{\text{c}}$, we can rewrite the stability condition, (3.45), to obtain a relation between thermal noise and density fluctuations of wave number $\boldsymbol{\kappa}$:

$$\eta = \frac{T}{T_{\text{c}}} > \frac{1}{1 + \frac{D_0}{k_0}\kappa^2}\,. \tag{3.47}$$

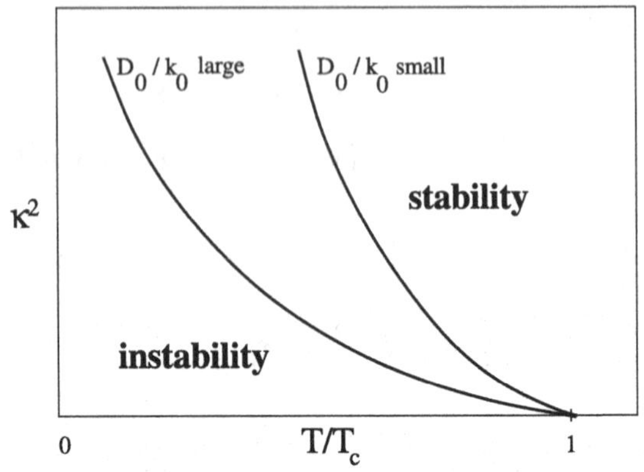

Fig. 3.5. Instability region of the homogeneous system for density fluctuations of wave number $\boldsymbol{\kappa}$ vs. reduced temperature $\eta = T/T_\mathrm{c}$, (3.47) [458]

Fluctuations below a critical wave number $\boldsymbol{\kappa}$ determined by (3.47) result in establishing of inhomogeneities in the distributions $n(\boldsymbol{r},t)$ and $h_0(\boldsymbol{r},t)$ (see Fig. 3.5).

We can also interpret the critical condition, (3.45), (3.47), with respect to parameters α and ε that characterize the agents' response and sensitivity. With $\varepsilon = k_\mathrm{B}T$, we can rewrite (3.45), and the *stability condition* for the homogeneous state then reads

$$\alpha\, s_0\, \bar{n} < \varepsilon\, (k_0 + \kappa^2\, D_0)\,. \tag{3.48}$$

This relation between parameters α and ε can be used to distinguish between different types of agent motion in the mean field limit (see Fig. 3.6). Here, "free motion" means that agents more or less ignore the attraction of the

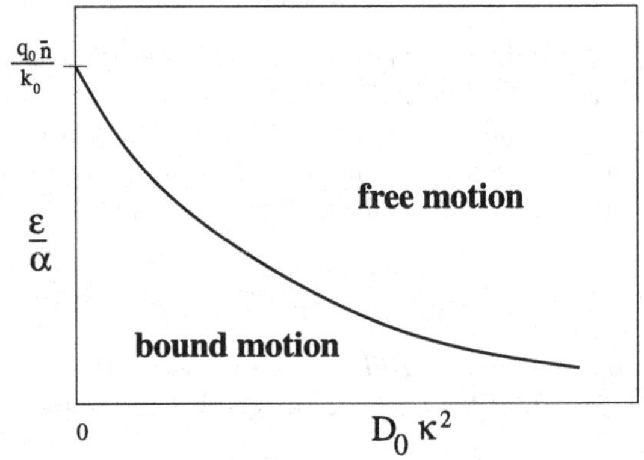

Fig. 3.6. Diagram indicating the transition from free motion of Brownian agents into bounded motion, in dependence on the "sensitivity" ($1/\varepsilon$) and the "response" (α). D_0 is the diffusion constant and k_0 the decay rate of the field [444]

field, thus behaving like random agents that move around. On the other hand, "bounded motion" means that the agents, on average, follow the gradient of the field that restricts their movement to the maxima of the field. Hence, agents will be trapped in certain areas, as already shown in the time series of the computer simulations, Fig. 3.4.

To describe processes of structure formation or collective behavior among Brownian agents, the parameter range of the "bounded motion" seems to be the more interesting with applications, e.g., in biological aggregation (see Sect. 3.2.6). In the other limit, all cooperative actions become impossible either because of an overrunning diffusion or an overcritical temperature, which means subcritical sensitivity. Although this insight results from a mean field analysis, it might basically hold also when agents have "individual" parameters, as discussed in the following sections.

3.2.3 Estimation of an Effective Diffusion Coefficient

To gain more insight into the aggregation process of Brownian agents shown in Fig. 3.4, we investigate the dynamics by an adiabatic approximation [457, 458]. Let us assume that the chemical field $h_0(\boldsymbol{r},t)$ relaxes faster, compared to the distribution of the agents, $n(\boldsymbol{r},t)$, into its stationary state and the diffusion coefficient of the field, D_0 is very small. With these assumptions, we get from (3.33),

$$\frac{\partial h_0(\boldsymbol{r},t)}{\partial t} \approx 0 \;, \quad h_0(\boldsymbol{r},t) = \frac{s_0}{k_0}\, n(\boldsymbol{r},t) \;, \quad \text{if} \quad D_0 \to 0 \,. \tag{3.49}$$

Equation (3.49) means that the spatiotemporal distribution of the chemical field quickly follows the spatiotemporal distribution of agents. Then, the Fokker–Planck equation, (3.32), for agent density $n(\boldsymbol{r},t)$ can be rewritten in terms of a usual diffusion equation by introducing an *effective diffusion coefficient* D_{eff}:

$$\frac{\partial n(\boldsymbol{r},t)}{\partial t} = \frac{\partial}{\partial \boldsymbol{r}}\left[D_{\text{eff}}\frac{\partial n(\boldsymbol{r},t)}{\partial \boldsymbol{r}}\right] , \quad D_{\text{eff}} = D - \frac{\alpha}{\gamma_0}\frac{\partial h_0(\boldsymbol{r},t)}{\partial n(\boldsymbol{r},t)}\, n(\boldsymbol{r},t) \,. \tag{3.50}$$

D_{eff} depends on the agent distribution, $n(\boldsymbol{r},t)$ and on the relation between $h_0(\boldsymbol{r},t)$ and $n(\boldsymbol{r},t)$. In the quasi-stationary limit, given by (3.49), the effective diffusion coefficient for agents reads explicitly

$$D_{\text{eff}} = D - \frac{\alpha}{\gamma_0}\frac{s_0}{k_0}\, n(\boldsymbol{r},t) = \frac{1}{\gamma_0}\Big[k_{\mathrm{B}}T - \alpha\, h_0(\boldsymbol{r},t)\Big] \,. \tag{3.51}$$

We note that an effective diffusion coefficient is not necessarily always larger than zero, which means spreading of agents over the whole surface. It can be also less than zero resulting in a lump of agents concentrating themselves only in certain regions. Depending on the spatiotemporal density $h_0(\boldsymbol{r},t)$, an effective diffusion coefficient at the same time can have quite the opposite effect at different places.

The transition from $D_{\text{eff}} > 0$ to $D_{\text{eff}} < 0$ is driven by the *local chemical production* of agents, i.e., D_{eff} becomes negative if the density $h_0(\boldsymbol{r}, t)$ locally exceeds a certain equilibrium value. This *equilibrium density* $h_0^{\text{eq}}(T)$, which depends only on temperature, is determined by the *phase separation line*, defined by

$$D_{\text{eff}} = 0 \ , \quad h_0^{\text{eq}}(T) = \frac{k_B T}{\alpha} \ . \tag{3.52}$$

The above relation has an obvious analogy to *nucleation theory*. Consider, for example, the usual phase diagram of a liquid-gas system, where the density is plotted versus temperature. As the initial state, we may assume a gaseous system at a temperature $T < T_c$ (see Fig. 3.7). The ratio $\varrho/\varrho^{\text{eq}}$ defines *supersaturation* of the gaseous system. A phase transition from the gaseous to the liquid state necessarily requires supersaturation larger than one, i.e., initial densities are inside the coexistence curve, which divides the homogeneous system from the two-phase region. As long as the supersaturation is not too great, phase separation can occur via nucleation processes, i.e., the formation of small droplets that can grow after they have reached supercritical size. Otherwise, the separation is most likely via "spinodal decomposition" [516].

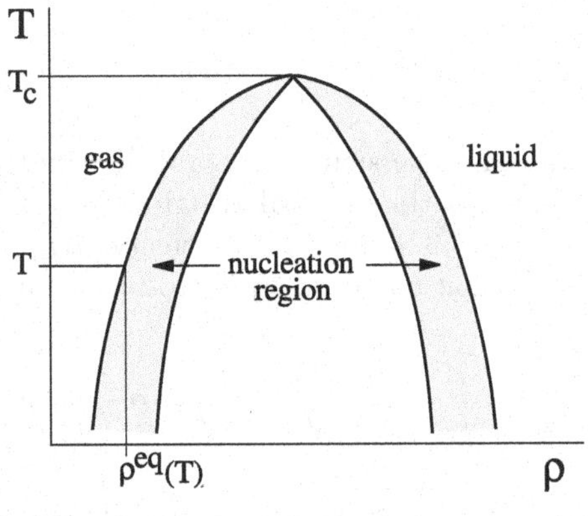

Fig. 3.7. Phase diagram of a liquid–gas system, where the density ϱ is plotted vs. temperature T. The shaded area, which is between the coexistence line (*outer curve*) and the spinodal (*inner curve*) indicates the region where phase transition via nucleation processes is possible

The gaseous phase is usually homogenized very quickly by diffusion processes; hence, supersaturation is a global parameter that characterizes the whole system. On the other hand, if these diffusion processes are rather slow, for example, in melts or porous media, we could also have locally different supersaturations at the same time [11, 441]. The effective diffusion coefficient,

(3.51), plays the role of such a *local critical parameter*, and therefore, in the following, it should be discussed in terms of *local supersaturation.*

The ratio between the normal diffusion coefficient D of the agents and the effective diffusion coefficient D_{eff} is given by

$$\sigma(\boldsymbol{r},t,T) = \frac{D_{\mathrm{eff}}}{D} = 1 - \frac{h_0(\boldsymbol{r},t)}{h_0^{\mathrm{eq}}(T)}\,, \tag{3.53}$$

where $h_0(\boldsymbol{r},t)/h_0^{\mathrm{eq}}(T) > 1$ defines *local supersaturation.* To determine the equilibrium value $h_0^{\mathrm{eq}}(T)$, we use the critical temperature T_{c}, (3.46), where D_{eff} always has to be positive because the system is homogeneous, $T > T_{\mathrm{c}}$, and no agglomeration occurs. Inserting (3.46) in (3.52), we find:

$$h_0^{\mathrm{eq}}(T_{\mathrm{c}}) = \frac{k_{\mathrm{B}}T_{\mathrm{c}}}{\alpha} = \frac{H_{\mathrm{tot}}}{A} = \frac{s_0}{k_0}\frac{N}{A}\,, \tag{3.54}$$

which matches with the mean density $\bar{h}_0$, (3.38). This leads to the expression for the equilibrium value:

$$h_0^{\mathrm{eq}}(T) = \frac{s_0}{k_0}\frac{N}{A}\frac{T}{T_{\mathrm{c}}} = \frac{s_0}{k_0}\,\bar{n}\,\eta\,, \tag{3.55}$$

and finally, we get the effective diffusion coefficient $\sigma(\boldsymbol{r},t,T)$, (3.53), in dependence on the reduced temperature:

$$\sigma(\boldsymbol{r},t,\eta) = 1 - \frac{h_0(\boldsymbol{r},t)}{\eta}\frac{k_0}{s_0\,\bar{n}}\,. \tag{3.56}$$

Figure 3.8 shows the *spatiotemporal distribution* of the effective diffusion coefficient, (3.56), for the simulations shown in Fig. 3.4 for times $t > \tau$, (3.37).

The inhomogeneously distributed black areas indicate a negative effective diffusion coefficient, resulting in bounded motion of agents and eventually in aggregation in that area. We note again that the transformation from free motion to bounded motion is genuinely driven by the actions of agents, i.e., the local production of the field, $h_0(\boldsymbol{r},t)$. If, as a result of these actions, agents increase the local density of the field above the equilibrium value, $h_0^{\mathrm{eq}}(T)$, the system turns locally into a supersaturated state which leads to the aggregation of agents. The negative effective diffusion coefficient D_{eff} locally prevents the agents from spreading out over the whole system; they are trapped in the attraction areas instead.

As shown in Fig. 3.8, for times $t > \tau$, the attraction area is decreasing with time in diversity as well as in area, indicating a selection process among the attraction areas, which will be discussed in detail in the following section.

3.2.4 Competition of Spikes

In the previous sections, we focused on the aggregation process of Brownian agents, but now we will shift the discussion to the spatiotemporal evolution

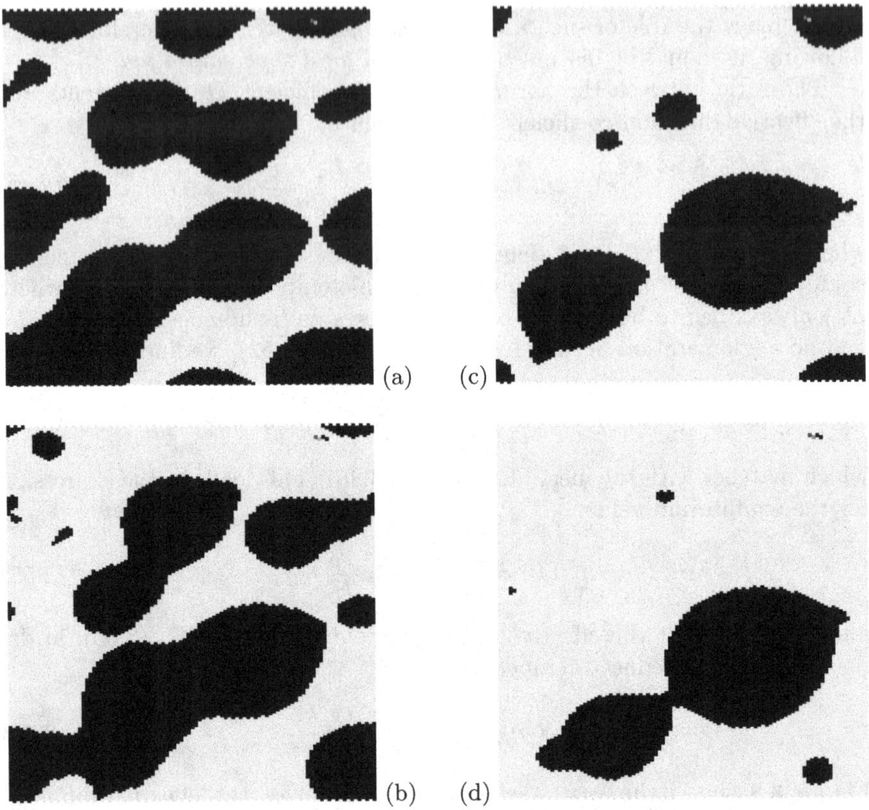

Fig. 3.8. Spatiotemporal evolution of the effective diffusion coefficient $\sigma(\boldsymbol{r}, t, \eta)$, (3.56), for the simulations shown in Fig. 3.4. The *black area* indicates $\sigma < 0$, which means an attraction area for agents; the *gray area* indicates $\sigma > 0$. Time in simulation steps: (**a**) $t = 5000$; (**b**) $t = 10,000$; (**c**) $t = 25,000$; and (**d**) $t = 50,000$. For parameters, see Fig. 3.4 [457]

of the field, $h_0(\boldsymbol{r}, t)$, which is generated by agents. It is obvious that a local concentration of agents will also result in a local growth of $h_0(\boldsymbol{r}, t)$, which in turn reamplifies the agent concentration. Computer simulations shown in Fig. 3.9 elucidate the evolution of the field for the aggregation process shown in Fig. 3.4.

As the time series of Fig. 3.9 illustrates, the spatiotemporal evolution of the field occurs in two steps. The *first stage* is characterized by *independent growth* of the spikes of the field, as presented in the left part of Fig. 3.9. The random density spots, produced by agents at the very beginning (see Fig. 3.9a), evolve into a very jagged profile characterized by a reduced number of peaks of comparable height (see Fig. 3.9c) at about $t = 1000$.

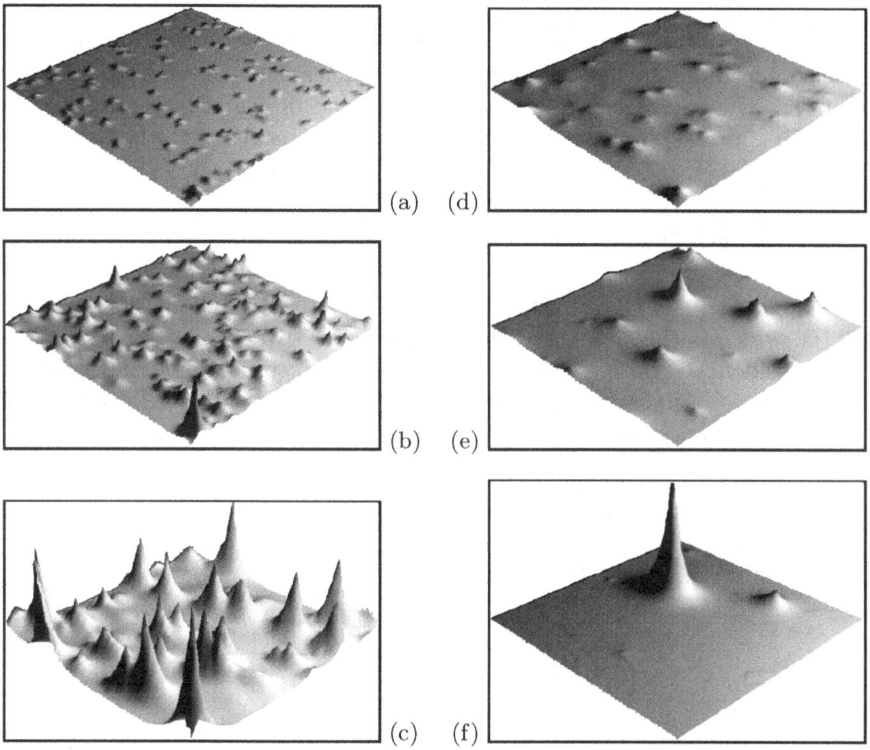

Fig. 3.9. Evolution of $h_0(\boldsymbol{r}, t)$ generated by $N = 100$ agents. Time in simulation steps. (*left side*) Growth regime: (**a**) $t = 10$, (**b**) $t = 100$, (**c**) $t = 1000$; (*right side*) competition regime: (**d**) $t = 1000$, (**e**) $t = 5000$, (**f**) $t = 50,000$. The scale of the right side is 10 times the scale of the left side. Hence, (**d**) is the same as (**c**). For parameters, see Fig. 3.4 [457]

The transition into the *second stage* occurs if the total production of the field becomes stationary, i.e., the decay at rate k_0 compensates for the production of the field by Brownian agents. We can conclude from (3.37) that the crossover is approximately at $t \approx \tau$, (3.37), which means for the decay rate $k_0 = 0.001$ used during the simulation, that after $t = 5000$ time steps, the total production of the field has reached its stationary value.

The second stage is characterized by a *competitive process* among the different spikes, which have bounded the Brownian agents. The competition leads to a decrease of the number of spikes, as presented in the right part of Fig. 3.9. Compared to the left part, the scale had to be changed by a factor of 10, to show the further evolution of the field. Figure 3.9d shows the same distribution as Fig. 3.9c, but only on the enlarged scale; Fig. 3.9e gives the distribution at the beginning of the competitive process, which relaxes after a long time into a one-peak distribution, as already indicated in Fig. 3.9f.

3.2.5 Derivation of a Selection Equation

The competition among the different spikes of the field can be well described in terms of a selection equation [457]. For the derivation, we rewrite the reaction–diffusion equation for the field, (3.33), by means of the homogeneous solutions, (3.34), (3.38):

$$\frac{\partial}{\partial t} h_0(\boldsymbol{r},t) = k_0\,\bar{h}_0 \left[\frac{n(\boldsymbol{r},t)}{\bar{n}} - \frac{h_0(\boldsymbol{r},t)}{\bar{h}_0}\right] + D_0\,\Delta h_0(\boldsymbol{r},t)\,. \tag{3.57}$$

The application of (3.38) implies that (3.57) is restricted to times $t > \tau$, (3.37), which is the timescale where the competitive process occurs. Let us now consider the limiting case that $n(\boldsymbol{r},t)$ relaxes faster, compared to $h_0(\boldsymbol{r},t)$, into a quasi-stationay state. The stationary solution for $n(\boldsymbol{r},t)$ results from the Fokker–Planck equation, (3.32):

$$D\,\frac{\partial n(\boldsymbol{r},t)}{\partial \boldsymbol{r}} - \frac{\alpha}{\gamma_0}\,\frac{\partial h_0(\boldsymbol{r},t)}{\partial \boldsymbol{r}}\,n(\boldsymbol{r},t) = \text{const}\,, \tag{3.58}$$

leading to the canonical distribution,

$$n^{\text{stat}}(\boldsymbol{r},t) = \bar{n}\,\frac{\exp\left[\frac{\alpha}{k_B T}\,h_0(\boldsymbol{r},t)\right]}{\left\langle \exp\left[\frac{\alpha}{k_B T}\,h_0(\boldsymbol{r},t)\right]\right\rangle_A} \tag{3.59}$$

with the mean value defined as

$$\left\langle \exp\left[\frac{\alpha}{k_B T}\,h_0(\boldsymbol{r},t)\right]\right\rangle_A = \frac{1}{A}\int\limits_A \exp\left[\frac{\alpha}{k_B T}\,h_0(\boldsymbol{r}',t)\right]\,dr'\,. \tag{3.60}$$

That means, in the limiting case of fast relaxation of $n(\boldsymbol{r},t)$, the distribution of agents follows in a quasi-stationary manner the slowly varying density $h_0(\boldsymbol{r},t)$. After inserting $n(\boldsymbol{r},t) = n^{\text{stat}}(\boldsymbol{r},t)$, (3.59), in (3.57), the dynamics of $h_0(\boldsymbol{r},t)$ is given by

$$\begin{aligned}\frac{\partial h_0(\boldsymbol{r},t)}{\partial t} &= \frac{k_0\,\bar{h}_0}{\left\langle \exp\left[(\alpha/k_B T)\,h_0(\boldsymbol{r},t)\right]\right\rangle_A}\,h_0(\boldsymbol{r},t) \\ &\quad\times\left\{\frac{\exp\left[(\alpha/k_B T)\,h_0(\boldsymbol{r},t)\right]}{h_0(\boldsymbol{r},t)} - \frac{\left\langle \exp\left[(\alpha/k_B T)\,h_0(\boldsymbol{r},t)\right]\right\rangle_A}{\bar{h}_0}\right\} \\ &\quad+D_0\,\Delta h_0(\boldsymbol{r},t)\,.\end{aligned} \tag{3.61}$$

Neglecting the diffusion term, this equation has an obvious analogy to the selection equations of the Fisher–Eigen type [130, 145] (see also Sect. 9.1.3):

$$\frac{dx_i}{dt} = x_i\left[E_i - \langle E_i\rangle\right]\,, \quad \langle E_i\rangle = \frac{\sum_i E_i\,x_i}{\sum_i x_i}\,. \tag{3.62}$$

Here, x_i denotes the size of a subpopulation i, E_i is the fitness of species i, and $\langle E_i \rangle$ is the mean fitness representing the global selection pressure. Selection equations like (3.62) also describe the process of Ostwald ripening among clusters of different sizes that can occur in the late stage of a first-order phase transition [434, 516].

For the system considered here, we identify the first term in brackets $\{\cdot\}$ in (3.61) as the *local fitness* and the second term as the *global fitness.* To guarantee local growth of the field $h_0(\boldsymbol{r}, t)$, the local fitness of the spikes has to be larger than the global fitness, which depends exponentially on the total distribution of spikes, (3.60).

As a result of the competitive process, we find an increasing inhomogeneity in the field, but, on the other hand, the growth of spikes leads in turn to an increase in the global selection pressure and a slowing down of the kinetics, as indicated by the prefactor of (3.61). This selection results in a decreasing number of spikes and eventually in the establishment of only one large peak, which is also shown in the computer simulations, Fig. 3.9.

For the *late stage* of the selection process, where only a few large and well separated spikes exist, (3.61) allows an estimation of the timescale of their survival or decay. Neglecting again the diffusion of the field, the decision between growth or decay of a spike is given by the sign of the term in brackets $[\cdot]$ in (3.62). Because the ratio in the late stage, (3.63), for the largest spikes has an amount of the order 1,

$$\frac{\exp\left[(\alpha/k_{\rm B}T)\, h_0(\boldsymbol{r},t)\right]\, \bar{h}_0}{\left\langle \exp\left[(\alpha/k_{\rm B}T)\, h_0(\boldsymbol{r},t)\right]\right\rangle_A \, h_0(\boldsymbol{r},t)} \sim 1 \tag{3.63}$$

we get as an estimation for the growth or decay of the spikes,

$$\begin{aligned}
&h_0(\boldsymbol{r},t) \sim \exp\left(k_0 t\right) \quad \text{(survival)}\\
&\quad \text{if } \exp\left[(\alpha/k_{\rm B}T)\, h_0(\boldsymbol{r},t)\right]\, \bar{h}_0 > \left\langle \exp\left[(\alpha/k_{\rm B}T)\, h_0(\boldsymbol{r},t)\right]\right\rangle_A h_0(\boldsymbol{r},t)\,;\\
&h_0(\boldsymbol{r},t) \sim \exp\left(-k_0 t\right) \quad \text{(decay)}\\
&\quad \text{if } \exp\left[(\alpha/k_{\rm B}T)\, h_0(\boldsymbol{r},t)\right]\, \bar{h}_0 < \left\langle \exp\left[(\alpha/k_{\rm B}T)\, h_0(\boldsymbol{r},t)\right]\right\rangle_A h_0(\boldsymbol{r},t)\,.
\end{aligned}$$

The estimated exponential growth and decay is, of course, modified by the diffusion of the field.

We want to underline the striking similarities of spike formation discussed here to pattern formation in active media [338]. Similarly, gliding bugs and crawling droplets on surfaces in physicochemical systems can also be considered [340]. Further amplification of growth by aggregating agents reminds us, to some extent, also of the gravitational collapse of stars without an energy balance [59]. If, different from the above model, the growth of the field is bounded by a maximum value, $h_0^{\max}$, this would cause *saturation effects.*

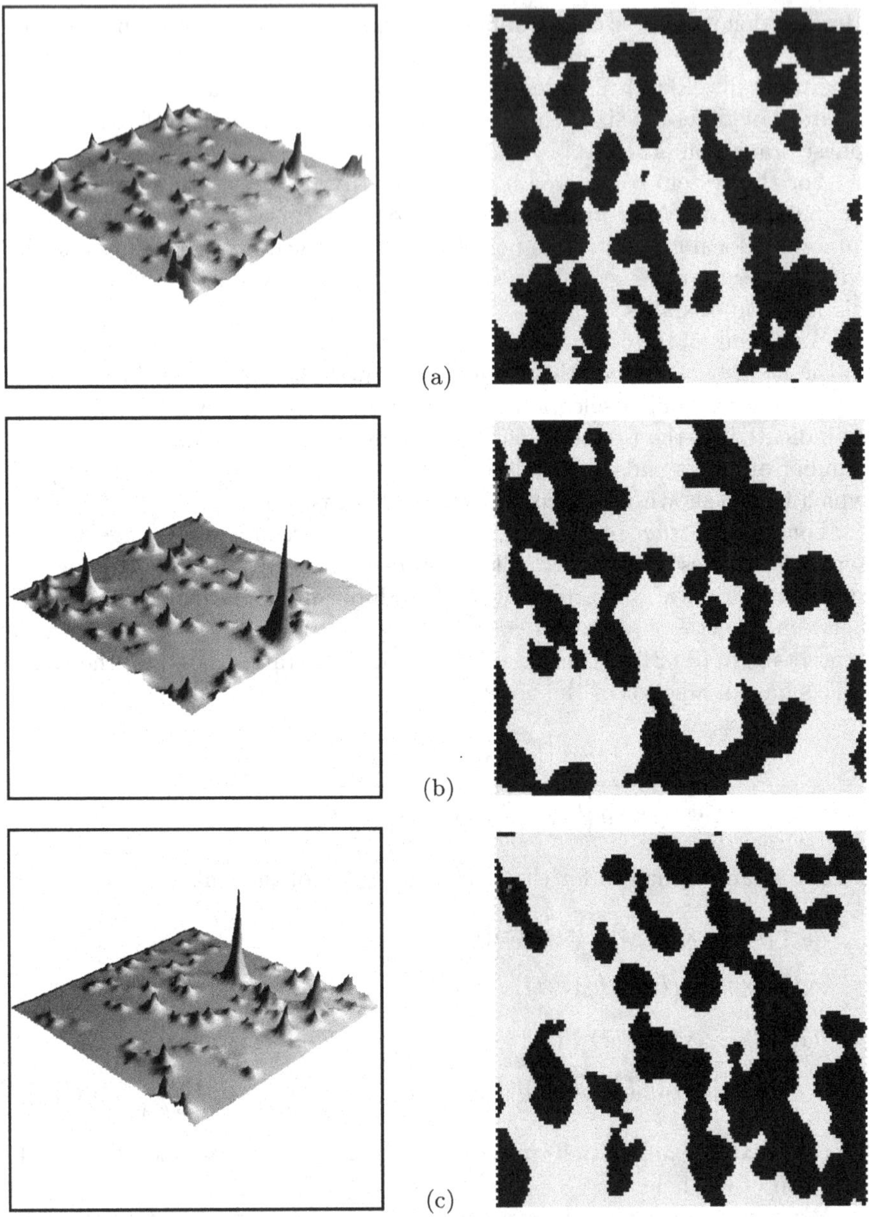

Fig. 3.10. Evolution of the field $h_0(\boldsymbol{r}, t)$ (*left*) and the related effective diffusion coefficient $\sigma(\boldsymbol{r}, t, \eta)$, (3.56) (*right*), during the competitive regime for a decay rate $k_0 = 0.01$. The scale of the field and the other parameters are the same as in Figs. 3.4, 3.8, 3.9. Time in simulation steps: (**a**) $t = 500$; (**b**) $t = 10,000$; (**c**) $t = 50,000$ [457]

Hence, further local growth of the field results in *spatially extended domains* instead of spikes after the maximum value is reached [432, 433].

To conclude our discussion of the aggregation of Brownian agents, a final note is devoted to the influence of the parameters. In the previous sections, the aggregation process has been described from two different perspectives: (i) from the perspective of agents by an effective diffusion coefficient, (3.53), (3.56); (ii) from the perspective of a self-consistent field by a selection equation for the growth of spikes, (3.61). As shown in the stability analysis, Sect. 3.2.2, a critical temperature T_c, (3.46), (3.47), exists for the aggregation process and for the formation of spikes, respectively. Only for $T < T_c$, local supersaturation, $h_0(\boldsymbol{r},t)/h_0^{eq}(\boldsymbol{r},t)$, (3.53), can reach values larger than one in the course of time. A larger value of the decay rate, k_0, for instance, decreases the equilibrium value, $h_0^{eq}(\boldsymbol{r},t)$, (3.55), and therefore may increase local supersaturation, which makes the local aggregation process easier.

On the other hand, a larger value of k_0 also decreases time τ, (3.37), where the independent growth of spikes turns into the selection process, so the competition starts earlier. This situation is shown in the time series of Fig. 3.10. The initial situation for the simulation is the same as that for Figs. 3.8 and 3.10, respectively; however, the decay rate k_0 is 10 times larger than before. Hence, the competition already starts at $t = 500$, and the spatial distribution of the field $h_0(\boldsymbol{r},t)$ and of the effective diffusion coefficient $\sigma(\boldsymbol{r},t,\eta)$ show a much larger diversity at this time. Therefore, the whole process of selection and agglomeration of agents takes a much longer time, which can be seen by comparing Fig. 3.8d, Fig. 3.9f with the related Fig. 3.10c, which is taken after the same number of simulation steps.

3.2.6 Comparison to Biological Aggregation

The aggregation model introduced in the previous section is based on nonlinear feedback between Brownian agents and a (chemical) field. The agents are active in that they produce the field, on the one hand, and respond to the gradient of the field, on the other hand. This can be considered a simple kind of communication. In fact, the interaction between Brownian agents can be described as a *nonlinear* and *indirect communication process* [445], in which all particles are involved (see also Sect. 1.2.2). The communication acts as a special type of *global coupling* between agents, which feeds back to their individual actions.

Communication is based on the exchange of information, and therefore needs a medium. In the model of Brownian agents, this medium is described as a space- and time-dependent field $h_0(r,t)$. By means of the field, the information generated is stored and distributed through the system via diffusion. On the other hand, the information only has a certain lifetime, expressed by the decay rate of the field. With respect to agents, communication consists of three subsequent processes:

- "writing": agents generate information locally by contributing to the field,
- "reading": agents receive information locally by measuring the gradient of the field,
- "acting": agents change the direction of their movement locally based on both the response to the information received and to erratic circumstances.

As already noticed in Sect. 3.1.1, chemical communication is widely found in *biological species* which gather guided by chemical signals originated by the individuals. There are many particular (biological) reasons for gathering processes; for instance, to give each other shelter, to reproduce, to explore new regions, and to feed or to endure starvation conditions. In the case of *myxobacteria* [112], it is known that they glide cooperatively and aggregate under starvation conditions. During gliding, they prefer to use paths that

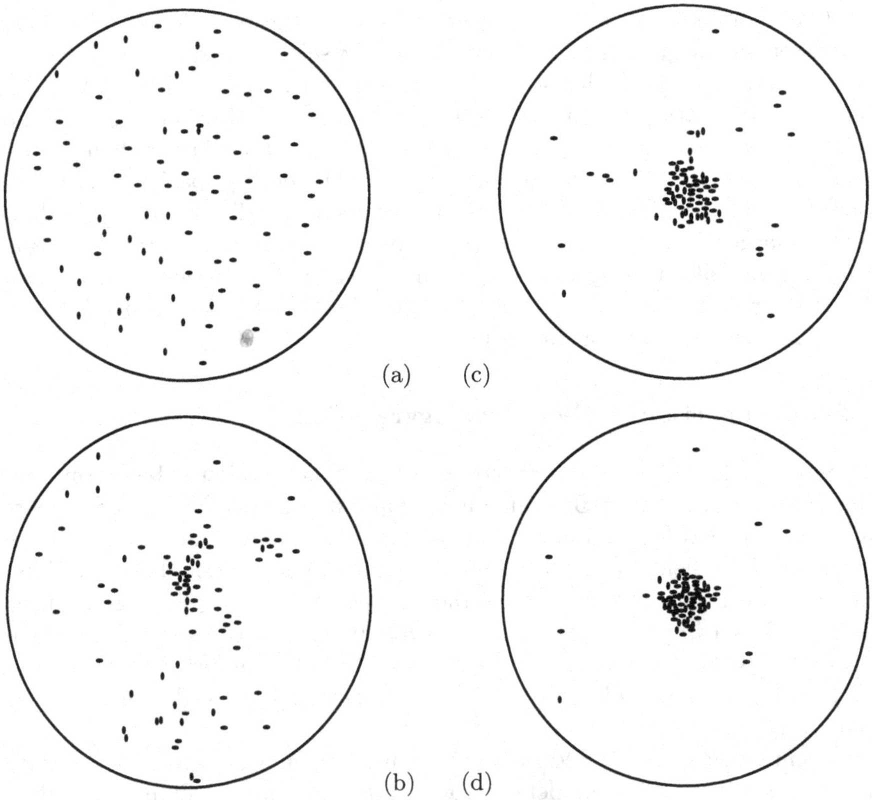

Fig. 3.11. Aggregation of larvae of the bark beetle *Dendroctonus micans*. Total population: 80 larvae; density: 0.166 larvae/cm^2; time in minutes: (**a**) $t = 0$, (**b**) $t = 5$, (**c**) $t = 10$, (**d**) $t = 20$. The pictures are redrawn from experimental results in [103]

they laid down. When the final aggregation takes place, they glide in streams toward developing mounds that later grow to form so-called fruiting bodies.

Another example of collective aggregation which should be compared with the computer simulations of Fig. 3.4, is shown in Fig. 3.11. Here, larvae of the bark beetle *Dendroctonus micans* (Coleoptera: Scolytidae) aggregate and feed side by side [103]. Group feeding has been found in many insect species. It often improves individual survival, growth, and development [172, 362, 514]; allows better exploitation of food resources [515]; and provides better protection against enemies by reinforcing the effects of chemical defense and aposematic warning signals [384].

In *Dendroctonus micans*, group feeding is mediated by the production of and the response to aggregation pheromones, *trans-* and *cis*-verbenol, verbenone, and myrtenol. The experiments done by Deneubourg et al. show the emergence of a single aggregate from a homogeneous distribution of larvae; the dependence of the initial larva density and the influence of small preformed clusters on the dynamics have been investigated [103].

A different example from insects, which is also based on chemical communication, is shown in Fig. 3.12. When ants die, workers carry the corpses out of the nest and place them in a pile, a behavior common to many ant species. In experiments done by Deneubourg et al. with ants of the species *Pheidole pallidula*, a large number of ant corpses was spread out on a surface. Very quickly, the workers collected them into a number of small clusters, which after a long period of time merged into larger clusters. The workers simply follow the rule to pick up the corpses, which emit an odor, i.e., a chemical, and to carry them to places with a higher concentration of the chemical. So, basically the corpses move toward the gradient of a chemical field, which is generated by them by means of some carriers. The mechanism of this aggregation process has been applied by Deneubourg et al. [102] to a sorting mechanism for robots that rearrange and collect different types of objects on a plane.

Chemotaxis [297, 310, 472, 491, 511, 512, 557, 570] is one of the major communication mechanisms in insect societies and has also been found in *slime mold amoebae* [270] and in the aggregation of cells [188, 510]. For example, Gruler et al. [186] investigated chemical communication between *migrating human leukocytes* (granulocytes) (see Fig. 3.13). Here, every cell transmits signals and guides its movement according to the received signals which, in the case of an attractive signal, results in the formation of aggregates. Gruler et al. found a threshold behavior, i.e., the aggregation process starts only above a *critical cell density*. Interestingly, the *cell response* can be *altered reversibly* by an external stimulus, in the particular case by bivalent ions such as calcium and magnesium. At high calcium concentrations, the migratory behavior of a single granulocyte is not affected by other granulocytes, but at low concentration, the granulocytes attract each other. Noteworthy,

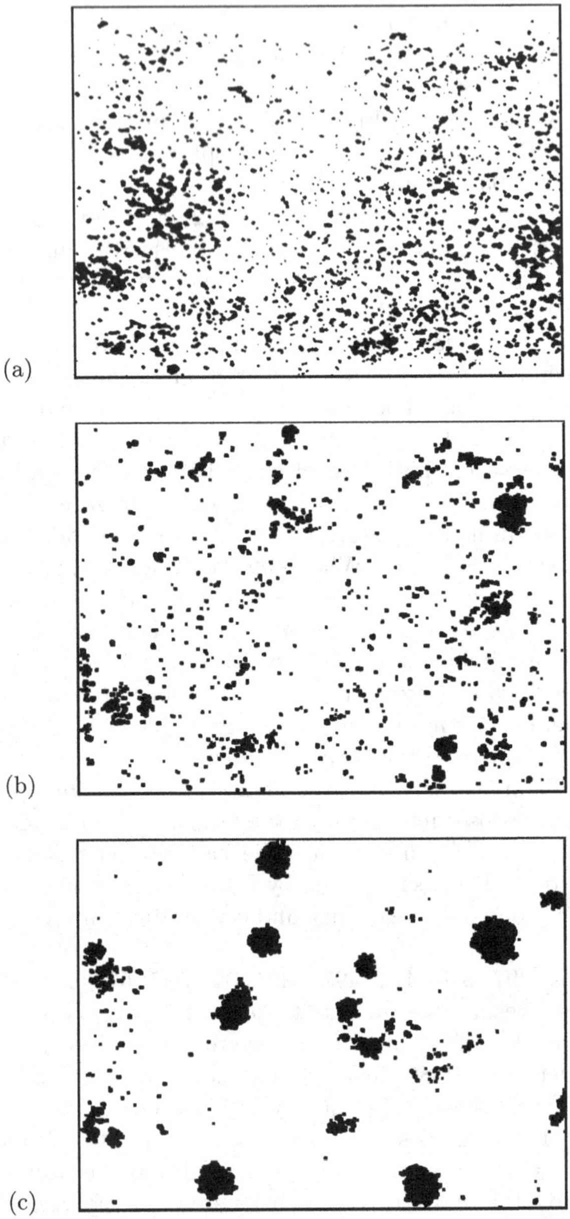

Fig. 3.12. Aggregation of ant corpses in a colony of *Pheidole pallidula*. Density: 1.6 corpses/cm^2; time in hours: (**a**) $t = 0$, (**b**) $t = 20$, (**c**) $t = 68$. The pictures are redrawn from experimental results in [102]

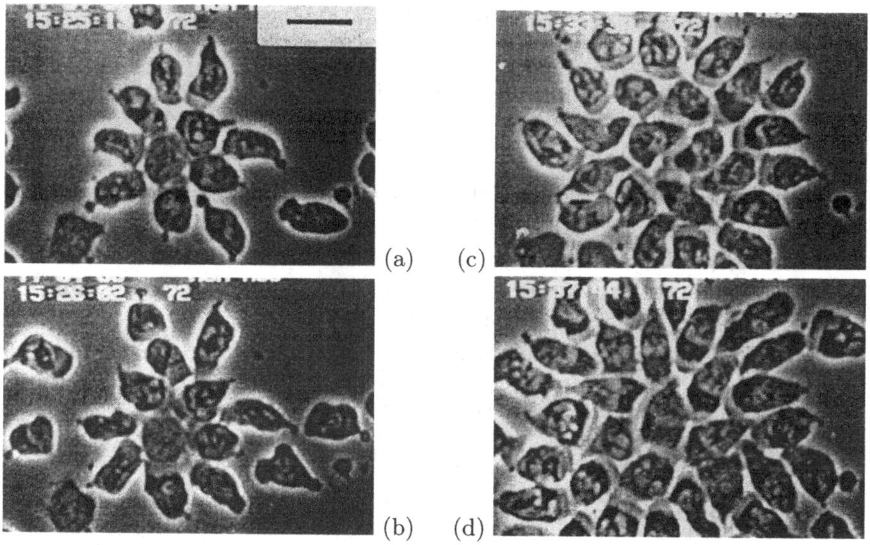

Fig. 3.13. Microscopic pictures of migrating granulocytes showing the formation of a cluster at different times. Cell density is approximately 1000 cells/mm^2. The bar in (**a**) corresponds to 20 μm. Time intervals after starting the experiment: (**a**) 2 min; (**b**) 2 min 50 s; (**c**) 4 min 50 s; (**d**) 14 min 30 s [186]

the guiding field produced by the migrating cells diffuses very fast by a factor of about 10^2 compared to the migration of granulocytes.

If we compare the conditions for a real aggregation process in cells with the assumptions for our model of aggregating Brownian agents, we find a number of interesting analogies:

1. The alteration of the cell response to external stimuli is covered in our model by the internal parameter θ_i, which can be changed and thus allows different types of response to the field.
2. The fast diffusion of chemotactic molecules compared to the migration of cells is reflected in our model by the adiabatic approximation of a quasi-stationary field.
3. The threshold behavior, i.e., the existence of a critical density for the aggregation process, is considered in our model via the critical parameters obtained from stability analysis, which expresses the relation between the initial density, the diffusion constant, and the production and decay rates, to allow an aggregation process.

Thus, we may conclude that the model of Brownian agents in Sect. 3.2.1 describes essential features of *real* aggregation processes found in biological systems. However, the chemotactic interaction in real biological processes still displays a number of possible variations, i.e., with respect to the production of the chemical field, the diffusion of the substance, etc. Some possible modifications will be discussed in Sect. 5.3.

At the very end, we shall also mention analogies to a *"real" agent model*, the so-called "heatbug" model, one of the standard examples to demonstrate the capabilities of the Swarm simulation platform:

> "Each agent in this model is a heatbug. The world has a spatial property, heat, which diffuses and evaporates over time. Each heatbug puts out a small amount of heat, and also has a certain ideal temperature it wants to be. The system itself is a simple time stepped model: each time step, the heatbug looks moves to a nearby spot that will make it happier and then puts out a bit of heat. One heatbug by itself can't be warm enough, so over time they tend to cluster together for warmth."[1]

In the specification of Sect. 3.2.1, the Brownian agent model captures the basic features of the "heatbug" model. The "spatial property," heat, is formally described by (3.29) for the spatiotemporal field $h_0(\boldsymbol{r}, t)$. The amount of heat that each agent puts out is a constant s_0 in our model, whereas in the original heatbug simulation, it can vary between a lower and an upper value. Additionally, the agent in the heatbug simulation may have a minimum and a maximum ideal temperature which is not considered here. Apart from these minor differences, the Brownian agent model of Sect. 3.2.1 provides a formal description of the heatbug simulation and, as has been shown in the previous sections, further allows *quantitative analysis* of the dynamics together with an estimation of the relevant *parameters.*

3.3 Pattern Formation in Reaction–Diffusion Systems

3.3.1 Coexistence of Spikes

In Sect. 3.2, we discussed a one-component (chemical) field, $h_0(\boldsymbol{r}, t)$ that influences the motion of agents. The coupled dynamics of the Langevin equation, (3.30), and the reaction–diffusion equation, (3.29), for the field results in the formation of spatiotemporal patterns. In particular, for a constant number of agents and a closed system, the formation and competition of density spikes of the chemical field have been shown.

Based on these investigations, in this section the model will be extended to simulate a variety of *physicochemical patterns*, which are known in reaction–diffusion systems. The new and interesting point is not the existence of these patterns, but the rather simple and most effective way they are generated and simulated [429, 431]. For this purpose, Brownian *agents* in this section

[1] `http://www.swarm.org/examples-heatbugs.html`

are identified with specific *chemical particles* (molecules, etc.) that play a role in physicochemical structure formation.

In a first step, we have to include *chemical reactions* in the model of Brownian agents. For the first example, we assume *adsorption* and *desorption* of particles on the surface. Hence, the total number of particles is not conserved but could be changed by a *homogeneous influx* of particles $N \to N+1$ at a rate Φ (e.g., by adsorption on the surface due to vapor deposition) and an *inhomogeneous outflux* $N \to N-1$ at a rate k_n due to local disappearence of particles (e.g., by desorption processes on the surface or reevaporation). Further, the particles on the surface generate a one-component chemical field $h_0(\boldsymbol{r}, t)$ again (cf. Fig. 3.14). Hence, the motion of a single particle can be described by the Langevin equation, (3.30), whereas the dynamics of the field is given by (3.29) with a nonconstant particle number N.

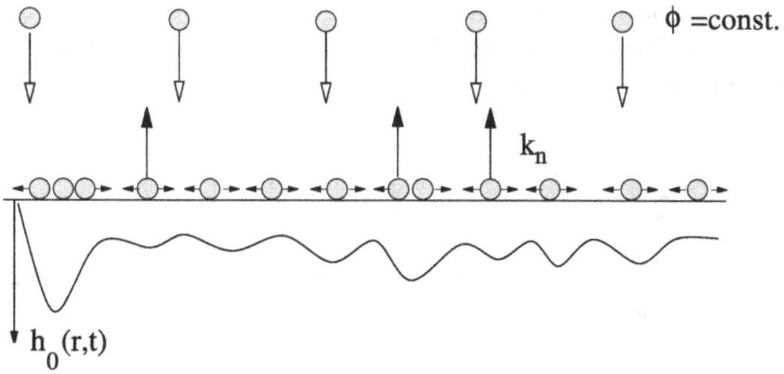

Fig. 3.14. In a physicochemical example, particles on a surface generate a chemical field, $h_0(\boldsymbol{r}, t)$ that can diffuse and decay again. The number of particles changes due to a homogeneous influx at a rate Φ and an inhomogeneous outflux at a rate k_n

The probability that a new particle with random position will appear on the surface during the next time interval Δt is $\Phi \Delta t\, A$, where A is the surface area, and the probability that one of the existing particles will disappear, is $k_n \Delta t\, N$ [see (3.28) and Sect. 3.1.4]. In general, adsorption and desorption rates could also be functions of the field $h_0(\boldsymbol{r}, t)$ and therefore depend on time and space coordinates. But here, we assume constant rates for simplicity.

For a nonconstant number of particles, the complete dynamics for the ensemble of particles can be formulated in terms of the *grand-canonical* N-particle distribution function $P_N(\boldsymbol{r}_1, \dots, \boldsymbol{r}_N, t)$ which gives the probability density of finding N particles in the vicinity of $r_1, \dots, r_N$ on the surface A at time t. Considering birth and death processes, such as the chemical reactions described above, the master equation for P_N reads in the limit of strong damping, $\gamma_0 \to \infty$ [142, 429, 431],

$$\begin{aligned}\frac{\partial}{\partial t}P_N(\boldsymbol{r}_1,\dots,\boldsymbol{r}_N,t) = &-(k_n\,N+\Phi\,A)\,P_N(\boldsymbol{r}_1,\dots,\boldsymbol{r}_N,t)\\ &+k_n\,(N+1)\int d\boldsymbol{r}_{N+1}\,P_{N+1}(\boldsymbol{r}_1,\dots,\boldsymbol{r}_{N+1},t)\\ &+\frac{\Phi}{N}\sum_{i=1}^{N}P_{N-1}(\boldsymbol{r}_1,\dots,\boldsymbol{r}_{i-1},\boldsymbol{r}_{i+1},\dots,\boldsymbol{r}_N,t)\\ &-\sum_{i=1}^{N}\left[\nabla_i\,(\mu\,\nabla_i h_0)\,P_N-D_n\,\Delta_i P_N\right], \qquad (3.64)\end{aligned}$$

where $\mu=\alpha/\gamma_0$ is the mobility of the particles and $D_n=\varepsilon/\gamma_0$ is the spatial diffusion coefficient. The first three terms on the right-hand side describe the loss and gain of particles with the coordinates $\boldsymbol{r}_1,\dots,\boldsymbol{r}_N$ due to chemical reactions. The last term describes the change in the probability density due to the motion of particles on surface A.

The mean field limit is obtained by introducing the density of particles in the grand canonical ensemble:

$$n(\boldsymbol{r},t)=\sum_{N=1}^{\infty}N\int d\boldsymbol{r}_1\dots d\boldsymbol{r}_{N-1}\,P_N(\boldsymbol{r}_1,\dots,\boldsymbol{r}_{N-1},\boldsymbol{r},t)\,. \qquad (3.65)$$

Integrating (3.64) due to (3.65) and neglecting higher correlations, we obtain the following reaction–diffusion equation for $n(\boldsymbol{r},t)$:

$$\frac{\partial}{\partial t}\,n(\boldsymbol{r},t)=-\nabla\,n\,(\mu\,\nabla h_0)+D_n\,\Delta n-k_n\,n+\Phi\,. \qquad (3.66)$$

With $n(\boldsymbol{r},t)$, the mean-field equation for the chemical field is again given by (3.33). The homogeneous solutions of both equations now read

$$\bar{n}=\frac{\Phi}{k_n}\,,\qquad \bar{h}_0=\frac{s_0}{k_0}\frac{\Phi}{k_n}=\frac{s_0}{k_0}\bar{n}\,. \qquad (3.67)$$

The stability analysis for homogeneous solutions can be carried out in the same way as described in Sect. 3.2.2. We find from the coupled mean field equations (3.33), (3.66) after linearization,

$$\begin{aligned}\frac{\partial\delta n}{\partial t}&=-\mu\bar{n}\,\Delta\delta h\;-\;k_n\,\delta n\;+\;D_n\,\Delta\delta n\,,\\ \frac{\partial\delta h}{\partial t}&=-k_0\,\delta h+s_0\,\delta n\;+\;D_0\,\Delta\delta h\,. \qquad (3.68)\end{aligned}$$

Using the ansatz (3.41), we find from (3.68) the following dispersion relation for small inhomogeneous fluctuations with wave numbers $\boldsymbol{\kappa}$:

$$\lambda_{1,2} = -\frac{1}{2}\left[k_0 + k_n + \kappa^2(D_n + D_0)\right] \pm \Bigg\{\frac{1}{4}\Big[k_0 + k_n + \kappa^2(D_n + D_0)\Big]^2 - k_0 k_n - \kappa^2\Big(D_n\, k_0 + D_0 k_n + \kappa^2 D_n\, D_0 - \mu\, s_0\, \bar{n}\Big)\Bigg\}^{1/2} . \tag{3.69}$$

The stability condition requires that both $\lambda_1(\boldsymbol{\kappa}) < 0$ and $\lambda_2(\boldsymbol{\kappa}) < 0$. This results in a fourth-order inequality for wave number $\boldsymbol{\kappa}$:

$$\kappa^2\Big(D_n\, k_0 + D_0\, k_n + \kappa^2 D_n\, D_0 - \mu\, s_0\, \bar{n}\Big) + k_0 k_n \geq 0\,, \tag{3.70}$$

which yields the following two *critical conditions* for stability:

$$\mu\, s_0\, \bar{n} > \left(\sqrt{D_n\, k_0} + \sqrt{D_0\, k_n}\right)^2 , \tag{3.71}$$

$$\kappa^2_{\text{crit}} = \frac{\mu\, s_0\, \bar{n} - D_n\, k_0 - D_0\, k_n}{2\, D_n\, D_0} . \tag{3.72}$$

From condition (3.72), we can derive again, in the limit $\boldsymbol{\kappa} \to 0$, a critical temperature T_c, which is now different from (3.46) because of the additional chemical reactions considered:

$$T_c = \frac{\alpha}{k_B}\,\frac{s_0}{k_0}\,\frac{N}{A} - \frac{D_0 \gamma_0}{k_B}\,\frac{k_n}{k_0} . \tag{3.73}$$

For $T < T_c$, the homogeneous state is unstable to periodic fluctuation with wave number κ^2_{crit}, which results in the formation of spatiotemporal structures in the field, h_0, as already discussed for the generation of spikes in Sect. 3.2.4. This time, however, we find *stable coexistence* of spikes instead of competition. This is clearly caused by the chemical reactions, which now control the unlimited growth of the existing spikes and are sources of a permanent formation of new ones.

Computer simulations with 8000 Brownian agents representing chemical particles, presented in Fig. 3.15, show the coexistence of spikes, which results in a nearly hexagonal pattern for the chemical field $h_0(\boldsymbol{r}, t)$. The distribution still looks noisy; however, it indicates already a stationary inhomogeneous pattern. Simulations with larger particle numbers result in very regular patterns (see also Fig. 3.16).

The model outlined can be generalized by considering that particles, denoted by A, are injected at the origin at a constant rate. These particles perform Brownian motion, which results in an inhomogeneous flux induced by the injection procedure. It is assumed that the A particles react with particles of a second species B, equally distributed on a surface, to form particles of species C:

$$A + B \to C\,. \tag{3.74}$$

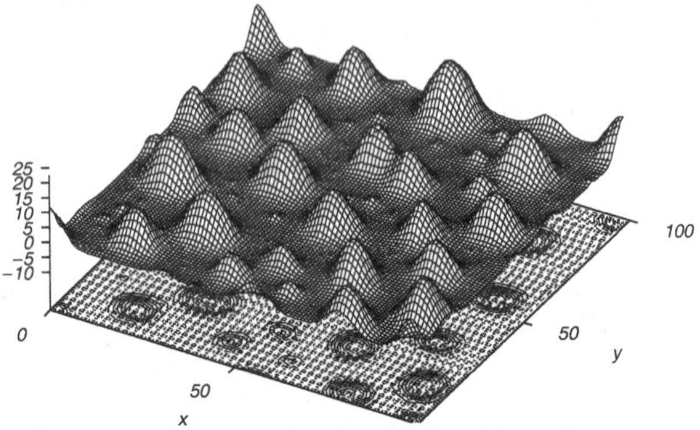

Fig. 3.15. A stationary, spatial inhomogeneous field $h_0(\boldsymbol{r}, t)$, generated by 8000 Brownian agents. In addition to nonlinear feedback between the agents and the field, the model also considers chemical reactions, such as adsorption and desorption of particles, which results in stable coexistence of spikes instead of a collapse into one single spike [429, 431]

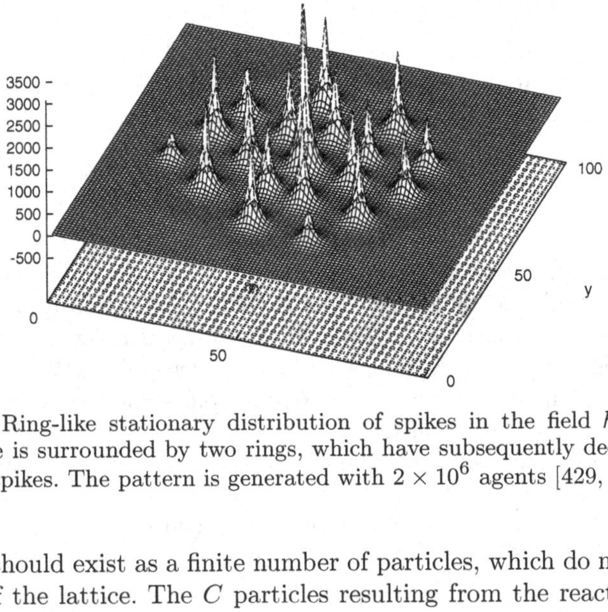

Fig. 3.16. Ring-like stationary distribution of spikes in the field $h_0(\boldsymbol{r}, t)$. The central spike is surrounded by two rings, which have subsequently decomposed to form small spikes. The pattern is generated with 2×10^6 agents [429, 431]

Species B should exist as a finite number of particles, which do not move, at each site of the lattice. The C particles resulting from the reaction, (3.74), are Brownian agents, i.e., they perform Brownian motion and generate a field $h_0(\boldsymbol{r}, t)$, which also influences their motion, as described by (3.29), (3.30). As a new element, we assume that the particles of species C can precipitate, i.e., become immobile. Then, the model produces *periodic precipitations*, similar to experiments on Liesegang rings [84, 231, 264, 271, 359, 521] (see also Fig. 10.12).

As a result of computer simulations, a recurring pattern is shown in Fig. 3.16 in the generated field $h_0(\boldsymbol{r}, t)$. The peaks indicate the existence of particle aggregations of different sizes. Here, a central aggregate is surrounded by two rings of precipitated material, which have subsequently decomposed into a sequence of smaller aggregates. The simulation was performed with 2,000,000 agents; thus the pattern looks much less noisy than that in Fig. 3.15.

3.3.2 Spiral Waves and Traveling Spots

In a next example, we investigate *excitable systems* with small particle numbers, which can form stable spiral-waves. To simulate this behavior, we use the piecewise linear Rinzl–Keller model [338]:

$$\frac{\partial}{\partial t}\, n(\boldsymbol{r},t) = -k_n\, n + (1-m)\, \Theta[n-a] + D_n\, \Delta n\,, \tag{3.75}$$

$$\frac{\partial}{\partial t}\, m(\boldsymbol{r},t) = s_0\, n - k_m\, m + D_m\, \Delta m\,. \tag{3.76}$$

$\Theta[n-a]$ is the Heaviside function, which is equal to 1, if $n > a$, and 0, if $n < a$. $n(\boldsymbol{r},t)$ should be the concentration of an *activator*. If it locally exceeds the excitation value a, *replication* of the activator starts. On the other hand, generation of the activator is controlled by an *inhibitor*, $m(\boldsymbol{r},t)$, which is generated by the activator at a rate s_0. The higher the inhibitor concentration, the lower the replication of n. Additionally, the activator and the inhibitor will decay at rates k_n and k_m, respectively. Both components can diffuse, however. Because the inhibitor influences the actual activator distribution only with some delay, $1/s_0$, the activator will be able to diffuse into regions where no inhibitor exists yet. As the result, a directed motion of the activator "cloud" in front of the inhibitor distribution occurs. In one spatial dimension, these clouds would be pulses of excited regions moving at constant velocity, whereas in two dimensions, spirals will be formed.

The activator–inhibitor dynamics described above can be reformulated in terms of the stochastic model of Brownian agents. Because the inhibitor is generated by the activator, it will be identified with the field $h_0(\boldsymbol{r},t)$, which in the mean-field limit follows the same reaction–diffusion dynamics, (3.33). The activator is described by Brownian agents, and its concentration is replaced by the microscopic density, (3.8). The motion of both types of particles is modeled by *simple* Brownian motion; all nonlinear dynamics is given by their chemical behavior. Note that Brownian agents this time do not respond to gradients of the field; however, their *replication* is determined by the local value of the field.

According to the dynamics above, chemical reactions are simulated as follows: Brownian agents that represent the activator are permanently desorbed with the probability $k_n \Delta t$ during the time interval Δt. If the local agent number $N(\boldsymbol{r},t)$ is above a critical value N_c, an input of particles different from zero occurs, i.e., the medium is excited at these space coordinates. The

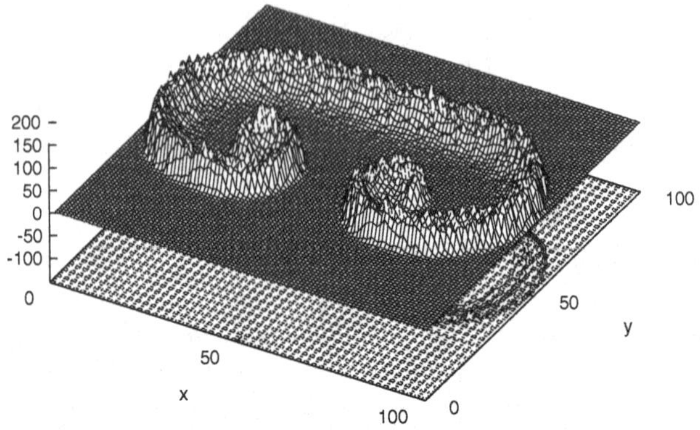

Fig. 3.17. Double spiral of 10^4 activator particles. Initially a target wave with two open ends was assumed which evolves into the two rotating spirals [429, 431]

local value of the input rate depends on the actual inhibitor concentration, $[1 - m(\boldsymbol{r}, t)]$. In nonexcited regions, the input of particles is zero.

In computer simulations, waves of excitations as well as single and double spirals have been found [429, 431]. A double spiral starting from an inhomogeneous activator distribution with two open ends is shown in Fig. 3.17. The well-developed structure shown is formed by only about 10^4 agents. An additional advantage of our approach is that the activator dynamics has to be calculated only in regions where particles are present, in contrast to PDEs which often are integrated on the whole lattice.

The same model can be also used to simulate the spatiotemporal dynamics of *traveling spots*. Recently, experimental results of traveling, interacting, and replicating spots [298, 558] were explained by localized moving excitations. The complex impact behavior of these spots has been discussed in [279]. To obtain similar structures, we now introduce a *global coupling* of the excitation value N_c, as an extension of the model described above. In our simulations, the excitation value N_c increases linearly with the global number of agents on the surface. Thus, the spirals cannot grow to their normal shape; only their mobility to move at constant velocity in one direction remains. Such a moving spot is shown Fig. 3.18. Moreover, Fig. 3.19 presents one of the possible *interactions between spots.* Two colliding spots are reflected nearly perpendicularly to their former motion, a behavior which was also discussed in [279].

Further investigations have shown that the character of the motion of these spots depends strongly on the number of particles forming a spot. For small particle numbers, the spots perform nearly Brownian motion, whereas for larger particle numbers, the motion of the spots becomes more or less ballistic [160].

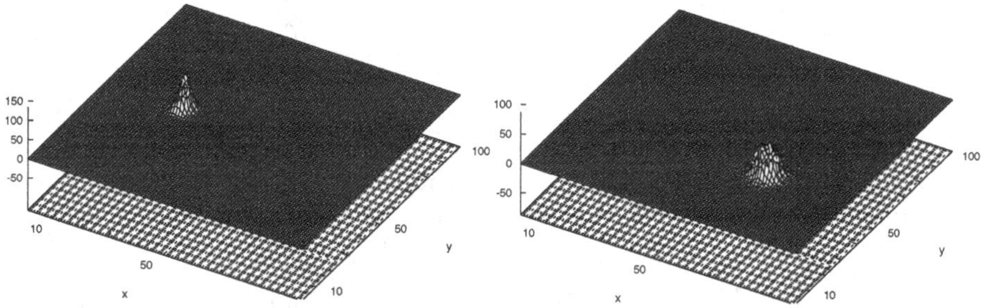

Fig. 3.18. Spot of about 3000 activator particles moving as a localized region of excitation at constant velocity from left to right [429, 431]

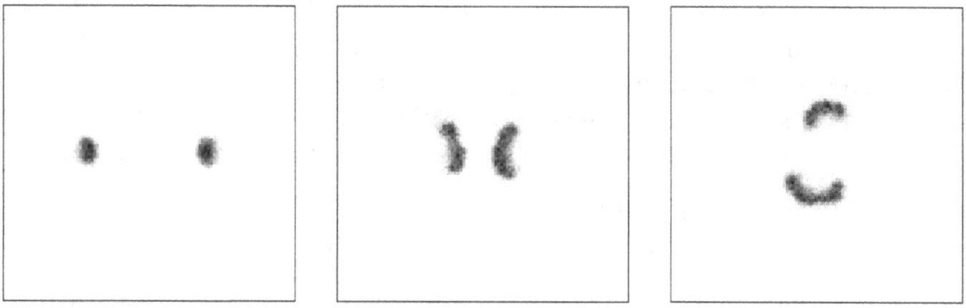

Fig. 3.19. Impact of two traveling spots with reflection perpendicular to the former motion. Black color indicates regions of large particle numbers of the activator [429, 431]

3.3.3 Traveling Waves

Finally, we discuss an example where Brownian agents *with different internal degrees of freedom* are used to simulate physicochemical structure formation. Let us assume that the action of Brownian agents may depend on an internal parameter θ_i, which can be either 0 or 1, i.e., there are two different types of agents. Only agents with the internal state $\theta_i = 1$ can contribute to the field $h_0(\boldsymbol{r}, t)$. Additionally, both kinds of agents can undergo a transition into the opposite state by changing their internal parameter. The total number of agents should be constant. The example can be summarized using the following symbolic reactions:

$$\xrightarrow{\beta} C_1 \xrightarrow{k(T)} C_0 \xrightarrow{\beta} \quad , \qquad N = N_0 + N_1 = \text{const} . \tag{3.77}$$

Here, C_θ denotes an individual agent C with internal state θ. β is the transition rate from state 0 to 1.

As (3.77) indicates, an outflux of C_0 is compensated for by an influx of C_1. This situation is similar to a *cross-flow reactor*. As an example, we investigate an exothermic reaction within this reactor [567]; hence, the contribution of C_1 agents to the field can be specified as follows:

$$s_i(\theta = 1) = \eta\, k(T) = \eta\, k_0 \exp\left(\frac{T}{T_0 + T}\right) \; , \qquad s_i(\theta = 0) = 0 \, . \tag{3.78}$$

Here, η is the heat released during one reaction, and $k(T)$ is the temperature-dependent reaction rate for the transition from state 1 to 0. The field $h_0(\boldsymbol{r}, t)$, to which the particles contribute, can, in this example, be identified as a *temperature field*: $h_0(\boldsymbol{r}, t) \to T(\boldsymbol{r}, t)$ which obeys an equation similar to (3.29):

$$\frac{\partial}{\partial t} T(\boldsymbol{r}, t) = \sum_{i=1}^{N} s_i(\theta = 1)\, \delta\Big[\boldsymbol{r} - \boldsymbol{r}_i(t, \theta = 1)\Big] - k_T\, T + \chi\, \Delta T \, . \tag{3.79}$$

The decay of the field results from coupling to a thermal bath outside the reactor; the diffusion of the field is replaced by heat conduction. In this example, nonlinear feedback between the motion of particles and the field is not given by the response to the gradient of the field, but by the *intensity of fluctuations*, $\varepsilon_i = k_{\mathrm{B}} T$. Hence, the resulting Langevin equation for the motion of C_0 and C_1 agents in the one-dimensional case reads as follows:

$$\frac{d\boldsymbol{r}_i(\theta_i)}{dt} = \boldsymbol{v}_i \quad , \qquad \frac{d\boldsymbol{v}_i(\theta_i)}{dt} = -\gamma_0 \Big[\boldsymbol{v}_i(\theta_i) \; - \; \boldsymbol{v}_0\Big] + \sqrt{2\, \varepsilon_i\, \gamma_0}\, \xi(t) \, . \tag{3.80}$$

In the cross-flow reactor, on average, all agents move at velocity $\boldsymbol{v}_0$ relative to the temperature field $T(\boldsymbol{r}, t)$ and can undergo the transitions specified in (3.77), which may increase the temperature locally. With probability $k[T(\boldsymbol{r}, t)]\Delta t$, the transition $\theta_i = 1 \to \theta_i = 0$ is realized, resulting in a local increase in temperature. Otherwise $\theta_i = 0 \to \theta_i = 1$ occurs with probablity $\beta \Delta t$ during the time interval Δt. As a result, a traveling periodic pattern occurs in the temperature field, as shown in Fig. 3.20. The results are in good agreement with those obtained in [567].

In the system considered, temperature plays the role of an *activator*, which increases locally due to the reactions and in turn amplifies the reactions

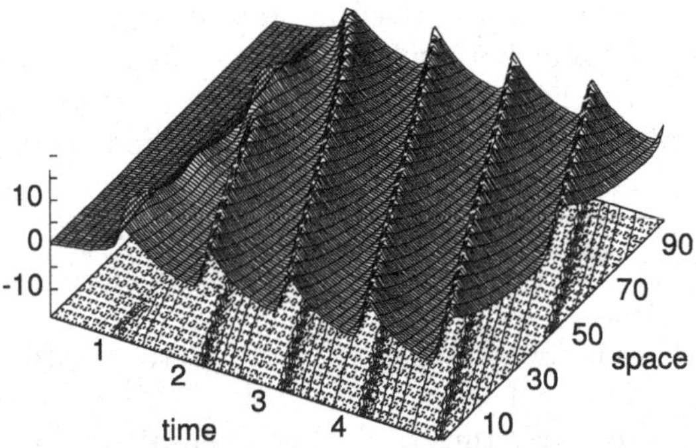

Fig. 3.20. Space–time plot of traveling periodic structures that occur in the temperature field. The simulation is carried out with only 5000 particles and is very stable and fast [431]

again. It has been shown [418], that activator–inhibitor systems that can generate Turing structures can also generate periodic *moving* structures. If the inhibitor moves with an overcritical velocity $v_0 > v_0^{\mathrm{crit}}$, relative to the activator, a differential flow-induced chemical instability (DIFICI) occurs. This instability leads to a periodic pattern, resulting from the fast moving inhibitor that violates the local equilibrium. As a specific example, we find a moving periodic structure in the temperature field for the exothermic reaction considered here.

We conclude this section with some notes in favor of the proposed agent-based or particle-based algorithm. Usually, reaction–diffusion problems are solved by integrating the related equations on a lattice. Hence, the system of partial differential equations corresponds to a large number of coupled ordinary differential equations. The time step required for the integration is determined mainly by nonlinearities of the equations. Considering, for example, (3.66), the allowed time step Δt should be less than $[\nabla n(\boldsymbol{r},t)]^{-2}$ if we suppose that $n(\boldsymbol{r},t)$ and $h_0(\boldsymbol{r},t)$ are of the same order. As large gradients come into play, the time step should be decreased according to the *square* of ∇n. On the other hand, if we solve the corresponding Langevin equation, (3.30), the gradient appears only linearly, and therefore, much larger time steps are allowed for the integration. Hence, simulation of a large number of particles does not necessarily cause larger simulation times because in the example considered, the equations are linearized.

4. Self-Organization of Networks

4.1 Agent-Based Model of Network Formation

4.1.1 Basic Assumptions and Equations of Motion

The previous chapter showed how the model of Brownian agents with an internal parameter θ_i can be used to simulate structures in reaction–diffusion systems. In this chapter, we want to explore this idea further: the internal parameter $\theta_i \in \{-1, 0, +1\}$ shall describe three different responses to the components of an effective field that can be further changed by Brownian agents, depending on their actual value of $\theta_i(t)$. The internal energy depot as another internal degree of freedom shall not be considered here. The model shall be applied to the generation of a specific pattern, which can be denoted as a network of links among arbitrary nodes. The *self-organized* formation of links among a set of nodes is of interest in many different fields. In electronic engineering, for instance, one is interested in *self-assembly* and *self-repair* of electrical circuits, and in biology, models for self-wiring and *neuronal networks* are investigated. On the social level, the self-organization of trail networks among different destinations is a similar problem, which will be discussed in detail in Sects. 5.4 and 6.2. The establishment of connections on demand in telecommunication or logistics is also related to the problem discussed here.

A desirable feature of self-organized networks is their *adaptivity*. This means that new nodes can be linked to the existing network or linked nodes can be disconnected from the network if this is required, e.g., by changing some external conditions. Noteworthy, such behavior should not be governed by a "supervisor" or "dispatcher"; it should rather result from the adaptive capabilities of the network itself.

Such problems become even more complicated if one considers that the nodes which should be linked to the networks are "unknown" in the sense, that they *first* have to be *discovered* and only *then* can be *connected*. In this chapter, we will show that Brownian agents can solve such a complex problem, and in the next chapter, the basic ideas will be applied to modeling how ants can discover food sources unknown to them and then can link them to their nest by generating a trail.

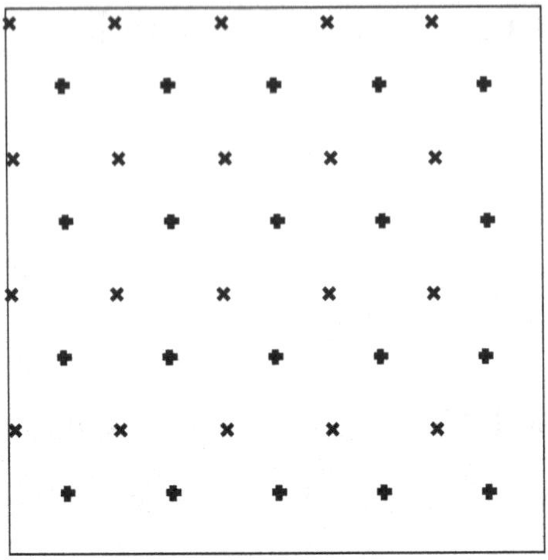

Fig. 4.1. Example of a regular distribution of 40 nodes on a 100×100 lattice. For the computer simulations, periodic boundary conditions have been used: $z_+ = 20$, $z_- = 20$. {x} indicates nodes with a potential $V_j = -1$; {+} indicates nodes with a potential $V_j = +1$ [468, 505]

In the following, we consider a two-dimensional surface, where a number of $j = 1, \ldots, z$ nodes are located at positions $\boldsymbol{r}_j^z$ (cf. Fig. 4.1). A number of z_+ nodes should be characterized by a positive potential, $V_j = +1$, whereas $z_- = z - z_+$ nodes have a negative potential, $V_j = -1$. The number of positive and negative nodes can be different; however, neutrality is obtained only, if

$$\sum_{j=1}^{z} V_j = 0 \;, \quad \text{if } z_- = z_+ \,. \tag{4.1}$$

It is the (twofold) task of Brownian agents, first to *discover* the nodes and then to *link* nodes with an opposite potential, in this way forming a self-organized network among the set of nodes. We note explicitly, that this task is quite complicated because the nodes do *not* have any *long-range effect* on agents, such as attraction or repulsion. Their effect is restricted to their locations, $\boldsymbol{r}_j^z$.

We further assume N Brownian agents moving on the surface, their actual positions denoted by $\boldsymbol{r}_i$, and their internal degree of freedom denoted by $\theta_i(t)$ that could have one of the following values: $\theta \in \{0, -1, +1\}$. Initially, $\theta_i(t_0) = 0$ holds for every agent. The parameter θ_i can be changed in the course of time by an interaction between moving agents and nodes as follows:

$$\Delta\theta_i(t) = \sum_{j=1}^{z} (V_j - \theta_i)\, \frac{1}{A} \int_A \delta\left[\boldsymbol{r}_j^z - \boldsymbol{r}_i(t)\right] d\boldsymbol{r} \,. \tag{4.2}$$

The delta function is equal to one only for $\boldsymbol{r}_j^z = \boldsymbol{r}_i$ and zero otherwise. So, (4.2) indicates, that an agent changes its internal state θ_i, only if it hits one of

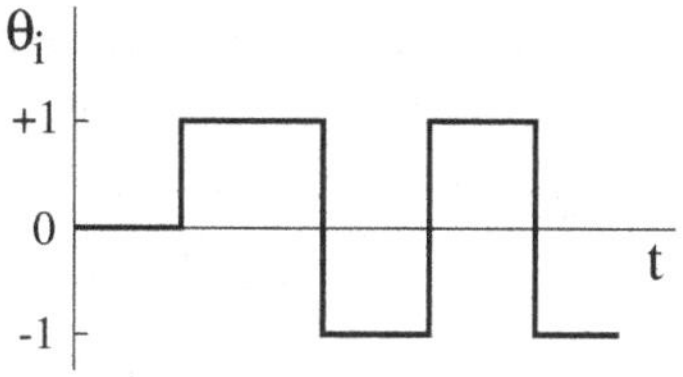

Fig. 4.2. Sketch of the change of internal parameter θ_i

the nodes. Then it takes over the value of the potential of the respective node, V_j, which means $\theta_i = \text{const}$, if $V_j = \theta_i$, and $\theta_i \to V_j$, if $V_j \neq \theta_i$ (see Fig. 4.2). To be mathematically correct, we note that the probability for a (pointlike) agent to hit a (pointlike) node is almost vanishing. However, the computer simulations discussed in the following section are carried out on a discrete lattice, so that the agent and node both have finite extensions, in that case, (4.2) makes sense.

If a Brownian agent hits one of the nodes, this impact may result in an active state for the agent – a *kick*, originated by the potential, which may change the internal parameter θ_i, due to (4.2). In the active state, it is assumed that the agent can produce a chemical, either component -1 or $+1$, depending on the actual value of the internal parameter. The production rate, $s_i(\theta_i, t)$, is assumed as follows:

$$s_i(\theta_i, t) = \frac{\theta_i}{2}\Big\{(1+\theta_i)\, s^0_{+1} \exp\left[-\beta_{+1}\,(t - t^i_{n+})\right] \\ - (1-\theta_i)\, s^0_{-1} \exp\left[-\beta_{-1}\,(t - t^i_{n-})\right]\Big\}\,. \tag{4.3}$$

Equation (4.3) means that the agent is not active, as long as $\theta_i = 0$, which means before it hits one of the nodes for the first time. After that event, the agent begins to produce either component $+1$, if $\theta_i = +1$, or component -1, if $\theta_i = -1$. This activity, however, declines with time, expressed in an exponential decrease of the production rate. Here, s^0_{+1}, s^0_{-1} are the initial production rates, and β_{+1}, β_{-1} are the decay parameters for the production of the chemical components $+1$ or -1. Respectively, t^i_{n+}, t^i_{n-} are the times when agent i hits either a node with a positive or a negative potential.

The spatiotemporal concentration of chemicals shall be described by a self-consistent field $h_\theta(\boldsymbol{r}, t)$, which in the case considered is a *chemical field*, consisting of either component $+1$ or -1. The field generated by Brownian agents is assumed again to obey a reaction equation, as given in (3.3), where k_θ is the decomposition rate, but diffusion is *not* considered here ($D_\theta = 0$):

$$\frac{\partial h_\theta(\boldsymbol{r}, t)}{\partial t} = -k_\theta\, h_\theta(\boldsymbol{r}, t) + \sum_{i=1}^{N} s_i(\theta_i, t)\, \delta_{\theta;\theta_i}\, \delta\left[\boldsymbol{r} - \boldsymbol{r}_i(t)\right] \quad , \quad \theta \in \{-1, +1\}\,. \tag{4.4}$$

The movement of the Brownian agent should be described again by a Langevin equation, (3.2), which in the *overdamped limit* reads,

$$\frac{d\boldsymbol{r}_i}{dt} = \frac{\alpha}{\gamma_0} \left. \frac{\partial h^e(\boldsymbol{r},t)}{\partial \boldsymbol{r}} \right|_{r_i,\theta_i} + \sqrt{\frac{2k_\mathrm{B}T}{\gamma_0}} \boldsymbol{\xi}_i(t) \, . \tag{4.5}$$

Here, $h^e(\boldsymbol{r},t)$ means an *effective* field, discussed already in Sect. 3.1.2, which is a specific function of the different components of the field, (3.4). It should influence the movement of agents in dependence on their internal parameter θ_i, as follows:

$$\frac{\partial h^e(\boldsymbol{r},t)}{\partial \boldsymbol{r}} = \frac{\theta_i}{2} \left[(1+\theta_i) \frac{\partial h_{-1}(\boldsymbol{r},t)}{\partial \boldsymbol{r}} - (1-\theta_i) \frac{\partial h_{+1}(\boldsymbol{r},t)}{\partial \boldsymbol{r}} \right] . \tag{4.6}$$

Equation (4.7) and Fig. 4.3 summarize the nonlinear feedback between the field and agents, as given by (4.3), (4.6):

$$\begin{array}{lll} \theta_i = 0: & \boldsymbol{\nabla}_i \, h^e(\boldsymbol{r},t) = 0, & s_i(\theta_i,t) = 0, \\ \theta_i = +1: & \boldsymbol{\nabla}_i \, h^e(\boldsymbol{r},t) = \boldsymbol{\nabla}_i \, h_{-1}(\boldsymbol{r},t), & s_i(\theta_i,t) = s_i(+1,t), \\ \theta_i = -1: & \boldsymbol{\nabla}_i \, h^e(\boldsymbol{r},t) = \boldsymbol{\nabla}_i \, h_{+1}(\boldsymbol{r},t), & s_i(\theta_i,t) = s_i(-1,t) \, . \end{array} \tag{4.7}$$

Our model assumes that agents with an internal state $\theta_i = 0$ do not contribute to the field and are not affected by the field. They simply move like Brownian particles. Agents with an internal state $\theta_i = +1$ contribute to the field by producing component $+1$, while they are affected by the part of the field that is determined by component -1. On the other hand, agents with an internal state $\theta_i = -1$ contribute to the field by producing component -1 and are affected by the part of the field, which is determined by component $+1$ (see also Fig. 4.3). Moreover, if an agent hits one of the nodes, the internal state can be switched due to (4.2). Hence, the agent begins to produce a different chemical, while affected by the opposite potential. Precisely, at one time, the agent does *not* respond to the gradient of the same field, to which it contributes via producing a chemical.

As the result of this nonlinear feedback between Brownian agents and the chemical field generated by them, we can observe the formation of macroscopic structures shown in the following section.

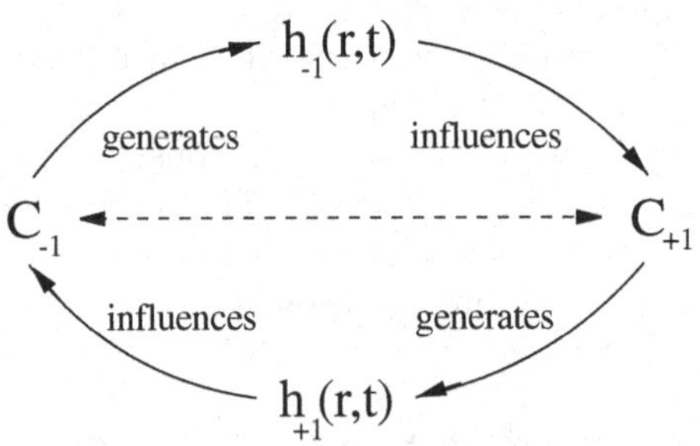

Fig. 4.3. Circular causation between Brownian agents in different internal states, C_{-1}, C_{+1}, and the two-component chemical field $h_\theta(\boldsymbol{r},t)$

4.1.2 Results of Computer Simulations

For the computer simulations, a triangular lattice with periodic boundary conditions has been used. Further, we have assumed that the parameters describing the production and decay of the chemical are the same for both components:

$$s^0_{+1} = s^0_{-1} = s_0 \ , \quad k_{+1} = k_{-1} = k_h \ , \quad \beta_{+1} = \beta_{-1} = \beta \ . \tag{4.8}$$

If not otherwise noted, agents start initially at random positions. For the evolution of the network, we evaluate the sum $\hat{h}(\boldsymbol{r}, t)$ of the two field components generated by the agents. For the plots, however, we have to match these values with a *gray scale* of 256 values, which is defined as follows:

$$c(\boldsymbol{r}, t) = 255 \left\{ 1 - \log \left[1 + 9 \, \frac{\hat{h}(\boldsymbol{r}, t) - \hat{h}_{\min}(t)}{\hat{h}_{\max}(t) - \hat{h}_{\min}(t)} \right] \right\} ,$$
$$\hat{h}(\boldsymbol{r}, t) = h_{+1}(\boldsymbol{r}, t) + h_{-1}(\boldsymbol{r}, t) \ . \tag{4.9}$$

This means that the highest actual value, $\hat{h}_{\max}(t)$, always refers to *black* ($c = 0$ in PostScript), whereas the actual minimum value, $\hat{h}_{min}(t)$, encodes *white* ($c = 255$). Both extreme values change, of course, in time; therefore each snapshot of the time series presented has its own value mapping.

As a first example, we show the evolution of the connections among four nodes (Fig. 4.4).

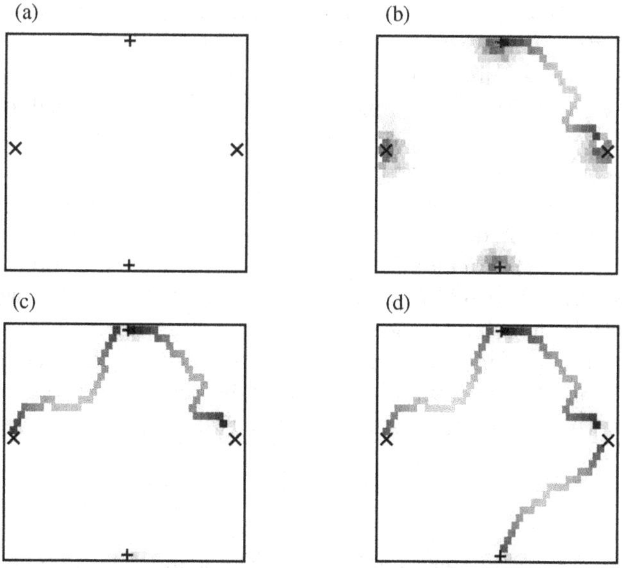

Fig. 4.4. Formation of links among four nodes: (**a**) initial state; (**b**) after 100 simulation steps; (**c**) after 1000 simulation steps; (**d**) after 4500 simulation steps. Lattice size 30 × 30, 450 agents. Parameters: s_0 = 25,000, k_h = 0.01, β = 0.2, k_h = 0.01 [468]

Figure 4.4 suggests that in the course of time all nodes with an opposite potential should be connected. This, however, is not true because the existing connections cause a *screening effect*, which forces agents to move along existing connections rather than making up new ones. This screening effect becomes more obvious, when the number of nodes is increased. Figure 4.5 shows the temporal evolution of a network, which should connect 40 nodes. Here, we see that in the course of time, the agents aggregate along the connections, which results in higher agent concentrations and in higher fields along the connections.[1]

In Fig. 4.5, the connections among the nodes exist as a two-component chemical field generated by Brownian agents. This self-assembling network is created very fast and remains stable after the initial period. Patterns, like the network shown, are intrinsically determined by the history of their creation. This means that irreversibility and early symmetry breaks play a considerable role in determining the final structure. These structures are unique due to their evolution and therefore can be understood only within a stochastic approach.

We note that network formation is not restricted to regular or symmetrical distributions of nodes. Figure 4.6 shows a simulation, where different nodes are connected with a center. An extension of the model has been applied to simulate the trunk trail formation in ants connecting a nest to different food sources [456] and will be discussed in Sect. 5.3.

To conclude the computer simulations, we note that our model shows the *self-assembling of networks* among arbitrary nodes based on the nonlinear interaction of Brownian agents. Different from a *circuit diagram*, for instance, that determines the different links among nodes in a *top-down* approach to hierarchical planning, connections here are created by agents *bottom-up* in a process of self-organization. The model turned out to be very flexible regarding the geometry of the nodes to be connected. The locations of the different nodes act as a boundary condition for structure formation but do not determine the way of connecting the different nodes. If, for example, a particular link breaks down, agents could *repair* it by reestablishing the field or by creating a new one. In Sect. 4.3, we will show that the switching behavior between a connected and a disconneced state can be very short, which would allow the construction of a dynamic switch. In Sect. 5.3, on the other hand, we will show that the model is also flexible in connecting additional nodes to the network or disconnecting obsolete ones.

---

[1] A video of these computer simulations can be found at `http://www.ais.fhg.de/~frank/network.html`. The author would like to thank Benno Tilch for his collaboration.

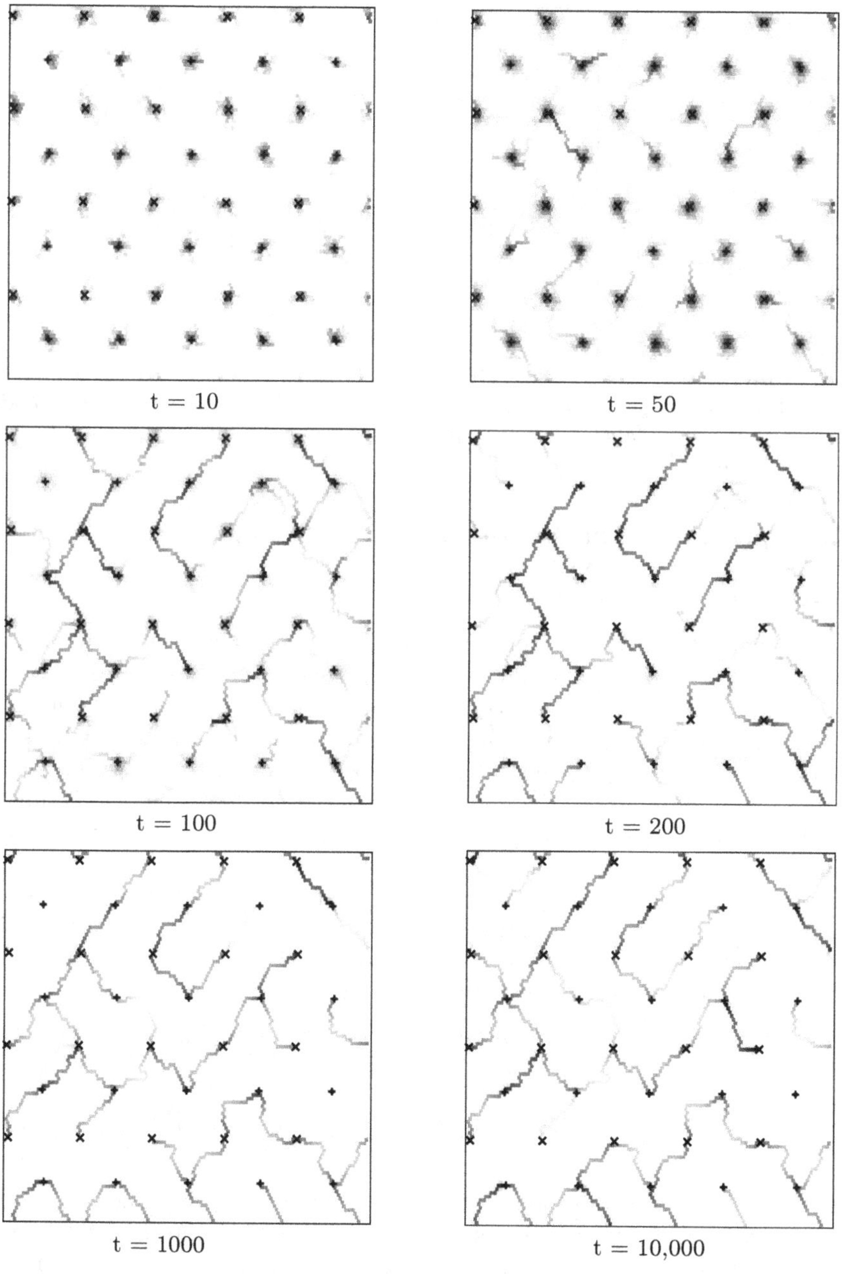

Fig. 4.5. Time series of the evolution of a network (time in simulation steps). Lattice size 100 × 100, 5000 agents, 40 nodes, $z_+ = 20$, $z_- = 20$. Parameters: $s_0 = 10,000$, $k_h = 0.03$, $\beta = 0.2$ [468, 505]

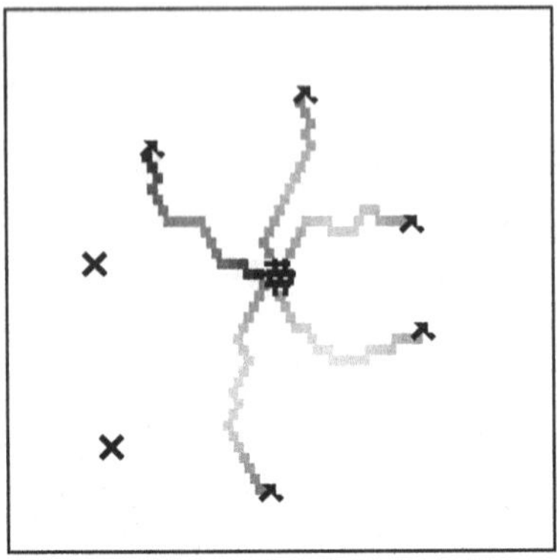

Fig. 4.6. Formation of links between a center ($z_- = 1$) and surrounding nodes ($z_+ = 7$) after 10,000 simulation steps. Lattice size: 50×50, 2000 agents. Parameters: $s_0 = 20,000$, $k_h = 0.02$, $\beta = 0.2$ [468]

4.2 Estimation of Network Connectivity

4.2.1 Critical Temperature

In our model of Brownian agents, stochastic forces are represented by thermal noise, which keeps agents moving. Hence, temperature T is an appropriate measure for *stochastic influences.* It is obvious from the Langevin equation, (4.5), that in the limit $T \to 0$, agents initially do not move and therefore do not find a node to start generating a field and to establish a network. On the other hand, for $T \to \infty$, the influence of the gradients of the field is almost negligible; hence agents do not turn to ordered motion among the nodes, and a network does not occur either.

To estimate the appropriate range of temperatures for the formation of networks, we turn to a continuous description and introduce the spatio-temporal density $n_\theta(\boldsymbol{r}, t)$, (3.9). As discussed in detail in Sect. 3.1.3, we obtain the following *reaction–diffusion* equation, (3.11), for $n_\theta(\boldsymbol{r}, t)$:

$$\begin{aligned} \frac{\partial n_\theta(\boldsymbol{r},t)}{\partial t} &= \frac{\partial}{\partial \boldsymbol{r}} \left[-\frac{\alpha}{\gamma_0} \frac{\partial h^e(\boldsymbol{r},t)}{\partial \boldsymbol{r}}\, n_\theta(\boldsymbol{r},t) + D_n \frac{\partial n_\theta(\boldsymbol{r},t)}{\partial \boldsymbol{r}} \right] \\ &\quad + \sum_{\theta' \neq \theta} \Big[w(\theta|\theta')\, n_{\theta'}(\boldsymbol{r},t) - w(\theta'|\theta)\, n_\theta(\boldsymbol{r},t) \Big] \,. \end{aligned} \tag{4.10}$$

$D_n = k_\mathrm{B} T/\gamma_0$ is the diffusion coefficient of Brownian agents. The summation term considers the gain and loss of agents with a given internal state θ. For the transition rates $w(\theta'|\theta)$, in accordance with (4.2), we have to consider four possible "reactions":

$$\begin{aligned} &C_0 \xrightarrow{w(+1|0)} C_{+1}, \qquad C_{+1} \xrightarrow{w(-1|+1)} C_{-1}, \\ &C_0 \xrightarrow{w(-1|0)} C_{-1}, \qquad C_{-1} \xrightarrow{w(+1|-1)} C_{+1} \,. \end{aligned} \tag{4.11}$$

Here, C_θ stands for an agent with internal parameter θ. We simply assume that the probability of changing the internal parameter is proportional to the mobility of the agents, expressed by D_n, and the spatial density of the respective nodes that cause the change. Hence,

$$\begin{aligned} w(+1|\theta) &= D_n \frac{z_+}{A}, \qquad w(\theta|\theta) = 0, \\ w(-1|\theta) &= D_n \frac{z_-}{A}, \qquad w(0|\theta) = 0 \qquad \theta \in \{-1, 0, +1\} . \end{aligned} \tag{4.12}$$

The total fraction of agents with internal state θ results from

$$x_\theta(t) = \frac{1}{N} \int_A n_\theta(\boldsymbol{r}, t)\, d\boldsymbol{r} . \tag{4.13}$$

With respect to (4.10), the equation for the change of $x_0(t)$ reads explicitly

$$\frac{dx_0(t)}{dt} = \sum_{\theta \neq 0} w(\theta|0)\, x_0(t) = -\sum_{\theta \neq 0} \frac{dx_\theta(t)}{dt} . \tag{4.14}$$

Using (4.12) for transition rates, (4.14) can be integrated with the initial condition, $x_0(0) = 1$, which yields

$$x_0(t) = \exp\left(-D_n \frac{z_+ + z_-}{A}\, t\right) . \tag{4.15}$$

This result can be compared with the simulations shown in Fig. 4.5. Figure 4.7 presents the corresponding fraction $x_\theta(t)$ of agents, which are currently in internal state θ. Starting with the *initial condition* $x_0 = 1, x_{\pm 1} = 0$, Fig. 4.7

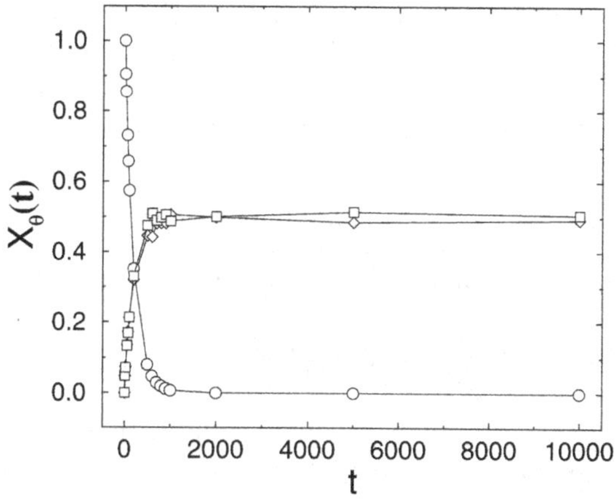

Fig. 4.7. Fraction $x_\theta(t) = N_\theta/N$ of agents with internal states $\theta \in \{-1, 0, +1\}$ vs. time in simulation steps. (○): $\theta = 0$, (◇): $\theta = +1$, (□): $\theta = -1$. The values are obtained from the simulation, Fig. 4.5. Initial conditions: $x_0 = 1, x_{\pm 1} = 0$

shows that after 1000 simulation steps, almost all agents have already found one of the nodes, thus changing their internal parameter to either $\{+1\}$ or $\{-1\}$. The equal distribution between these two states, of course, fluctuates slightly around $x_\theta = 0.5$, depending on the actual positions of agents.

If we compare (4.15) with the computer simulations presented in Fig. 4.7, then the best fit of the appropriate curve yields exactly (4.15), with $D_n = 1$. The fractions for the remaining two internal parameters can be described accordingly:

$$x_\theta(t) = \frac{z_\theta}{z_+ + z_-}\left[1 - \exp\left(-D_n \frac{z_+ + z_-}{A}\, t\right)\right] \quad , \quad \theta \in \{-1, +1\} \, . \tag{4.16}$$

As also indicated by the computer simulations, in the asymptotic limit, the fraction x_θ is determined by the appropriate *number of nodes*, which change the internal state of the agents to θ.

For further investigations, we restrict ourselves to times where $x_\theta(t) \to x_\theta^{\mathrm{stat}}$, which is after about 2000 time steps in the computer simulations. Then, we have to consider only *two* different densities, $n_{+1}(\boldsymbol{r}, t)$, $n_{-1}(\boldsymbol{r}, t)$, where the corresponding reaction–diffusion equations, (4.10), have an additional indirect coupling via the effective field, $h^e(\boldsymbol{r}, t)$, (4.4), (4.6).

For further discussions, we assume that the spatiotemporal field *relaxes faster* than the related distribution of agents into a quasistationary equilibrium. The field components $h_\theta(\boldsymbol{r}, t)$ should still depend on time and space coordinates but due to the fast relaxation, there is a fixed relation to the agent distribution. From (4.4), we find with $\dot{h}_\theta(\boldsymbol{r}, t) = 0$,

$$h_\theta(\boldsymbol{r}, t) = \frac{1}{k_h} \sum_{i=1}^{N} s_i(\theta_i, t)\, \delta_{\theta, \theta_i(t)}\, \delta\left[\boldsymbol{r} - \boldsymbol{r}_i(t)\right] \quad , \quad \theta \in \{-1, +1\} \, . \tag{4.17}$$

The δ sum in (4.17) cannot be easily replaced by the related densities because of $s_i(\theta_i, t)$, (4.3), that decreases exponentially in time. We assume that a minimum production rate s_{min} exists, which is, in the given model, the smallest possible amount of chemical released by the agent (naturally, it could be a molecule, if s is measured in molecule numbers). With t_0 as time, when the agent hits the node, we get from

$$s_{\mathrm{min}} = s_0 \exp\left[-\beta(t - t_0)\right] \, , \tag{4.18}$$

the maximum time t_{max} after which the production is *negligible*:

$$t_{\mathrm{max}} - t_0 = \frac{1}{\beta} \ln\left(\frac{s_0}{s_{\mathrm{min}}}\right) \, . \tag{4.19}$$

In a mean-field approximation, we may assume that the time-dependent production rate $s_i(\theta_i, t)$ can be replaced by an average rate $\bar{s}$, in which case (4.17) yields

$$h_\theta(\boldsymbol{r}, t) = \frac{\bar{s}}{k_h}\, n_\theta(\boldsymbol{r}, t) \, . \tag{4.20}$$

Using (4.3), the average production rate can be defined as

$$\bar{s} = \frac{1}{t_{\max} - t_0} \int_{t_0}^{t_{\max}} s_0 \exp\Big[- \beta\,(t - t_0)\Big] \,. \tag{4.21}$$

With (4.19), we get

$$\bar{s} = \frac{s_0 \;-\; s_{\min}}{\ln s_0 \;-\; \ln s_{\min}} \,. \tag{4.22}$$

With these approximations, we can rewrite the reaction–diffusion equation, (4.10), for $\theta \in \{-1, +1\}$:

$$\begin{aligned} \frac{\partial}{\partial t}\, n_\theta(\boldsymbol{r}, t) = &- \frac{\partial}{\partial \boldsymbol{r}} \left[n_\theta(\boldsymbol{r}, t)\, \frac{\alpha}{\gamma_0}\, \frac{\bar{s}}{k_h}\, \frac{\partial n_{-\theta}(\boldsymbol{r}, t)}{\partial \boldsymbol{r}} \right] + D_n\, \frac{\partial^2}{\partial \boldsymbol{r}^2} n_\theta(\boldsymbol{r}, t) \\ &- D_n \frac{z_{-\theta}}{A}\, n_\theta(\boldsymbol{r}, t) + D_n \frac{z_\theta}{A}\, n_{-\theta'}(\boldsymbol{r}, t) \,. \end{aligned} \tag{4.23}$$

In the following, we assume that $z_- = z_+ = z/2$, as used for the computer simulations presented in Fig. 4.5.

The homogeneous solution for $n_\theta(\boldsymbol{r}, t)$ is given by the mean densities:

$$\bar{n}_\theta = \frac{1}{A} \int_A n_\theta(\boldsymbol{r}, t)\, dr = \frac{\bar{n}}{2} \;, \quad \bar{n} = \frac{N}{A} \,. \tag{4.24}$$

To investigate the stability of the homogeneous state, we consider small fluctuations around $\bar{n}_\theta$:

$$n_\theta(\boldsymbol{r}, t) = \bar{n}_\theta + \delta n_\theta \;; \qquad \left| \frac{\delta n_\theta}{\bar{n}_\theta} \right| \ll 1 \,. \tag{4.25}$$

Inserting (4.25) in (4.23), linearization gives

$$\frac{\partial \delta n_\theta}{\partial t} = D_n\, \Delta \delta n_\theta \;-\; \left[\frac{\alpha\, \bar{s}\, \bar{n}}{2 \gamma_0 k_h} \right] \, \Delta \delta n_{-\theta} \;-\; \frac{2z}{A} \Big(\delta n_\theta - \delta n_{-\theta} \Big) \,. \tag{4.26}$$

With the ansatz

$$\delta n_\theta \sim \exp\left(\lambda t + i \boldsymbol{\kappa} \boldsymbol{r} \right) \,, \tag{4.27}$$

we find from (4.26) the dispersion relation $\lambda(\boldsymbol{\kappa})$ for small inhomogeneous fluctuations with wave vector $\boldsymbol{\kappa}$. This relation yields two solutions:

$$\lambda_1(\boldsymbol{\kappa}) = -\kappa^2 \left(D_n - \frac{\alpha\, \bar{s}\, \bar{n}}{2 \gamma_0 k_h} \right) \,, \quad \lambda_2(\boldsymbol{\kappa}) = -\kappa^2 \left(D_n + \frac{\alpha\, \bar{s}\, \bar{n}}{2 \gamma_0 k_h} \right) - \frac{2z}{A} \,. \tag{4.28}$$

For homogeneous fluctuations, we obtain from (4.28),

$$\lambda_1 = 0, \quad \lambda_2 = -\frac{2z}{A} < 0, \quad \text{for } \boldsymbol{\kappa} = 0 \; , \tag{4.29}$$

which means that the homogeneous system is marginally stable to homogeneous fluctuations. Stability to inhomogeneous fluctuations ($\boldsymbol{\kappa} \neq 0$) is obtained only as long as $\lambda_1(\boldsymbol{\kappa}) < 0$ and $\lambda_2(\boldsymbol{\kappa}) < 0$. Although this relation is always satisfied for $\lambda_2(\boldsymbol{\kappa})$, it holds for $\lambda_1(\boldsymbol{\kappa})$ only as long as the following inequality is satisfied:

$$D_n = \frac{k_{\mathrm{B}}T}{\gamma_0} > D_{\mathrm{crit}} = \frac{\alpha\,\bar{s}\,\bar{n}}{2\gamma_0 k_h} \; . \tag{4.30}$$

In analogy to the discussion in Sect. 3.2.2, this relation defines a critical temperature:

$$T_{\mathrm{c}} = \frac{\alpha}{2}\,\frac{\bar{s}\,\bar{n}}{k_{\mathrm{B}}\,k_h} \; . \tag{4.31}$$

A structure formation process originated by agents can occur only below the critical temperature, $T < T_{\mathrm{c}}$, whereas for $T > T_{\mathrm{c}}$, the homogeneous state $n_\theta(\boldsymbol{r}, t) = \bar{n}/2$ is the stable state, and all localized structures, such as the links between the nodes, disappear. This result allows us to determine the *range of parameters* which have to be chosen to observe network formation. In the following section, we want to apply (4.31) to investigate the connectivity of the network in dependence on the temperature.

4.2.2 Network Connectivity and Threshold

To characterize a network, one of the most important questions is whether or not two nodes k and l are connected. In the model considered, a connection is defined in terms of the chemical field $\hat{h}(\boldsymbol{r}, t)$ produced by agents. During the first stage of network formation, agents have randomly visited almost every lattice site before their motion turned into a bounded motion among the nodes. Therefore, the field $\hat{h}(\boldsymbol{r}, t)$ has a non-zero value for almost every $\boldsymbol{r}$, which exponentially decays but never vanishes. Hence, to *define a connection* in terms of $\hat{h}(\boldsymbol{r}, t)$, we have to introduce a *threshold value* h_{thr}, which is the minimum value considered for a link. More precisely, a *connection* between two nodes k and l should exist only, if there is a path $a \in A$ between k and l along which the actual value of the field is larger than the threshold value:

$$\hat{h}(a, t) > h_{\mathrm{thr}}, \quad \text{for } a \in A \; . \tag{4.32}$$

Such a definition does not necessarily assume that the connection has to be a *direct link*. Instead, it could be any path a, which may also include other nodes, as long as the value $\hat{h}(a, t)$ along the path is above the threshold.

We want to define *local connectivity* E_{lk} as follows:

$$E_{lk} = \begin{cases} 1 & \text{if } k \text{ and } l \text{ are connected by a path } a \in A, \\ & \text{along which } \hat{h}(a,t) > h_{\text{thr}} \\ 0 & \text{otherwise} . \end{cases} \tag{4.33}$$

We note that connectivity E_{lk} does not change if two nodes k and l are connected by more than one path.

If we consider a number of z nodes, then *global connectivity* E that refers to the whole network is defined as follows:

$$E = \frac{\sum\limits_{k=1}^{z} \sum\limits_{l>k}^{z} E_{lk}}{\sum\limits_{k=1}^{z} \sum\limits_{l>k}^{z} 1} = \frac{2}{z(z-1)} \sum_{k=1}^{z} \sum_{l>k}^{z} E_{lk} . \tag{4.34}$$

Depending on the configuration of nodes, there may be numerous different realizations for connections, which result in the same connectivity E.

To use the definition for connectivity to evaluate the simulated networks, first we have to define the threshold value h_{thr}. This should be the *minimum value* of $\hat{h}(\boldsymbol{r},t)$ along a *stable connection* between two nodes. For our estimations, we treat the connection between two nearest neighbor nodes k and l as a *one-dimensional* structure, where x is now the space coordinate and L the linear distance between the two nodes k and l. The node at $x = 0$ should have a positive potential $V = +1$, and the node at $x = L$ has a negative potential $V = -1$:

$$0 \leq x \leq L\,, \quad V(0) = +1\,, \quad V(L) = -1\,. \tag{4.35}$$

We assume that a stable connection exists, if both field components $h_\theta(x,t)$ have reached their stationary values:

$$\frac{\partial h_\theta(x,t)}{\partial t} = -k_\theta\, h_\theta(x,t) + \sum_i s_i(\theta,t)\, \delta(x - x_i) = 0\,. \tag{4.36}$$

Of course, we do not know how many agents are actually on the connection between k and l. For our estimates, we have to bear in mind that h_{thr} should determine the *lower limit* of the possible values of $\hat{h}$. Therefore, it is justified to assume the worst case, which means that the local number of agents at a specific location is given just by the *average agent density* $\bar{n} = N/A$, where N is the total agent number and A is the surface area. Further, we found in the computer simulations (cf. Fig. 4.7), that in the long-time limit, agents are equally distributed between the two internal states, $\theta = \{+1, -1\}$. Hence, we assume that on any location along the connection, there are $n_\theta = \bar{n}/2$ agents in state θ. Using this lower limit for the agent number, the delta function in (4.36) can be replaced by $\bar{n}/2 = N/2A$ in the continuous limit.

Further, we consider that agents move along the x coordinate at constant velocity (which also matches the assumption $\dot{v} \approx 0$ of the overdamped Langevin equation):

$$v = |\boldsymbol{v}| = \frac{|x|}{t} \,, \quad 0 \leq x \leq L \,. \tag{4.37}$$

This allows us to replace t in the time-dependent production rate, $s_i(\theta, t)$. With these simplified assumptions and the conventions (4.8), (4.36) reads for component $\theta = +1$,

$$\frac{\partial h_{+1}(x,t)}{\partial t} = -k_h\, h_{+1}(x,t) + \frac{\bar{n}}{2}\, s_0 \exp\left(-\beta \frac{x}{v}\right) \,. \tag{4.38}$$

Integrating (4.38) yields, with $h_{+1}(x, t = 0) = 0$,

$$h_{+1}(x,t) = \frac{\bar{n}}{2} \frac{s_0}{k_h} \exp\left(-\beta \frac{x}{v}\right) [1 - \exp(-k_h\, t)] \,. \tag{4.39}$$

Finally, for $t \to \infty$, we find from (4.39) the stationary solution,

$$h_{+1}(x) = \frac{\bar{n}}{2} \frac{s_0}{k_h} \exp\left(-\frac{\beta}{v} x\right) \,. \tag{4.40}$$

The corresponding field component $h_{-1}(x', t)$ should have the same stationary solution as (4.40), with $x' = L - x$. The resulting total field $\hat{h}(x, t)$ reads, in the stationary limit (see Fig. 4.8),

$$\begin{aligned} \hat{h}(x) &= h_{+1}(x) + h_{-1}(L - x) \qquad\qquad (4.41)\\ &= \frac{\bar{n}}{2} \frac{s_0}{k_h} \left\{ \exp\left(-\frac{\beta}{v} x\right) + \exp\left[-\frac{\beta}{v}(L - x)\right] \right\} \,, \quad 0 \leq x \leq L \,. \end{aligned}$$

The threshold h_{thr} should be defined as the minimum of $\hat{h}(x)$, which holds for $L/2$. As a result, we find

$$h_{\text{thr}} = \hat{h}\left(\frac{L}{2}\right) = \frac{N}{A} \frac{s_0}{k_h} \exp\left(-\frac{\beta}{v} \frac{L}{2}\right) \,. \tag{4.42}$$

Here, the threshold value is a function of the mean agent density $\bar{n}$, the parameters s_0, k_h, and β, and the distance L between the two nodes. However, due to the decay of the field (k_h) and the decreasing production rate with time (β), agents can link only nodes that are at a distance closer than $L^\star$; thus (4.42) makes sense only for $L < L^\star$.

To estimate $L^\star$, we consider that the production rate s_i decreases exponentially in the course of time, and therefore a maximum time $t_{\max}$, (4.19), results after which production is negligible. We can now discuss the case

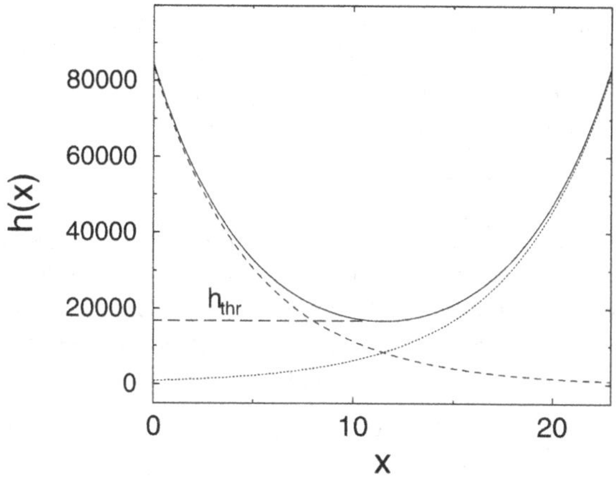

Fig. 4.8. Stationary solutions for $h_{+1}(x)$, (4.40) ($\cdots$), $h_{-1}(x)$ ($--$), and $\hat{h}(x)$, (4.41) (——). $L = L_{\max}/2$, (4.43). The threshold h_{thr} is defined as the minimum value of $\hat{h}(\boldsymbol{r}, t)$ in the stationary limit. For the parameters, see Fig. 4.5 [468, 505]

where the agent moves straight at constant velocity, without changing its direction. Then the maximum distance crossed would be

$$L_{\max} = v\, t_{\max} = \frac{v}{\beta} \ln\left(\frac{s_0}{s_{\min}}\right) . \tag{4.43}$$

To the contrary, if we assume that the agent moves like a random agent, the average distance reached after t steps is given by the mean displacement, $\Delta R = \sqrt{2d\, D_n\, t}$, (1.16), which yields for $d = 2$,

$$L_{av} = \sqrt{4D_n\, t_{\max}} = \sqrt{\frac{4\, D_n}{\beta} \ln\left(\frac{s_0}{s_{\min}}\right)} . \tag{4.44}$$

The *real maximum distance* which can be connected by one agent in the model considered, is, of course, between these limits. We found during the computer simulations (see also Sect. 4.3.1 and Fig. 4.15) that $L_{\max}/2$ is a reasonable estimate for $L^{\star}$:

$$\sqrt{\frac{4\, D_n}{v}\, L_{\max}} < L^{\star} \approx \frac{L_{\max}}{2} < L_{\max} . \tag{4.45}$$

Using this approximation, we find with (4.42), (4.43), finally, the estimate for the threshold [468]:

$$h_{\text{thr}} = \frac{N}{A} \frac{s_0}{k_h} \left(\frac{s_{\min}}{s_0}\right)^{1/4} . \tag{4.46}$$

Provided with the set of parameters used for the simulations, we find for the threshold approximately $h_{\text{thr}} \approx 2 s_0$. We note again that this is an estimate that might give a rather high value because of the assumed worst conditions. On the other hand, we can now be sure that values for $\hat{h}(a,t)$ above the threshold *really* represent a *stable* connection a.

4.2.3 Numerical Results

After these theoretical considerations, we can now calculate the connectivity E, (4.34), for the network, simulated in Fig. 4.5. Figure 4.9 shows the increase in the connectivity of the given network in the course of time. In agreement with the visible evolution of the network presented in Fig. 4.5, three different stages can be clearly distinguished:

1. an *initial* period ($t < 10^2$), where no connections yet exist;
2. a *transient* period ($10^2 < t < 10^4$), where the network is established;
3. a *saturation* period ($t > 10^4$), where almost all nodes are connected, and only small fluctuations in the connectivity occur.

Figure 4.9 results only from a single realization of the network, but due to stochastic influences that affect the formation of the network, on average, we find certain fluctuations in connectivity E. This can be seen, for instance, in Fig. 4.10 which shows the connectivity for the same setup as used for Fig. 4.5, averaged over 200 simulations. In Fig. 4.10, mean connectivity $\langle E \rangle$ is plotted versus reduced temperature T/T_c, (4.31), which describes how much agents are affected by stochastic influences. Figure 4.11 shows some simulated realizations of networks for different reduced temperatures.

Figure 4.10 clearly indicates that an optimal range of temperature, $0.3 \leq T/T_c \leq 0.5$, exists where the average connectivity reaches a maximum. For the given setup of nodes and the given set of parameters, within this optimal range, on average, more than 90% of the nodes are connected. For $T > 0.5\ T_c$, the connectivity breaks down drastically, which means that the

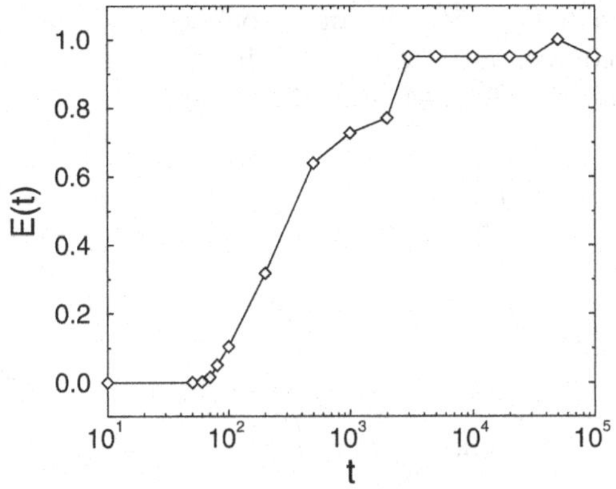

Fig. 4.9. Network connectivity E, (4.34), vs. time t in simulation steps, calculated from the series of Fig. 4.5 [468, 505]

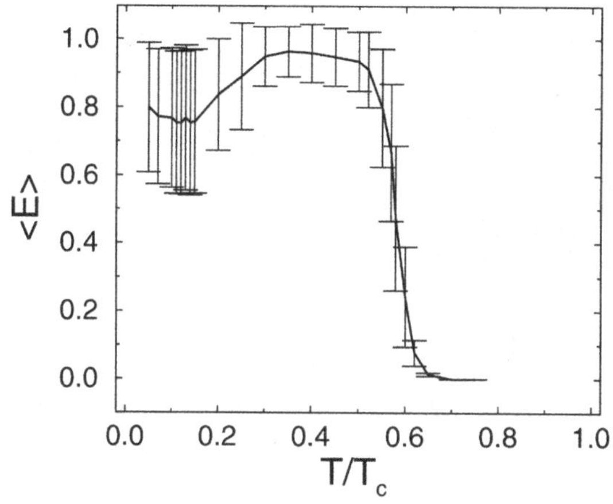

Fig. 4.10. Network connectivity E, (4.34), averaged over 200 simulations vs. reduced temperature T/T_c, (4.31). For the setup, see Fig. 4.5 [467]

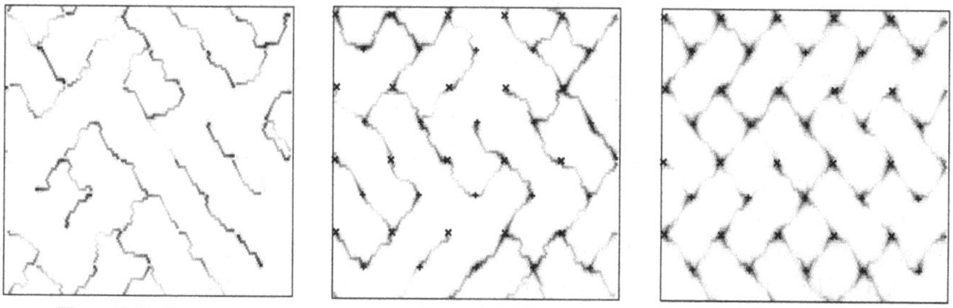

T = 0.2 T_c T = 0.4 T_c T = 0.6 T_c

Fig. 4.11. Network formation dependent on reduced temperature T/T_c, (4.31). Shown is the value $c(\boldsymbol{r}, t)$, (4.9), after 5000 simulation steps. For other parameters, see Fig. 4.5 [467]

motion of agents is determined mainly by stochastic forces, and no attention is paid to the gradient of the field. On the other hand, for $T < 0.3\ T_c$, the connectivity decreases because during the transient period of establishing the network, the agents have paid too much attention to the gradient and therefore are trapped in the first established links, instead of moving around.

The average connectivity $\langle E \rangle$ displays the largest fluctuations in the range of lower temperatures, so almost every realization between $E = 0.5$ and $E = 1$ has a certain likelihood. Here, the early history in creating the network plays an important role in the eventual pattern and connectivity. However, in the optimal temperature range, the effect of the early symmetry breaks is weakened to a certain extent, thus leading to smaller fluctuations in E.

We also want to note that in the range of smaller temperatures, of course, fewer connections among nodes occur, but these connections provide a much stronger links in terms of the field $\hat{h}(\boldsymbol{r}, t)$ than at higher temperatures. One

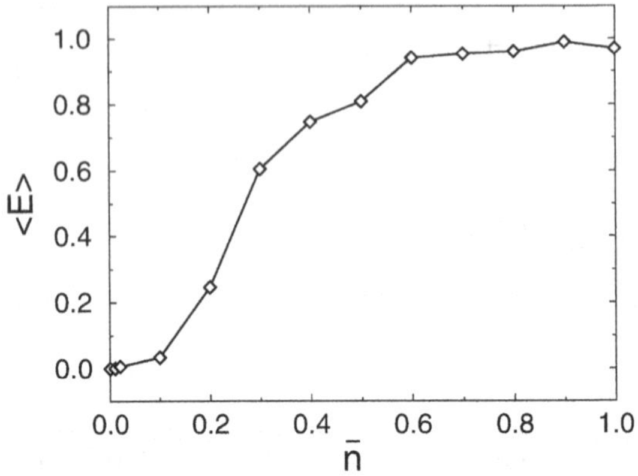

Fig. 4.12. Network connectivity $\langle E \rangle$, (4.34), averaged over 5 simulations vs. mean density of agents, $\bar{n}$. $\eta = 0.1$; for further parameters, see Fig. 4.5 [468]

of the reasons for the breakdown of connectivity at higher temperatures, shown in Fig. 4.10, results from the fact that many of the links established have values below the threshold $h_{\rm thr}$ and therefore are not considered stable connections. This can be seen by comparing the pictures in Fig. 4.11. Here, for $T = 0.6\, T_{\rm c}$, the nodes still seem to be connected; the value of the related field, however, is below the threshold $h_{\rm thr}$.

The estimate of an optimal temperature has been carried out for a fixed density of Brownian agents. However, we also investigated the average connectivity $\langle E \rangle$ at a fixed temperature in dependence on the density of agents $\bar{n}$, which is shown in Fig. 4.12.

Here, we clearly see that below a critical density, the connectivity is almost zero because not enough agents are available to establish connections. On the other hand, above a certain density, the connectivity has reached its saturation value, which could also be below one, as Fig. 4.12 shows. Hence, an increase in the number of agents does not necessarily result in establishing more links. This is caused by the *screening effect*, already mentioned in Sect. 4.1.2, which eventually concentrates all agents to move along established links.

4.3 Construction of a Dynamic Switch

4.3.1 Setup for the Switch

The computer simulations discussed in Sect. 4.1 have proved the very fast and flexible establishment of links between nodes of opposite potential by Brownian agents. Here, we want to apply the basic idea to construct a *dynamic switch*, which is either in the state ON or OFF, depending on external influences. The situation is sketched in Fig. 4.13.

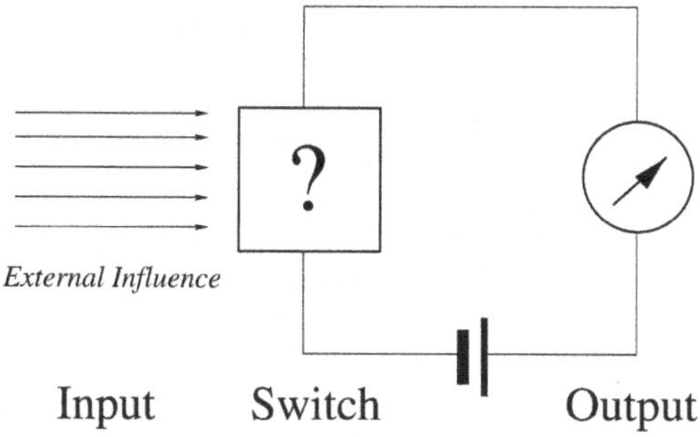

Fig. 4.13. Sketch of the situation considered for a dynamic switch, which should operate dependent on the presence of an external influence

Let us assume a situation, where we need an indicator of an external perturbation, for instance, an electromagnetic field or a poisonous chemical vapor. That means the switch that controls the power supply of the output device should be ON if the external influence is present and OFF if it disappers again. In the following, we will realize this switch by means of Brownian agents. To simulate a dynamic switch, which fits into the black box in Fig. 4.13, we consider only two nodes with opposite potential. Figure 4.14 shows the initial situation (*left*) and a typical connection of nodes (*right*). Hence, we define the states of the dynamic switch as follows:

$$\begin{array}{ll} \text{switch:} & \\ \text{ON} & E = 1 \;\; \text{if} \;\; \exists a \in A \;\; \text{with} \;\; \hat{h}(a,t) > h_{\text{thr}} \, ; \\ \text{OFF} & E = 0 \;\; \text{if} \;\; \nexists a \in A \;\; \text{with} \;\; \hat{h}(a,t) > h_{\text{thr}} \, . \end{array} \qquad (4.47)$$

In agreement with (4.33), the switch is assumed ON if a connection a between the two nodes exists along which every value of the field is above the threshold. Otherwise, the switch should be OFF. The switch between the two states is caused by the existence, or influence, of an external perturbation.

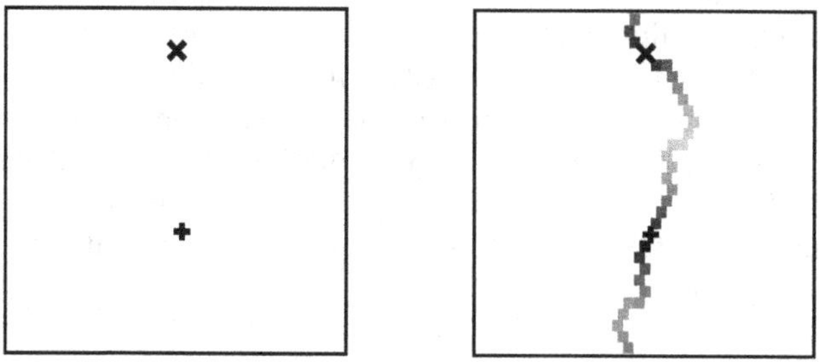

Fig. 4.14. Setup for the switch consisting of two nodes. {x} indicates the node with potential $V_j = -1$; {+} indicates the node with potential $V_j = +1$. (*left*) Unconnected situation (OFF); (*right*) example of a connected situation (ON). Shown is the value $c(\boldsymbol{r},t)$, (4.9). Lattice size: 30×30, 450 agents. Parameters: $L = 15$, $s_0 = 10{,}000$, $k_h = 0.02$, $\beta = 0.3$ [466]

With respect to our model, we can consider different possibilities for such an influence:

1. The external influence changes the behavior of the *agents*. This is possible in different ways:
   - a change in the response parameter α that determines the influence of gradient ∇h^e, as described by (4.5);
   - a change in the noise level $\varepsilon = k_B T$ that determines the "sensitivity" of the agents;
   - a change in the production rate $s_i(\theta_i, t)$, (4.3) that determines the generation of the effective field, either via a change in s_θ^0 or β_θ.
2. The external influence changes the dynamics of the *effective field*, (4.4), generated by the agents. This is possible via the decay rate k_θ. On the other hand, changes in s_θ^0 or β_θ would also influence the magnitude of the chemical field.

In the following, we restrict the discussion to two of these possibilities. First, we may assume that the external influence affects the rate β_θ that describes the exponential decay in the production rate $s_i(\theta_i, t)$, after the agent has hit a node, (4.3). In Sect. 4.3.3, we will also discuss the case (2) that the decay rate of the field itself will be affected by the external influence.

A change in rate β simply means that agents will generate *stronger* or *weaker* links between fixed nodes. In the latter case, the probability increases that the chemical concentration will fall below the threshold, $\hat{h}(a,t) < h_{\text{thr}}$, which means that the nodes are disconnected, (4.33). The threshold value h_{thr}, on the other hand, is a function of the distance between the nodes, (4.42). As pointed out in Sect. 4.2.2, due to the decay of the field (k_h) and the decreasing production rate with time (β), agents will be able to link only nodes that are at a distance closer than a *critical value* $L^\star$. An estimate given in (4.45) indicates that $L^\star$ should be between $\sqrt{4D_n L_{\text{max}}/v}$ and L_{max}, which is given by (4.43). A closer estimate can be obtained by computer simulations.

Figure 4.15 shows the mean connection time t_{con}, which has been calculated for different values of L, using an ensemble of Brownian agents. We found that t_{con} increases exponentially with L if $L < 18$, but overexponentially for larger values of L, so that the connection time finally diverges at the critical distance $L^\star$. From Fig. 4.15, we could find that $L_{\text{max}}/2$ is a reasonable estimate for $L^\star$ in the two-dimensional case. We mention that, as a result of increasing connection time in Fig. 4.15, the probability of establishing a connection, also decreases drastically.

4.3.2 Simulations of the Dynamic Switch

As Figs. 4.14 and 4.15 demonstrate, Brownian agents can connect two nodes, if their distance is well below the critical value $L^\star$, (4.45). In addition to that, for application in a dynamic switch, it is most important that the switch

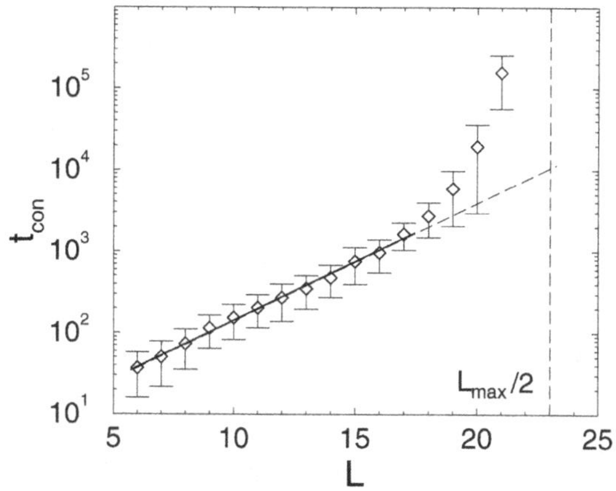

Fig. 4.15. Mean connection time t_{con} (in simulation steps) needed to connect two nodes at a distance L (in lattice sites). The simulated values ($\diamond$) are obtained from an average over 10,000 simulations. For $L < 18$, the values can be fitted by the exponential relation $t_{\text{con}} = 5.19 \exp(0.33\, L)$. The connection times tend to diverge for $L \to L^\star$, which can be estimated as $L^\star = L_{\text{max}}/2$, (4.43). For parameters, see Fig. 4.14 [466, 505]

respond almost immediately to changes in external conditions. Hence, we need to know *how fast* the two nodes are connected or disconnected if external perturbation occurs. Figure 4.16 shows the results of computer simulations of the dynamic switch.

The top part of Fig. 4.16 shows an example of the time-dependent change in external influences, which are assumed to affect decay rate β. Hence, the lower value of β corresponds to ON and the upper value to OFF, respectively. The middle part of Fig. 4.16 shows the corresponding behavior of the switch for three different simulations, whereas the bottom part of Fig. 4.16 shows the behavior of the switch averaged over 400 simulations.

At the beginning of the simulation, the external influence is assumed ON. As the three realizations and the averaged connectivity in Fig. 4.16 show, there is an initial delay in the response of the switch to external change, which results from the fact that the agents have to be "initialized." They first have to discover the nodes via random movement, thus changing their internal state $\theta = 0$ into $\theta = \pm 1$. Therefore, fluctuations in the first connection time are rather large. After that, however, switching behavior is very *synchronized.* The bottom part of Fig. 4.16 shows that the switch is very reliable. For the set of parameters considered, more than 95% of the simulations created a connection between the nodes after the external influence is switched on.

Figure 4.16 further shows that the switch responds very fast but with a certain *delay* to changes in the external influence. This is caused by the fact that it takes some time for agents to increase the chemical concentration above the threshold value after the external influence is ON. On the other

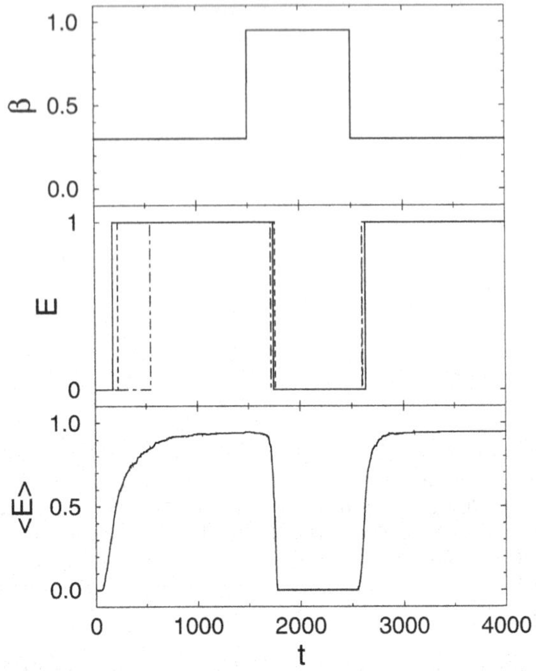

Fig. 4.16. (*top*): Change in decay rate β, (4.3), in time. The lower value of β corresponds to ON in the external influences, the upper value corresponds to OFF. (*middle*): Switch behavior obtained from three different simulations, indicated by (–·–), (– –), and (—). (*bottom*): Switch behavior averaged over 400 simulations. For parameters, see Fig. 4.14 [466]

hand, it also takes some time for the chemical field to decay at rate k_h below the threshold after the external influence is OFF. These delays, of course, depend on the parameters chosen, in particular, on k_h, β, and the threshold value $h_{\rm thr}$, which have to fulfill certain relations to have a reliable switch. This requires proper adjustment of the threshold value and also of the gradient of the field.

One of the basic features of our network model is the idea that the agent is guided by the *gradient* of a field, which it has previously generated. This, however, implies that the gradient of this field still points in the *right direction* when the agent is on its way back to the previously visited node. Obviously, this holds only for a certain range of parameters, k_h, β, which determine the actual value of the field. Using the initial condition $h_\theta(0) = 0$, (4.4), for the field, $h_\theta(\boldsymbol{r}, t)$ can be formally integrated:

$$h_\theta(\boldsymbol{r}, t) = \int\limits_0^t dt' \, \exp\left[-k_\theta \, (t - t')\right] \sum_i s_i(\theta_i, t') \, \delta_{\theta,\theta_i} \, \delta\left[\boldsymbol{r} - r_i(t')\right] \, . \tag{4.48}$$

For our further estimations, we treat the connection between the two nodes as a *one-dimensional* structure, as in Sect. 4.2.2. Hence, x is now the space coordinate, and L the linear distance between the two nodes k and l. The node at $x = 0$ should have a positive potential $V = +1$, and the node at $x = L$ has a negative potential $V = -1$, (4.35). Further, we consider *only one* agent with the initial conditions:

$$x(t=0) = 0 \ , \quad v = |\boldsymbol{v}| = |x|/t = \text{const} \ , \quad \theta(t=0) = +1 \ , \tag{4.49}$$

which should move only between the two nodes, $0 \le x \le L$. With these assumptions and the simplifications (4.8), (4.48) can be rewritten:

$$\begin{aligned} h_{+1}(x,t) &= \int\limits_0^t dt' \, \exp\left[-k_h\,(t-t')\right] \, s_0 \, \exp(-\beta\, t') \, \delta(x - vt') \\ &= \frac{s_0}{v} \, \exp(-k_h\, t) \int\limits_0^t dt' \, \exp\left[(k_h - \beta)\, t'\right] \, \delta\left(t' - \frac{x}{v}\right) \, . \end{aligned} \tag{4.50}$$

The solution of (4.50) yields then,

$$h_{+1}(x,t) = \frac{s_0}{v} \, \exp(-k_h t) \, \exp\left[(k_h - \beta)\frac{x}{v}\right] \ , \quad \text{if} \ \ \frac{x}{v} \le t \le t_{\max} \, . \tag{4.51}$$

Here, we have considered that an agent contributes to the field h_{+1} only as long as it has not reached the node at $x = L$. If we assume that $L = L_{\max}$, then the maximum time is given by $t = t_{\max}$, (4.19). In particular, for $t = t_{\max}$, the spatial dependence of the field can be explicitly written, using (4.19):

$$h_{+1}(x, t_{\max}) = \frac{s_0}{v} \left(\frac{s_{\min}}{s_0}\right)^{\frac{k_h}{\beta}} \exp\left[(k_h - \beta)\, \frac{x}{v}\right] \, . \tag{4.52}$$

We already see from (4.51), (4.52) that the gradient of $h_{+1}(x,t)$ points back to the node at $x = 0$ only if $k_h < \beta$. This, however, holds for times $t \le t_{\max}$. Much more important are the times $t \ge t_{\max}$, that is, after the agent has reached the node at $x = L$, and now tries to find its way back to $x = 0$, guided by the gradient of $h_{+1}(x,t)$.

For $t \ge t_{\max}$, the time evolution of the field $h_{+1}(x,t)$ is simply given by

$$\begin{aligned} \frac{\partial h_{+1}(x,t)}{\partial t} &= -k_h \, h_{+1}(x,t) \ , \quad \text{if}\, t_{\max} \le t \le 2t_{\max} \, , \\ h_{+1}(x,t) &= h(x, t_{\max}) \, \exp\left[-k_h\,(t - t_{\max})\right] \ , \end{aligned} \tag{4.53}$$

because the agent on its way back does not contribute to the field h_{+1}. If we express the time by $t = 2t_{\max} - x(t)/v$, where $x(t)$ is the actual position of the agent, the exponential function in (4.53) can be rewritten:

$$\begin{aligned} \frac{x(t)}{v} &= 2\,t_{\max} - t\,, \\ \exp\left[-k_h\,(t-t_{\max})\right] &= \exp\left(-k_h\,t_{\max}\right)\,\exp\left[k_h\,\frac{x(t)}{v}\right]\,, \\ &= \left(\frac{s_{\min}}{s_0}\right)^{\frac{k_h}{\beta}}\,\exp\left[k_h\,\frac{x(t)}{v}\right]\,. \end{aligned} \tag{4.54}$$

Inserting (4.52), (4.54) in (4.53), we find for the field h_{+1} at the actual position of the agent, $x(t)$,

$$h_{+1}[x(t)] = \frac{s_0}{v}\left(\frac{s_{\min}}{s_0}\right)^{\frac{2k_h}{\beta}}\,\exp\left[(2k_h-\beta)\,\frac{x}{v}\right]\,, \tag{4.55}$$

and for the gradient that guides the agent for times $t_{\max} \leq t \leq 2t_{\max}$,

$$\frac{\partial h_{+1}[x(t)]}{\partial x} = h_{+1}[x(t)]\,\frac{2k_h-\beta}{v}\,, \quad t_{\max} \leq t \leq 2t_{\max}\,. \tag{4.56}$$

Hence, we find as the result that the gradient of $h_{+1}(x)$ points back to $x=0$ for $t_{\max} \leq t \leq 2t_{\max}$, only if the following relation between the decay rate k_h of the field and the decay rate β of the production function holds:

$$\beta \geq 2\,k_h\,. \tag{4.57}$$

Although the above considerations are carried out for the movement of only one agent, (4.57) gives us an important restriction for the choice of the parameters to allow connections between nodes.

4.3.3 Estimation of the Switch Delay

In the previous section, we assumed that the external influence affects the decay rate β that describes the declining production of the chemical in time. Now we want to discuss the second case mentioned in Sect. 4.3.1, that external influences affect only the field $\hat{h}(\boldsymbol{r},t)$ generated by agents, in terms of the *decay rate* k_h. If we think, e.g., of light or radiation as external perturbations, then it is known that they can certainly change the decay rate of a chemical field. On the other hand, the behavior of agents now is not at all affected by the external influence.

To estimate the response of the field to the switch under the external influence, we (i) restrict the discussion to the one-dimensional case discussed in the previous section and (ii) apply the approximation used in Sect. 4.2.2 to determine the threshold value $h_{\rm thr}$. This means that the (unknown) number of agents along the link between two nodes is approximated by the mean density $\bar{n}$, which is a lower limit. Then, the change of the field component $h_{+1}(x,t)$ is given by (4.38), which yields the time-dependent solution $h_{+1}(x,t)$, (4.39), and the stationary solution $h_{+1}(x)$, (4.40). Both $h_{+1}(x,t)$ and $h_{+1}(x)$ are equal only in the asymptotic limit, $t \to \infty$. For practical purposes, it is more interesting to find out when the difference is smaller than a certain value ε:

$$h_{+1}(x) - h_{+1}(x,t) \leq \varepsilon \,. \tag{4.58}$$

With (4.40) and (4.39), this leads to an estimate for the time needed to reach a "nearly" stationary solution of the field:

$$t_{\mathrm{stat}} \geq \frac{1}{k_h} \ln \left[\frac{N}{2A} \frac{s_0}{k_h\, \varepsilon} \exp\left(-\frac{\beta}{v}\, x \right) \right] . \tag{4.59}$$

To find the maximum estimate for t_{stat}, we choose $x = 0$, which gives the highest value, and assume that ε should be equal to the smallest increase in the field, s_{min}, (4.18), which can be produced by $\bar{n}/2$ agents:

$$\varepsilon = \frac{N}{2A}\, s_{\mathrm{min}} \,. \tag{4.60}$$

Then, for the time needed to reach the stationary value of the field, we find

$$t_{\mathrm{stat}} \geq \frac{1}{k_h} \ln \left(\frac{1}{k_h} \frac{s_0}{s_{\mathrm{min}}} \right) . \tag{4.61}$$

We note that t_{stat} does not consider the time needed for agents to find a node, but only the time that a previously established connection needs to become a *stable* connection.

In Sect. 4.2.2, we already showed that the *minimum value* for the *total field*, $\hat{h}(x) = h_{+1}(x) + h_{-1}(L - x)$, in the stationary limit, is given by (4.42), which yields for $L/2$,

$$\hat{h}\left(\frac{L}{2}\right) = \frac{\bar{n}\, s_0}{k_h} \exp\left(-\frac{\beta}{v} \frac{L}{2} \right) . \tag{4.62}$$

Here, the minimum value depends on the local agent number $\bar{n}$, the parameters s_0, k_h, and β, and the distance between the two nodes L, which is a fixed value for the switch. From the computer simulations, Fig. 4.15, we know that $L < L^\star$, (4.45). The smallest possible value $\hat{h}(L/2)$, (4.62), is obtained for $L = L^\star$, which has been used to define the threshold h_{thr}, (4.42).

In the case considered, we have assumed that the external influence changes the value of the decay rate k_h of the field; therefore, now we have to consider two different values, k_{OFF} and k_{ON}. If the switch is ON, the *minimum value* $\hat{h}(L/2)$ has to be *above* the threshold h_{thr}, whereas for the state OFF, at least the minimum value of $\hat{h}(L/2)$ has to be *below* the threshold to disconnect the nodes. Equation (4.62) defines the relation between the minima in both states:

$$\hat{h}(L/2)_{\mathrm{OFF}} = \frac{k_{\mathrm{ON}}}{k_{\mathrm{OFF}}}\, \hat{h}(L/2)_{\mathrm{ON}} \,, \tag{4.63}$$

whereas the threshold value, (4.42), (4.46), is given by

$$h_{\mathrm{thr}} = \hat{h}\left(\frac{L^\star}{2}\right) = \frac{\bar{n}\, s_0}{k_{\mathrm{ON}}} \exp\left(-\frac{\beta}{v}\frac{L^\star}{2}\right) = \frac{\bar{n}\, s_0}{k_{\mathrm{ON}}} \left(\frac{s_{\mathrm{min}}}{s_0}\right)^{1/4} . \tag{4.64}$$

Because k_{ON} also determines the threshold value, we need only to adjust k_{OFF} properly to ensure that the value of $\hat{h}(L/2)$ in the OFF state is below the threshold.

If the external influence is switched between ON and OFF, the response of the switch occurs only with a certain *delay* because it takes some time before the field $\hat{h}(L/2,t)$ reaches the threshold value. This delay time can be estimated using the time-dependent equation for the total field $\hat{h}(x,t)$ at position $L/2$. From

$$\frac{\partial \hat{h}(L/2,t)}{\partial t} = -k_h\, \hat{h}(L/2,t) + \bar{n}\, s_0 \exp\left(-\frac{\beta}{v}\frac{L}{2}\right) , \tag{4.65}$$

we find the following general solutions for the two possibilities:

$$\begin{aligned}
&\mathrm{ON} \to \mathrm{OFF} : \\
&\hat{h}(L/2, t > t_0) = h^0_{\mathrm{ON}} \exp\left[-k_{\mathrm{OFF}}\,(t-t_0)\right] \\
&\qquad + \frac{s_0\,\bar{n}}{k_{\mathrm{OFF}}} \exp\left(-\frac{\beta}{v}\frac{L}{2}\right) \{1 - \exp\left[-k_{\mathrm{OFF}}\,(t-t_0)\right]\} \,; \\
&\mathrm{OFF} \to \mathrm{ON} : \\
&\hat{h}(L/2, t > t_0) = h^0_{\mathrm{OFF}} \exp\left[-k_{\mathrm{ON}}\,(t-t_0)\right] \\
&\qquad + \frac{s_0\,\bar{n}}{k_{\mathrm{ON}}} \exp\left(-\frac{\beta}{v}\frac{L}{2}\right) \{1 - \exp\left[-k_{\mathrm{ON}}\,(t-t_0)\right]\} \,.
\end{aligned} \tag{4.66}$$

Here, t_0 is the time when the external influence switches, and h^0 is the value of $\hat{h}(L/2, t = t_0)$, which should be either above (ON) or below (OFF) the threshold to allow the respective switch.

Figure 4.17 shows the solutions of (4.66). It is obvious that after the initial time delay needed to establish the field $\hat{h}(a,t)$, the response of the switch to external changes is rather fast, expressed in a steep increase/decay of the field.

The *longest possible delay* of the switch to respond to external changes can be expected, if $h^0(t)$ has already reached its stationary value, which is given by (4.62). Using this approximation for h^0, we find from the condition,

$$\hat{h}(L/2, t > t_0) = h_{\mathrm{thr}} \,, \tag{4.67}$$

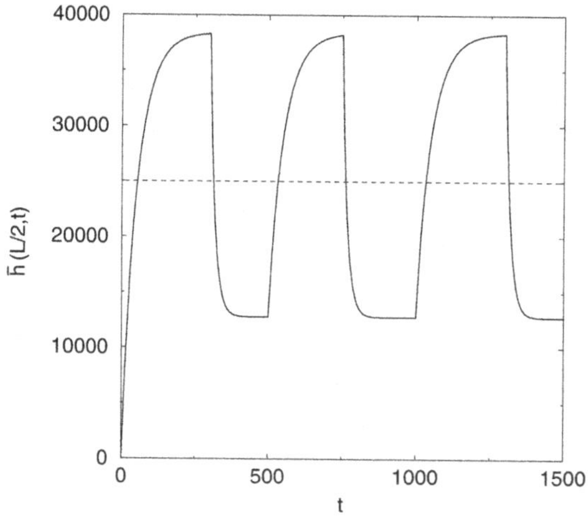

Fig. 4.17. Change in the minimum value $\hat{h}(L/2,t)$, (4.66), in time. The *dashed* line marks the threshold value h_{thr}, (4.46). $\hat{h}(L/2,t) > h_{\text{thr}}$ corresponds to ON; $\hat{h}(L/2,t) < h_{\text{thr}}$ corresponds to OFF. Parameters: $\bar{n} = 0.5$, $s_0 = 10,000$, $s_{\min} = 1$, $k_{\text{ON}} = 0.02$, $k_{\text{OFF}} = 0.06$, $\beta = 0.25$, $L = 15$

the time needed to reach the threshold value after the external influences have changed at t_0:

$$\begin{aligned}
&\text{ON} \to \text{OFF}\ : \\
&t_{\text{OFF}} - t_0 = \frac{1}{k_{\text{OFF}}} \ln \left[\frac{1 - \dfrac{k_{\text{ON}}}{k_{\text{OFF}}}}{\exp\left(\dfrac{\beta}{v}\dfrac{L}{2}\right)\left(\dfrac{s_{\min}}{s_0}\right)^{1/4} - \dfrac{k_{\text{ON}}}{k_{\text{OFF}}}} \right] ; \\
&\text{OFF} \to \text{ON}\ : \\
&t_{\text{ON}} - t_0 = \frac{1}{k_{\text{ON}}} \ln \left[\frac{1 - \dfrac{k_{\text{ON}}}{k_{\text{OFF}}}}{1 - \exp\left(\dfrac{\beta}{v}\dfrac{L}{2}\right)\left(\dfrac{s_{\min}}{s_0}\right)^{1/4}} \right] .
\end{aligned} \tag{4.68}$$

The values $t_{\text{ON}} - t_0$ and $t_{\text{OFF}} - t_0$ give the time delay for the switch to respond to the external change. For the set of parameters given in Fig. 4.17, we find that $t_{\text{ON}} - t_0 = 32$ time steps and $t_{\text{OFF}} - t_0 = 12$ time steps. Consequently, it cannot be guaranteed that perturbations that occur for a shorter time than given by $t_{\text{ON}} - t_0$ and $t_{\text{OFF}} - t_0$ are indicated by the switch suggested here. On the other hand, if the perturbations exist longer than the given switch time, we could expect a very reliable indicator function for the switch.

5. Tracks and Trail Formation in Biological Systems

5.1 Active Walker Models

5.1.1 Master Equation Approach to Active Walkers

In this chapter, the agent model of network formation discussed in the previous chapter is extended to applications in biology to describe the formation of tracks and aggregates in myxobacteria and trail formation in ants. We will again exploit the basic features of the Brownian agent model, (i) discrete internal degrees of freedom θ_i to cope with different behavioral responses and (ii) the generation of a self-consistent effective field $h^e(\boldsymbol{r}, t)$ by agents. However, different from other chapters, we will describe the motion of Brownian agents continuous in time and space in this chapter by a *discrete approximation.* In this approximation, the simple Brownian particle (performing an overdamped motion) would become a *random walker* and the Brownian *agent* with its different *activities* would then become an *active walker* – a term that from now on shall be used equivalently.

The discretization of space and time variables has been already used for *numerical* treatment (see Sect. 3.1.4); it will be reflected now also in the equations of motion of agents. If we consider a *discrete lattice* instead of a continuous space, the moving agent performs a *walk* on the lattice. It moves one step of constant length l per time unit Δt, i.e., the speed, $|v| = l/\Delta t$, remains constant. Different from the continuous description, the walker *cannot* choose *every* direction, but only a discrete number of directions in accordance with the lattice geometry. Figure 5.1 shows an example of a two-dimensional triangular lattice with periodic boundary, which means that each lattice site has six nearest neighbor sites at equal distance l. The set of nearest neighbor sites $\boldsymbol{r}'$ with respect to position $\boldsymbol{r}$ shall be denoted by $\mathcal{L}$. If we assume that the random walker moves from $\boldsymbol{r}$ to $\boldsymbol{r}'$ during time step Δt, the possible positions $\boldsymbol{r}'$ are specified for the triangular lattice as follows:

$$\boldsymbol{r}' = \boldsymbol{r} + \boldsymbol{\Delta r} \; , \quad \boldsymbol{\Delta r} = l\left(\cos\tau\, \boldsymbol{e_x} + \sin\tau\, \boldsymbol{e_y}\right) \; , \quad \tau = \frac{n\,\pi}{3} \; , \quad n = \{1, 2, \dots, 6\} \; . \tag{5.1}$$

Considering for the moment rather small time steps, $\Delta t \to 0$ and *only one* walker (i.e., we omit index i), the probability density $p(\boldsymbol{r}, \theta, t)$ of finding

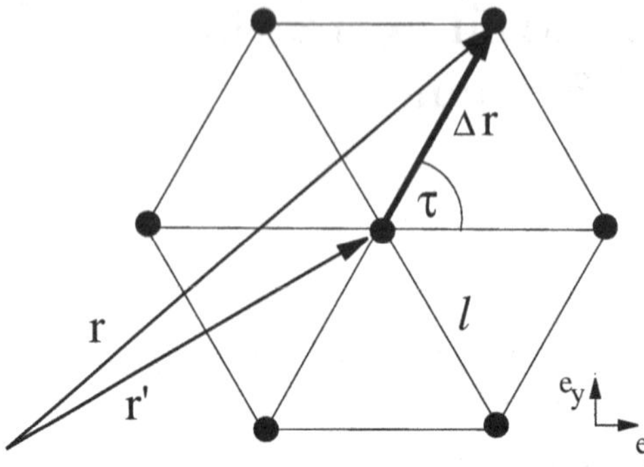

Fig. 5.1. Definition of the positions $\boldsymbol{r}'$ on a triangular lattice

a walker with internal parameter θ at position $\boldsymbol{r}$ at time t may change due to the (continuous in time) master equation, (1.42) (see Sect. 1.3.3):

$$\begin{aligned}\frac{\partial}{\partial t}p(\boldsymbol{r},\theta,t) = &\sum_{\boldsymbol{r}'\in\mathcal{L}} w(\boldsymbol{r}|\boldsymbol{r}')\,p(\boldsymbol{r}',\theta,t) - w(\boldsymbol{r}'|\boldsymbol{r})\,p(\boldsymbol{r},\theta,t)\\ &+\sum_{\theta'\neq\theta} w(\theta|\theta')\,p(\boldsymbol{r},\theta',t) - w(\theta'|\theta)\,p(\boldsymbol{r},\theta,t)\;. \end{aligned} \tag{5.2}$$

In the simplest case of a *nonbiased* random walker, the transition rates to change position $\boldsymbol{r}$ are given by

$$w(\boldsymbol{r}'|\boldsymbol{r}) = \frac{\omega}{d}\;,\quad w(\boldsymbol{r}|\boldsymbol{r}) = 0\;. \tag{5.3}$$

In general, ω [1/s] is the jump frequency per unit time, which is set to one here. d is the number of possible directions, which is six on the triangular lattice. Hence, normalization of the transition rates reads $\sum_d w(\boldsymbol{r}'|\boldsymbol{r}) = \omega = 1$, which means that *one* step in one of the possible directions is realized with certainty during the time step Δt.

Let us now consider, as before, that a walker can generate an *effective field* $h^e(\boldsymbol{r},t)$. Hence, it is an *active walker* in the same sense as *Brownian agents*, and the above considerations also apply if a discrete space is used. The effective field should again influence further movement of the walker. If it responds to the *gradient* of an effective field, the transition rates, (5.3), change to

$$w(\boldsymbol{r}+\boldsymbol{\Delta r}|\boldsymbol{r}) = \frac{\omega}{d}\left[1+\beta\,\frac{h^e(\boldsymbol{r}+\boldsymbol{\Delta r},t)-h^e(\boldsymbol{r}-\boldsymbol{\Delta r},t)}{|(\boldsymbol{r}+\boldsymbol{\Delta r})-(\boldsymbol{r}-\boldsymbol{\Delta r})|}\right]\;, \tag{5.4}$$

where β acts as a dimensional constant. Here, we have used the discrete version of the derivative:

$$\frac{\partial h^e(\boldsymbol{r},t)}{\partial r} = \frac{h^e(\boldsymbol{r}+\boldsymbol{\Delta r},t)-h^e(\boldsymbol{r}-\boldsymbol{\Delta r},t)}{|2\,\boldsymbol{\Delta r}|}\;. \tag{5.5}$$

Linearization implies that the gradient in opposite directions is assumed to be the same with only the sign changed. In the absence of the field $h^e(\boldsymbol{r})$ or for a vanishing gradient, $h^e(\boldsymbol{r}+\boldsymbol{\Delta r},t) = h^e(\boldsymbol{r}-\boldsymbol{\Delta r},t)$, the walker changes its position like the random walker with equal probabilities for the possible directions. However, in the presence of a positive gradient of the field, $\boldsymbol{\nabla} h^e(\boldsymbol{r},t)$, the probability of going in the direction of the gradient increases.

If we assume that the jump distance l per step is a sufficiently small value, $l \to 0$, then the differences in h^e can be replaced by the first derivative, $\partial h^e/\partial \boldsymbol{r}$. Further, in the limit of small jump sizes, the discrete version in space of the master equation, (5.2), can be transformed into a second-order partial differential equation in continuous space [169], which reads in a general form (see Sect. 1.3.3),

$$
\begin{aligned}
\frac{\partial}{\partial t} p(\boldsymbol{r},\theta,t) = & -\frac{\partial}{\partial \boldsymbol{r}}\, a_1(\boldsymbol{r})\, p(\boldsymbol{r},\theta,t) + \frac{1}{2}\frac{\partial^2}{\partial \boldsymbol{r}^2}\, a_2(\boldsymbol{r})\, p(\boldsymbol{r},\theta,t) \\
& + \sum_{\theta' \neq \theta} w(\theta|\theta')\, p(\boldsymbol{r},\theta',t) - w(\theta'|\theta)\, p(\boldsymbol{r},\theta,t)\,. \qquad (5.6)
\end{aligned}
$$

Here, the coefficients $a_n(\boldsymbol{r})$ are defined as follows:

$$
\begin{aligned}
a_1(\boldsymbol{r}) &= \sum_{\boldsymbol{r}' \in \mathcal{L}} (\boldsymbol{r}'-\boldsymbol{r})\, w(\boldsymbol{r}'|\boldsymbol{r})\,, \\
a_2(\boldsymbol{r}) &= \sum_{\boldsymbol{r}' \in \mathcal{L}} (\boldsymbol{r}'-\boldsymbol{r})^2\, w(\boldsymbol{r}'|\boldsymbol{r})\,. \qquad (5.7)
\end{aligned}
$$

With the transition rates, (5.4), these coefficients yield

$$
a_1(\boldsymbol{r}) = \omega\, l\, \beta\, \frac{\partial h^e(\boldsymbol{r},t)}{\partial \boldsymbol{r}}\,, \quad a_2(\boldsymbol{r}) = \omega\, l^2\,. \qquad (5.8)
$$

If we use the abbrevations for the constants,

$$
\omega\, l\, \beta = \frac{\alpha}{\gamma_0}\,, \quad \frac{1}{2}\omega\, l^2 = D\,, \qquad (5.9)
$$

then (5.6) can be explicitly rewritten in the form,

$$
\begin{aligned}
\frac{\partial}{\partial t} p(\boldsymbol{r},\theta,t) = & -\frac{\partial}{\partial \boldsymbol{r}}\, \frac{\alpha}{\gamma_0}\, \frac{\partial h^e(\boldsymbol{r},t)}{\partial \boldsymbol{r}}\, p(\boldsymbol{r},\theta,t) + D\, \frac{\partial^2}{\partial \boldsymbol{r}^2}\, p(\boldsymbol{r},\theta,t) \\
& + \sum_{\theta' \neq \theta} w(\theta|\theta')\, p(\boldsymbol{r},\theta',t) - w(\theta'|\theta)\, p(\boldsymbol{r},\theta,t)\,. \qquad (5.10)
\end{aligned}
$$

We can generalize the result of (5.10) for N walkers. Using again the probability density $P(\underline{r},\underline{\theta},t) = P(\boldsymbol{r}_1,\theta_1,\ldots,\boldsymbol{r}_N,\theta_N,t)$, (3.5), we then find the master equation for $P(\underline{r},\underline{\theta},t)$ in the form of (3.7).

Thus, we conclude our different levels of description introduced so far. On the level of an *individual* agent i, we will either use the Langevin equations, (3.1), (3.2), which are continuous in time and space, or the discrete description in terms of active walkers. In the limit of an overdamped motion of agents and with the assumption of small jump sizes and small time steps, in first-order approximation both descriptions result in the same master equation, (3.7), for the *ensemble* of agents or walkers, respectively. Consequently, on the level of an ensemble, we find the same density equations in the mean-field limit, where the probability density for agents is replaced by the macroscopic density. In this chapter, we will prefer the discrete picture in terms of active walkers. The discrete model will be extended in Sect. 5.2 by considering also a *bias* of the walk and a spatially restricted response to the gradient of the effective field. But first, we will give some examples of active walker models applied to different problems in the next section.

5.1.2 Active Walker Models of Fractal Growth Patterns

The model of Brownian agents or active walkers outlined above tries to substantiate the dynamics by considering Langevin equations for agents, and reaction–diffusion equations for the dynamics of a self-consistent effective field $h^e(\boldsymbol{r}, t)$. Of course, there are many different possibilities for defining the equations of motion for agents and for the dynamics of the field, as shown in the following for the simulation of fractal patterns, which occur, for instance, during dielectrical breakdown, in diffusion limited aggregation, in retinal growth of neurons, and during bacterial aggregation.

For instance, it is possible to replace a rather complex self-consistent field, which may consist of different components, (3.4), by a simpler time-dependent potential, $U(\boldsymbol{r}, t)$, which consists of two parts:

$$U(\boldsymbol{r}, t) = U_b(\boldsymbol{r}, t) + U_0(\boldsymbol{r}, t) \, .$$

Here, $U_b(\boldsymbol{r}, t)$ is a *background* potential that can be changed, e.g., by external dynamics, and $U_0(\boldsymbol{r}, t)$ is the part of the potential that can be changed by active walkers:

$$U_0(\boldsymbol{r}, t+1) = U_0(\boldsymbol{r}, t) + \sum_{i=1}^{N} W_i \left[\boldsymbol{r} - \boldsymbol{r}_i(t - t_0)\right] \, . \tag{5.11}$$

$\boldsymbol{r}_i$ is the position of walker $i = 1, \dots, N$ on a discrete lattice, W_i represents the change in potential per time step, which can occur also with a certain *time lag* t_0. Two different examples for W_i, used by Lam et al. in a different context [159, 269, 288, 291], are shown in Fig. 5.2.

The essential ingredient of the active walker model is again nonlinear feedback between walkers and the potential $U(\boldsymbol{r}, t)$, which influences the further motion of the walkers. In the case of a Boltzmann *active walker*, the

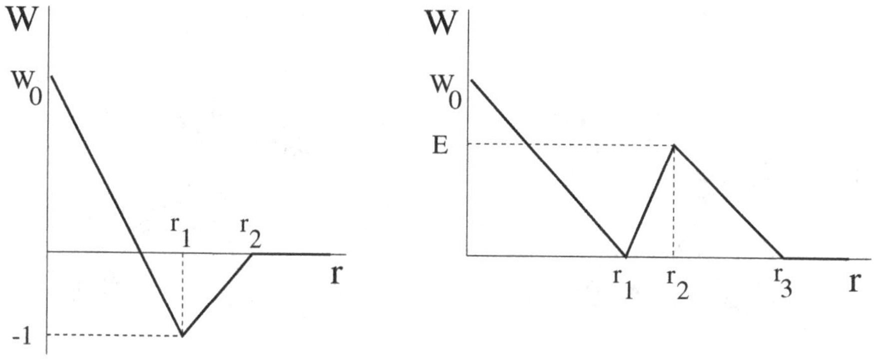

Fig. 5.2. Two types of functions $W_i(|\boldsymbol{r}|)$ (5.11) [291]

probability for the walker i to move from site $\boldsymbol{r}_i$ to one of the neighboring sites $\boldsymbol{r}_i + \boldsymbol{\Delta r}$, is assumed as [389]

$$p(\boldsymbol{r}_i \to \boldsymbol{r}_i + \boldsymbol{\Delta r}, t) \propto \exp\left[\frac{U(\boldsymbol{r}_i, t) - U(\boldsymbol{r}_i + \boldsymbol{\Delta r}, t)}{T}\right] . \tag{5.12}$$

Consequently, walkers move with a larger probability toward local *minima* of the potential. The parameter T, which may be interpreted as a temperature, can be chosen to scale local differences in the potential.

Another case, named a *probabilistic active walk* [291], assumes for the probability,

$$p(\boldsymbol{r}_i \to \boldsymbol{r}_i + \boldsymbol{\Delta r}, t) \propto \begin{cases} \Big[U(\boldsymbol{r}_i, t) - U(\boldsymbol{r}_i + \boldsymbol{\Delta r}, t)\Big]^{\eta} & \text{if } U(\boldsymbol{r}_i, t) > U(\boldsymbol{r}_i + \boldsymbol{\Delta r}, t) \\ 0 & \text{otherwise} . \end{cases} \tag{5.13}$$

The parameter η may be used to adjust the compactness of fractal clusters (see also Sect. 8.3.1), as discussed, e.g., in [289, 291].

So far, the model described simulates simple track patterns. The situation becomes more interesting if *branching* of tracks is introduced. Suppose that a walker is located at $\boldsymbol{r}_i$ and will move to $\boldsymbol{r}_j$. The remaining (not chosen) nearest neighbor sites are denoted by $\boldsymbol{r}_k$. If the following inequality is satisfied,

$$U(\boldsymbol{r}_i, t) - U(\boldsymbol{r}_k, t) > \kappa\Big[U(\boldsymbol{r}_i, t) - U(\boldsymbol{r}_j, t)\Big] , \tag{5.14}$$

the site $\boldsymbol{r}_k$ will be occupied by a *new* active walker, and *branching* occurs. Parameter κ is the branching factor.

Figure 5.3 shows computer simulations of a *dielectric breakdown pattern* in a liquid. These characteristic discharge patterns are known from experiments, e.g., in nematic liquid crystals [269] in the presence of an electrical field between the center and the boundary of a Hele–Shaw cell. Provided with

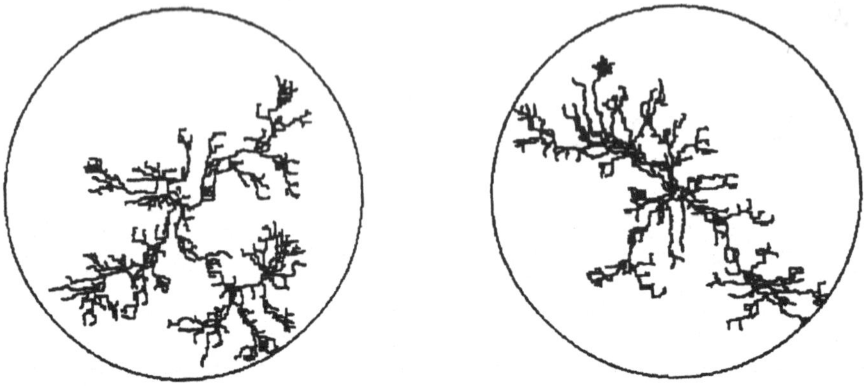

Fig. 5.3. Two different realizations of a dielectric breakdown pattern simulated with an active walker model [269]. For the initial setup, see the text. Parameters: $W_0 = 5$, $E = 5$, $r_1 = r_2 = 10$, $r_3 = 15$, $L = 252$, $\kappa = 0.8$, $t_0 = 0$, $\eta = 1$

a supercritical field, the creation of burned filaments between the center and the boundary of the cell can be observed during discharge. The computer simulations in Fig. 5.3 assume radial symmetry, with $\boldsymbol{r}_0 = (0, 0)$ as the center of the Hele–Shaw cell and $|\boldsymbol{r}_{\text{max}}| = L/2$ as the boundary. Here, $L \times L$ denotes the size of a square lattice. Further, a background potential is assumed that decreases linearly between $U_b(\boldsymbol{r}_0) = 600$ and $U_b(L/2) = 0$. The function $W_i(\boldsymbol{r} - \boldsymbol{r}_i)$ is chosen in accordance with Fig. 5.2 (*right*); (5.13) is used for the movement of walkers. Initially, four walkers are placed outside the center of the lattice at positions, $(0, \pm 2)$, $(\pm 2, 0)$, branching, i.e., the creation of new walkers in the neighborhood is allowed due to (5.14). The fractal dimensions of the typical dendritic patterns shown in Fig. 5.3, are in good agreement with experimental results [269].

A related fractal growth phenomenon is known from *electrochemical deposition* [202, 289, 331], where, for instance, Zn deposits form growing "trees" in a $ZnSO_4$ solution in the presence of an external electrical field. This has been simulated by Lam et al. with an active walker model, using the function $W_i(\boldsymbol{r} - \boldsymbol{r}_i)$ in accordance with Fig. 5.2 (*left*), and the same initial positions for the four active walkers, as in Fig. 5.3. The result of the simulation, which can be compared with experimental observation (see Fig. 5.4), shows a rather dense radial morphology of the branching trees.

5.1.3 Active Walker Models of Bacterial Growth

Patterns that are very similar to those shown in the previous section have also been reported in the growth of *bacterial colonies*, e.g., in *Bacillus subtilis*, strain 168 [50]. The colonies adopt various shapes of fractal growth patterns ranging from rather compact to very ramified, if conditions are varied. The structure formation can be modeled with a more refined ac-

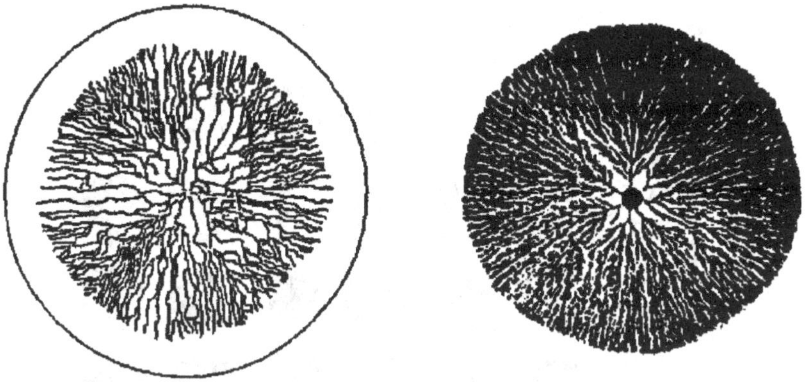

Fig. 5.4. Dense radial morphology obtained from an active walker model (*left*) and in experiments on Zn deposits (*right*). Parameters for the computer simulations: $W_0 = 5$, $r_1 = 12$. $r_2 = 15$, $L = 100$, $\kappa = 0.99999$, $t_0 = 0$, $\eta = 1$; for initial conditions, see Fig. 5.3 [289, 291]

tive walker model, suggested by Ben–Jacob et al., which is denoted as the *communicating walkers model* [47, 49–52]. Here, each walker represents about 10^4 bacteria, i.e., within a coarse graining of the colony; it should be viewed as a rather *mesoscopic* unit, not as an individual bacterium. The walkers perform an off-lattice random walk within a well-defined envelope (defined on a triangular lattice). This envelope grows outwards from the center of the lattice by moving each segment of the envelope after it has been hit N_c times by walkers. This requirement represents local communication or cooperation in the behavior of the bacteria. In a first approximation, the level of N_c represents the *agar concentration*, as more collisions are needed to push the envelope on a harder substrate. Another varying condition is given by the *initial nutrient concentration* $P = c(\boldsymbol{r}, 0)$. In the course of time, the nutrient concentration $c(\boldsymbol{r}, t)$ changes by *diffusion* with constant D_c and *consumption* of food by the walkers, which can be described by the following reaction–diffusion equation:

$$\frac{\partial c(\boldsymbol{r},t)}{\partial t} = D_c\, \Delta c(\boldsymbol{r},t) - \sum_i \delta(\boldsymbol{r}-\boldsymbol{r}_i)\ \min\left[c_r, c(\boldsymbol{r},t)\right] . \qquad (5.15)$$

Here, the consumption rate, $\min\left[c_r, c(\boldsymbol{r},t)\right]$, means that each walker consumes nutrients at a fixed rate c_r, if sufficient food is available, and otherwise consumes the available amount.

The nutrient is used to increase the *internal energy* w_i of a walker, which decreases at a fixed rate e again. Hence, the internal energy of walker i follows the dynamics,

$$\frac{dw_i}{dt} = \min\left[c_r, c(\boldsymbol{r},t)\right] \ - \ e\ . \qquad (5.16)$$

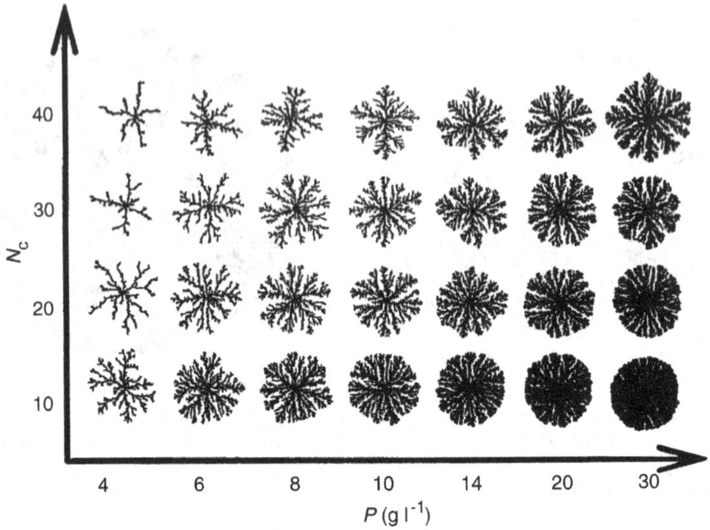

Fig. 5.5. Results of numerical simulations of the communicating walker model, where P denotes the initial value of the nutrient concentration and N_c corresponds to the agar concentration. Typical lattice size 600×600; typical number of walkers 10^5 [50]

Depending on the level of internal energy, a walker can switch its *activity* as follows: for $w_i = 0$, the walker becomes *stationary*, i.e., it is not moving, and for $w_i \geq w_{\text{thr}}$, where w_{thr} is some (upper) threshold value, the walker divides into two *(reproduction)*.

Figure 5.5 shows the variety of fractal patterns resulting from the simulation model, depending on the initial level of nutrient P and the agar concentration N_c. As in the growth of bacterial colonies, the patterns are compact at high P and become fractal with decreasing food level. For a given P, the patterns are more ramified as the agar concentration increases.

Different from the above simulations, bacteria can also develop dense organized patterns at *very low* levels of P. A possible explanation of this phenomenon counts on *chemotactic interaction* between bacteria, i.e., additional chemical communication in low nutrient concentrations, which is also known from *Dictyostelium* amoeba. To be specific, it is assumed that stationary walkers, i.e., walkers that have been exposed to a low level of food, produce a chemical at a fixed rate s_r, whereas active, nonstationary walkers consume the chemical at a fixed rate c_c, in the same way as the nutrient. Further, the chemical is allowed to diffuse, which leads to the following reaction–diffusion equation for the concentration of the chemical, $h(\boldsymbol{r}, t)$:

$$\frac{\partial h(\boldsymbol{r}, t)}{\partial t} = D_h \, \Delta h(\boldsymbol{r}, t) - \sum_i \delta(\boldsymbol{r} - \boldsymbol{r}_i) \; \min\left[c_c, h(\boldsymbol{r}, t)\right] + \sum_j \delta(\boldsymbol{r} - \boldsymbol{r}_j) \; s_r \; . \tag{5.17}$$

Here, the summation over i refers to *active* walkers, whereas the summation over j refers to *stationary* walkers. Finally, it is assumed that active walkers change their movements according to the chemical signal received, i.e., instead of a pure random walk, they move with a higher probability in the direction of the chemical.

As the result of such additional communication, the ramified aggregation patterns obtained in the simulations at a low level of P change into a dense structure with thin branches and a well-defined circular envelope. The results of computer simulations with and without chemotactic interactions carried out by Ben–Jacob et al. can be compared in Fig. 5.6a,b, where part (b) agrees with experimental observations of bacterial growth at a low level of nutrient.

The active walker models discussed above and in Sect. 5.1.2 are only some specific realizations out of the many possible. In general, active walker models are very flexible in different respects: (i) the way a "potential" or a "field" can be changed by walkers; (ii) the eigendynamics of the potential or the field; (iii) the rules for the motion of walkers dependent on the potential or the field; and (iv) possible transitions in the activities of walkers, depending on internal or external influences. Hence, active walker models have been used to simulate a broad variety of pattern formation in complex systems [159, 217, 222, 269, 288, 291, 389, 456–458, 486, 492].

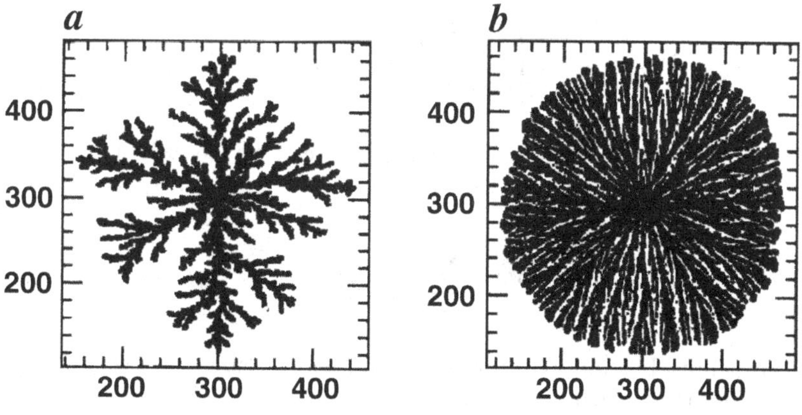

Fig. 5.6. Effect of chemotactic signaling on the communicating walkers model. (**a**) In the absence of chemotaxis for $P = 10\,gl^{-1}$ and $N_c = 40$ (see Fig. 5.5). (**b**) In the presence of chemotaxis for the same values of P and N_c. The axes refer to the triangular lattice size [50]

5.2 Discrete Model of Track Formation

5.2.1 Biased Random Walks

To apply the active walker model to the formation of tracks, trails, and aggregates in biology, as a first step we want to discuss some biological features of movement, such as the persistence of the walk, in more detail. We start with the discrete random walk model, as already introduced in Sect. 5.1.1 for a two-dimensional triangular lattice with periodic boundary conditions.

Let $p_i(\boldsymbol{r}, g_i, t)$ denote the probability density of finding random walker i on the lattice site with space vector $\boldsymbol{r}$ at time t. The *orientation* of walker i at time t should be denoted by the discrete value $g_i(t)$, that is, the direction it has moved in the previous time step. $g_i(t)$ can be regarded as another *agent specific state variable*, similar to internal parameter θ_i. On a lattice, each lattice site has d nearest neighbor sites at equal distance l (see Fig. 5.1), which can be numbered from 1 to d. Hence, g_i could have one of the possible values, $1, 2, \dots, d$. It is assumed that the walker is moving at constant space velocity $v_s = l/\Delta t$, i.e., it is moving *one lattice site* per time interval Δt.

The set of nearest neighbor sites $\boldsymbol{r}'$ with respect to position $\boldsymbol{r}$ shall be denoted by $\mathcal{L}$. If we assume that the walker was at lattice site $\boldsymbol{r}' = \boldsymbol{r} - \boldsymbol{\Delta r}_{g_i}$ at time $t - \Delta t$ and is moving at velocity v_s, vector $\boldsymbol{\Delta r}_{g_i}$ can be expressed as

$$\boldsymbol{\Delta r}_{g_i} = \Delta t\, v_s \begin{pmatrix} \cos \frac{2\pi}{d} g_i \\ \sin \frac{2\pi}{d} g_i \end{pmatrix} , \quad g_i = 1, 2, 3, \dots, d . \tag{5.18}$$

The evolution of the probability $p_i(\boldsymbol{r}, g_i, t)$ can be described in terms of a Frobenius–Perron equation which is discrete in time and space:

$$p_i(\boldsymbol{r}', g_i', t + \Delta t) = \sum_{g_i=1}^{d} K(\boldsymbol{r}', g_i' | \boldsymbol{r}, g_i)\, p_i(\boldsymbol{r}, g_i, t) . \tag{5.19}$$

The Frobenius–Perron operator $K(\boldsymbol{r}', g_i' | \boldsymbol{r}, g_i)$ describes the average over all possible paths to reach $\boldsymbol{r}'$ from $\boldsymbol{r}$:

$$K(\boldsymbol{r}', g_i' | \boldsymbol{r}, g_i) = \big\langle \delta\big[\boldsymbol{r}' - \boldsymbol{r}'(\boldsymbol{r}, g_i), g_i' = g_i'(\boldsymbol{r}, g_i)\big] \big\rangle . \tag{5.20}$$

Here, $\delta\big(x' - x\big)$ is Dirac's delta function used for the continuous variable $\boldsymbol{r}$. The function $\boldsymbol{r}'(\boldsymbol{r}, g_i)$ is given by (5.18); hence, on the lattice considered with d direct neighbor sites, K reads

$$K(\boldsymbol{r}', g_i' | \boldsymbol{r}, g_i) = w(g_i' | g_i)\, \delta\big[\boldsymbol{r}' - (\boldsymbol{r} + \boldsymbol{\Delta r}_{g_i'})\big] , \tag{5.21}$$

which results in the Frobenius–Perron equation:

$$p_i(\boldsymbol{r}', g_i', t + \Delta t) = \sum_{g_i=1}^{d} w(g_i'|g_i)\, p_i(\boldsymbol{r}' - \boldsymbol{\Delta r}_{g_i'}, g_i, t) \,. \tag{5.22}$$

The term $w(g_i'|g_i)$ denotes the transition rate for the walker to choose the orientation (or direction) g_i' in the next time step, provided it had orientation g_i during the previous time step. For a simple, nonbiased random walk, the probability of each orientation is equally distributed; hence, we simply find $w(g_i'|g_i) = 1/d$ (see Sect. 5.1.1).

Different from the simple random walk, the motion of microscopic biological entities usually displays a certain persistence, which has already been mentioned in Sect. 2.1.3. It means that instead of an equal probability of moving in every direction, the probability of moving in the same direction, as in the previous step, is larger than those of the other directions. Because $g_i(t)$ describes the orientation of a walker at time t, the *preference for the moving direction* is described by [492]

$$z_i\{\boldsymbol{r}, \boldsymbol{r}', g_i(t)\} = \begin{cases} b \geq 1.0\,, & \text{if } \boldsymbol{r}' \text{ is neighbor in direction } g_i(t) \text{ of the orien-} \\ & \text{tation of the walker, which is located at } \boldsymbol{r} \\ 1.0\,, & \text{else}\,. \end{cases} \tag{5.23}$$

The persistence results in a certain bias b for the random walk, but the direction of the bias is not constant as it would be, e.g., in the presence of an external field, but depends on the "history" of the last step, i.e., on $g_i(t)$.

Using the *bias* $b \geq 1$, we find for the probabilities to move into the different directions,

$$w(g_i'|g_i = g_i') = \frac{b}{d}\,,$$

where $b \leq d$,

$$w(g_i'|g_i \neq g_i') = \frac{1}{d-1}\Big[1 - w(g_i'|g_i')\Big] = \frac{d-b}{d(d-1)}\,. \tag{5.24}$$

We note that for the discrete lattice, bias is limited to values less than d. Obviously, for $b = 1$, a nonbiased random walk is obtained, and for $b \to d$, the motion of a walker degenerates into a straight line. Based on these considerations, transition rates for the Frobenius–Perron equation will be defined as

$$w(g_i'|g_i) = \frac{1}{d(d-1)}\Big[(d-b) + d(b-1)\,\delta_{g_i',g_i}\Big]\,, \quad \sum_{g_i=1}^{d} w(g_i'|g_i) = 1\,. \tag{5.25}$$

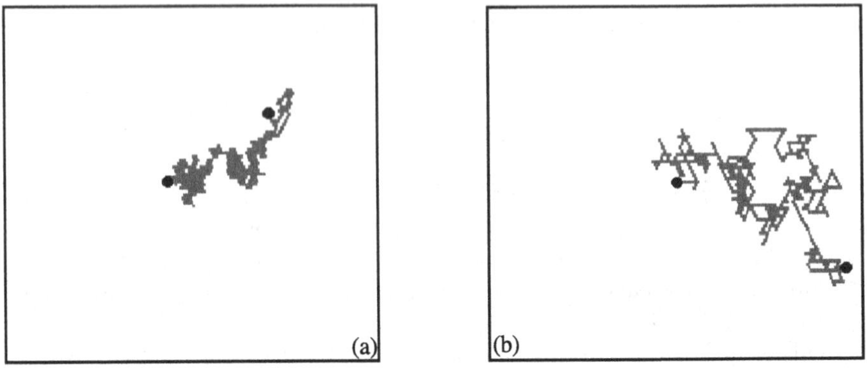

Fig. 5.7. (**a**) Simple random walk ($b = 1$, $a = 1$), (**b**) biased random walk ($b = 4$, $a = 0.4$) of a single walker. The starting position of the walker (*center of each picture*) and the end position after 676 steps are marked with large dots (lattice size: 100×100)

Here, δ_{g_i', g_i} is the Kronecker delta used for discrete values g_i, which is one only if $g_i = g_i'$, and zero otherwise. The $d \times d$ matrix of the possible transitions then reads explicitly

$$w(g_i'|g_i) = \frac{1}{d(d-1)} \begin{pmatrix} b(d-1) & (d-b) & (d-b) & \cdots \\ (d-b) & b(d-1) & (d-b) & \cdots \\ (d-b) & (d-b) & b(d-1) & \cdots \\ \vdots & \vdots & \vdots & \ddots \end{pmatrix} . \tag{5.26}$$

As Fig. 5.7 shows, a biased random walk, compared to a simple random walk, displays a certain persistence in the moving direction and allows a walker to reach out farther.

The bias that represents the persistence or the inertia of the walker, respectively, can be given by a simple geometric illustration (which holds only in continuous space). If we imagined the walker as a moving two-dimensional geometric object, then a circle would represent a nonbiased motion, where each possible direction is equally probable. However, if the object has a more elliptic form (cf. Fig. 5.8), then we intuitively realize that its motion occurs more likely in the direction of the least effort, which is the direction of the larger half axis. Insects like ants, for example, have an oblong body with bilateral symmetry, which causes bias in their motion (see Fig. 5.9). Of course, the object can change its moving direction from time to time by turning around, but the motion is still more likely in the new direction of the larger half axis. Thus, the ratio of the two half axes of the ellipse, $a = x/y$, can be regarded as an alternative expression for the bias b. We propose the following simple linear relation:

$$(b-1) = (d-1)(1-a) , \tag{5.27}$$

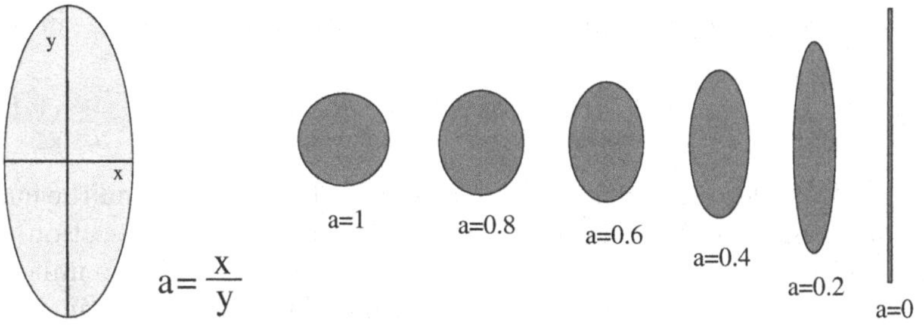

Fig. 5.8. Graphical illustration of bias by assuming a walker as an ellipse. The geometric asymmetry ratio, $a = x/y$, then represents the bias, $b \geq 1$, according to (5.27)

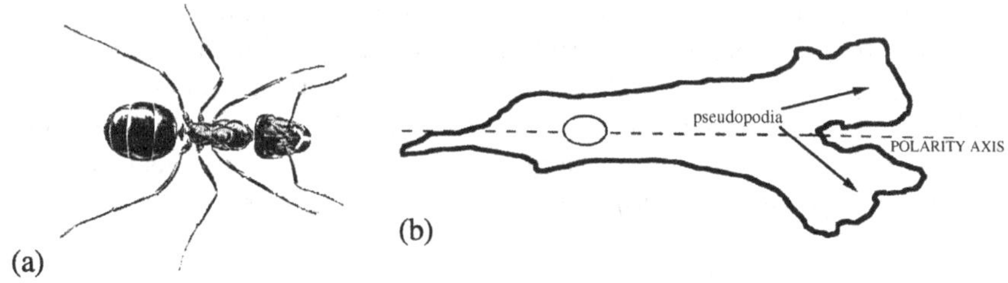

Fig. 5.9. (**a**) Sketch of a worker of the ant species *Formica polyctena* [239]. The antennae on the head are clearly visible. (**b**) Sketch of a tissue cell (top view) in active locomotion on a surface (projected cell outline from a fibroblast) [105]. Both pictures indicate the oblong shape and the bilateral symmetry of the biological entity

which means that $b = 1$ and $a = 1$ for an unbiased random walk. In the limit, where the object's motion would be only a straight line, $b = 6$, and $a = 0$.

As already mentioned, the biased random walk model provides a suitable approximation of the motion of *microscopic biological entities*, such as infusoria, amoebia, bacteria, and cells. It has been shown [4, 109, 110, 377, 385] that the diffusion coefficient for these species can be expressed as

$$D = \frac{v_s^2}{n} \tau , \tag{5.28}$$

where v_s^2, the second moment of the speed distribution, is a constant, n is the spatial dimensionality, and τ is the *persistence time* [109], i.e., the time during which the entity moves in the same direction, before it turns. In this form, (5.28) agrees with (2.26), which explains the equivalence between the "physical" and the "biological" definition of the diffusion coefficient. Due to [377], the persistence time τ can be calculated from

$$\tau = \frac{1}{\lambda (1 - \psi)} , \tag{5.29}$$

where λ is the *turning rate* and ψ means the *persistence index* or mean cosine,

$$\psi = \int_{-\pi}^{\pi} f(\varphi)\, \cos\varphi \, d\varphi \ , \quad \text{for } n = 2 \ . \tag{5.30}$$

Here, $\varphi \in [0, \pi]$ describes the cone angle between the previous and the current direction of motion. The distribution $f(\varphi)$ is related to the selection of direction. Assuming a uniform random selection of direction on the unit circle, $f(\varphi) = 1/2\pi$ yields $\psi = 0$. It has been shown [377, 385] that, in general, $\psi = I_1(k)/I_0(k)$, where the $I_z(k)$ are Bessel functions of order z. For $k = 0$, the uniform random selection of direction is obtained, and as $k \to \infty$, the new direction of motion tends to be the same as the previous direction, and $\psi \to 1$.

It is obvious that the persistence index ψ is directly related to the bias b with the notable difference that the above relations hold for a continuous variation of φ, whereas on the lattice, only a discrete number of possible directions exists. However, we may use (5.30) to obtain a relation between ψ and b for the discrete case. With the assumptions for the transition rates for a biased random walk, $w(g_i'|g_i)$, (5.25), and using further the definition of φ and cyclic values,

$$\varphi = \frac{2\pi}{d}(g_i - g_i') \ , \quad g_i \pm d \equiv g_i \ , \quad 2g_i \pm d \equiv -2g_i \ , \tag{5.31}$$

(5.30) results in

$$\begin{aligned} \psi &= \sum_{g_i'=1}^{d} w(g_i'|g_i)\ \cos\frac{2\pi}{d}(g_i - g_i') \\ &= \frac{1}{d(d-1)}\{[d - b + d(b-1)] - (d-b)\} \\ &= \frac{b-1}{d-1} = (1-a) \ . \end{aligned} \tag{5.32}$$

The relation $\psi = (1-a)$ reproduces both the limit of a *uniform random selection* of direction, i.e., $a = 1$, $\psi = 0$, and the limit of a *straight motion*, i.e., $a = 0$, $\psi = 1$. Obviously, in the latter case, the random walk *degenerates* and is no longer described by a diffusion approximation. For $a \to 0$, the persistence time τ goes to infinity, and the diffusion coefficient, (5.28), diverges. Hence, the diffusion approximation of a discrete biased random walk should be restricted to a range of $0.2 \leq a \leq 1$. Then, the diffusion constant for the biased random walk, D_a, with (5.28), (5.29) and (5.32) yields:

$$D_a = \frac{D_n}{a} \ , \quad D_n = \frac{v_s^2}{2\,\lambda} \ , \tag{5.33}$$

where D_n is the diffusion coefficient of a nonbiased random walk.

We note that (5.30) may yield values for $\psi \in [-1, +1]$; hence, negative values are also possible, which are not reproduced within our simplified approach. For instance, the so-called "telegraph process", i.e., the reverse of the direction at every step, yields $\psi = -1$. The motion of microscopic biological entities, however, is usually best described by small positive values of ψ. Data from the two-dimensional locomotion of *Dictyostelium* amoeba, for instance, yield $\psi \approx 0.7$ [200], whereas for human granulocytes, $\psi \approx 0.3$ has been found [201]. For the three-dimensional random walk of bacteria, $\psi = 0.33$ has also been obtained [55]. These values suggest that the value of $a = 0.4$ for the bias of a two-dimensional random walk is a good estimate for computer simulations.

5.2.2 Reinforced Biased Random Walks

Let us now assume in accordance with the investigations in Sect. 5.1.1 that walkers are *active walkers* in the same manner as Brownian agents, i.e., the walkers can generate a *one-component chemical field*, $h_0(\boldsymbol{r}, t)$ that follows reaction–diffusion dynamics. If $h_0(\boldsymbol{r}, t)$ changes during the next time step (i) by decay at rate k_0; (ii) by production, described by the function $Q[h_0(\boldsymbol{r}, t), N(\boldsymbol{r}, t)]$; and (iii) by diffusion with a diffusion constant D_0, the discrete dynamics in time and space can be approximated by (3.18) (cf. Sect. 3.1.4):

$$h_0(\boldsymbol{r}, t + \Delta t) = (1 - k_0\, \Delta t)\, h_0(\boldsymbol{r}, t) + \Delta t\; Q[h_0(\boldsymbol{r}, t), N(\boldsymbol{r}, t)] \\ - \frac{\Delta t\, D_0}{(\Delta r)^2} \left[d\, h_0(\boldsymbol{r}, t) - \sum_{\boldsymbol{r}' \in \mathcal{L}} h_0(\boldsymbol{r}', t) \right] . \tag{5.34}$$

$N(\boldsymbol{r}, t)$ denotes the number of active walkers covering the point $\boldsymbol{r} \in A^2$ at time t. For production, we simply assume a linear relation,

$$Q[h_0(\boldsymbol{r}, t), N(\boldsymbol{r}, t)] = s_0\, N(\boldsymbol{r}, t) \; ; \tag{5.35}$$

more complex functions will be discussed in Sect. 5.3. $\mathcal{L}$ defines the set of nearest neighbor sites, which can be obtained from

$$\boldsymbol{r}' = \boldsymbol{r} + l \begin{pmatrix} \cos\nu \\ \sin\nu \end{pmatrix} , \quad \nu = \frac{2\pi\, g}{d} , \quad g = 1, 2, 3, \ldots, d , \tag{5.36}$$

where the discrete number g denotes again the direction of the nearest neighbor, which corresponds to angle ν. $\Delta r = l$ is the absolute value of the distance between $\boldsymbol{r}$ and the nearest neighbor sites, $\boldsymbol{r}'$.

Nonlinear feedback between the chemical field and the further movement of active walkers should again result from the fact that the walker prefers to move toward the ascent of the field. In the discrete formulation, the walker is able to recognize the chemical, if it is in the close vicinity of its site, namely on the nearest neighbor sites $\boldsymbol{r}' \in \mathcal{L}$ – which means a short-range or local recognition rather than a "view" over a long distance. In the discrete version, the gradient of the field can be written as

$$\frac{\partial h_0(\boldsymbol{r},t)}{\partial \boldsymbol{r}} = \frac{h_0(\boldsymbol{r}+\boldsymbol{\Delta r},t) - h_0(\boldsymbol{r}-\boldsymbol{\Delta r},t)}{2\,\boldsymbol{\Delta r}} \,. \tag{5.37}$$

With $\boldsymbol{\Delta r} = \boldsymbol{r}' - \boldsymbol{r}$, (5.36), the gradient can also be expressed as a function $f(\boldsymbol{r},g)$, which depends on the current position $\boldsymbol{r}$ and the direction g, which is a cyclic value, (5.31). $f(\boldsymbol{r},g)$ may be defined as

$$f(\boldsymbol{r}_i,g) = \alpha \left.\frac{\partial h_0(\boldsymbol{r},t)}{\partial \boldsymbol{r}}\right|_{\boldsymbol{r}_i} \,, \quad \sum_{g=1}^{d} f(\boldsymbol{r}_i,g) = 0 \,. \tag{5.38}$$

The response parameter $\alpha > 0$ (see Sect. 3.1.1), which acts as a dimensional constant, can also consider a threshold parameter h_{thr}, which is of relevance, in particular, for biological systems:

$$\alpha = \Theta\Big[h_0(\boldsymbol{r}_i,t) - h_{\mathrm{thr}}\Big] \,. \tag{5.39}$$

h_{thr} is assumed of small magnitude, indicating that a walker can detect even small amounts of chemical.

Now, the movement of a walker is subject to two influences. On one hand, the walker performs a biased random walk and, on the other hand, it tends to move toward the gradient of a chemical field. The simplest assumption would be a superposition of these two influences. However, investigations of the aggregation of active walkers in Sect. 3.2 already revealed the existence of critical parameters for the response to a chemical field. In particular, if the reduced temperature η, (3.47), which refers to thermal noise, is equal or larger than one, a walker is not at all affected by the existence of spatial gradients of the chemical field. For $\eta \le 1$, a walker is subject to both deterministic and random influences, whereas for $\eta \to 0$, only deterministic influences resulting from the gradient of the chemical field affect the walker's motion.

We may use this result to weight the two influences on the movement of the walker by means of η. Thus, we define the probability of choosing different possible orientations g_i', provided that the walker i has orientation g_i at position $\boldsymbol{r}_i$, as

$$w(g_i'|g_i) = \frac{1}{c(g_i)}\Big\{(1-\eta)\, f(\boldsymbol{r}_i,g_i') + \eta\,\Big[(d-b) + d(b-1)\,\delta_{g_i',g_i}\Big]\Big\} \,, \tag{5.40}$$

with $0 \le \eta \le 1$. The *normalization constant* $c(g_i)$ results from the condition

$$\sum_{g_i=1}^{d} w(g_i'|g_i) = 1 \,. \tag{5.41}$$

Because of the definition of $f(\boldsymbol{r}_i, g_i)$ we find for this case,

$$c(g_i) = \eta\, d(d-1) = c(\eta, d) \ . \tag{5.42}$$

Equation (5.40) assumes that for $0 < \eta < 1$, an active walker is affected by the gradient of the chemical field in *every* direction, or in terms of a biological species, that the organism can measure the gradient in every direction. This assumption may hold for some species, but it does not seem to be valid for others. It is known that insects like ants can measure concentration gradients of pheromones [76, 127, 386], but only in a certain range of space and if the concentration locally exceeds a certain threshold. For the detection of chemical markings, these species use specific receptors located at their antennae; hence, the length and the angle between the antennae (cf. Fig. 5.9a) determine the angle of perception for chemical gradients.

A similar picture holds for cells such as fibroblasts (cf. Fig. 5.9b) that persistently move in the direction of the leading lamella via an extension of the cell membrane and cytoplasm (pseudopod) in the anterior region and by a contraction of the posterior region [105]. This process is assumed to be mediated via a transfer of receptors to the anterior region, where they exist in a larger concentration. Consequently, the response to external gradients comes mostly from the leading lamella.

We can count on these facts by considering *broken symmetry* of an active walker, i.e., it can check for a gradient of the chemical only in a certain range of space. Different from a simple physical agent that feels a gradient in every direction, a walker feels a gradient only within a certain *angle of perception*, which should be 2φ (see Fig. 5.10).

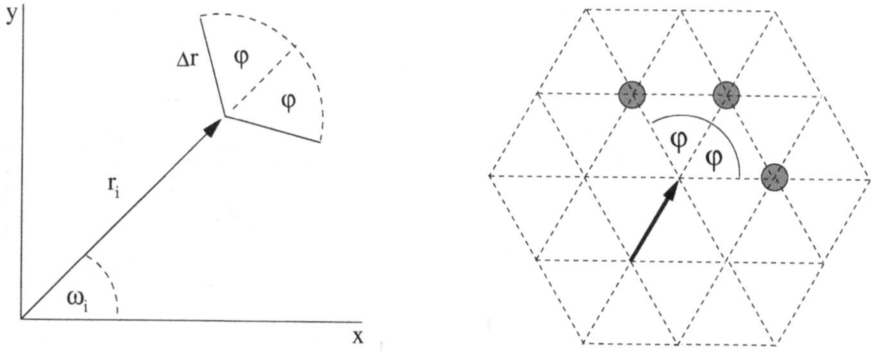

Fig. 5.10. Illustration of the angle of perception in the continuous (*left*) and the discrete (*right*) model. The current position/orientation of a walker is described by the arrow. The spatial range of perception is determined in the continuous model by the length Δr (which can be imagined as the length of the antennae) and the angle 2φ. In the discrete model, Δr is expressed by the distance l of the nearest neighbor sites, and 2φ is chosen so that only the sites in front of walkers (marked with large dots) are "visible"

On a lattice, the spatially restricted perception of a chemical gradient means that an active walker counts on $f(\boldsymbol{r}_i, g_i')$ only for a few selected values of g_i'. We may assume the following three values $g_i' = (g_i, g_i+1, g_i-1)$, which means that a walker is affected by gradients that are in the direction of its current orientation g_i, or close to it, $g_i \pm 1$, but not, for example, behind it. We note again the cyclic values for g_i, (5.31). Consequently, the probability of choosing an orientation, (5.40), reads now

$$w(g_i'|g_i) = \frac{1}{c(g_i)}\Big\{(1-\eta)\,f(\boldsymbol{r}_i, g_i')(\delta_{g_i',g_i} + \delta_{g_i',g_i\pm 1}) + \eta\Big[(d-b) + d(b-1)\,\delta_{g_i',g_i}\Big]\Big\}\,. \tag{5.43}$$

The normalization constant $c(g_i)$ can be calculated from the normalization of the transition probabilities, (5.41), which yields now

$$c(g_i) = \eta\, d(d-1) + (1-\eta)\,\Big[f(\boldsymbol{r}_i, g_i) + f(\boldsymbol{r}_i, g_i - 1) + f(\boldsymbol{r}_i, g_i + 1)\Big]\,. \tag{5.44}$$

The effect of the restricted perception angle can be clearly seen in Fig. 5.11, which displays the biased random walk of a single active walker, either with full perception, (5.40), or with restricted perception, (5.43), of the chemical gradient. In both cases, the diffusion constant D_0 of the chemical, (5.34), is set to zero. In the first case, the walker is always affected by the chemical it just produced, which strongly determines its decision about the next step. Because the largest gradient points back mainly to the previously visited site, a back and forth motion results, which only slightly changes because of random influences. Thus, the walker is locally trapped by the chemical.

(a) (b)
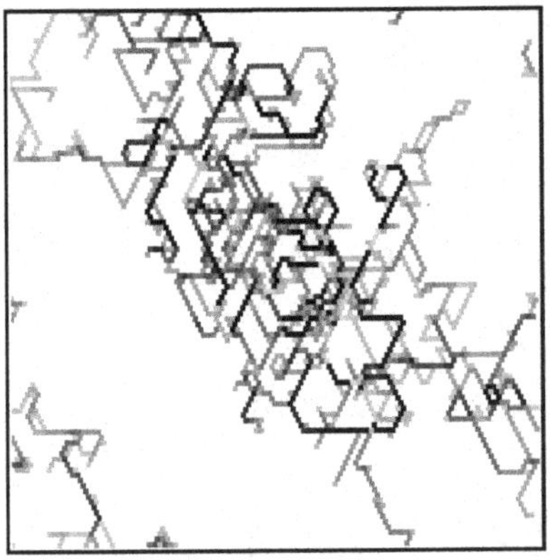

Fig. 5.11. Reinforced biased random walk of a single active walker (**a**) with full perception, (5.40); (**b**) with spatially restricted perception, (5.43), of the chemical gradient. The chemical concentration $h_0(\boldsymbol{r}, t)$ (coded on a gray scale) is shown after $t = 10{,}000$ simulation steps. Parameters: $a = 0.4$, $D_0 = 0$, $k_0 = 10^{-4}$, $\eta = 0.15$, $s_0 = 80$, $h_{\mathrm{thr}} = 5$. Lattice size: 100×100, periodic boundary conditions

For a restricted angle of perception, a walker is *not* immediately affected by the previously produced chemical, simply because it does not recognize the backward gradient. This will give the walker the chance to leave the area, but once it comes back by chance, it might still find chemical markings, which force it to use (and to reamplify) a previous track. In the following section, we will discuss the influence of the decay rate and the diffusion constant of the chemical on the formation of tracks in more detail.

5.2.3 Formation of Tracks

The previous section has shown that the consideration of an angle of perception for an active walker can be crucial in obtaining track patterns because it prevents the walker from getting trapped. However, as we have learned from the investigations of Sect. 3.2.2, this kind of "localization" holds only for a certain range of parameters. Above a critical temperature T_c, (3.46), a walker always performs an unbounded motion (see Fig. 3.6), even if it is responding to the gradient in every direction.

In this section, we want to investigate the appearence of track patterns resulting from the interaction of many walkers. So, the initial condition assumes that at $t = 0$, N active walkers start from the *same* lattice site, in analogy to a "nest" or a "source", with random initial directions. Every walker performs a biased random walk and produces a chemical, which follows the dynamics of (5.34). As before, the chemical does *not* diffuse, $D_0 = 0$. If the chemical is sensed within the angle of perception, the walker chooses the direction of its next step in accordance with the gradient of the chemical, as described by (5.43). Noteworthy, all walkers use the same kind of chemical markings.

A characteristic track pattern, which result from 100 active walkers is shown in Fig. 5.12.

The main conclusion to be drawn from Fig. 5.12 is the transition of the initially random movement of the walker population into a movement along distinct tracks that are used with different frequencies. We can distinguish between a *major track*, which appears in black, and a number of *minor tracks*, which are less used and therefore appear in lighter gray. Figure 5.12 also shows that large areas of the simulated surface are left empty because the walker community is bound mainly to existing tracks.

The track pattern remains as a *quasi-stationary structure*, that means, after the structure is established, only slight shifts of the less frequently used parts occur. This is shown in the time series of Fig. 5.13, which indicates that the main structure builds up relatively quickly and then is stable over a very long period. Due to fluctuation in the movements of walkers, some new tracks are always added to the structure, and others less used decay again – but the main structure remains the same.

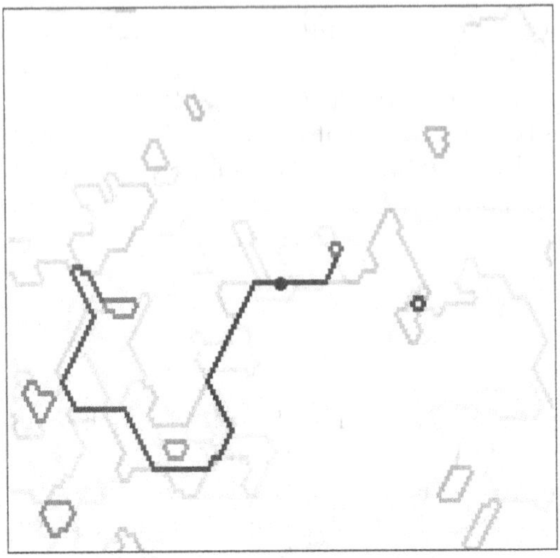

Fig. 5.12. Track pattern generated by 100 active walkers after $t = 5000$ simulation steps. At the beginning, the walkers have been released at once in the middle (nest) of the triangular lattice (size 100×100). The gray scale of the tracks encodes the concentration of $h_0(\boldsymbol{r}, t)$. Parameters: $a = 0.4$, $D_0 = 0$, $k_0 = 0.005$, $\eta = 0.05$, $q^0 = 80$, $h_{\text{thr}} = 1$ [456]

Track patterns, as shown in Figs. 5.12 or 5.13, are always unique due to fluctuations that affect the formation of tracks, especially during the initial stage. The chemical markings generated at different times during the evolution contribute also with different weights to the present state of the track pattern. Due to nonlinear feedback between the existing and the newly produced markings, this process occurs in a nontrivial way. The markings generated in the early stages of the system's evolution, certainly disappear – but, on the other hand, if walkers have used the track again in the course of evolution, the early markings have been brushed up and reinforced. In this way, the early markings stamp the pattern because of early symmetry breaks, as can be clearly seen in the time series of Fig. 5.13. However, tracks that are not reused, that fade out in the course of time and do not influence the further evolution of the pattern.

As a consequence, only those tracks that are consecutively used and therefore are renewed by walkers survive in the course of evolution, in a process of competition similar to that described in Sect. 3.2.5. Due to fluctuations, new tracks can be created at any time; the question, however, is whether a new track can compete with existing tracks. Here, the *enslaving principle* [198] of already existing tracks becomes important: the more these tracks are carved into the surface, the more chemical markings are confined to specific "areas," the more difficult it would be to establish a new track. We will come back to this interesting phenomenon later, when discussing trail formation by ants and pedestrians.

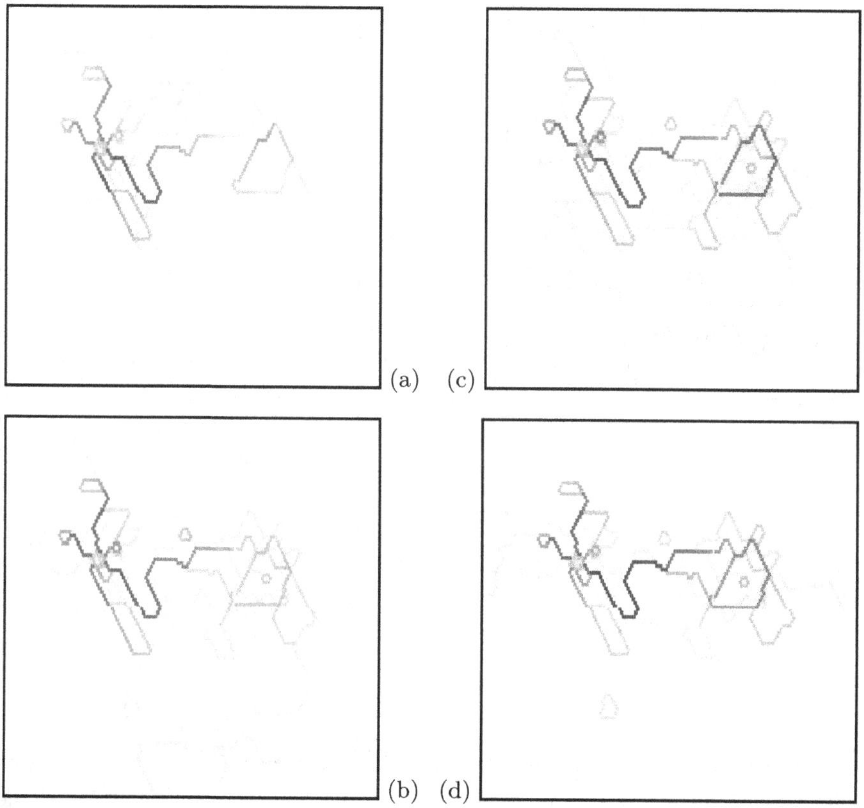

Fig. 5.13. Evolution of a track pattern generated by 10 active walkers. Time in simulation steps: (a) $t = 1000$, (b) $t = 5000$, (c) $t = 10,000$, (d) $t = 60,000$. Parameters: $a = 0.4$, $D_0 = 0$, $k_0 = 10^{-4}$, $\eta = 0.05$, $s_0 = 80$, $h_{\mathrm{thr}} = 5$. Lattice size: 100×100, periodic boundary conditions

Track patterns like those of Fig. 5.12 have been also obtained by Stevens [487, 489, 490] who simulated gliding myxobacteria that produce slime tracks commonly used for movement and aggregation. Similar computer simulations of Edelstein–Keshet et al. [127, 128] investigated, in a specific biological context, the influence of individual properties of ants in trails following the resulting pattern.

We restrict the discussion here to the influence of two parameters, which describe the dynamics of the chemical field, (5.34), the decay rate k_0 and the diffusion constant D_0. Figure 5.14 presents the chemical field generated by active walkers for four different sets of parameters after the same number of simulation steps. The comparison shows that for large values of the decay rate, no commonly used track patterns occur, simply because the local decay of the chemical markings cannot be compensated for by the local chemical production of the (few) walkers in time. We know from the investigations in

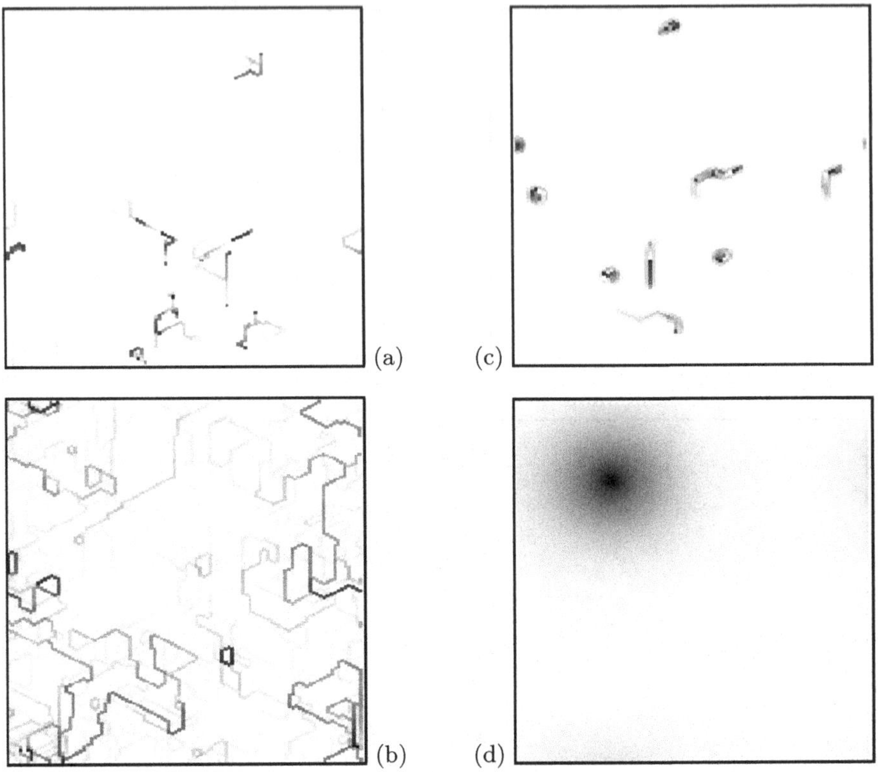

Fig. 5.14. Influence of the decay rate k_0 and the diffusion coefficient D_0 of the chemical field, (5.34). The structures generated by 10 active walkers are shown in terms of the chemical field after $t = 10,000$ simulation steps. (**a**) $k_0 = 10^{-1}$, $D_0 = 0$; (**b**) $k_0 = 10^{-4}$, $D_0 = 0$; (**c**) $k_0 = 10^{-1}$, $D_0 = 10^{-2}$; (**d**) $k_0 = 10^{-4}$, $D_0 = 10^{-2}$. Other parameters: $\eta = 0.1$, $s_0 = 80$, $h_{\rm thr} = 5$. Lattice size: 100×100, periodic boundary conditions

Sect. 3.2.2 that the *total amount* of chemical produced by the walkers asympotically approaches a constant value, (3.36). But the question is whether the probability of forming a *common* track, would be large enough, especially if there is no guiding force such as an extended gradient, to attract walkers to a common region. So, we find only small pieces of tracks (see Fig. 5.14a,c).

As we have learned from the investigations in Sect. 3.2, which to a large extent also apply to the situation discussed here, diffusion of the chemical allows establishing an extended gradient. Diffusion, however, destabilizes the track, which is a very narrow structure. So, especially for small decay rates, we find an aggregation of walkers instead of movement along distinct tracks (see Fig. 5.14d).

For *finite time intervals*, the formation of track patterns is most likely observed in the range of *small decay rates* and *negligible diffusion* of the chemical, as Fig. 5.14b shows. We note that the decay of the chemical is still crucial for *commonly* used tracks. Without decay, no selection pressure for the tracks exists, which would result in a surface almost totally covered with chemical markings. A *track*, however, as the above simulations suggest, should be a *distinct, quasi-stationary structure* on which the *majority* of walkers can be found during their movements.

5.3 Track Formation and Aggregation in Myxobacteria

5.3.1 Modification of the Active Walker Model

In this section, we want to discuss a variation of the discrete model investigated in the previous section, which elucidates the formation of aggregates under certain circumstances [492]. Active walkers represent a biological species now, i.e., *myxobacteria*, which move on surfaces or at air–water interfaces at a velocity ranging from 1 to 13 μm/min [265, 403]. Like most gliding bacteria, myxobacteria produce extracellular slime that marks the paths used by the bacteria. These slime trails play an important role in the guidance of the motion of other following myxobacterial cells [488]. When a myxobacterium glides on an untrailed area and encounters another slime trail, it will typically glide onto it. Myxobacteria can also use the trails of other gliding bacteria [75]. On the slime trail, the bacterium increases its gliding velocity [404].

Common gliding soon turns into a local aggregation of bacteria. In the example of the myxobacteria *Myxococcus xanthus*, the bacteria first begin to associate in organized rafts of 4 to 6 cells gliding together as a unit, which then are extended to arrays larger than 10 cell units in width and length [374]. The aggregation is accompanied by subsequent pattern formation, such as coherent motion around centers, formation of spirals, and oscillating waves, that finally results in the formation of so-called "fruiting bodies," but the details are not important here. A similar kind of aggregation is also known from the cellular slime mold, *Dictyostelium discoideum* [235].

For a description of the myxobacterial aggregation, we will use the discrete active walker model introduced in Sect. 5.2.2, with only a few modifications [492]. The similarities and differences from the previous model are summarized in the following:

- The movement of a walker is again described as a *biased random walk* on a two-dimensional lattice, where A^2 denotes the lattice area. For the simulations, discussed in the next section, a *square lattice* with periodic boundary conditions is used, i.e., $d = 4$.
- The concentration of the substance, to which a walker is sensitive, is described again by the field $h_0(\boldsymbol{r}, t)$, which follows the reaction–diffusion dynamics given by (5.34). Different from previous investigations, we assume a homogeneous initial concentration $h_0(\boldsymbol{r}, 0) = 1.0$.

- A walker shall not be characterized by a restricted angle of perception as before; instead it is sensitive to the field in every direction. Further, the response is not characterized by a threshold value h_{thr} or weighed by a value η, i.e., the walker is always sensitive to every magnitude of concentration.
- We do not assume a response to the gradient of the field, but to the magnitude of the field, i.e., the transition probability should be proportional to $h_0(\boldsymbol{r}, t)$.

The probability that a walker i, located at latticle site $\boldsymbol{r}$ at time t, will be found at one of the neighboring sites $\boldsymbol{r}' \in \mathcal{L}$ at time $t + \Delta t$ is now simply assumed as [492]

$$p_i(\boldsymbol{r}', g_i', t + \Delta t) = \frac{h_0(\boldsymbol{r}', t) z[\boldsymbol{r}, \boldsymbol{r}', g_i(t)]}{\sum\limits_{\boldsymbol{r}' \in \mathcal{L}} h_0(\boldsymbol{r}', t) z[\boldsymbol{r}, \boldsymbol{r}', g_i(t)]} . \tag{5.45}$$

Here, $z[\boldsymbol{r}, \boldsymbol{r}', g_i(t)]$, (5.23), denotes again the preference for the moving direction *without* respect to a chemical field, which is mainly determined by bias b.

The dynamics for the substance is described by (5.34), but for the production function $Q[h_0(\boldsymbol{r}, t), N(\boldsymbol{r}, t)]$, now a more complex assumption is used:

$$Q[h_0(\boldsymbol{r}, t), N(\boldsymbol{r}, t)] = \left[\beta^{N(\boldsymbol{r},t)} - 1\right] h_0(\boldsymbol{r}, t) + s_0 \, \frac{\beta^{N(\boldsymbol{r},t)} - 1}{\beta - 1} . \tag{5.46}$$

$N(\boldsymbol{r}, t)$ denotes the number of walkers covering point $\boldsymbol{r} \in A^2$ at time t. Due to (5.46), walkers produce a fixed amount s_0 at their current location and *additionally* β times the existing concentration of substance, with $s_0, \beta \geq 0$. This assumption is based on biological investigations of myxobacteria. For $\beta = 1$, which agrees with the basic model for chemotactic response, a linear production term results again:

$$Q[h_0(\boldsymbol{r}, t), N(\boldsymbol{r}, t)] = s_0 \, N(\boldsymbol{r}, t) \ , \quad \frac{\beta^{N(\boldsymbol{r},t)} - 1}{\beta - 1} = N(\boldsymbol{r}, t) \ \text{ if } \ \beta \to 1 . \tag{5.47}$$

The model outlined describes again a *reinforced random walk* due to nonlinear coupling between the concentration of the substance, (5.34), and the movement of walkers, (5.45), which may result in an aggregation of walkers. For myxobacteria, we will investigate both a linear ($\beta = 1$) and a superlinear ($\beta > 1$) production of slime. We note that a more realistic model of myxobacterial aggregation is given by Stevens [490], where both slime trail following and the response to a diffusing chemoattractant are needed to get stable centers of aggregation. Here, we restrict ourselves to the simpler model of only one substance.

Before we show computer simulations that elucidate the relevance of different parameters, we want to point out that the discrete interacting model, introduced here, can be approximated again by a continuous model. Assuming the simplest case of a reinforced random walk of *one* walker in two dimensions, i.e., with four nearest neighbor sites on the lattice, a diffusion approximation can be carried out [378]. For a nonbiased random walk, i.e., $b = 1.0$, from (5.34), (5.45), the following *chemotactic system* of coupled partial differential equations can be derived:

$$\partial_t p = D\,\nabla\left(\nabla p - \chi\,\frac{p}{h_0}\,\nabla h_0\right)\,, \tag{5.48}$$

$$\partial_t h_0 = D_0 \Delta h_0 + Q(h_0, p) - k_0 h_0\,. \tag{5.49}$$

Here, $p(\boldsymbol{r}, t)$ denotes the probability density that a walker will be located at point $\boldsymbol{r}$ at time t, $h_0(\boldsymbol{r}, t)$ denotes the density of the chemical substance, and $\chi = 2$ is the *chemotactic coefficient.* $Q(h_0, p)$ is a suitable function for the growth of the chemical substance, for instance, $Q(h_0, p) = s_0 p$.

For $D_0 = 0$ and $k_0 = 0$ and with an initial peak for $p(\boldsymbol{r}, 0)$, only a high production rate for the substance accounts for the expansion of p in finite time, whereas for a low production rate, the initial peak of p breaks down. This is closely related to a model by Davis [96], who investigated a *one-dimensional* reinforced random walk of one walker without diffusion of the substance. In this case, the approximation yields (5.48) with $\chi = 1$. He found that the expansion of p occurs only for superlinear growth of the substance, i.e., the walker localizes at a random place if the substance is produced superlinearly and does not localize if it is produced only linearly.

If the decision of the walker for the next move is proportional to the gradient ∇h_0 instead of h_0, (5.45), i.e., if the probability density has the form,

$$p_i(\boldsymbol{r}', g_i, t + \Delta t) \sim a_1 + a_2\Big[h_0(\boldsymbol{r}, t) - h_0(\boldsymbol{r}', t)\Big] \;\; ; \quad \boldsymbol{r}' \in \mathcal{L} \;\; ; \;\; a_1,\, a_2 \in \mathbb{R}\,, \tag{5.50}$$

one obtains, instead of (5.48) [378],

$$\begin{aligned} \partial_t p &= D'\,\nabla\Big(a_1 \nabla p - 2\,a_2 p \nabla h_0\Big) \\ &= D\,\Delta p - \nabla p\Big(\chi \nabla h_0\Big) \\ \text{with} \quad D &= D'\,a_1\,, \;\; \chi = 2D\,a_2/a_1\,. \end{aligned} \tag{5.51}$$

In this form, (5.51) agrees with the density (3.32) for active Brownian agents; thus, for a further discussion of the continuous equation, we refer to Sect. 3.2. A rigourous approach to deriving density equations from an interacting many-particle system can be found in [489].

We finally note that chemotaxis equations (5.48), (5.49) have been investigated in different respects. The stability analysis for a quite general situation is done in [423]; expansion results are given in [59, 232, 256] for a *diffusing substance*, sometimes with $D_0 \gg 1$. For chemotaxis equations with a nondiffusing substance, $D_0 = 0$, qualitative results are given in [378, 399]. In [378], an expansion of p in finite time, finite stable peaks, and the collapse of developing peaks are discussed.

5.3.2 Simulation of Myxobacterial Aggregation

In our model of interacting walkers, there is interplay between the parameters describing the performance of walkers, such as persistence b or the production of substance s_0, β, and parameters that describe the evolution of the chemical substance, such as decay k_0 or diffusion D_0. Another parameter of interest is the initial population density. The influence of these parameters on the aggregation process will be elucidated by computer simulations [492].

We consider a square lattice A^2 of 70×70 grid points with periodic boundary conditions. This grid size is chosen to avoid too strong boundary effects and at the same time guarantee a clear output. Initially, $N = 1000$ walkers are randomly distributed on the inner square lattice A_1^2 of 30×30 grid points (see Fig. 5.15). At each time step, walkers move to one of their four nearest neighbors and interact with the surface, as described by (5.34), (5.45). The initial concentration of the substance is set at $h_0(\boldsymbol{r}, 0) = 1$.

In the following, we will discuss different sets of paramters with respect to their influence on the aggregation process.

(i) $D_0 = 0$, $k_0 = 0$ *(no diffusion and no decay of the substance)*, $s_0 = 0.1$, $\beta = 1$ *(linear production of the substance)*, $b = 1$ *(no persistence of the walk)*

Fig. 5.15. Initial conditions: $N = 1000$ walkers are randomly distributed on the inner square lattice A_1^2 of 30×30 grid points. The total size of the square lattice A^2 is 70×70 grid points. Black squares mark a single walker; the squares of different gray levels mark two to nine walkers, and white squares mark 10 and more walkers [492]

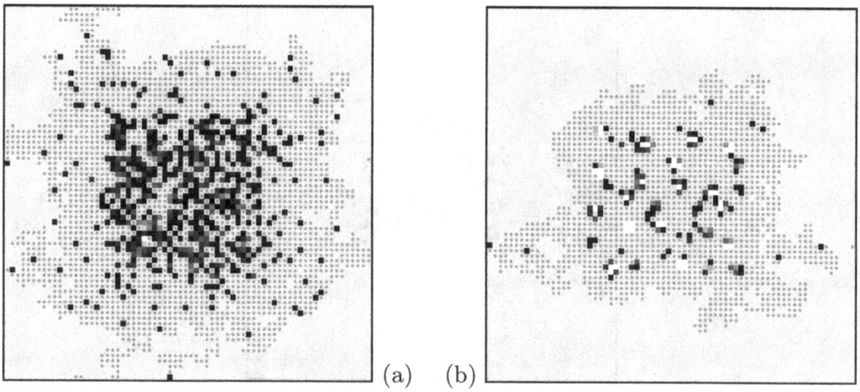

Fig. 5.16. Reinforced random walk of 1000 walkers: (a) The substance is laid down and measured in between the grid points. $t = 200$ simulation steps. (b) The substance is laid down and measured on the grid points. $t = 1000$ simulation steps. Gray dots mark the paths that the walkers have used. Black squares mark a single walker; the squares of different gray levels mark two to nine walkers; white squares mark 10 and more walkers. Parameters: $D_0 = 0$, $k_0 = 0$, $s_0 = 0.1$, $\beta = 1$, $b = 1$ [492]

The simulations shown in Fig. 5.16 refer to two different situations. Figure 5.16a displays a situation where the chemical substance is laid down and measured *in between* the grid points; in Fig. 5.16b, the chemical substance is laid down and measured *directly on* the grid points, as also described by the model (5.45), (5.34). In the first case, *no aggregation* occurs, which also agrees with the result in [96] for a single walker. In the second case, however, a stronger taxis effect results. Hence, if the initial walker density, N/A_1^2, exceeds a critical value, the swarm remains more local, and the walkers form small clusters. This can be seen by comparing Fig. 5.16a with Fig. 5.16b (see also [378]).

(ii) $D_0 = 0.05$, $k_0 = 0$ *(diffusion, but no decay of the substance)*, $s_0 = 0.1$, $\beta = 1$ *(linear production of the substance)*, $b = 1$ *(no persistence of the walk)*

The simulation in Fig. 5.17 shows that if the substance diffuses, fewer clusters appear, but more walkers are trapped, compared to the situation without diffusion, Fig. 5.16b. In this case, initial conditions play a less important role because the diffusion of the attractive substance effaces initial clusters and guides walkers from regions with only few walkers toward regions with developing aggregation centers. If the diffusion constant D_0 is increased, the aggregation centers interfere even more, and eventually only a single cluster is formed instead of different separated clusters. This agrees with the competition described in Sect. 3.2.4. To counteract the destabilization of existing clusters, production of the substance has to be increased, and a higher initial concentration of walkers has to be chosen.

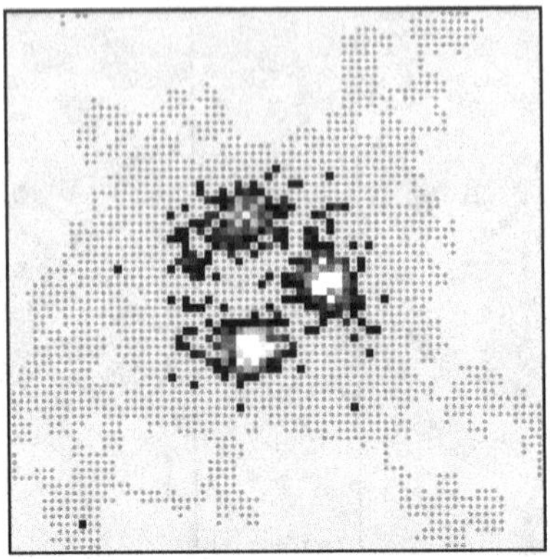

Fig. 5.17. Reinforced random walk of 1000 walkers. Same situation as in Fig. 5.16b, but with diffusion of the substance. $t = 1000$ simulation steps. Parameters: $D_0 = 0.05$, $k_0 = 0$, $s_0 = 0.1$, $\beta = 1$, $b = 1$ [492]

(iii) $D_0 = 0$, $k_0 = 0$ *(no diffusion and no decay of the substance)*, $s_0 = 0.1$, $\beta = 1.01$ *(superlinear production of the substance)*, $b = 1$ *(no persistence of the walk)*

The simulation in Fig. 5.18a shows that for superlinear production of the substance, well separated aggregates are formed very quickly, even if there is no diffusion of the substance. This effect can be amplified by high initial density of walkers. On the other hand, if walkers are more distant initially, they are trapped in many very small clusters.

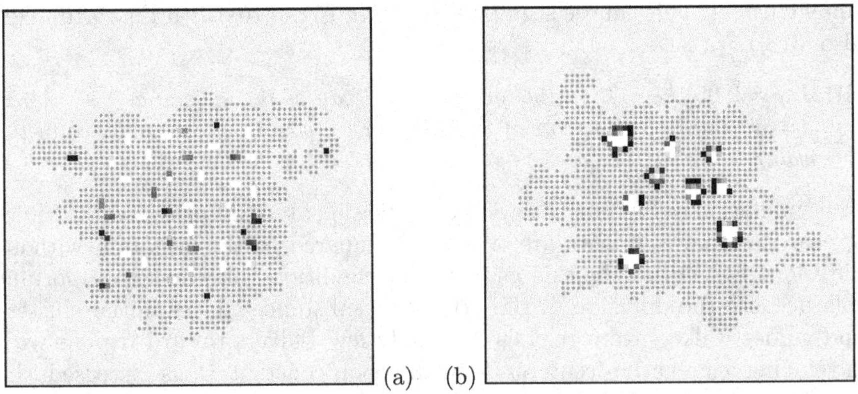

(a) (b)

Fig. 5.18. Reinforced random walk of 1000 walkers, considering superlinear production of the substance. The results are shown after $t = 200$ simulation steps (**a**) without diffusion of the substance, $D_0 = 0$; (**b**) with diffusion of the substance, $D_0 = 0.05$. Other parameters: $k_0 = 0$, $s_0 = 0.1$, $\beta = 1.01$, $b = 1$ [492]

(iv) $D_0 = 0.05$, $k_0 = 0$ *(diffusion, but no decay of the substance)*, $s_0 = 0.1$, $\beta = 1.01$ *(superlinear production of the substance)*, $b = 1$ *(no persistence of the walk)*

If diffusion of the substance is considered for the situation described in (iii), streaming of walkers toward clusters becomes stronger; clusters grow to a larger size and become well separated, which can be seen by comparing Fig. 5.18a with Fig. 5.18b.

(v) $D_0 = 0.05$, $k_0 = 0.05$ *(diffusion and decay of the substance)*, $s_0 = 0.1$, $\beta = 1.01$ *(superlinear production of the substance)*, $b = 3$ *(persistence of the walk)*

Decay of the chemical substance would amplify the effect of aggregation; however, walkers far away from the centers have a tendency to jump back and forth between two grid points. This effect can be smoothed out by an increase in diffusion D_0, which then results again in streaming of walkers toward the clusters, as already shown in Fig. 5.18b.

To keep close to myxobacterial behavior, the simulations are finally carried out considering additional persistence of a walker's motion. Then, walkers swarm out and return to the aggregates more easily, once they have chosen the correct direction, a behavior found close to reality. An increase of the diffusion of the substance, $D_0 \neq 0$, can support this "behavior." As the simulation in Fig. 5.19 shows, diffusion and decay of the substance can stabilize the different aggregation centers.

We can conclude from the computer simulations that in our simple model, fine-tuning of parameters can account for swarming of walkers, aggregation, and stabilization of aggregation centers. We found that a large production

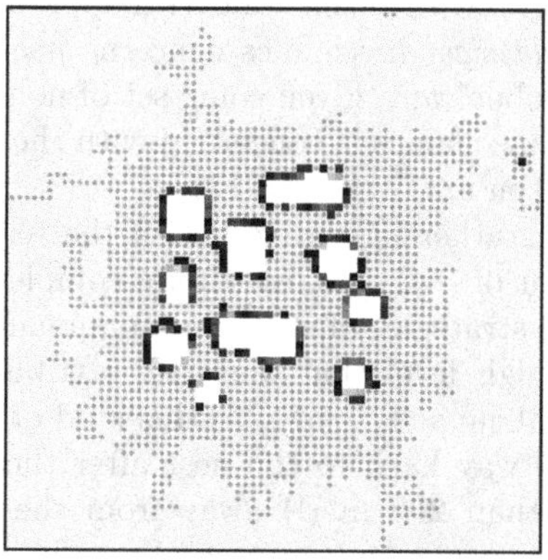

Fig. 5.19. Reinforced presistent random walk of 5000 walkers, considering superlinear production, diffusion, and decay of the substance. The results are shown after $t = 200$ simulation steps. Parameters: $D_0 = 0.05$, $k_0 = 0.05$, $s_0 = 0.1$, $\beta = 1.01$, $b = 3$ [492]

rate of slime, combined with a supercritical initial concentration of walkers, results in the formation of aggregation centers, but, on the other hand, prevents swarming. Diffusion of the slime enforces the aggregation effect but effaces the centers a little bit. Decay of the slime stabilizes the aggregation centers but increases the chance that single walkers are trapped in certain regions. The persistence of walker movement does not change this qualitative behavior; however, it makes the simulations more realistic. We note that in the more complex model for myxobacterial aggregation described in [490], superlinear slime production does not affect aggregation in the way discussed here. Further research needs to be done to understand this, as well as to find the critical conditions that separate different dynamic regimes.

5.4 Trunk Trail Formation of Ants

5.4.1 Biological Observations

In this section, we try to combine the different investigations into the formation of networks, Sect. 4.1, and tracks, Sect. 5.2, to achieve a model suitable for describing the formation of trail systems in biological systems, i.e., ants. The physical perspective on "ants" is rather reductionistic because it basically refers to the assumed *random walk* of ants, which is apparently not even in accordance with biological observations.

A quote by Stauffer and Stanley concludes the physical reductionistic view on ants in a pertinent manner: "The use of the term *ant* to describe a random walker is used almost universally in the theoretical physics literature – perhaps the earliest reference to this colorful animal is a 1976 paper of de Gennes that succeeded in formulating several general physics problems in terms of the motion of a "drunken" ant with appropriate rules for motion. Generally speaking, *classical* mechanics concerns itself with the prediction of the position of a "sober" ant, given some set of non-random forces acting on it, while *statistical* mechanics is concerned with the problem of predicting the position of a drunken ant." [481, p. 158]

Different from this rather figurative use of the term "ant," this section deals with the behavior of *real ants*, which are complex biological creatures. Searching and homing strategies of small-brain animals, like ants, have been developed to such a high level that one may attribute terms like "intelligence" [392] or "mental capacity" [561] to this kind of behavior. *Desert ants* for example find their way back to the nest after their foraging excursions have led them more than 250 m (!) away from the start [528]. This impressive performance may depend on several capabilities cooperating, such as (i) *geocentric navigation*, which uses landmarks (for desert ants, genus *Cataglyphis*, see [531]; with respect to cognitive maps see also [530]); and (ii) *egocentric navigation*, which is based on route integration (dead reckoning) and provides the animal with a continuously updated homeward-based vector (see [358, 532], for desert ants, genus *Cataglyphis*).

Both mechanisms are based on visual navigation and on the capability of internal storage of information (memory). On the other hand, different ant species are also capable of external storage of information, e.g., by setting chemical signposts (pheromones) that provide additional olfactory orientation. These chemicals are used for orientation and also for communication between individuals in a rather complex way (for a review of chemical communication in ants, see, for instance, [239, 494]). External storage of information by means of chemicals plays an important role in establishing the impressive trail systems for which *group-raiding ant species* are known. These trails provide the basic orientation for foraging and homing of the animals. A more detailed description of this phenomenon is given below.

With respect to the model of interacting walkers discussed in this book, we want to focus on the question whether these trail patterns could also be obtained under the restrictions that (i) no visual navigation and internal storage of information are provided; and (ii) at the beginning, no chemical signposts exist that lead the animals to the food sources and afterward back to the nest.

Certainly, these restrictions do not describe ants that can use the different levels of orientation mentioned. For example, the influence of memory and space perception on foraging ants, *Pogonomyrmex*, is investigated in [90, 196]. Our approach starts from quite a different perspective [443, 444, 456]. We do not try to reveal the biological constitutions of ants for solving this problem but propose a simple model to reproduce the characteristic trail patterns observed in ants by means of active walkers that have far less complex capabilities than the biological creatures. Our model therefore may serve as a *toy model* to test what kind of interaction between individuals may lead to a trail system and what are the minimal conditions for its existence.

In this section, the active walker model is used to simulate the formation of directed trails, as observed in group-raiding ants. The behavioral observations of trunk trail formation in ants are summarized in a most succinct manner by Hölldobler and Wilson [239]. As pointed out on page 285 of their comprehensive treatise, *The Ants*, "a great many species in the *Myrmicinae*, *Dolichoderinae*, and *Formicinae* lay trunk trails, which are traces of orientation pheromones enduring for periods of days or longer. In the case of leafcutter ants of the genus *Atta*, harvesting ants of the genus *Pogonomyrmex*, the European shining black ant *Lasius fuliginosus*, and the polydomous wood and mound-building ants in the *Formica exsecta* and *rufa* groups, the trails can last for months at a time (...) Trunk trails used for foraging are typically dendritic in form. Each one starts from the nest vicinity as a single thick pathway that splits first into branches and then into twigs (see Fig. 5.20). This pattern deploys large numbers of workers rapidly into foraging areas. In the species of *Atta* and the remarkable desert seed-eating *Pheidole militicida*, tens

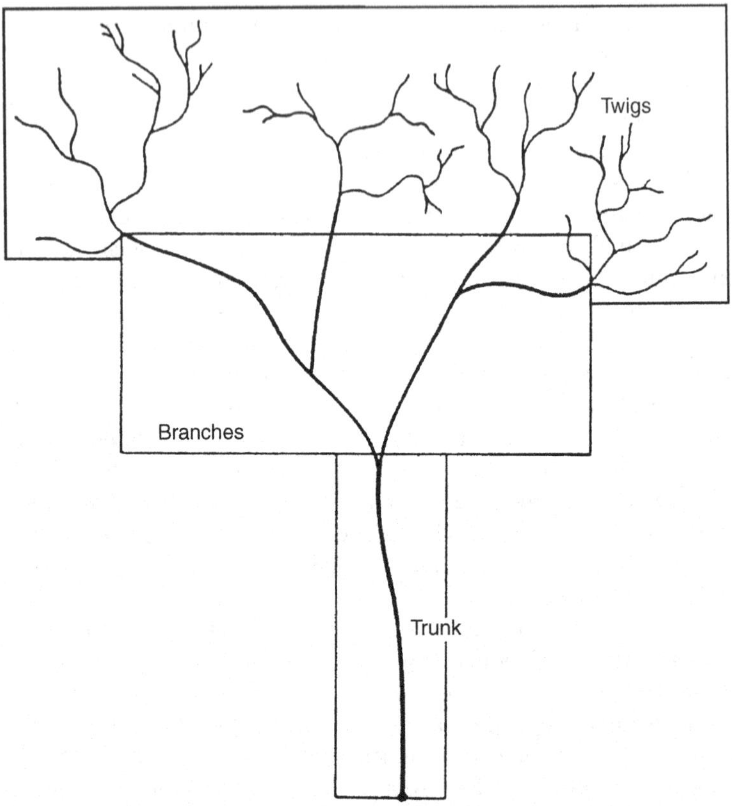

Fig. 5.20. Schematic representation of the complete foraging route of *Pheidole milicida*, a harvesting ant of the southwestern U.S. deserts. Each day tens of thousands of workers move out to the dendritic trail system, disperse singly, and forage for food (from Hölldobler and Möglich [238]; reprinted in [239, p. 285])

or even hundreds of thousands of workers move back and forth on a daily base. A few workers drift away from the main trunk route, but most do not disperse on a solitary basis until they reach the terminal twigs. When small food items are encountered, the workers carry them back into the outer branches of the system and then homeward. The twigs and branches can now be envisioned as tributaries of ant masses flowing back to the nest. When rich deposits of food are found, on the other hand, the foragers lay recruitment trails to them. In time new deposits of orientation pheromone accumulate along which foragers move with the further inducement of recruitment pheromones. By this process the outer reaches of the trunk-trail system shift subtly from day to day."

We note that the foraging behavior described above is, to a certain extent, different from that of the army ants (e.g., *Eciton burchelli* and other species of *Eciton* and *Dorylus*), althrough the foraging patterns of different *Eciton* species also show a dendritic structure [239, 494]. But *Eciton burchelli* is a swarm raider, whose foraging workers spread out into a fan-shaped swarm with a broad front, whereas most other army ant species are column raiders, whose mass moves outward along narrow trails (see the pioneering works of Schneirla [436, 437] and Rettenmeyer [408]). Recently, Deneubourg et al. [101] (see also [158]), presented a computer simulation model showing how these patterns could be generated from interactions between identical foragers. This model is based on trail-laying and trail-following behavior with an adjustable speed of the ants. It counts on a *single pheromone* and does not include additional recruitment.

The phenomena, discussed here, are based on *two different stages*; (i) *exploration* of the food sources by solitary scouts and (ii) *recruitment* and *exploitation* of the food sources. As reported, e.g., for the tropical ponerine ant, *Leptogenys ocellifera* [327] (see also the review in [529]) during the first stage, individual "scout" ants search solitarily for arthropod prey. On their way out of the nest, they deposit a trail pheromone consisting of poison-gland secretions, providing orientational cues. Once successful, the scout, on its way back to the nest, lays a pheromone trail consisting of both poison-gland and pygidial-gland secretions; the latter stimulate recruitment [23, 24]. Along this recruitment trail, up to 200 nest mates will move in single file, one behind the other (and not in a swarm raid) from the nest to the newly discovered food source and transport it to the nest.

As pointed out, trunk trails are used to bring a large number of foragers in a most effective way into the foraging area and to recruit new nest mates to the new food. The trails are used with considerable fidelty both for foraging and return trips. But trunk trails are also used as connecting routes by ants that occupy multiple nests, as reported for *Crematogaster* and species in the dolichoderine genera *Azteca*, *Hypoclinea*, and *Iridomyrex* [239].

5.4.2 Active Walker Model of Trail Formation in Ants

In this section, the active walker model of track formation introduced in Sect. 5.2.2 will be extended to simulate the formation of directed trails, as observed in group-raiding ants. There is a notable difference between tracks and trails. The tracks shown, for example, in Fig. 5.12, are *undirected* tracks used for movement. No destination points, such as nest or food sources, are present. A *directed trail*, however, is a specific link between a source and a destination.

To simulate trunk trail formation in ants, as described in the previous section, we consider active walkers that move on a surface according to (5.43), which is repeated here:

$$\begin{aligned} w(g_i'|g_i) = \frac{1}{c(g_i)} \Big\{ & (1-\eta_i)\, f(\boldsymbol{r}_i, g_i')\, (\delta_{g_i',g_i} + \delta_{g_i',g_i\pm 1}) \\ & + \eta_i \left[(d-b) + d(b-1)\, \delta_{g_i',g_i} \right] \Big\} \,. \end{aligned} \tag{5.52}$$

The normalization constant $c(g_i)$ is given by (5.44). Equation (5.52) describes a biased random walk, which additionally considers the response to chemical gradients within an angle of perception. We note that the parameter η, which refers to the noise in the systems, now becomes an "individual" parameter, η_i, similar to the individual *sensitivity* introduced in Sect. 3.1.1. The use of η_i will be explained below.

Initially, all walkers are concentrated in one place, the "nest." Further, every walker is characterized by an internal parameter $\theta_i = \pm 1$, which describes whether or not a walker has found food yet. If the walker starts from the nest, the internal parameter is always set to $\theta_i = +1$. Only if the walker has detected a food source, is θ_i changed to $\theta_i = -1$. Depending on the value of θ_i, the walker can produce one of two different chemicals at rate $s_i(\theta_i, t)$, which is obtained from the network model, discussed in Sect. 4.1.1:

$$\begin{aligned} s_i(\theta_i, t) = \frac{\theta_i}{2}\Big\{ &(1+\theta_i)\, q_{+1}^0 \, \exp[-\beta_{+1}\,(t - t_n^i)] \\ &- (1-\theta_i)\, q_{-1}^0 \, \exp[-\beta_{-1}\,(t - t_f^i)]\Big\} \,. \end{aligned} \tag{5.53}$$

Due to (5.53), a walker produces chemical of component $(+1)$, when it leaves the nest $(\theta_i = +1)$ at time t_n^i, and starts to produce chemical of component (-1) instead, after it has found food $(\theta_i = -1)$ at time t_{f}^i. The model assumes further that, as a question of disposal, the quantity of chemical produced by a specific walker after leaving the nest or the food source, decreases exponentially in time. s_θ^0 are the initial drop rates, and β_θ are the decay parameters for the drop rates.

The dynamics of two chemical components is given by the following reaction equation in accordance with (5.34):

$$h_\theta(\boldsymbol{r}, t + \Delta t) = (1 - k_\theta)\, h_\theta(\boldsymbol{r}, t) + Q[h_\theta(\boldsymbol{r}, t), N_\theta(\boldsymbol{r}, t)] \,. \tag{5.54}$$

Here, $h_\theta(\boldsymbol{r}, t)$ is the space- and time-dependent concentration of chemical θ that can decay at rate k_θ, but is not assumed to diffuse. The production function $Q[h_\theta(\boldsymbol{r}, t), N_\theta(\boldsymbol{r}, t)]$ is defined as follows:

$$Q[h_\theta(\boldsymbol{r}, t), N_\theta(\boldsymbol{r}, t)] = \sum_{i=1}^{N(t)} s_i(\theta_i, t)\, \delta_{\theta_i,\theta}\, \delta(\boldsymbol{r} - \boldsymbol{r}_i) \,. \tag{5.55}$$

Note that the total number of walkers, $N(t)$, is not a constant now and can change in time. We assume in accordance with biological observations, that walkers which have found food can recruit *additional walkers*. The recruitment rate is given by N_r, which is the additional number of walkers that move out, *if* one successful walker $(\theta_i = -1)$ was able to return to the nest. The maximum number of walkers in the simulations is limited to N_{m}, which denotes the *population size*.

Let us add some notes in favor of the assumptions with respect to the biological phenomenon to be described. Active walkers introduced should represent ants which continuously mark their way with a pheromone. This does not hold for all of the about 18,000 different ant species, but some species, such as army ants *Eciton* or Argentine ant *Iridomyrmex humilis*, behave that way. In these cases, chemical signposts are always present, and deciding which trail to follow depends only on the actual pheromone concentration [76, 180]. We have further assumed that walkers produce two kinds of chemicals (or pheromones). Which kind of chemical is used by a walker will be specified depending on whether or not it has found food. Only in the first case, recruitment of additional walkers occurs. As reported in Sect. 5.4.1, ants can surely use different kinds of pheromones for orientation and recruitment; the orientation pheromones comprising the trunk trails are often secreted by glands different from those that produce the recruitment pheromones.

To complete the model, we now need to specify how walkers are affected by different kinds of chemicals. In agreement with the biological observations in Sect. 5.4.1, we assume that initially a group of N_0 walkers, which we call the *scouts*, leaves their nest. Along the way, they drop a trail pheromone, denoted as chemical (+1). Because active walkers act only locally and determine their way with no additional information, a scout can hit a food source only by chance. In that case, this particular scout begins to drop a different pheromone, chemical (−1), indicating the discovery of a food source. But it continues to be sensitive to chemical component (+1), which provides local orientation, and therefore increases the chance that the scout is guided back to the nest.

Hence, the scouts are sensitive only to chemical (+1), and the response function $f(\boldsymbol{r}_i, g_i)$ in (5.52) consequently reads for *scouts*,

$$\begin{aligned} f(\boldsymbol{r}_i, g) &= \Theta\Big[h_{+1}(\boldsymbol{r},t) - h_{\text{thr}}\Big]\; \frac{\partial h_{+1}(\boldsymbol{r},t)}{\partial \boldsymbol{r}}\bigg|_{\boldsymbol{r}_i} \\ &= \Theta\Big[h_{+1}(\boldsymbol{r},t) - h_{\text{thr}}\Big]\; \frac{h_{+1}(\boldsymbol{r}+\boldsymbol{\Delta r}(g),t) - h_{+1}(\boldsymbol{r}-\boldsymbol{\Delta r}(g),t)}{2\,\boldsymbol{\Delta r}}\;. \end{aligned} \tag{5.56}$$

Here, it is assumed again that walkers respond to *gradients* in the chemical field, if the amount is above a certain threshold h_{thr}.

When a successful scout ($\theta_i = -1$) returns to the nest, it recruits a number N_r of walkers with the internal parameter $\theta_i = +1$ to move out. The *recruits* are different from the scouts only in that they are sensitive to chemical component (−1) and not to (+1), when they start from the nest, but they also drop chemical (+1) as long as they have not found any food. If a recruit hits a food source, it switches its internal parameter to $\theta_i = -1$. Hence, it begins to drop chemical (−1) and becomes sensitive to chemical component (+1), which should guide it back to the nest where it, indicated by $\theta_i = -1$, can recruit new walkers.

Because recruits are sensitive to both chemicals, depending on their current situation, the response function for *recruits* reads

$$f(\boldsymbol{r}_i, g) = \frac{\theta_i}{2} \left\{ \Theta\Big[h_{-1}(\boldsymbol{r}_i, t) - h_{\text{thr}}\Big] \, (1 + \theta_i) \left. \frac{\partial h_{-1}(\boldsymbol{r}, t)}{\partial \boldsymbol{r}} \right|_{\boldsymbol{r}_i} \right.$$
$$\left. - \, \Theta\Big[h_{+1}(\boldsymbol{r}_i, t) - h_{\text{thr}}\Big] \, (1 - \theta_i) \left. \frac{\partial h_{+1}(\boldsymbol{r}, t)}{\partial \boldsymbol{r}} \right|_{\boldsymbol{r}_i} \right\} . \quad (5.57)$$

In accordance with biological observations, we now have a two-step scenario characterized by (i) exploration of food sources and (ii) recruitment and exploitation of food sources. We want to point out again that neither the scouts nor the recruits have any information about the location of the nest or the food sources. They use only the local orientation provided by the concentration of that chemical to which they are sensitive. They make a local decision only about the next step and cannot store their way in an individual memory.

On could argue that to simulate the foraging trails of ants, the most simple assumption could be that *only* individuals that found food create a pheromone trail, while turning straight back to the nest. These trails can be used by other ants to find food sources, once they find the trails. However, those simulations have to overcome the problem of where the first ant which found food got the information to find its way back home. In biological systems, geocentric or egocentric navigation as described in Sect. 5.4.1, could provide this additional information needed for successfully returning to the nest.

Because our model simulates the formation of trunk trails without counting on navigation and internal storage of information, our main assumption is that a simple chemotactic response would be sufficient to generate trails between the nest and food sources. However, we have to distinguish trails that have been created during an (unsuccessful) search for food from trails that really lead to a food source. In our model, this additional information is encoded by assuming that walkers now produce, and respond to, two different chemical fields.

To complete our model, we have to address an additional problem. It is known from central foragers like ants that they can leave a place where they don't find food and reach out to other areas. This indicates that they can at least increase their mobility because they have a certain aim, finding food. Active walkers, on the other hand, do not have aims and do not reflect a situation. So, they stick to their local markings even if they did not find any food source. To increase the mobility of active walkers in those cases, we assume that every walker has individual sensitivity to follow a chemical gradient. This would be the factor $(1-\eta_i)$ in the probability, $w(g_i'|g_i)$, (5.52), for choosing the next direction. As long as a walker does not hit a food source, this sensitivity is continuously decreasing via an increase in η_i, which means that the walker more and more ignores the chemical detected and thus also becomes able to choose sites not visited so far.

However, if the walker does not find any food source after a certain number of steps, the sensitivity $(1 - \eta_i)$ goes to zero, and the walker behaves like a (biased) random walker. From the utilitarian perspective, such a walker becomes "useless" for the exploitation of food sources, if its individual noise level η_i exceeds a maximum value $\eta_{\max}$. So we simply assume that the walker "dies" at this level and is removed from the system. On the other hand, if the walker hits a food source, its sensitivity is set back to the initial high value and is kept constant to increase the chance that the walker finds its way back along the gradient. The switch of the level of sensitivity is accompanied by the use of the two different chemical markers, and in this particular sense reminds us of *chemokinesis* (cf. Sect. 3.1.1), where the level of activity can be changed due to chemical conditions in the surroundings.

Using the internal parameter θ_i again, we may assume that the sensitivity of a walker changes according to

$$1 - \eta_i = 1 - \frac{\theta_i}{2} \left\{ (1 + \theta_i) \left[\eta_0 + \delta\eta \, (t - t_n^i) \right] + (1 - \theta_i) \, \eta_0 \right\} . \qquad (5.58)$$

Here, $\delta\eta$ is the increase in the individual noise level η_i per time step if the walker has not found any food source.

In this way, we have completed our model for the formation of trunk trails in ants. The basic ingredient is the chemotactic response to a chemical gradient which is produced by walkers themselves. Hence, the basic equations are given by the probability of choosing the next step, (5.52), which is coupled with equations for the two chemical fields, (5.53), (5.54), and (5.55). Other features, such as recruitment or the adjustment of sensitivity, are included to make the simulations more realistic in accordance with biological observations.

We note that, with respect to biology, there are different parameters that may influence trail following in addition to sensitivity, such as trail fidelity, traffic density, detection distance, endurance of the trail, navigation capabilities [128, 196]. However, our model considers only minimal assumptions for trail formation. Here, the formation of trail patterns is based solely on simple *local* chemical communication between walkers, with no additional capabilities of orientation or navigation.

The directed trails, which should connect the nest with the food sources, do not exist from the very beginning; they have to be generated by the walker community in a process of self-organization. Hence, walkers have to perform two quite different tasks, which are referred to each other: first, they have to detect food places – unknown to them, and then they have to link these places to their original starting point by forming a trail, with local chemical orientation as the only tool provided. The computer simulations outlined below should prove that the simple local rules assumed for the actions of walkers are sufficient to solve such a complex problem.

5.4.3 Simulation of Trunk Trail Formation in Ants

Different from the arbitrary *nondirected* track patterns, *directed trails* should link a starting point (e.g., a nest) to a destination point (e.g., a food source). To define start and destination, we assume a center (nest) on a two-dimensional surface where the walkers are initially concentrated, and a number of food sources unknown to the walkers. Both the nest and the food sources *do not attract* the walkers by a certain long-range attraction potential; they are just particular sites of a certain size (defined in lattice sites). To make the simulation more realistic, we may consider that the food sources can be exhausted by visting walkers, which, it is assumed, carry part of them back to the nest.

Our simulations are carried out on a triangular lattice of size 100×100 with periodic boundary conditions. For the food sources, two different distributions are assumed: (1) continous food distribution at the top/bottom lines of the lattice and (2) random distribution of five separated food sources. Both types of sources are simulated by N_F food clusters, each of a size of seven lattice sites, where F_0 is the total amount of food in each cluster. Because the sources could be exhausted by visting walkers, we introduce the parameter df as the amount taken away by a walker, so that all sources are exhausted after $n_\mathrm{ex} = N_F F_0/df$ visits.

To conclude the setup for the computer simulations, we give the parameters (in arbitrary units) chosen for the simulations discussed below:

$$N_0 = 20,\ N_r = 5,\ N_m = 100,\ s_\theta^0 = 80,\ h_\mathrm{thr} = 1,\ \beta_\theta = 0.0512,$$
$$\eta_0 = 0.01,\ \delta\eta = 0.001,\ \eta_\mathrm{max} = 0.5,\ b = 3\ (a = 0.6)\ .$$

Further, we have chosen

$$\text{for simulation (1):}\ \ k_\theta = 0.001,\ N_f = 50,\ F_0/df = 55\ ;$$
$$\text{for simulation (2):}\ \ k_\theta = 0.01,\ N_f = 5,\ F_0/df = 550\ .$$

(1) Trunk Trail Formation to Extended Forage Areas

First we discuss computer simulations for trail formation between a nest (in the center) and an extended food source, which is a longer distance from the nest. This is also the case, on which the biological observation of Fig. 5.20 is based. The time series of Fig. 5.21 shows the evolution of the trail system for exploiting the extended food source (at the top/bottom lines of the lattice) in terms of the spatial concentration of the chemical components (+1) (*left*) and (−1) (*right*). The concentration is coded in a gray scale; the deeper gray marks the higher concentration [on a logarithmic scale, see (4.9)].

Figure 5.21a,d gives the concentration of both chemicals after the initial period. We see that the food is already discovered by the walkers, but so far no trail exists between the food and the nest, indicated in the nonoverlapping concentration fields of both components. Therefore, during the simulation, we

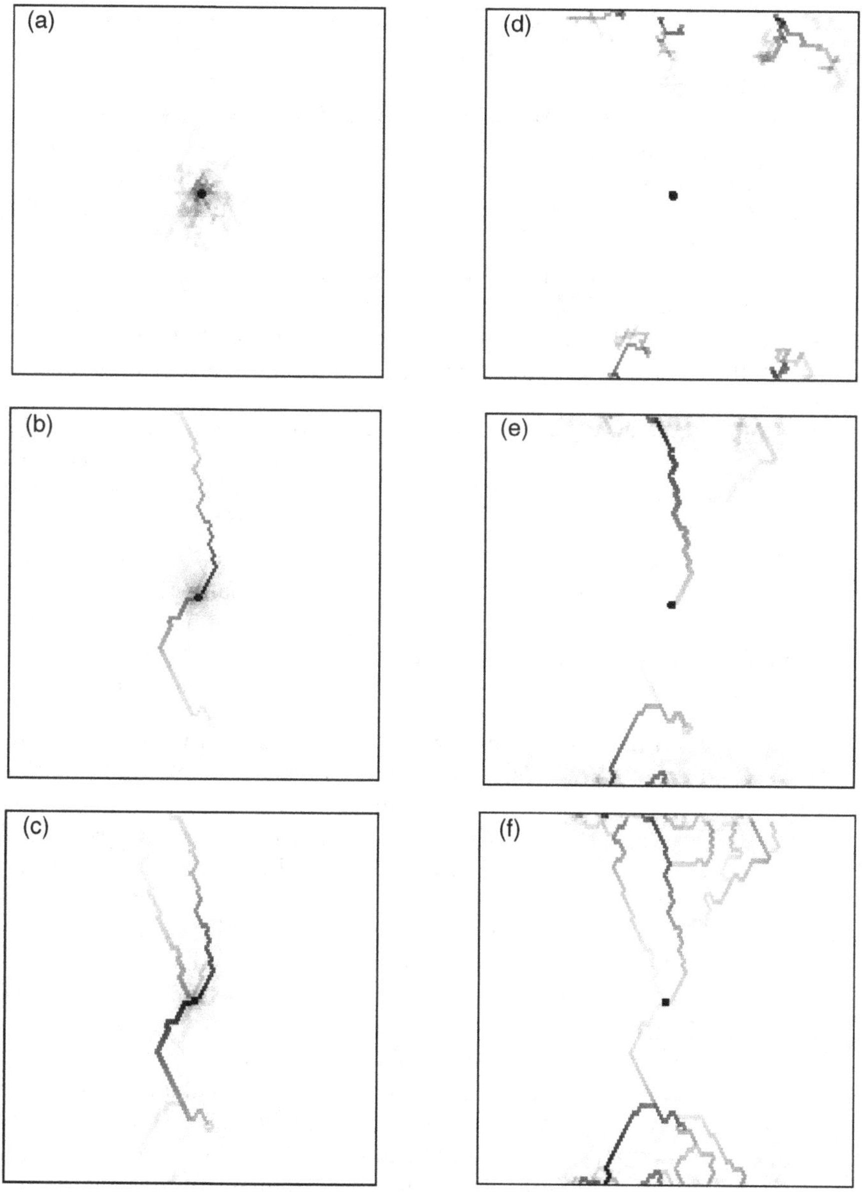

Fig. 5.21. Formation of trails from a nest (*middle*) to a line of food at the top and the bottom of a lattice. (**a**)–(**c**) show the distribution of chemical component (+1) (see text), and (**d**)–(**f**) show the distribution of chemical component (−1). Time in simulation steps: (**a**), (**d**) $t = 1000$; (**b**), (**e**) $t = 5000$, (**c**), (**f**) $t = 10,000$. For parameters, see text [456]

see walkers moving quite randomly. But, after a time lag of disorientation, the emergence of distinct trunk trails can be clearly observed in Fig. 5.21b,e and Fig. 5.21c,f.[1]

A match of the concentration fields for both chemical components for the main trails shows that the walkers use the same trunk trails for their movements toward the food sources and back home. The exhaustion of some food clusters and the discovery of new ones in the neighborhood results in branching of the main trails in the vicinity of the food sources (at the top or bottom of the lattice), leading to the dendritic structures expected. The trail system observed in Fig. 5.21c,f remains unchanged in its major parts – the trunk trails, although some minor trails in the vicinity of the food sources slightly shift in time – as has also been reported in the biological observations of trunk trail formation in ants.

The simulations presented in Fig. 5.21 show two separate trail systems, one from the nest to the top, the other to the bottom of the lattice. The emergence of both of them at the same time depends on the number of walkers available to maintain the whole trail system. Because our model is probabilistic, the trunk trails of the upper and lower trail system are not identical, but the dendritic structures are similar. Especially, the trail system to the food sources at the bottom is, with respect to its branching structure, very similar to that drawn in Fig. 5.20. Thus, we conclude that the basic features of trunk trail formation are represented by our simulation.

(2) Trunk Trail Formation to Separate Conspicuous Food Items

In a second example, we discuss computer simulations for trail formation between a nest (in the center) and a number of separated food sources at different distances from the nest. Figure 5.22 shows a time series of trunk trail formation in terms of the spatial concentration of chemical component (-1), which is coded again in a gray scale. In Fig. 5.22a, we see that two of the food sources randomly placed on the lattice are about the same distance to the nest, but the one, which – by chance – has been discovered first, will also be the one linked first to the nest. This reflects the influence of *initial symmetry-breaking effects.*

We assumed that the food sources could be exhausted by walkers carrying food to the nest. Once the food source vanished, the individual sensitivity of walkers coming from the nest to the food cannot be set back to the high initial value, and walkers further increase their mobility by ignoring the trail, they reach out again, and by chance discover new food sources. But because a trail already exists, those sources have a larger probability of being discovered that are close to the one that disappeared, and part of the "old" trail is reused now to link the new food source to the nest (compare Fig. 5.22a,b,c).

[1] The reader is invited to see a video of these computer simulations – with a different set of parameters – at `http://www.ais.fhg.de/~frank/ants.html`.

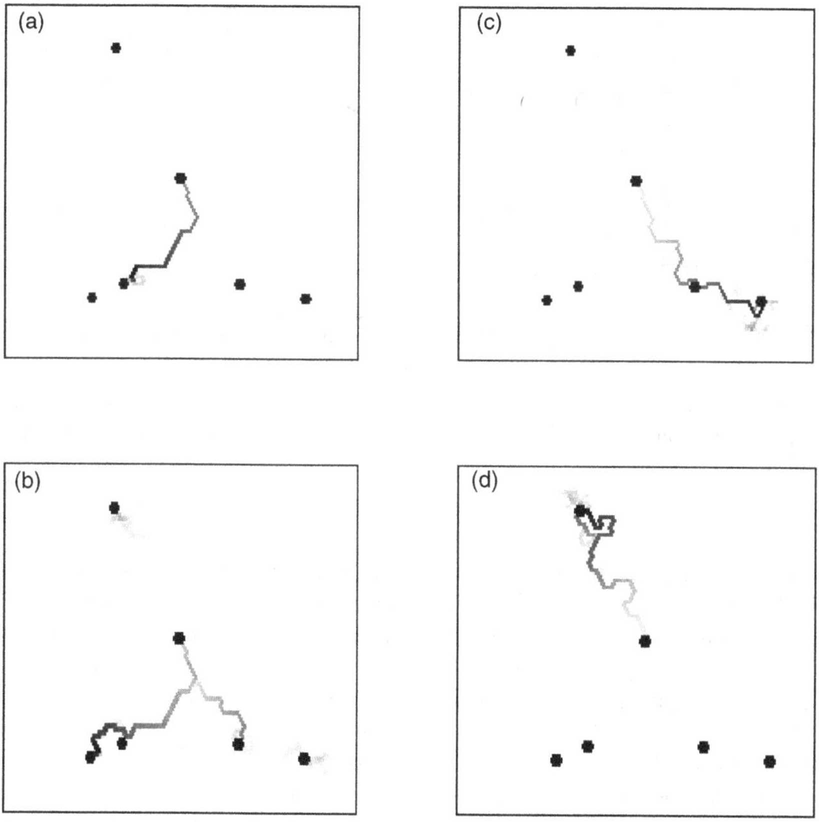

Fig. 5.22. Formation of trails from a nest (*middle*) to five randomly placed food clusters. The distribution of chemical component (−1) (see text) is shown after (**a**) 2000, (**b**) 4000, (**c**) 8500, and (**d**) 15,000 simulation time steps, respectively. For parameters, see text [456]

Figure 5.22b shows that at the same time also more than one source could be linked to the nest by trunk trails. However, all trails compete to be maintained, and the trails could survive only if the concentration of chemicals is above a certain critical value. To compare this fact with the simulations presented in Fig. 5.21 (where two trunk trail systems appeared at the same time), we have chosen for the simulations of Fig. 5.22, a decay rate of the chemicals ten times higher than that for the simulations of Fig. 5.21. This clearly increases the selection pressure, and therefore, the coexistence of different trails lasts only for a short time. Hence, the trunk trails to the different food sources will appear one after the other; and the old trails disappear again by decomposition because they are no longer maintained. Because we do not have extended sources now, the trunk trails do not branch in the vicinity of the food, exept in the coexistence state, when connecting different sources at the same time.

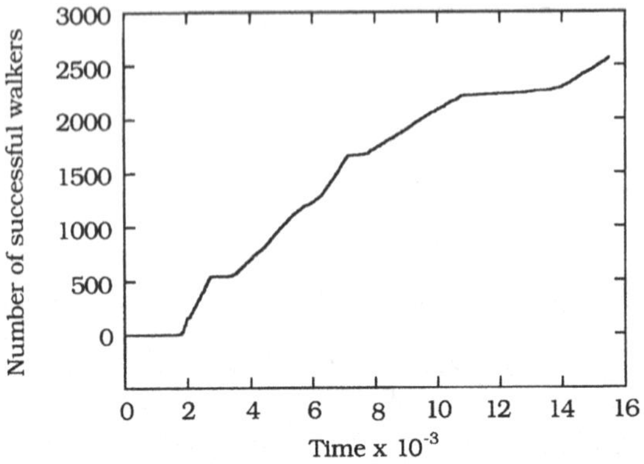

Fig. 5.23. Cumulative number of walkers that have come back to the nest with food as a function of time (in simulation steps). The data are obtained from the simulation presented in Fig. 5.22 [456]

The flexibility of the model is indicated by the fact that, even after a long period of trail formation in the lower part of the lattice, the walkers can link the food source on the upper left side of the lattice to the nest (Fig. 5.22d) and therefore finally have detected and linked all five sources to the nest.

For the simulation shown in Fig. 5.22, Fig. 5.23 gives the cumulative number of walkers which return to the nest with food (indicated by the internal parameter $\theta_i = -1$) as a function of time.

We see a large initial time lag indicating disorientation, where no walker finds the way home because of the lack of a trail. But the local interactions of walkers eventually lead to the emergence of the first trail, which is reinforced by recruits. Hence, the concentration of both chemical components on the trails is increased, and the process of food exploitation is self-amplified. This is indicated by an increasing number of walkers successfully returning to the nest. Once the source is exhausted, we see a new time lag needed to form the trail to the second source, and so on.

Thus, we can distinguish again between two alternating dynamic stages: (i) a stage of rather random motion of walkers (*exploration* of sources) and (ii) a stage of directed motion of walkers along the trail (*exploitation* of the sources). The disorientation stage before exploiting the last source is especially long, because walkers could not build on any previous part of a trail in that direction, whereas the disorientation stage before exploiting the third and the fourth sources is very short because of trails into that area that already exist.

Let us conclude the simulations above. In our model, active walkers that do not have the ability of visual navigation or storage of information, first have to discover different distributions of food sources and then have to link these sources to a central place by forming a trail, using no other guidance than the chemical markings that they produced. It turns out from the computer simulations that, for different kinds of food sources, the model generates a distinctive trail system to exploit food sources and is highly flexible in discovering and linking new sources. During the evolution of the trail system, we can distinguish between two different stages: the first is a rather random movement of walkers dropping chemical almost everywhere, and no trail exists. But during the second stage, a distinct major trail appears that is reamplified by the walkers moving forward and backward on the trail.

The spontaneous emergence of a collective trail system can be described as a self-organizing process. As in every evolutionary game, also for the self-organization of trail systems, critical parameters for the emergence of the structures exist. This is directly related to the discussion of critical parameters in Sects. 3.2.2 and 4.2. If, for instance, the decay rate for the decomposition of the chemicals is too high or the sensitivity of the walkers is too low, they lose their local orientation and get lost, so that no trail appears. If, on the other hand, the decay rate is too low, it takes a much longer time before a distinct trunk trail appears because the selection pressure is too low. Or, if the sensitivity of walkers is too high when they move out of the nest, their motion is restricted close to the nest because they always follow markings. Similar considerations hold for the local orientation of successful walkers which have to find their way back to the nest. But only if the number of recruits is large enough to maintain the whole trail system will it keep its distinctive overall structure intact and become a special kind of *quasi-stationary structure* – as long as the food sources last.

In the model provided, the formation of trail systems is based on interactions of walkers on a local or "microscopic" level that could lead to the emergence of the structure as a whole on a global or "macroscopic" level. Compared to the complex "individual-based" models in ecology [98, 252], the active walker model proposed here provides a very simple but efficient tool for simulating a specific structure with only a few adjustable parameters. The major difference from biology is that the active walkers used in the simulations do not have individual capabilities of storing and processing informations about the locations of food or the nest or counting on navigational systems or additional guidance. They rather behave like physical particles that respond to local forces quite simply without "implicit and explicit intelligence" [196]. With respect to the formation of trunk trails, this could indicate that visual navigation and information storage do not necessarily have to be indispensable presumptions to obtain advanced and efficient foraging patterns.

6. Movement and Trail Formation by Pedestrians

6.1 Movement of Pedestrians

6.1.1 The Social Force Model

In this chapter, modeling of biological motion, track formation, and trail formation is extended to social systems. It will be shown that within the framework of Brownian agents, the motion of humans and the generation of human trails can also be suitably described.

Quantitative approaches to the movement of pedestrians date back to the 1950s [332]. They range from queuing models [309], which do not explicitly take into account the effects of the specific geometry of pedestrian facilities, to models of route choice behavior [65] of pedestrians in dependence on their demands, entry points, and destinations. The last has the drawback of not modeling the effects of pedestrian interactions, except that pedestrians will take detours if their preferred way is crowded. Further, it has been suggested that pedestrian crowds can be described by the Navier–Stokes equations of fluid dynamics [228]. This approach, however, implicitly assumes that energy and momentum are collisional invariants which is obviously not the case for pedestrian interactions. Recently, Helbing proposed a fluid-dynamic pedestrian model. This is a pedestrian specific *gas kinetic* (Boltzmann-like) model [208, 209] that also considers the intentions, desired velocities, and pair interactions of pedestrians.

Because a numerical solution of fluid dynamic equations is very difficult, current research rather focuses on the *microsimulation* of pedestrian crowds. Here, active walker models, as discussed in a biological context in Sects. 5.3, 5.4 come into play. We will start with a continuous description in terms of Brownian agents, instead of the discrete description used in previous sections.

Brownian agent i represents a *pedestrian* now who moves on a two-dimensional surface and is described by the following Langevin equations:

$$\frac{d\boldsymbol{r}_i(t)}{dt} = \boldsymbol{v}_i(t) \ , \quad \frac{d\boldsymbol{v}_i(t)}{dt} = -\frac{1}{\tau_i}\boldsymbol{v}_i(t) + \boldsymbol{f}_i(t) + \sqrt{\frac{2\,\varepsilon_i}{\tau_i}}\,\boldsymbol{\xi}_i(t). \tag{6.1}$$

τ_i represents the *relaxation time* of velocity adaptation, which is related to the friction coefficient, $\gamma_0 \sim 1/\tau_i$. The term $\boldsymbol{f}_i$ in (6.1) represents deterministic influences on the motion, such as intentions to move in a certain direction at a certain desired velocity, or to keep a distance from neighboring agents. $\boldsymbol{f}_i(t)$ can be interpreted as a *social force* that describes the influence of the environment and other pedestrians on individual behavior. However, the social force is not *exerted* on a pedestrian, it rather describes concrete motivation to act [210, 301].

The *fluctuation term* takes into account random variations of behavior, which can arise, for example, when two or more behavioral alternatives are equivalent, or by accidental or deliberate deviations from the usual rules of motion. ε_i [cf. (3.2)] is again the intensity of the stochastic force $\boldsymbol{\xi}_i(t)$. The i dependence takes into account that different agents could behave more or less erratically, depending on their current situations.

There are some major differences between social forces and forces in physics:

1. For social forces, the Newtonian law *actio* = *reactio* does not hold.
2. Energy and momentum are *not collisional invariants*, which implies that there is no energy or momentum conservation.
3. Pedestrians (or, more generally, individuals) are *active systems* that produce forces and perform changes themselves.
4. The effect of social forces comes about by *information exchange* via complex mental, psychological, and physical processes.

In the following, we want to specify the social force model of pedestrian motion [208, 218] that, in a simpler form, has also been proposed in [177]. First, we assume that a pedestrian i wants to walk in a *desired direction* $\boldsymbol{e}_i$ (the direction of his/her next destination) at a certain *desired speed* v_i^0. The desired speeds within pedestrian crowds are Gaussian distributed [227, 229, 230, 363]. Hence, deviation of the *actual velocity* $\boldsymbol{v}_i$ from the *desired velocity* $\boldsymbol{v}_i^0 := v_i^0\boldsymbol{e}_i$ leads to a tendency $\boldsymbol{f}_i^0$ to approach $\boldsymbol{v}_i^0$ again within a certain *relaxation time* τ_i. This can be described by an *acceleration term* of the form

$$\boldsymbol{f}_i^0(v_i^0\boldsymbol{e}_i) := \frac{1}{\tau_i}v_i^0\boldsymbol{e}_i\,. \tag{6.2}$$

In a second assumption, we consider that pedestrians keep a certain distance from *borders* (of buildings, walls, streets, obstacles, etc.). This effect $\boldsymbol{f}_{i\mathrm{B}}$ can be described by a repulsive, monotonic decreasing potential V_B:

$$\boldsymbol{f}_{i\mathrm{B}}(\boldsymbol{r}_i - \boldsymbol{r}_\mathrm{B}^i) = -\nabla_{\boldsymbol{r}_i}V_\mathrm{B}(|\boldsymbol{r}_i - \boldsymbol{r}_\mathrm{B}^i|)\,. \tag{6.3}$$

Here, $\boldsymbol{r}_i$ is the actual *location* of pedestrian i, whereas $\boldsymbol{r}_\mathrm{B}^i$ denotes the location of the border that is nearest location $\boldsymbol{r}_i$.

Additionally, the motion of a pedestrian i is also influenced by other pedestrians j. These interactions cause him/her to perform *avoidance maneuvers* or to slow down to keep a situation-dependent distance from other pedestrians. A quite realistic description of pedestrian interactions results from the assumption that each pedestrian i respects the *"private spheres"* of other pedestrians j. These *territorial effects* $\boldsymbol{f}_{ij}$ can be modeled by *repulsive potentials* $V_j(\beta)$:

$$\boldsymbol{f}_{ij}(\boldsymbol{r}_i - \boldsymbol{r}_j) = -\nabla_{\boldsymbol{r}_i} V_j\Big\{\beta(\boldsymbol{r}_i - \boldsymbol{r}_j)\Big\}. \tag{6.4}$$

The sum of repulsive potentials V_j defines the *interaction potential* V_{int} which influences the behavior of each pedestrian:

$$V_{\text{int}}(\boldsymbol{r}_i, t) := \sum_j V_j\Big\{\beta[\boldsymbol{r}_i - \boldsymbol{r}_j(t)]\Big\}. \tag{6.5}$$

Additionally, the tendency of social groups (friends, family) to move together can be considered *attractive* interaction potentials very similar to (6.5), but with an opposite sign [350].

The *social force (total motivation)* $\boldsymbol{f}_i$ is given by the *sum* of all effects, (6.2), (6.3), (6.4), that influence a pedestrian's decision at the *same* time:

$$\begin{aligned}\boldsymbol{f}_i(t) &:= \boldsymbol{f}_i^0(v_i^0 \boldsymbol{e}_i) + \boldsymbol{f}_{i\text{B}}(\boldsymbol{r}_i - \boldsymbol{r}_\text{B}^i) + \sum_{j(\neq i)} \boldsymbol{f}_{ij}(\boldsymbol{r}_i - \boldsymbol{r}_j) \\ &= \frac{1}{\tau_i} v_i^0 \boldsymbol{e}_i - \nabla_{\boldsymbol{r}_i}\Big[V_\text{B}(|\boldsymbol{r}_i - \boldsymbol{r}_\text{B}^i|) + V_{\text{int}}(\boldsymbol{r}_i, t)\Big]. \end{aligned} \tag{6.6}$$

The Langevin equation, (6.1), together with the social force, (6.6), characterize the social force model of pedestrian motion. Hence, the description of pedestrian *crowds* consists of nonlinearly coupled stochastic differential equations of the form (6.1), (6.6).

6.1.2 Simulation of Pedestrian Motion

In the previous section, the dynamic behavior of pedestrian crowds has been formulated in terms of a social force model; that means that pedestrians behave as if they are subject to an accelerative force and to repulsive forces describing the reaction to borders and other pedestrians. For computer simulations, we need to specify the repulsive potential $V_j(\beta)$, which describes the "private spheres" of pedestrians.

Here, we simply assume [221] a monotonic decreasing function in β that has equipotential lines in the form of an ellipse that is directed in the direction of motion (see Fig. 6.1). The reason for this is that pedestrian j requires space for the next step, which is taken into account by pedestrian i. $\beta(\boldsymbol{r}_i - \boldsymbol{r}_j)$ denotes the *semiminor axis* of the ellipse and is given by

$$2\beta = \sqrt{(|\boldsymbol{r}_i - \boldsymbol{r}_j| + |\boldsymbol{r}_i - \boldsymbol{r}_j - v_j\,\Delta t\,\boldsymbol{e}_j|)^2 - (v_j\,\Delta t)^2}\,. \tag{6.7}$$

$l_j := v_j\,\Delta t$ is about the *step width* of pedestrian j. For more refined assumptions about the repulsive interaction potential $V_j(\beta)$, see [350].

The social force model of pedestrian dynamics, (6.1), (6.6), has been simulated by Helbing, Molnár [218–221, 350] for a large number of interacting

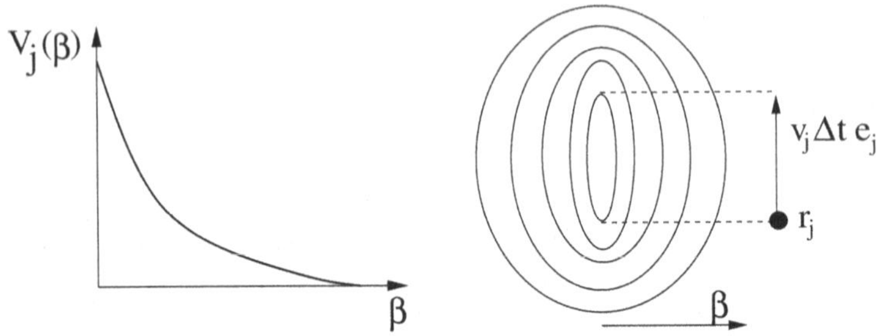

Fig. 6.1. (*left*) Decrease of the repulsive interaction potential $V_j(\beta)$ with β, (*right*) elliptical equipotential lines of $V_j[\beta(\boldsymbol{r} - \boldsymbol{r}_j)]$ for different values of β [221]

pedestrians confronted with different situations. Despite the fact that the proposed model is very simple, it describes a lot of observed phenomena very realistically. Under certain conditions, the emergence of *spatiotemporal patterns*, or *collective behavior*, in pedestrian crowds can be observed, for example, (i) the development of lanes (groups) consisting of pedestrians walking in the *same* direction, (ii) oscillatory changes of walking direction at narrow passages (for example, doors), and (iii) spontaneous formation of roundabout traffic at intersections.[1]

Figure 6.2 shows two examples of *trajectories* of pedestrians moving in different directions on a narrow walkway. Without obstacles, we find the *formation of lanes* indicated by the bundling of trajectories of pedestrians moving in the same direction. The movement in lanes is due to the fact that avoidance maneuvers are less frequently necessary if pedestrians walking in the same direction move close to each other. These lanes vary dynamically and depend on the width of the floor as well as on pedestrian density.

The lower part of Fig. 6.2 shows the formation of lanes in the presence of an obstacle, such as a double door. Pedestrian trajectories indicate that, in the case of two alternative passages, each door is occupied by one walking direction for a long period of time. As the result of a spontaneous symmetry break which is reinforced by nonlinear interaction of pedestrians, eventually almost all pedestrians moving in the same direction use the same door.

The phenomena shown in Fig. 6.2, which also agree with our individual experience, are the result of a self-organization process among pedestrians rather than of external forces or deliberative actions. The only assumptions of the social force model are that each pedestrian is characterized by a desired direction, a desired speed, and a desired distance from obstacles or other pedestrians.

---

[1] The reader is invited to check out the nice Java applets on `http://www.helbing.org/Pedestrians/` simulating different forms of collective behavior in pedestrians: (i) lane formation in a street, (ii) oscillation at a bottleneck, and (iii) interactions at a crossing.

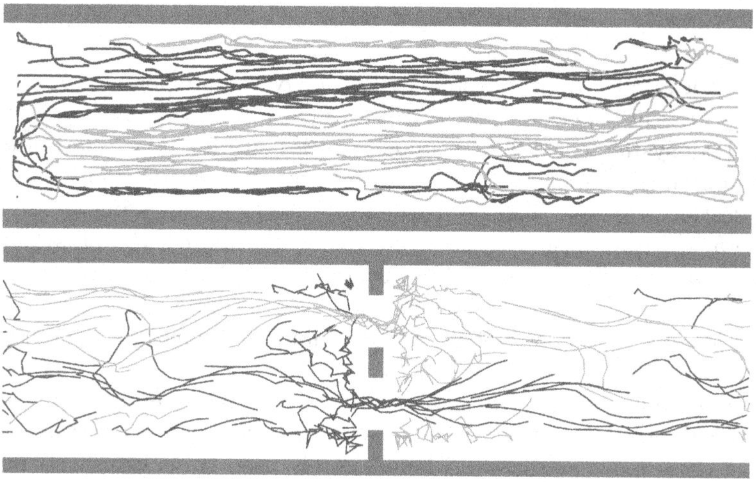

Fig. 6.2. Trajectories of pedestrians moving on a narrow walkway. Dark gray indicates pedestrians moving from left to right, and light gray indicates pedestrians moving in the reverse direction. (*upper part*): Spontaneous formation of lanes in the absence of obstacles [219]; (*lower part*): spontaneous symmetry break in the use of two doors. The picture indicates some pedestrian "jams" in front of the respective doors; further we see that pedestrians, who have chosen the "wrong" door, are trapped in the corners [220]

We note that the computer simulations of pedestrian crowds can be extended to optimization of pedestrian facilities by varying the geometric specification of floors, walkways, etc., and then calculating efficiency measures of pedestrian flow, as suggested in [220, 350]. Another important research area that also recently attracted the attention of physicists is simulation of pedestrian crowds in *panic* or *evacuation* situations[2] [213, 215, 438]. It has been shown by Helbing et al. that these phenomena can be simulated and explained by Brownian agent models rather similar to those discussed in this book [212, 214].

6.2 Trail Formation by Pedestrians

6.2.1 Model of Trail Formation

The trajectory patterns that occur in Fig. 6.2 as the result of self-organization in pedestrian crowds are *virtual patterns*, different from tracks or trails that may exist on their own – at least for a while – in time *and* space. But

---

[2] Java Applets for extensive computer simulations of panics and escape dynamics of pedestrians as well as related information can be found at `http://angel.elte.hu/~panic/pedsim/index.html`

in this section, we will extend the description of pedestrian motion also by considering of human trail formation.

The formation of trail systems is a widely spread phenomenon in ants and also in other species, such as hoof animals, and mice, and even humans [424]. Trail systems of different animal species and humans differ, of course, regarding their shape, duration, and extension. However, more striking is the question whether there is a common underlying dynamics that allows a generalized description of the formation and evolution of trail systems. In this section, we will derive a rather general model of trail formation based on the model of trail formation by ants (see Sect. 5.4) and a social force model of pedestrians that now will be specified for the formation of human trail systems.

As our experience tells us, human trails adapt to the requirements of their users. In the course of time, frequently used trails become more developed, making them more attractive, whereas rarely used trails vanish. Trails with large detours become optimized by creating shortcuts. New destinations or entry points are connected to an existing trail system. These dynamic processes occur basically without any common planning or direct communication among users. Instead, the adaptation process can be understood as a self-organization phenomenon, resulting from nonlinear feedback between users and trails [442].

For the derivation of the general model, we start with a continuous description in terms of Brownian agents instead of the discrete description used in Sect. 5.4. The agents are not specified as pedestrians or animals; rather they are considered arbitrary moving *agents*, which can be described by a Langevin equation similar to (6.1). Again, these agents continuously change their environment by leaving *markings* while moving. These markings can, for example, be imagined as damaged vegetation on the ground (for hoofed animals or pedestrians) or as chemical markings (for ants).

The spatiotemporal distribution of existing markings will be described by a *ground potential* $G_\theta(\boldsymbol{r},t)$. Trails are characterized by particularly large values of $G_\theta(\boldsymbol{r})$. The subscript θ allows us to distinguish different *kinds* of markings. Due to weathering or chemical decay, the markings have a certain *lifetime* $T_\theta(\boldsymbol{r}) = 1/k_\theta$ that characterizes their local *durability.* k_θ denotes the decay rate used in (4.4). Therefore, existing trails tend to fade, and the ground potential would exponentially adapt to *natural ground conditions*, $G^0_\theta(\boldsymbol{r})$, if the production of markings is stopped. However, the creation of new markings by agent i is described by the term $Q_i(\boldsymbol{r}_i,t)\,\delta(\boldsymbol{r}-\boldsymbol{r}_i)$, where Dirac's delta function $\delta(\boldsymbol{r}-\boldsymbol{r}_i)$ gives only a contribution at the actual position $\boldsymbol{r}_i(t)$ of the agent. The quantity $Q_i(\boldsymbol{r}_i,t)$ represents the strength of new markings and will be specified later. In summary, we obtain the following equation for the spatiotemporal evolution of the ground potential:

$$\frac{dG_\theta(\boldsymbol{r},t)}{dt} = \frac{1}{T_\theta(\boldsymbol{r})}\Big[G^0_\theta(\boldsymbol{r}) - G_\theta(\boldsymbol{r},t)\Big] + \sum_i Q_i(\boldsymbol{r}_i,t)\,\delta\Big[\boldsymbol{r}-\boldsymbol{r}_i(t)\Big]\,. \tag{6.8}$$

The motion of agent i on a two-dimensional surface will be described again by the Langevin equation, (6.1). Due to the social force ansatz, (6.6), the term $\boldsymbol{f}_i$ represents deterministic influences on the motion, such as intentions to move in a certain direction at a certain desired velocity or to keep a distance from neighboring agents. Because we will focus on rare direct interactions, the term $\boldsymbol{f}_{ij}$ in (6.6) can be approximately neglected here. Further, we will also neglect effects, $\boldsymbol{f}_{i\mathrm{B}}$, resulting from the avoidance of borders, etc. Then, $\boldsymbol{f}_i$ is specified as follows:

$$\boldsymbol{f}_i(t) = \frac{v_i^0}{\tau_i}\boldsymbol{e}_i(\boldsymbol{r}_i, \boldsymbol{v}_i, t) \,. \tag{6.9}$$

Here, v_i^0 describes the *desired velocity*, and $\boldsymbol{e}_i$ is the *desired direction* of the agent. With (6.9), the Langevin equation, (6.1), becomes

$$\frac{d\boldsymbol{v}_i(t)}{dt} = \frac{v_i^0\boldsymbol{e}_i(\boldsymbol{r}_i, \boldsymbol{v}_i, t) - \boldsymbol{v}_i(t)}{\tau_i} + \sqrt{\frac{2\,\varepsilon_i}{\tau_i}}\,\boldsymbol{\xi}_i(t) \,, \tag{6.10}$$

where the first term reflects an *adaptation* of the *actual walking direction* $\boldsymbol{v}_i/|\boldsymbol{v}_i|$ to the desired walking direction $\boldsymbol{e}_i$ and an acceleration toward the desired velocity v_i^0 with a certain relaxation time τ_i. Assuming that the time τ_i is rather short compared to the timescale of trail formation which is characterized by durability T_θ, $\boldsymbol{v}_i$, (6.10), can be adiabatically eliminated. This leads to the following equation of motion for agents:

$$\frac{d\boldsymbol{r}_i}{dt} = \boldsymbol{v}_i(\boldsymbol{r}_i, t) \approx v_i^0\boldsymbol{e}_i(\boldsymbol{r}_i, \boldsymbol{v}_i, t) + \sqrt{2\,\varepsilon_i\tau_i}\,\boldsymbol{\xi}_i(t) \,. \tag{6.11}$$

To complete our trail formation model, we finally specify the *orientation relation*

$$\boldsymbol{e}_i(\boldsymbol{r}_i, \boldsymbol{v}_i, t) = \boldsymbol{e}_i\left\{\left[G_\theta(\boldsymbol{r}, t)\right], \boldsymbol{r}_i, \boldsymbol{v}_i\right\} \,, \tag{6.12}$$

which determines the desired walking direction in dependence on ground potentials $G_\theta(\boldsymbol{r}, t)$. Obviously, the concrete orientation relation for pedestrians will differ from that for ants. Thus, before discussing trail formation by pedestrians, we will reformulate the problem of trail formation by ants, discussed in Sect. 5.4.2, to elucidate the applicability of the current approach.

For *ants*, markings are chemical signposts, *pheromones*, which also provide the basic orientation for foraging and homing of the animals. To distinguish those trails that lead to a food source, the ants, *after* discovering a food source, use *another* pheromone to mark their trails, which stimulates the recruitment of additional ants to follow that trail. In our agent model, we count on that fact by using the two different chemical markings (+1), (−1), as discussed in detail in Sect. 5.4.2. Which of these markings is produced by agent i depends on its internal parameter $\theta_i \in \{-1, +1\}$. Hence, it makes sense to identify the production term for the ground potential in (6.8) with the production rate, $s_i(\theta_i, t)$, (5.53), for ants:

$$Q_i(\boldsymbol{r}_i, t) \equiv s_i(\theta_i, t)\,. \tag{6.13}$$

Due to the two chemical markings, we have two different ground potentials $G_{-1}(\boldsymbol{r}, t)$ and $G_{+1}(\boldsymbol{r}, t)$ here, which provide orientation for the agents. In the following, we need to specify how they influence the motion of the agents, especially their desired directions $\boldsymbol{e}_i(\boldsymbol{r}_i, \boldsymbol{v}_i, t)$. At this point, we take into account that the behavior of agent i will not be *directly* affected by the ground potentials $G_\theta(\boldsymbol{r}, t)$, which reflect the pure existence of markings of type θ at place $\boldsymbol{r}$. It will rather be influenced by the *perception* of its environment from its actual position $\boldsymbol{r}_i$, which for ants can be restricted by the *angle of perception* φ and the length of their antennae Δr (see also Fig. 5.10 for the continuous case). The effective influence of ground potentials on a specific agent will be described by the *trail potentials*

$$V_\theta^{\mathrm{tr}}(\boldsymbol{r}_i, \boldsymbol{v}_i, t) = V_\theta^{\mathrm{tr}}\left\{\left[G_\theta(\boldsymbol{r}, t)\right], \boldsymbol{r}_i, \boldsymbol{v}_i\right\}\,. \tag{6.14}$$

In agreement with the considerations in Sect. 5.4.2 for the discrete description, we choose here the assumption,

$$V_\theta^{tr}(\boldsymbol{r}_i, \boldsymbol{v}_i, t) = \int_0^{\Delta r} dr' \int_{-\varphi}^{+\varphi} d\varphi'\, r'\, G_\theta\Big\{\boldsymbol{r}_i + r'\Big[\cos(\omega_i + \varphi'), \sin(\omega_i + \varphi')\Big], t\Big\}\,, \tag{6.15}$$

where the angle ω_i is given by the current moving direction (see Fig. 5.10)

$$\boldsymbol{e}_i^*(t) = \frac{\boldsymbol{v}_i(t)}{|\boldsymbol{v}_i(t)|} = \Big[\cos\omega_i(t), \sin\omega_i(t)\Big]\,. \tag{6.16}$$

The perception of already existing trails will have an *attractive effect* $\boldsymbol{f}_i^{tr}(\boldsymbol{r}_i, \boldsymbol{v}_i, t)$ on agents. This can be defined by the gradients of the trail potentials, similar to (5.57):

$$\boldsymbol{f}_i^{tr}(\boldsymbol{r}_i, \boldsymbol{v}_i, t) = \frac{\theta_i}{2}\,\Big[\,(1+\theta_i)\,\boldsymbol{\nabla} V_{-1}^{tr}(\boldsymbol{r}_i, \boldsymbol{v}_i, t) \;-\; (1-\theta_i)\,\boldsymbol{\nabla} V_{+1}^{tr}(\boldsymbol{r}_i, \boldsymbol{v}_i, t)\,\Big] \tag{6.17}$$

Here, it is again considered that agents that move out of the nest to reach a food source ($\theta_i = +1$) orient by chemical (-1), whereas agents that move back from the food ($\theta_i = -1$) orient by chemical ($+1$).

We complete our model of trail formation by specifying the *orientation relation* of agents. Assuming

$$\boldsymbol{e}_i(\boldsymbol{r}_i, \boldsymbol{v}_i, t) = \frac{\boldsymbol{f}_i^{tr}(\boldsymbol{r}_i, \boldsymbol{v}_i, t)}{|\boldsymbol{f}_i^{tr}(\boldsymbol{r}_i, \boldsymbol{v}_i, t)|}\,, \tag{6.18}$$

the desired walking direction $\boldsymbol{e}_i(\boldsymbol{r}_i, \boldsymbol{v}_i, t)$ points in the direction of the steepest increase of the relevant trail potential $V_\theta^{tr}(\boldsymbol{r}_i, \boldsymbol{v}_i, t)$. However, this ansatz does not take into account the ants' persistence to keep the previous direction of motion, as discussed in Sect. 5.4.2. A bias would reduce the probability of changing to the opposite walking direction by fluctuations. Therefore, we modify (6.18) to

$$\boldsymbol{e}_i(\boldsymbol{r}_i, \boldsymbol{v}_i, t) = \frac{\boldsymbol{f}_i^{tr}(\boldsymbol{r}_i, \boldsymbol{v}_i, t) + \boldsymbol{e}_i^*(t - \Delta t)}{\mathcal{N}_i(\boldsymbol{r}_i, \boldsymbol{v}_i, t)}\,, \tag{6.19}$$

where $\mathcal{N}_i(\boldsymbol{r}_i, \boldsymbol{v}_i, t) = |\boldsymbol{f}_i^{tr}(\boldsymbol{r}_i, \boldsymbol{v}_i, t) + \boldsymbol{e}_i^*(t - \Delta t)|$ is a normalization factor. That means that on ground without markings, the moving direction tends to agree with that at the previous time $t - \Delta t$, but it can change by fluctuations.

Finally, we may also include the adjustment of the sensitivity in our continuous description. Assuming again a linear increase in the individual noise level of the agent by the amount $\delta\varepsilon$, as long as it did not find any food, we may choose, in accordance with (5.58), the following ansatz for ε_i in (6.11):

$$\varepsilon_i(t) = \frac{\theta_i}{2}\left\{ (1+\theta_i)\left[\varepsilon_0 + \delta\varepsilon\,(t - t_n^i)\right]^2 + (1-\theta_i)\,\varepsilon_0^2 \right\}. \tag{6.20}$$

t_n^i again is the starting time of agent i from the nest, and ε_0 is the initial noise level.

Hence, we have shown that the basic features of the discrete model of trunk trail formation by ants, Sect. 5.4, can also be included in the continuous description based on the Langevin equation. The results of computer simulations have already been discussed in Sect. 5.4.3. Thus, we can move forward to apply the general model introduced above to trail formation by pedestrians.

6.2.2 Human Trail Formation

Quantitative models of trail formation by pedestrians have been derived only recently [217, 222]. Human trail formation can be interpreted as a complex interplay between human motion, human orientation, and environmental changes. For example, pedestrians, on one hand, tend to take the shortest way to their destinations. On the other hand, they avoid walking on bumpy ground, because it is uncomfortable. Therefore, they prefer to use existing trails, but they would build a new shortcut, if the relative detour would be too long. In the latter case they generate a new trail because footprints clear some vegetation. Examples of resulting trail systems can be found in green areas, such as public parks (see Fig. 6.3).

In this section, we want to adapt the previous model of trail formation to human trail patterns. We restrict the description to the most important

Fig. 6.3. Between the straight, paved ways on the university campus in Stuttgart-Vaihingen, a trail system has evolved (*center of the picture*). Two types of nodes are observed: intersections of two trails running in a straight line and junctions of two trails that merge smoothly into one trail [222]

factors, keeping in mind that pedestrians can also show much more complicated behavior than described here. We start again with (6.11), which can also be applied to the motion of pedestrians. But, we have to specify the *orientation relation*, $\boldsymbol{e}_i(\boldsymbol{r}_i, \boldsymbol{v}_i, t)$, (6.12), which describes how moving pedestrians are affected by their environment. In turn, we also have to specify now how pedestrians change their environment by leaving footprints.

In the case considered, we do *not* have to distinguish between different kinds of markings. Thus, we will need only one ground potential $G(\boldsymbol{r}, t)$, and the subscript θ can be omitted. The value of G is a measure of the *comfort of walking*. Therefore, it can depend considerably on weather conditions, which are not discussed here any further. For the strength $Q_i(\boldsymbol{r}, t)$ of the markings produced by footprints at place $\boldsymbol{r}$, we assume that

$$Q_i(\boldsymbol{r}, t) = I(\boldsymbol{r}) \left[1 - \frac{G(\boldsymbol{r}, t)}{G_{\max}(\boldsymbol{r})}\right] , \tag{6.21}$$

where $I(\boldsymbol{r})$ is the location-dependent intensity of clearing vegetation. The saturation term $[1 - G(\boldsymbol{r}, t)/G_{\max}(\boldsymbol{r})]$ results from the fact that the clarity of a trail is limited to a maximum value $G_{\max}(\boldsymbol{r})$.

On plain, homogeneous ground without any trails, the desired direction $\boldsymbol{e}_i$ of pedestrian i at place $\boldsymbol{r}_i$ is given by direction $\boldsymbol{e}_i^*$ of the next destination $\boldsymbol{d}_i$:

$$\boldsymbol{e}_i(\boldsymbol{r}_i, \boldsymbol{v}_i, t) = \boldsymbol{e}_i^*(\boldsymbol{d}_i, \boldsymbol{r}_i) = \frac{\boldsymbol{d}_i - \boldsymbol{r}_i}{|\boldsymbol{d}_i - \boldsymbol{r}_i|} = \boldsymbol{\nabla} U_i(\boldsymbol{r}_i) , \tag{6.22}$$

where the *destination potential* is

$$U_i(\boldsymbol{r}_i) = -|\boldsymbol{d}_i - \boldsymbol{r}_i| . \tag{6.23}$$

However, the perception of already existing trails will have an *attractive effect* $\boldsymbol{f}_{\rm tr}(\boldsymbol{r},t)$ on a pedestrian, which will again be defined by the gradient of the trail potential $V_{tr}(\boldsymbol{r},t)$, specified later:

$$\boldsymbol{f}_{tr}(\boldsymbol{r}_i,t) = \boldsymbol{\nabla} V_{\rm tr}(\boldsymbol{r}_i,t)\,. \qquad (6.24)$$

Because the potentials U and $V_{\rm tr}$ influence the pedestrian at the same time, it seems reasonable to introduce an *orientation relation* similar to (6.18), by taking the sum of both potentials:

$$\begin{aligned}\boldsymbol{e}_i(\boldsymbol{r}_i,\boldsymbol{v}_i,t) &= \frac{\boldsymbol{f}_{tr}(\boldsymbol{r}_i,t)+\boldsymbol{e}_i^*(\boldsymbol{d}_i,\boldsymbol{r}_i)}{\mathcal{N}(\boldsymbol{r}_i,t)} \\ &= \frac{1}{\mathcal{N}(\boldsymbol{r}_i,t)}\boldsymbol{\nabla}[U_i(\boldsymbol{r}_i)+V_{tr}(\boldsymbol{r}_i,t)]\,. \qquad (6.25)\end{aligned}$$

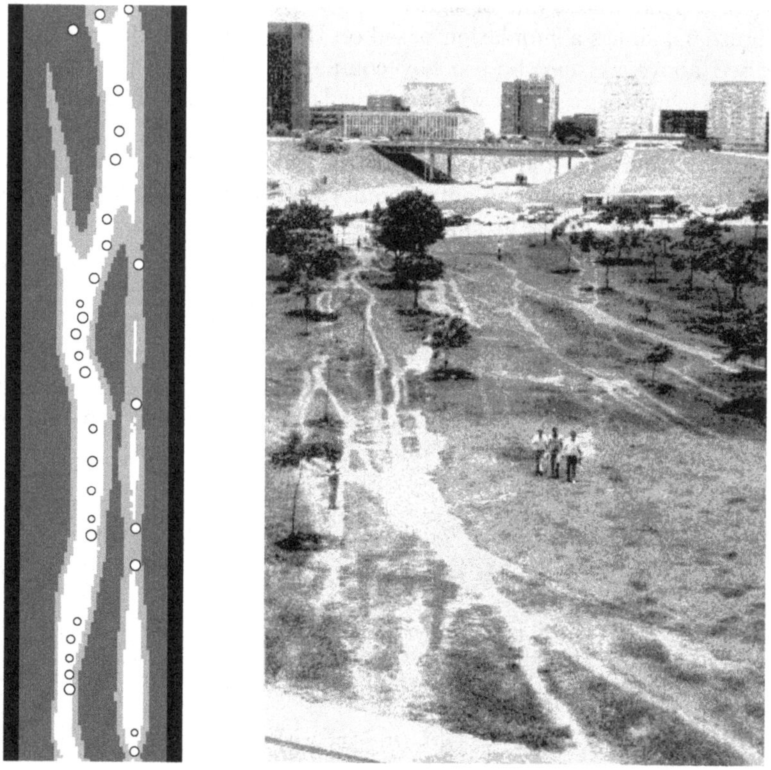

Fig. 6.4. When pedestrians leave footprints on the ground, trails will develop, and only parts of the ground are used for walking (in contrast to paved areas). The similarity between the simulation result (*left*) and the trail system on the university campus of Brasilia (*right*, reproduction with kind permission of K. Humpert) is obvious [211, 217, 222]

Here, $\mathcal{N}(\boldsymbol{r},t) = |\boldsymbol{\nabla}[U_i(\boldsymbol{r}) + V_{tr}(\boldsymbol{r},t)]|$ serves as a normalization factor. Equation (6.25) ensures that the vector $\boldsymbol{e}_i(\boldsymbol{r}_i,t)$ points in a direction that is a *compromise* between the *shortness* of the direct way to the destination and the *comfort* of using an existing trail.

Finally, we need to specify the trail potential V_{tr} for pedestrians. Obviously, a trail must be *recognized* by a pedestrian and near enough to be used. Whereas the ground potential $G(\boldsymbol{r},t)$ describes the *existence* of a trail segment at position $\boldsymbol{r}$, the *trail potential* $V_{\text{tr}}(\boldsymbol{r}_i,t)$ reflects the *attractiveness* of a trail from the actual position $\boldsymbol{r}_i(t)$ of the pedestrian. Because this will decrease with the distance $|\boldsymbol{r}-\boldsymbol{r}_i|$, we have applied the relation

$$V_{tr}(\boldsymbol{r}_i,t) = \int d^2r\, \mathrm{e}^{-|\boldsymbol{r}-\boldsymbol{r}_i|/\sigma(\boldsymbol{r}_i)} G(\boldsymbol{r},t)\,, \qquad (6.26)$$

where $\sigma(\boldsymbol{r}_i)$ characterizes the line of sight, i.e., the range of visibility. In analogy to (6.15), this relation could be generalized to include conceivable effects of a pedestrian's line of sight.

Figure 6.4 shows a simulation based on the human trail formation model described above. As can be seen by comparison with the photograph, the results are in good agreement with empirical observations. The simulation has started with the assumption of plain, homogeneous ground. All agents have their own destinations and entry points from which they start at a randomly chosen time. In Fig. 6.4, the entry points and destinations are distributed over the small ends of the ground.

6.2.3 Simulation of Pedestrian Trail Systems

The use of existing trails depends on the visibility, as given by (6.26). Assuming that the sight parameter σ is approximately space-independent, an additional simplification of the equations of trail formation can be reached by introducing dimensionless variables:

$$\boldsymbol{x} = \frac{\boldsymbol{r}}{\sigma}\,, \qquad (6.27)$$

$$\tau(\boldsymbol{x}) = \frac{t}{T(\sigma\boldsymbol{x})}\,, \qquad (6.28)$$

$$G'(\boldsymbol{x},\tau) = \sigma G(\sigma\boldsymbol{x},\tau T)\,, \qquad (6.29)$$

$$V'_{tr}(\boldsymbol{x},\tau) = \int d^2x'\, \mathrm{e}^{-|\boldsymbol{x}'-\boldsymbol{x}|}\, G'(\boldsymbol{x}',\tau)\,, \qquad (6.30)$$

$$U'_i(\boldsymbol{x}) = -|\boldsymbol{d}_i/\sigma - \boldsymbol{x}|\,, \qquad (6.31)$$

etc. Neglecting fluctuations in (6.11) for the moment, this implies the following scaled equations [222] for pedestrian motion:

$$\frac{d\boldsymbol{x}_i(\tau)}{d\tau} = \boldsymbol{v}'_i(\boldsymbol{x}_i,\tau) \approx \frac{v_i^0 T(\sigma\boldsymbol{x}_i)}{\sigma}\boldsymbol{e}'_i(\boldsymbol{x}_i,\tau)\,, \qquad (6.32)$$

for human orientation,

$$\boldsymbol{e}_i'(\boldsymbol{x},\tau) = \frac{\boldsymbol{\nabla}[U_i'(\boldsymbol{x}) + V_{tr}'(\boldsymbol{x},\tau)]}{|\boldsymbol{\nabla}[U_i'(\boldsymbol{x}) + V_{tr}'(\boldsymbol{x},\tau)]|}, \tag{6.33}$$

and for environmental changes,

$$\begin{aligned} \frac{dG'(\boldsymbol{x},\tau)}{d\tau} &= \Big[G_0'(\boldsymbol{x}) - G'(\boldsymbol{x},\tau)\Big] \\ &\quad + \left[1 - \frac{G'(\boldsymbol{x},\tau)}{G_{\max}'(\boldsymbol{x})}\right] \sum_i \frac{I(\sigma\boldsymbol{x})T(\sigma\boldsymbol{x})}{\sigma}\, \delta\Big[\boldsymbol{x} - \boldsymbol{x}_i(\tau)\Big]. \end{aligned} \tag{6.34}$$

Therefore, we find the surprising result that the dynamics of trail formation is (apart from the influence of the number and places of entering and leaving pedestrians) already determined by two local parameters, κ and λ, instead of four, namely, the products

$$\begin{aligned} \kappa(\boldsymbol{x}) &= \frac{I(\sigma\boldsymbol{x})\, T(\sigma\boldsymbol{x})}{\sigma}, \\ \lambda(\boldsymbol{x}) &= \frac{V_0\, T(\sigma\boldsymbol{x})}{\sigma}. \end{aligned} \tag{6.35}$$

Herein, V_0 denotes the mean value of the desired velocities v_i^0. The influence of κ, which is a measure of the attractiveness of a trail, is elucidated in Fig. 6.5, where pedestrians move between all possible pairs of three fixed places (such as shops, houses, underground stations, or parking lots). If trail attractiveness is small, a direct way system develops (see the left part of Fig. 6.5); if it is large, a minimal way system is formed, which is the shortest way system that connects all entry points and destinations (see the *right* part of Fig. 6.5). Otherwise a minimal detour system will result that looks similar to the trail system in the *center* of Fig. 6.3.

Another example is shown in Fig. 6.6, where pedestrians move between all possible pairs of four fixed places. Starting with plain, spatially homogeneous ground, the pedestrians take *direct ways* to their respective destinations. However, by continuously leaving markings, they produce trails that have an attractive effect on nearby pedestrians because this is more comfortable than clearing new ways. Thus, pedestrians begin to use already existing trails after some time. In this way, a kind of *selection process* between trails occurs (cf. also Sect. 3.2.4). Frequently used trails are reinforced, which makes them even more attractive, whereas rarely used trails may vanish. Therefore, even pedestrians with different entry points and destinations use and maintain common parts of the trail system.

We note that the tendency of *trail bundling* can be interpreted as some kind of *agglomeration process*, which is *delocalized* due to the directedness of agents' motion. Similar to the explanation in Sect. 5.4.3, this agglomeration is based on three processes: (i) the attractive effect of existing trails, which

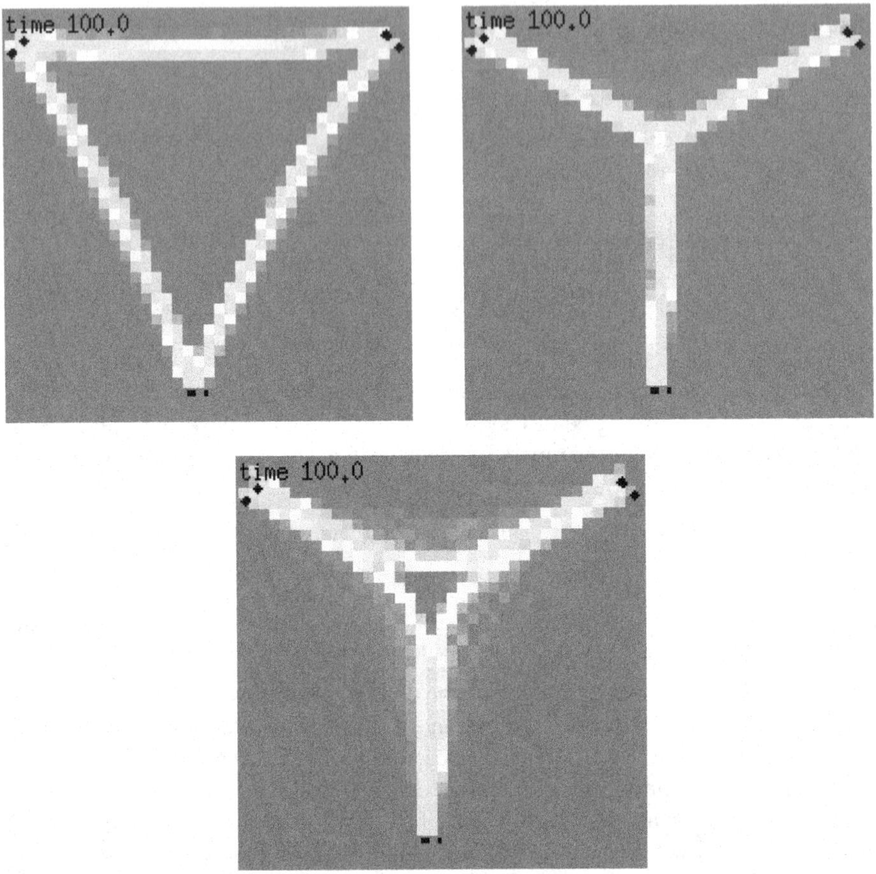

Fig. 6.5. The structure of the emerging trail system (light gray) essentially depends on the attractiveness of the trails (i.e., on the parameter $\kappa = IT/\sigma$): (*top: left*) direct trail system (small value of κ), (*top: right*) minimal trail system (large value of κ), (*bottom*) minimal detour system. The gray scale, which encodes the ground potential, allows us to reconstruct the temporal evolution of the trail system before its final state is reached [211, 222]

cause a tendency to approach the trail; (ii) the reinforcement of trails because of their usage; and (iii) the vanishing of rarely chosen trails in the course of time.

If the attractiveness of emerged trails is large, the final trail system would be a *minimal way system*. However, because of pedestrians' dislike for taking detours, the evolution of the trail system normally stops before this state is reached. In other words, a so-called *minimal detour system* develops if model parameters are chosen realistically. As Fig. 6.6 shows, the resulting trails can considerably differ from the direct ways that pedestrians would use if they were equally comfortable.

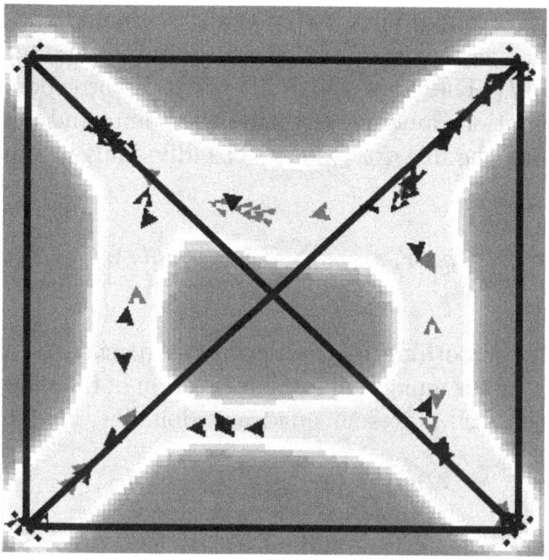

Fig. 6.6. Comparison of different types of way systems between four places: The *direct way system* (which is represented by the black lines) provides the shortest connections between all entry points and destinations, but it covers a lot of space. In real situations, pedestrians will produce a *'minimal detour system'* as the best compromise between a direct way system and a *minimal way system* (which is the shortest way system that connects all entry points and destinations). The illustration shows a simulation result which could serve as a planning guideline. Its asymmetry is caused by differences in the frequency of trail usage. Note that the above figure, in contrast to Figs. 6.5 and 6.7, does not display the ground potential, but the trail potential. The latter appears considerably broader because it takes into account the range of visibility of the trails. Arrows represent the positions and walking directions of pedestrians. Therefore, they indicate the ways actually taken [222]

For planning purposes, model parameters λ and κ must be specified in a realistic way. Then, one needs to simulate the exptected flows of pedestrians that enter the system considered at certain entry points with the intention to reach certain destinations. Already existing ways can be taken into account by the function $G_0'(\boldsymbol{x})$. According to our model, a trail system will evolve that minimizes overall detours and thereby provides an optimal compromise between a direct and a minimal way system.

6.2.4 Macroscopic Equations of Trail Formation

The formation of trail systems can be described as a process of self-organization. Based on interactions of agents on a local or "microscopic" level, the emergence of a global or "macroscopic" structure occurs. From the above microscopic, or agent-based model of trail formation, we can derive related *macroscopic equations* [222]. For this purpose, we need to distinguish different subpopulations a of individuals i. By $a(\tau)$, we denote the time-dependent set of individuals i who have started from the same entry point $\boldsymbol{p}_a$ with the same destination $\boldsymbol{d}_a$. Therefore, the different sets a correspond to the possible (directed) combinations among existing entry points and destinations.

Next, we define the *density* $\varrho_a(\boldsymbol{x},\tau)$ of individuals of subpopulation a at place $\boldsymbol{x}$ by

$$\varrho_a(\boldsymbol{x},\tau) = \sum_{i\in a(\tau)} \delta\Big[\boldsymbol{x} - \boldsymbol{x}_i(\tau)\Big] . \tag{6.36}$$

Note that spatial smoothing of the density is reached by discretization of space that is needed for numerical implementation of the model. For example, if the discrete places $\boldsymbol{x}_i$ represent quadratic domains,

$$\mathcal{A}(\boldsymbol{x}_i) = \{\boldsymbol{x} : |\boldsymbol{x} - \boldsymbol{x}_i|_\infty \le L\} , \tag{6.37}$$

with area $|\mathcal{A}| = L^2$, the corresponding density is

$$\varrho_a(\boldsymbol{x}_i,\tau) = \frac{1}{|\mathcal{A}|} \int\limits_{\mathcal{A}(\boldsymbol{x}_i)} d^2x \sum_{i\in a(\tau)} \delta\Big[\boldsymbol{x} - \boldsymbol{x}_i(\tau)\Big] . \tag{6.38}$$

However, for simplicity, we will treat the continuous case. The quantity,

$$N_a(\tau) = \int d^2x\, \varrho_a(\boldsymbol{x},\tau) = \sum_{i\in a(\tau)} \int d^2x\, \delta\Big[\boldsymbol{x} - \boldsymbol{x}_i(\tau)\Big] , \tag{6.39}$$

describes the number of pedestrians of subpopulation a, who are walking on the ground at time τ. It changes by pedestrians entering the system at entry point $\boldsymbol{p}_a$ at rate $R_a^+(\boldsymbol{p}_a,\tau)$ and leaving it at destination $\boldsymbol{d}_a$ at a rate $R_a^-(\boldsymbol{d}_a,\tau)$.

Due to the time dependence of the sets $a(\tau)$, we will need to distinguish among the following sets:

$$\begin{aligned} a_\cap(\tau) &= a(\tau+\Delta) \cap a(\tau) , \\ a_+(\tau) &= a(\tau+\Delta) \setminus a_\cap(\tau) , \\ a_-(\tau) &= a(\tau) \setminus a_\cap(\tau) , \end{aligned} \tag{6.40}$$

where $a_\cap(\tau)$ is the set of pedestrians remaining in the system, $a_+(\tau)$ is the set of entering pedestrians, and $a_-(\tau)$ is the set of leaving pedestrians. The following relations hold for these sets:

$$\begin{aligned} a_+(\tau) \cap a_-(\tau) &= \emptyset , \\ a_\cap(\tau) \cup a_+(\tau) &= a(\tau+\Delta) , \\ a_\cap(\tau) \cup a_-(\tau) &= a(\tau) . \end{aligned} \tag{6.41}$$

Therefore, (6.36) implies

$$
\begin{aligned}
\frac{\partial \varrho_a(\boldsymbol{x},\tau)}{\partial \tau} &= \lim_{\Delta\to 0}\frac{1}{\Delta}\left\{\sum_{i\in a(\tau+\Delta)}\delta\Big[\boldsymbol{x}-\boldsymbol{x}_i(\tau+\Delta)\Big]-\sum_{i\in a(\tau)}\delta\Big[\boldsymbol{x}-\boldsymbol{x}_i(\tau)\Big]\right\}\\
&= \lim_{\Delta\to 0}\sum_{i\in a_\cap(\tau)}\frac{1}{\Delta}\Big\{\delta\Big[\boldsymbol{x}-\boldsymbol{x}_i(\tau+\Delta)\Big]-\delta\Big[\boldsymbol{x}-\boldsymbol{x}_i(\tau)\Big]\Big\}\\
&\quad+\lim_{\Delta\to 0}\frac{1}{\Delta}\sum_{i\in a_+(\tau)}\delta\Big[\boldsymbol{x}-\boldsymbol{x}_i(\tau+\Delta)\Big]\\
&\quad-\lim_{\Delta\to 0}\frac{1}{\Delta}\sum_{i\in a_-(\tau)}\delta\Big[\boldsymbol{x}-\boldsymbol{x}_i(\tau)\Big]\,.
\end{aligned}
\tag{6.42}
$$

Taking into account

$$
\lim_{\Delta\to 0}\frac{1}{\Delta}\delta\Big[\boldsymbol{x}-\boldsymbol{x}_i(\tau+\Delta)\Big]=\lim_{\Delta\to 0}\frac{1}{\Delta}\delta\Big[\boldsymbol{x}-\boldsymbol{x}_i(\tau)\Big]+\frac{\partial}{\partial\tau}\delta\Big[\boldsymbol{x}-\boldsymbol{x}_i(\tau)\Big]\,, \tag{6.43}
$$

which follows by Taylor expansion, we obtain

$$
\begin{aligned}
\frac{\partial \varrho_a(\boldsymbol{x},\tau)}{\partial \tau} &= \lim_{\Delta\to 0}\sum_{i\in a(\tau+\Delta)}\frac{1}{\Delta}\Big\{\delta\Big[\boldsymbol{x}-\boldsymbol{x}_i(\tau+\Delta)\Big]-\delta\Big[\boldsymbol{x}-\boldsymbol{x}_i(\tau)\Big]\Big\}\\
&\quad+\lim_{\Delta\to 0}\frac{1}{\Delta}\sum_{i\in a_+(\tau)}\delta\Big[\boldsymbol{x}-\boldsymbol{x}_i(\tau)\Big]\\
&\quad-\lim_{\Delta\to 0}\frac{1}{\Delta}\sum_{i\in a_-(\tau)}\delta\Big[\boldsymbol{x}-\boldsymbol{x}_i(\tau)\Big]\,.
\end{aligned}
\tag{6.44}
$$

With

$$
\begin{aligned}
\lim_{\Delta\to 0}\frac{1}{\Delta}\Big\{\delta\Big[\boldsymbol{x}-\boldsymbol{x}_i(\tau+\Delta)\Big]-\delta\Big[\boldsymbol{x}-\boldsymbol{x}_i(\tau)\Big]\Big\} &= \frac{\partial}{\partial\tau}\delta\Big[\boldsymbol{x}-\boldsymbol{x}_i(\tau)\Big]\\
&= -\boldsymbol{\nabla}\delta\Big[\boldsymbol{x}-\boldsymbol{x}_i(\tau)\Big]\frac{d\boldsymbol{x}_i}{d\tau}
\end{aligned}
\tag{6.45}
$$

and the relations

$$
R_a^+(\boldsymbol{x},\tau)=\lim_{\Delta\to 0}\frac{1}{\Delta}\sum_{i\in a_+(\tau)}\delta\Big[\boldsymbol{x}-\boldsymbol{x}_i(\tau)\Big]\,, \tag{6.46}
$$

$$
R_a^-(\boldsymbol{x},\tau)=\lim_{\Delta\to 0}\frac{1}{\Delta}\sum_{i\in a_-(\tau)}\delta\Big[\boldsymbol{x}-\boldsymbol{x}_i(\tau)\Big]\,, \tag{6.47}
$$

for the rates of pedestrians joining and leaving subpopulation a, we finally arrive at

$$\frac{\partial \varrho_a(\boldsymbol{x},\tau)}{\partial \tau} = -\boldsymbol{\nabla}\cdot \sum_{i\in a(\tau)} \boldsymbol{v}_i'(\boldsymbol{x}_i,\tau)\,\delta\Big[\boldsymbol{x}-\boldsymbol{x}_i(\tau)\Big] + R_a^+(\boldsymbol{x},\tau) - R_a^-(\boldsymbol{x},\tau)\,, \tag{6.48}$$

where $R_a^+(\boldsymbol{x},\tau)$ is zero away from entry point $\boldsymbol{p}_a$ and the same holds for $R_a^-(\boldsymbol{x},\tau)$ away from destination $\boldsymbol{d}_a$.

Now, we define the *average velocity* $\boldsymbol{V}_a$ by

$$\boldsymbol{V}_a(\boldsymbol{x},\tau) = \frac{1}{\varrho_a(\boldsymbol{x},\tau)} \sum_{i\in a(\tau)} \boldsymbol{v}_i'(\boldsymbol{x}_i,\tau)\delta\Big[\boldsymbol{x}-\boldsymbol{x}_i(\tau)\Big]\,. \tag{6.49}$$

This gives us the desired *continuity equation*,

$$\frac{\partial \varrho_a(\boldsymbol{x},\tau)}{\partial \tau} + \boldsymbol{\nabla}\cdot\Big[\varrho_a(\boldsymbol{x},\tau)\boldsymbol{V}_a(\boldsymbol{x},\tau)\Big] = R_a^+(\boldsymbol{x},\tau) - R_a^-(\boldsymbol{x},\tau)\,, \tag{6.50}$$

describing pedestrian motion. Fluctuation effects can be taken into account by additional *diffusion terms*

$$\sum_b \boldsymbol{\nabla}\cdot\Big[D_{ab}(\{\varrho_c\})\boldsymbol{\nabla}\varrho_b(\boldsymbol{x},\tau)\Big] \tag{6.51}$$

on the right-hand side of (6.50) [211, 216]. This broadens the trails somewhat.

Next, we rewrite (6.34) for environmental changes in the form

$$\frac{dG'(\boldsymbol{x},\tau)}{d\tau} = \Big[G_0'(\boldsymbol{x}) - G'(\boldsymbol{x},\tau)\Big] + \left[1 - \frac{G'(\boldsymbol{x},\tau)}{G_{\text{max}}'(\boldsymbol{x})}\right]\sum_a \kappa(\boldsymbol{x})\varrho_a(\boldsymbol{x},t)\,. \tag{6.52}$$

Finally, the orientation relation becomes

$$\boldsymbol{e}_a(\boldsymbol{x},\tau) = \frac{\boldsymbol{\nabla}[U_a(\boldsymbol{x}) + V_{tr}'(\boldsymbol{x},\tau)]}{|\boldsymbol{\nabla}[U_a(\boldsymbol{x}) + V_{tr}'(\boldsymbol{x},\tau)]|}\,, \tag{6.53}$$

with

$$U_a(\boldsymbol{x}) = -|\boldsymbol{d}_a/\sigma - \boldsymbol{x}|\,. \tag{6.54}$$

Therefore, the average velocity is given by

$$\begin{aligned} \boldsymbol{V}_a(\boldsymbol{x},\tau) &\approx \frac{1}{\varrho_a(\boldsymbol{x},\tau)}\sum_{i\in a(\tau)} \frac{v_i^0 T(\sigma\boldsymbol{x})}{\sigma}\boldsymbol{e}_a(\boldsymbol{x},\tau)\delta\Big[\boldsymbol{x}-\boldsymbol{x}_i(\tau)\Big] \\ &\approx \frac{V^0 T(\sigma\boldsymbol{x})}{\sigma}\boldsymbol{e}_a(\boldsymbol{x},\tau) = \lambda(\boldsymbol{x})\boldsymbol{e}_a(\boldsymbol{x},\tau)\,, \end{aligned} \tag{6.55}$$

where V_0 is again the average desired pedestrian velocity. For *frequent* interactions (avoidance maneuvers) of pedestrians, V_0 must be replaced by suitable monotonically decreasing functions $V_a(\{\varrho_c\})$ of densities ϱ_c [208, 211, 216]. Moreover, fluctuation effects will be stronger, leading to larger diffusion functions $D_{ab}(\{\varrho_c\})$ and broader trails.

Summarizing our results [222], we have found a macroscopic formulation of trail formation that is given by (6.50)–(6.55), with (6.30). Apart from possible analytical investigations, this allows us to determine implicit equations for a stationary solution, if the rates $R_a^+(\boldsymbol{x},\tau)$ and $R_a^-(\boldsymbol{x},\tau)$ are time-independent. Setting the temporal derivatives to zero, we find the relations

$$G'(\boldsymbol{x}) = \frac{G'_0(\boldsymbol{x}) + \sum_a \kappa(\boldsymbol{x})\varrho_a(\boldsymbol{x})}{1 + \sum_a \kappa(\boldsymbol{x})\varrho_a(\boldsymbol{x})/G'_{\max}(\boldsymbol{x})} \tag{6.56}$$

and

$$\nabla \cdot \Big[\varrho_a(\boldsymbol{x})\boldsymbol{V}_a(\boldsymbol{x})\Big] = R_a^+(\boldsymbol{x}) - R_a^-(\boldsymbol{x})\,. \tag{6.57}$$

Together with (6.30) and (6.53)–(6.55), the relations (6.56), (6.57) allow us to calculate the trail system finally evolving. Again, we see that the resulting state depends on the two parameters, λ and κ. In addition, it is determined by the respective boundary conditions, i.e., the configuration and frequency of usage of the entry point–destination pairs that are characterized by the concrete form of entering rates $R_a^+(\boldsymbol{x})$ and leaving rates $R_a^-(\boldsymbol{x})$.

The advantage of applying the macroscopic equations is that the trail system finally evolving can be calculated much more efficiently because considerably less time is required for computing. The numerical solution can now be obtained by a simple iterative method that is comparable to the self-consistent field technique. Figure 6.7 shows an example for different values of κ that shall be compared with Fig. 6.5. As expected, the results obtained from the iterated macroscopic equations agree with those of related microsimulations.

We can conclude that the agent concept is suitable for modeling trail formation by pedestrians and animals. The structure of the resulting trail system can vary considerably with the species. Whereas our model ants find their destinations (the food sources) by chance, pedestrians can directly orient toward their destinations, so that fluctuations are not a necessary model component in this case. Thus, for certain ant species, a dendritic trail system is found, whose detailed form depends on random events, i.e., the concrete history of its evolution. Pedestrians, however, produce a minimal detour system, i.e., an optimal compromise between a direct way system and a minimal way system.

As a consequence, we can derive a macroscopic model for trail formation by pedestrians, but not for ants. This implies a self-consistent field method for very efficient calculation of the trail system finally evolving. This is

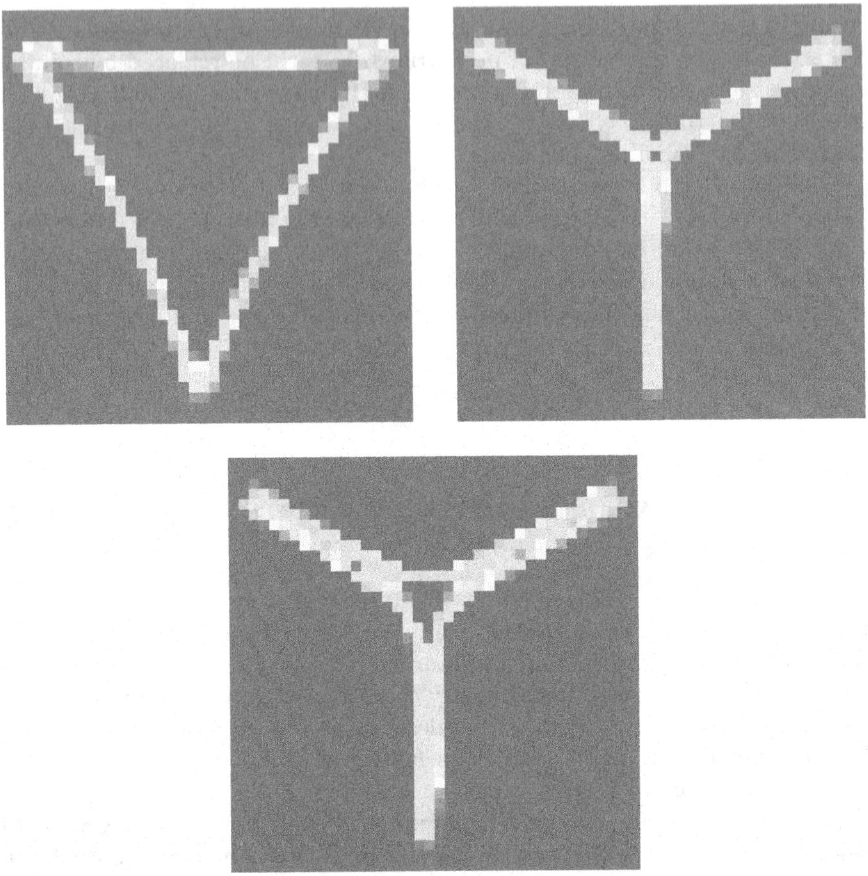

Fig. 6.7. Stationary solution of the macroscopic trail formation model, obtained by an iterative self-consistent field method. Because the boundary conditions were chosen as in Fig. 6.5, the results in dependence on parameter κ are almost identical with those of the corresponding microsimulations [217, 222]

determined by the locations of entry points and destinations (e.g., houses, shops, and parking lots) and the rates of choosing possible connections between them. Apart from this, it depends only on two parameters, which was demonstrated by scaling to dimensionless equations. These are related to trail attractiveness and the average velocity of motion.

7. Evolutionary Optimization Using Brownian Searchers

7.1 Evolutionary Optimization Strategies

7.1.1 Ensemble Search with Brownian Agents

In this chapter, we apply the approach of Brownian agents to complex search and optimization problems. It allows us also to introduce some new elements into the dynamics, such as the *reproduction* of (successful) agents at the cost of less successful ones.

An optimization process can be described as a special search for minima of a related potential in the configuration space. Evolutionary optimization strategies are based on the idea of an *ensemble of searchers* that move through the configuration space, searching for minima. In the case considered, these *searchers* shall be described as *Brownian agents* that move in a potential $U(\boldsymbol{x})$ due to the Langevin equation, (1.39), introduced in Sect. 1.3.2 [20, 444]. We will concentrate mainly on the overdamped limit described by Langevin equation, (1.40):

$$\frac{d\boldsymbol{x}_i}{dt} = -\frac{1}{\gamma_0}\left.\frac{\partial U(\boldsymbol{x})}{\partial \boldsymbol{x}}\right|_{\boldsymbol{x}_i} + \sqrt{\frac{2\,\varepsilon_i}{\gamma_0}}\,\boldsymbol{\xi}_i(t)\,. \tag{7.1}$$

The potential $U(\boldsymbol{x})$ has to describe the optimization landscape sufficiently. Different from a self-consistent field $h^{\mathrm{e}}(\boldsymbol{x},t)$, introduced in Sect. 3.1.2, $U(\boldsymbol{x})$ *cannot* be changed by agents. However, we consider again that agents are characterized by an *individual noise intensity* $\varepsilon_i \sim k_{\mathrm{B}}T$, which adjusts their susceptibility to stochastic influences.

As we have shown in Sect. 1.3.3, for $\varepsilon_i = \varepsilon = \mathrm{const}$, the Langevin equation, (7.1), corresponds to a Fokker–Planck equation, (1.55), for the probability density $P(\boldsymbol{x},t)$ of searchers in the form,

$$\begin{aligned}\frac{\partial P(\boldsymbol{x},t)}{\partial t} &= \boldsymbol{\nabla} D\Big[\boldsymbol{\nabla} P(\boldsymbol{x},t) \,+\, \beta\, P(\boldsymbol{x},t)\, \boldsymbol{\nabla} U(\boldsymbol{x})\Big] \\ &= \boldsymbol{\nabla}\Big[(1/\gamma_0)\, P(\boldsymbol{x},t)\boldsymbol{\nabla} U(\boldsymbol{x})\Big] + D\,\Delta P(\boldsymbol{x},t)\,,\end{aligned} \tag{7.2}$$

with

$$D = \frac{\varepsilon}{\gamma_0} = \frac{k_{\mathrm{B}}T}{\gamma_0}\;, \quad \beta = \frac{1}{k_{\mathrm{B}}T}\,. \tag{7.3}$$

If we apply (7.2) to the search problem, then $\boldsymbol{x}$ denotes the coordinate in the *search space* characterized by a *scalar potential function* $U(\boldsymbol{x})$, which is also denoted as a fitness function. Equation (7.2) is a special case of a general search dynamics of the form [18],

$$P(\boldsymbol{x}, t + \Delta t) = \mathcal{T}\{P(\boldsymbol{x}, t); \Delta t\}\,. \tag{7.4}$$

The dynamics $\mathcal{T}$ is considered *optimization dynamics* if any (or nearly any) initial distribution $P(\boldsymbol{x}, 0)$ converges to a target distribution, $\lim_{t\to\infty} P(\boldsymbol{x}, \tau)$, which is concentrated around the minimum $\boldsymbol{x}_0$ of $U(\boldsymbol{x})$. We restrict our analysis here to the case where $\mathcal{T}$ is given by a second-order partial differental equation, such as (7.2). Solvable cases of (7.2) can be extracted from the ansatz,

$$P(\boldsymbol{x}, t) = \exp\left[-\frac{\beta}{2}\,U(\boldsymbol{x})\right]\, y(\boldsymbol{x}, t)\,, \tag{7.5}$$

which leads to

$$\frac{dy(\boldsymbol{x}, t)}{dt} = -V(\boldsymbol{x})\, y(\boldsymbol{x}, t) + D\,\Delta y(\boldsymbol{x}, t)\,. \tag{7.6}$$

After separating of the time and space variables,

$$y(\boldsymbol{x}, t) = \exp\left(-\lambda\, t\right) \psi(\boldsymbol{x})\,, \tag{7.7}$$

we obtain the following eigenvalue equation [142, 412]:

$$H\,\psi(\boldsymbol{x}) = \lambda\,\psi(\boldsymbol{x}) = -D\,\Delta\psi(\boldsymbol{x}) + V(\boldsymbol{x})\,\psi(\boldsymbol{x}) \tag{7.8}$$

with the eigenvalue λ and the potential

$$V(\boldsymbol{x}) = \frac{\beta^2}{4}\, D\,\boldsymbol{\nabla} U(\boldsymbol{x}) \cdot \boldsymbol{\nabla} U(\boldsymbol{x}) - \frac{\beta}{2}\, D\,\Delta U(\boldsymbol{x})\,, \tag{7.9}$$

which can be interpreted as a "redefined" fitness [20]. Equation (7.8) is the well-known stationary Schrödinger equation from quantum mechanics. Considering a discrete spectrum, it has the general solution,

$$P(\boldsymbol{x}, t) = \exp\left[-\frac{\beta}{2}\,U(\boldsymbol{x})\right] \sum_{i=0}^{\infty} c_i\,\psi_i(\boldsymbol{x})\, \exp\left(-\lambda_i\, t\right)\,, \tag{7.10}$$

where the λ_i are eigenvalues and the $\psi_i(\boldsymbol{x})$ are eigenfunctions, respectively. A detailed discussion that also considers convergence criteria of the sum in (7.10) and the construction of a Ljapunov function and denotes the differences to quantum mechanics is given in [18]. Further, it is shown that the first eigenvalue, λ_0, of (7.10) vanishes, and the equilibrium distribution is given by

$$P^0(\boldsymbol{x}) = \text{const}\;\exp\left[-\beta\, U(\boldsymbol{x})\right]. \tag{7.11}$$

Hence, in the asymptotic limit, $t \to \infty$, the distribution $P(\boldsymbol{x}, t)$ converges to an equilibrium distribution, which is concentrated around the optimum because the minimum of $U(\boldsymbol{x})$ corresponds to the maximum of $P^0(\boldsymbol{x})$.

In the case considered, it means that a search strategy of Brownian agents described by (7.1), can find the mininum of a potential $U(\boldsymbol{x})$ in the asymptotic limit. This does not necessarily imply that such a search strategy would be *efficient* enough for *practical purposes.* Therefore, in this chapter, we discuss some extensions of the search strategy that are adapted from thermodynamic and biological optimization strategies.

As already mentioned, the dynamics of a search process has to ensure that the occupation probability of searchers in the course of time should increase in the minima of the potential. In a *simple case*, the optimization problem is characterized by only a few optimization criteria, which are weakly or not correlated. Then, usually a *perfect solution* of the problem exists, which corresponds to a clear minimum in the related potential (see Fig. 7.1a). On the other hand, a *complex optimization problem* is characterized by many optimization criteria that can be strongly correlated (see Fig. 7.1b).

Optimization problems like this are known as *frustrated problems* in statistical physics or *multiobjective problems* in optimization theory and have long been investigated from both theoretical and practical perspectives [12, 79, 243, 272, 348, 373, 474, 485, 549, 568]. Multiobjective optimization problems are characterized by a tremendous number of nearly evenly matched solutions which have to be found in a very rugged landscape of the related optimization function. To find the Pareto-optimal solutions, evolutionary algorithms can be applied [146, 178, 237, 355, 376, 402, 440]. These algorithms deal with *stochastic search strategies* in an ensemble of searchers that adapt certain features from natural evolution.

If a large quantity of evenly matched minima exists, the search strategy should avoid total locking of the searchers in the minima found so far to guarantee further search. In (7.2), the mobility of searchers is adjusted by the diffusion coefficient $D \sim T$, (7.3), that describes stochastic motion in the configuration space. Thus, an increase in mobility can be realized by a temporary increase in temperature. Different optimization routines, like

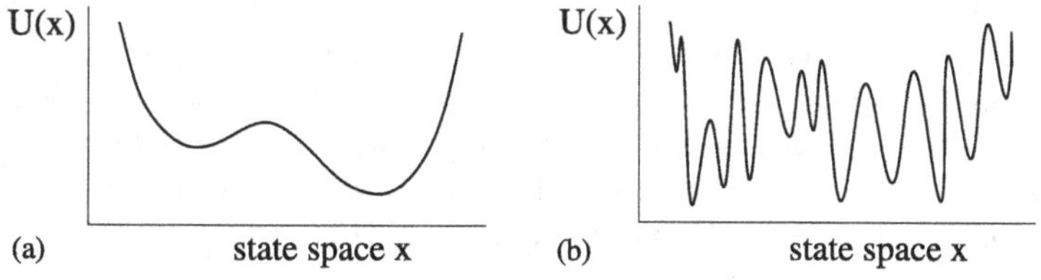

Fig. 7.1. (**a**) Simple optimization problem, characterized by a clear minimum of related potential $U(\boldsymbol{x})$ in state space $\boldsymbol{x}$. (**b**) Complex optimization problem, characterized by a number of nearly evenly matched minima in potential $U(\boldsymbol{x})$

the *simulated annealing approach*, count on such a temporary change in *global* temperature. On the other hand, an increase in temperature for *all* searchers could also result in the loss of appropriate minima which then have to be found again. Hence, the proper adjustment in the temperature of searchers is a major problem in optimization.

In the model of Brownian agents, this dilemma can be solved by using an *individual searching temperature* for every agent [444], i.e., an individual noise intensity $\varepsilon_i \sim T$. Agents that have already found a minimum should be very sensitive during the search, expressed in a very low ε_i, whereas agents that are far from the minima should increase their mobility to search further. We mention that the idea of adjustable sensitivity has been successfully applied to active random walkers simulating ants, which search for unknown food sources (see Sect. 5.4) [456]. Here, we may suggest $\varepsilon_i \propto U(\boldsymbol{x}_i)$, where $U(\boldsymbol{x}_i)$ is the current value of the potential at the location of agent i.

Moreover, *additional coupling* between agents can be assumed, which in terms of communication means instant information for all agents about the best minimum found so far. In this case, only the searcher with the best minimum will stay at rest, whereas other agents are forced to search further. Using global coupling, the individual noise level in the Langevin equation, (7.1), for Brownian searchers can be modified to [444]

$$\varepsilon_i \to \varepsilon_i\{U(\boldsymbol{x}_i)\} - \varepsilon_{\min} \ , \quad \varepsilon_{\min}(t) = \varepsilon_0 \cdot \min\{U(\boldsymbol{x}_i, (t))\} \, . \tag{7.12}$$

Here $\varepsilon_{\min}(t)$ is defined by the minimum of all potential values at the current positions of the agents. Noteworthy, every searcher counts on two pieces of information now: (i) *local* information, provided by the local value of the potential $U(\boldsymbol{x}_i)$ and its gradient $\boldsymbol{\nabla} U$ and (ii) *global* information, provided by $\varepsilon_{\min}(t)$, which means additional coupling among all agents.

7.1.2 Boltzmann Strategy and Darwinian Strategy

The search with Brownian agents is based on the assumption of a continuous state space, where searchers move at constant velocity. However, it is also possible to describe the search problem in discrete space, as shown in the following. Let us consider a numbered set of states $i = 1, \ldots, s$ in the configuration space, each of them characterized by a scalar U_i (the potential energy). Further, we assume a total number N of searchers participating. Then, $N_i(t)$ is the actual number of searchers occupying state i at time t. The occupation number should be an integer, but in the limit case $N \to \infty$ discussed first, the occupation fraction $N_i(t)/N$ may be replaced by the probability of occupation at time t, denoted by $p_i(t)$.

Dynamics that finds the minimum among a discrete set of scalar values, U_i, with $i = 1, \ldots, s$, is given by

$$\frac{dp_i(t)}{dt} = \sum_{i \neq j} \Big[A_{ij}\, p_j(t) - A_{ji}\, p_i(t)\Big]\,, \tag{7.13}$$

where A_{ij} denotes the transition rate for a searcher to move from state j to state i. If we require that the stationary solution of the occupation probability

$$p_i^0 = \lim_{t \to \infty} p_i(t) \tag{7.14}$$

has its maximum at the minimum value of U,

$$p_i^0 \sim \exp\left(-U_i/T\right), \tag{7.15}$$

this leads to the following condition for transition rates:

$$\frac{A_{ij}}{A_{ji}} = \exp\Big[-(U_i - U_j)/T\Big]\,. \tag{7.16}$$

A possible ansatz for A_{ij}, which fulfills this condition, reads

$$A_{ij} = A_{ij}^0 \times \begin{cases} 1 & \text{if } U_i < U_j \\ \exp\Big[-(U_i - U_j)/T(t)\Big] & \text{if } U_i \geq U_j\,. \end{cases} \tag{7.17}$$

These are the transition rates of the known Metropolis algorithm [335]. The analogy between equilibrium statistical mechanics and the Metropolis algorithm was first discussed in [272]. Equation (7.17) means that transitions $j \to i$ toward a lower value $U_i < U_j$ are always accepted, but transitions that lead to a deterioration, $U_i > U_j$, are accepted only at a rate related to the difference in the potential, scaled by the temperature. Thus, due to motion along the gradients, the steepest local descent of the potential will be reached; however, due to thermal fluctuations, locking in those local minima will be avoided.

The prefactor A_{ij}^0 in (7.17) is symmetrical, $A_{ij}^0 = A_{ji}^0$; it defines a set of possible states i that can be reached from state j by a searcher. The simplest definition might be

$$A_{ij}^0 = \begin{cases} 1 & \text{if } \; i \text{ is adjacent to } j \\ 0 & \text{if } \; i \text{ is not adjacent to } j\,. \end{cases} \tag{7.18}$$

The term *adjacent* here means that state i results only from a single elementary mutation of state j; in other words, a change between the different states can occur only in small steps. Another, but more complicated case, is given by $A_{ij}^0 = g(d_{ij})$, where g is a monotonously decreasing function of the minimal number of elementary mutations to reach state j from state i and vice versa.

The temperature T, which appears in (7.17), allows us to adjust transition rates for the deterioration case. For $T \to \infty$, all movements of searchers are equally accepted, whereas for $T \to 0$, only improvements in the search are accepted. The latter is equivalent to the gradient method in optimization theory.

For constant temperatures, the stationary solution, (7.15), is known as the canonical or Boltzmann distribution; therefore the strategy, (7.13), is also named the Boltzmann strategy [67, 120]. It is a *thermodynamic strategy* used during evolution to find minima of certain thermodynamic functions. In the Boltzmann strategy, similar to the *simulated annealing* approach [12, 272, 373, 421, 474], temperature $T(t)$ decreases during the search by a certain rule, e.g., by a power law. This decrease leads to the consequence that first the larger basins of potential minima are explored ("coarse-grained search") and later on a "fine-grained" search occurs within these minimum regions.

Hence, the Boltzmann strategy is based on three elements:

1. Due to motion along gradients, the steepest local ascent of entropy or the steepest local descent of a potential is reached.
2. Due to thermal fluctuations, locking in local minima of the potential is avoided.
3. Due to a continous decrease in temperature, the search becomes more precise in the course of time.

The Boltzmann process asymptotically finds the minimum in a given set of scalars U_i because the minimum of the potential has the highest probability. One can further show [142] that during the search process, the function

$$K(t) = \sum_i p_i(t) \log \frac{p_i(t)}{p_i^0} = \frac{F(t) - F_0}{T} \tag{7.19}$$

monotonically decreases. Here $F(t)$ means the *free energy* of the system with the equilibrium value F_0.

In biological evolution, some new elements occurred in the optimization process, namely,

1. self-reproduction of species whose fitness is above the average fitness,
2. mutation processes due to error reproduction,
3. increase of the precision of self-reproduction in the course of evolution.

We adapt these elements for our search strategy, which is then called the Darwinian strategy because it includes some biological elements known from population dynamics [67, 122, 141]. Biological strategies are different from the above thermodynamic strategies in that here the searchers do not remain constant but can be changed with respect to fundamental processes, such as *reproduction* and *mutation*.

Let us consider again the population of N searchers, which are now distributed in different subpopulations $x_i = N_i/N$; $(i = 1, ..., N)$, each characterized by a *replication rate* E_i which might be proportional to *fitness*. Then, the average replication rate, or average fitness, $\langle E \rangle$, is given by

$$\langle E \rangle = \frac{1}{N} \sum_{i=1}^{N} E_i \, N_i(t) \quad , \quad N = \sum_{i=1}^{N} N_i(t) \, . \tag{7.20}$$

According to Fisher–Eigen dynamics [119, 130, 145], the evolution of subpopulations is given by the equation,

$$\frac{dx_i}{dt} = \Big(E_i - \langle E \rangle\Big) \, x_i + \sum_{j \neq i} \Big(A_{ij}^D \, x_j - A_{ji}^D \, x_i\Big) \, . \tag{7.21}$$

The first term describes the reproduction of subspecies i with respect to its fitness, i.e., on average, only subpopulations whose fitness is above the mean fitness, $E_i > \langle E \rangle$, grow. Due to the mean value $\langle E \rangle$, there exists a global coupling between different subpopulations. The second term describes *mutations*, which means that searchers by chance can be transferred into a state with a better or worse fitness. The transition rates A_{ij}^D are assumed to be *symmetrical* because there are no directed mutations. The average size of the total population remains unchanged.

The effect of increasing precision in self-reproduction can be considered again by a temperature dependence of transition rates, where decreasing temperature leads to smaller probability of mutation. For $A_{ij}^D \to 0$, this evolutionary dynamics is known to approach asymptotically a final state where the average fitness $\langle E \rangle$ is equal to the maximal fitness, which means that only the (one) subpopulation with the best fitness value will survive. For finite mutation rates, $A_{ij}^D > 0$, the target of the search is the eigenvector of (7.21) corresponding to the highest eigenvalue, which for small mutations rates, is close to the maximal value, E_{max} [142].

For application to search processes, it is reasonable to choose fitness E_i of subspecies i as the negative of the potential U_i, indicating that the subspecies which has found the better minimum in the potential landscape, also has the higher reprodution rate. If we use the probability distribution $P(\boldsymbol{x}, t)$ for the *continuous description* again, the dynamics of (7.21) can be rewritten in the form,

$$\frac{\partial P(\boldsymbol{x}, t)}{\partial t} = \Big[\langle U \rangle - U(\boldsymbol{x})\Big] \, P(\boldsymbol{x}, t) + D \, \Delta P(\boldsymbol{x}, t) \, , \tag{7.22}$$

with the mean value for fitness,

$$\langle U \rangle \, (t) = \frac{\int U(\boldsymbol{x}) \, P(\boldsymbol{x}, t) \, d\boldsymbol{x}}{\int P(\boldsymbol{x}, t) \, d\boldsymbol{x}} \, . \tag{7.23}$$

Equation (7.22) is another representation of general search dynamics, (7.4), which has to be compared with (7.2) for the Boltzmann strategy. Using the ansatz,

$$P(\boldsymbol{x}, t) = \exp\left[\int_0^t \langle U\rangle\,(t')\,dt'\right]\,y(\boldsymbol{x}, t)\,, \tag{7.24}$$

one gets from (7.22),

$$\frac{dy(\boldsymbol{x}, t)}{dt} = -H\,y(\boldsymbol{x}, t) = -U(\boldsymbol{x})\,y(\boldsymbol{x}, t) + D\,\Delta y(\boldsymbol{x}, t)\,. \tag{7.25}$$

After separating time and space variables,

$$y(\boldsymbol{x}, t) = \sum_i a_i \exp\left(-\lambda_i\,t\right)\psi_i(\boldsymbol{x})\,, \tag{7.26}$$

we obtain the following stationary Schrödinger equation [17]:

$$\Big(\lambda_i - H\Big)\,\psi_i(\boldsymbol{x}) = D\,\Delta\psi_i(\boldsymbol{x}) + \Big[\lambda_i - U(\boldsymbol{x})\Big]\,\psi_i(\boldsymbol{x}) = 0\,, \tag{7.27}$$

with eigenvalues λ_i and eigenfunctions $\psi_i(\boldsymbol{x})$. A comparison between (7.8) and (7.27) leads to the notable difference, that in biological strategy, the eigenvalue λ_0 has a *nonzero* value. With respect to normalization of the solution, only relative eigenvalues $\lambda_i - \lambda_0$ are obtained [20], reflecting the fact that a shift in the fitness function should not influence the dynamics. In particular, the relaxation time is modified to [18]

$$t_0 = \frac{1}{\lambda_i - \lambda_0}\,. \tag{7.28}$$

To compare both strategies [17, 67], we note that the Boltzmann strategy can detect appropiate potential minima, even in an unknown, rugged optimization landscape, as long as potential barriers between local minima are not too high, which forces the locking in side minima. On the other hand, Darwinian strategy can cross high barriers by tunneling if the next minimum is close enough.

Based on the continuous description, Asselmeyer et al. [18, 20] calculated different velocities which describe how fast the algorithm reaches a particular fitness value, or finds an optimum. The convergence velocity can be defined by the time derivative of the mean value of the fitness:

$$v_p = -\frac{d}{dt}\,\langle U\rangle\,. \tag{7.29}$$

With respect to (7.2), (7.22), one finds for the Boltzmann strategy [20],

$$v_p = D\,\beta\,\Big\langle \boldsymbol{\nabla} U \cdot \boldsymbol{\nabla} U\Big\rangle - D\,\langle \Delta U\rangle\,, \tag{7.30}$$

which depends on the curvature and gradient of the potential. A sufficient condition for *positive* convergence velocity reads,

$$\beta\,(\nabla U)^2 > \Delta U\,. \tag{7.31}$$

For the Darwinian strategy, one obtains instead [20]

$$v_p = \langle U^2 \rangle - \langle U \rangle^2 \;-\; D\,\langle \Delta U \rangle\,. \tag{7.32}$$

Comparison of (7.30), (7.32) leads to the sufficient condition that the biological strategy is faster than the thermodynamic strategy, if the following inequality holds:

$$\Big(U - \langle U \rangle\Big)^2 > D\,\beta\,\Big(\nabla U\Big)^2\,. \tag{7.33}$$

This condition will be fulfilled in optimization landscapes with high curvature and many peaks. On the other hand, the thermodynamic strategy would be faster in landscapes with slight curvature and widely extended maxima, which agrees with the above conclusion.

Further, we can conclude from (7.32) that a *positive* convergence velocity is obtained only if the diffusion coefficient D is below a critical value [20]:

$$D < D_{crit} = \frac{\langle U^2 \rangle - \langle U \rangle^2}{\langle \Delta U \rangle}\,. \tag{7.34}$$

Because D is related to temperature, this condition gives us some estimate of a proper choice of parameters of the strategy. Because $\langle U \rangle$ changes in time according to (7.23), we can deduce from (7.34) a frame for the possible adjustment of temperature during the search.

Within a discrete description, the diffusion coefficient is related to transition rates A^D_{ij}, (7.21), which describe the *mutations* of searchers. Hence, the critical condition, (7.34), means that mutations with respect to fitness do not have to be too large to guarantee success during the search.

We finally note that the existence of a critical value of the diffusion coefficient has an analogy to the aggregation process of Brownian agents discussed in Sect. 3.2. There, we found that Brownian agents aggregate in space in the course of time only if their diffusion coefficient is below a certain threshold, related to a critical temperature. Here, a similar effect can be obtained in the *search space*. If the diffusion coefficient of searchers is below a critical value (which may change in time), then searchers can "aggregate" in the search space, i.e., they commonly occupy minima of the optimization potential.

7.1.3 Mixed Boltzmann–Darwinian Strategy

In addition to the Darwinian strategy introduced in the previous section, other evolutionary strategies based on biological dynamics have been suggested, e.g., the Lotka–Volterra strategy, which includes subpopulations interacting e.g., via predator–prey interactions, or the Haeckel strategy, which considers life cycles [66, 114]. More complex strategies based on *learning, exchange, and storage of information* or *sexual reproduction*, can also be considered.

However, a first and simple step to improve the optimization strategies could be to combine the advantages of the Boltzmann and Darwinian strategies. Therefore, a *mixed* Boltzmann–Darwinian *strategy* has been introduced [17, 20, 67]. From the Boltzmann strategy, asymmmetrical transition rates, (7.17), are adopted that favor transition toward the minimum. On the other hand, from Darwinian strategy, reproduction dynamics is adopted, and fitness E_i of the subspecies is chosen as the negative of potential U_i.

Here, we introduce a more general formulation for the mixed strategy [452], which covers different limit cases:

$$\frac{dx_i}{dt} = \sum_{j \neq i} \left[\kappa \, \mathcal{G}(U_j - U_i) \, x_j \, x_i + A_{ij} \, x_j - A_{ji} \, x_i \right] , \tag{7.35}$$

with transition matrices A_{ij} obtained from (7.17). The function $\mathcal{G}(x)$ can be any linear or nonlinear monotonically nondecreasing function. For the applications discussed here, we use two limiting cases for the choice of $\mathcal{G}(x)$:

$$\begin{aligned} &\text{(i) } \mathcal{G}(U_j - U_i) = U_j - U_i \,, && \text{difference selection} \\ &\text{(ii) } \mathcal{G}(U_j - U_i) = \Theta(U_j - U_i) = \begin{cases} 1 \text{ if } & U_j > U_i \\ 0 \text{ if } & U_j > U_i \end{cases} && \text{tournament selection} \,. \end{aligned} \tag{7.36}$$

The function $\mathcal{G}(U_j - U_i)$ allows us to describe different selection processes. The *tournament selection* (see also [27, 28, 178]) means a "hard" or k.o. selection, where the better one wins regardless of the value of the advantage. Compared to this, the *difference selection*, which also occurs in the Fisher–Eigen model, weights this advantage, and therefore is a "weak" selection. For $\mathcal{G}(U_j - U_i) = U_j - U_i$, (7.35) can be transformed into

$$\frac{dx_i}{dt} = \kappa \left(\langle U \rangle - U_i \right) x_i + \sum_{j \neq i} \left(A_{ij} \, x_j - A_{ji} \, x_i \right) . \tag{7.37}$$

We note, that in (7.37), as well as in (7.35), global coupling between different subpopulations exists. From the general equation (7.35), we can obtain different limit cases of evolutionary strategies, which are appropriate for solving the optimization problem of minimizing U_i. By changing parameters κ and T in the range $0 \leq \kappa \leq 1$, $0 < T \leq \infty$, (7.35) yields

$$\begin{aligned} &\kappa = 0,\, T > 0 && \text{Boltzmann strategy (7.13)}\,, \\ &\kappa = 1,\, T \to \infty,\, \mathcal{G}(U_j - U_i) = U_j - U_i && \text{Fisher–Eigen strategy (7.21)}\,, \\ &\kappa = 1,\, T \to \infty,\, \mathcal{G}(U_j - U_i) = \Theta(U_j - U_i) && \text{tournament strategy}\,. \end{aligned} \tag{7.38}$$

The dynamics of the mixed Boltzmann–Darwinian strategy can also be reformulated in terms of a continuous description, similar to the Darwinian strategy in Sect. 7.1.2. Asselmeyer et al. [17, 20] discussed the approximative version of the mixed strategy,

$$\frac{\partial P(\boldsymbol{x},t)}{\partial t} = \kappa \Big[\langle U \rangle - U(\boldsymbol{x}) \Big] P(\boldsymbol{x},t) + \beta D \boldsymbol{\nabla} \Big[P(\boldsymbol{x},t) \boldsymbol{\nabla} U(\boldsymbol{x}) \Big] + D \Delta P(\boldsymbol{x},t) . \tag{7.39}$$

Equation (7.39) has been treated with the ansatz,

$$P(\boldsymbol{x},t) = \exp \left[\kappa \int_0^t \langle U \rangle (t') \, dt' - \frac{\beta}{2} U(\boldsymbol{x}) \right] y(\boldsymbol{x},t) , \tag{7.40}$$

leading to

$$\frac{dy(\boldsymbol{x},t)}{dt} = -W(\boldsymbol{x},\kappa,\beta)\, y(\boldsymbol{x},t) + D \Delta y(\boldsymbol{x},t) , \tag{7.41}$$

with the refined potential [17],

$$W(\boldsymbol{x},\kappa,\beta) = \kappa U(\boldsymbol{x}) - \frac{\beta}{2} D \Delta U(\boldsymbol{x}) + \frac{\beta^2}{4} D \boldsymbol{\nabla} U(\boldsymbol{x}) \cdot \boldsymbol{\nabla} U(\boldsymbol{x}) . \tag{7.42}$$

Equation (7.42) yields $W(\boldsymbol{x},\kappa,\beta) = U(\boldsymbol{x})$ for the Darwinian case, $\beta = 0$, $\kappa = 1$, and $W(\boldsymbol{x},\kappa,\beta) = V(\boldsymbol{x})$, (7.9), for the Boltzmann case, $\kappa = 0$. Again using the separation ansatz, (7.26), the eigenfunctions result now from

$$\Big[\lambda_i - W(\boldsymbol{x},\kappa,\beta) \Big] \psi_i(\boldsymbol{x}) + D \Delta \psi_i(\boldsymbol{x}) = 0 . \tag{7.43}$$

The linearity of the differential (7.43) leads to simple relations between solutions. For example, the velocity of the Boltzmann strategy can be added to the velocity of the Darwinian strategy with respect to constant κ. The convergence velocity will be positive as long as the diffusion constant D satisfies the inequality [20],

$$D < D_{crit} = \frac{\kappa \left(\langle U^2 \rangle - \langle U \rangle^2 \right)}{\langle \Delta U \rangle - \beta \Big\langle \boldsymbol{\nabla} U \cdot \boldsymbol{\nabla} U \Big\rangle} . \tag{7.44}$$

Now, the magnitude of D_{crit} can be used to adapt parameters κ and β.

Numerous examples [17, 67, 114, 122] have shown that the mixed strategy is a tool well-suited for multicriteria optimization. Let us, for example, assume an optimization function consisting of two parts, $U(\boldsymbol{x}) = U_1(\boldsymbol{x}) + U_2(\boldsymbol{x})$, where $U_1(\boldsymbol{x})$ represents a more local requirement, as discussed in the next section. Then, the mixed strategy may use the Boltzmann strategy for local optimization of $U_1(\boldsymbol{x})$, and the Darwinian strategy for global optimization of the whole function $U(\boldsymbol{x})$. Mixing of both strategies increases the convergence velocity, and it also reduces the tendency of the search process to become trapped in local minima because of the Darwinian part, which allows a "tunneling" process.

Before we start to apply the different strategies to optimization of a complex, multicriteria problem, we need to discuss the simulation algorithm for a *finite number* of searchers. So far, we considered the case $N \to \infty$, where the discrete numbers of searchers $N_i(t)$ occupying state i, might be replaced by probabilities $p_i(t)$ or population densities $x_i(t)$. During the simulations, however, one always deals with a finite number of searchers; hence, the search process will be realized by a kind of the Ehrenfest urn game. For a stochastic approach to the search process, let us introduce the discrete distribution of searchers:

$$\boldsymbol{N}(t) = N_1(t), N_2(t), ..., N_s(t) \; , \quad N = \sum_{i=1}^{s} N_i(t) = \text{const} \, , \tag{7.45}$$

where N_i is the number of searchers that occupy state i. The probability of finding a particular distribution $\boldsymbol{N}$ at time t is described by $P(\boldsymbol{N},t) = P(N_1, N_2, ..., N_s, t)$. The change of this probability distribution in the course of time can be described by a *master equation* that considers all possible processes leading to the change in the particular distribution $\boldsymbol{N}$ (see Sect. 1.3.3):

$$\frac{\partial P(\boldsymbol{N},t)}{\partial t} = \sum_{\boldsymbol{N}' \neq \boldsymbol{N}} \Big[w(\boldsymbol{N}|\boldsymbol{N}') P(\boldsymbol{N}',t) - w(\boldsymbol{N}'|\boldsymbol{N}) P(\boldsymbol{N},t) \Big] \, . \tag{7.46}$$

The terms $w(\boldsymbol{N}'|\boldsymbol{N})$ are transition rates per unit time to change the distribution $\boldsymbol{N}$ into distributions $\boldsymbol{N}'$. For the strategies introduced above, we have now two different sets of transition probabilities.

The transition rate for *mutation* processes reads

$$w_M(N_1, \ldots, N_i + 1, N_j - 1, \ldots, N_s | N_1, \ldots, N_i, N_j, \ldots, N_s) = A_{ij} \; N_j \, , \tag{7.47}$$

which means that the probability that a searcher occupying state j moves to state i is proportional to the number of searchers in state j and transition rates for the Boltzmann process, (7.17). With respect to computer simulations, (7.47), is consistent with these rules:

1. Select one among the N searchers (say searcher j).
2. Choose a possible mutation A_{ij}.
3. Decide according to 7.17 whether or not the mutation is accepted.

For strategies including Darwinian elements, we have additionally to consider the transition rate for *selection* processes, which reads

$$\begin{aligned} w_S(N_1, \ldots, N_i + \omega_{ij}, N_j - \omega_{ij}, \ldots, N_s | N_1, \ldots, N_i, N_j, \ldots, N_s) \\ = \kappa \Big| \mathcal{G}(U_j - U_i) \Big| \, N_i N_j / N \, , \end{aligned} \tag{7.48}$$

with

$$\omega_{ij} = \begin{cases} 1 & \text{if } U_i < U_j \\ -1 & \text{if } U_i > U_j \,. \end{cases}$$

Equation (7.48) means that the probability that searchers in state j compete with searchers in state i depends on the occupation number (or size of the subpopulation) and the fitness, which is related to the potential of the occupied state. For the selection function $\mathcal{G}(x)$, two possibilities have been introduced in (7.36). For the computer simulations discussed below, we prefer the *tournament selection* (k.o. selection) which in most cases is more effective than the difference selection. It is consistent with these rules:

1. Select a pair of searchers (say i and j) from the $N\,(N-1)/2$ possibilities.
2. Compare potentials U_i and U_j.
3. Transform the searcher with the higher value of U to the class of searchers with the lower U (which is related to better fitness).

During stochastic simulation of the search, transition rates for all possible changes of the distribution $\boldsymbol{N}$ are calculated for every time step. The probability that a particular transition occurs during the next time step is given by $w(\boldsymbol{N}'|\boldsymbol{N})/\sum_{\boldsymbol{N}'} w(\boldsymbol{N}'|\boldsymbol{N})$. The actual transition is chosen randomly with respect to the different weights of the possible transitions (see also Sect. 3.1.4 for the stochastic simulation technique).

It can be proved analytically and by computer simulations [142] that the stochastic game based on transition probabilities, (7.47), (7.48), on average and for large numbers of N, results in the equation for the mixed Boltzmann–Darwinian strategy, (7.35), and therefore has the same search properties.

7.2 Evaluation and Optimization of Road Networks

7.2.1 Road Networks

The optimization of networks that connect a given set of nodes is of common interest in many different areas, among them electrical engineering, telecommunication, road construction, and trade logistics. The problem turned out to be a challenge for both practical applications and theoretical investigations [13, 243, 259, 483]. As one *example*, we consider here a *road network*; however the optimization method as well as the results can be generalized for similar network problems.

In Sect. 5.4 and Sect. 6.2, the focus was on the *formation* of trail systems between sources and destinations as a process of *self-organization*. However, if these connections are used, e.g., for transportation, then they also have to satisfy some economic requirements. The optimization of a road network has to consider, for example, two different requirements: (i) to minimize the total cost of constructing and maintaining the roads, which should be proportional to the total length of the roads between the different points of interest (nodes) and (ii) to minimize the effort (detour) to reach any given node from any other node, which should be proportional to the lengths of the roads along the shortest existing connection between the two nodes. Both demands cannot be completely satisfied at the same time because a minimized detour means a direct connection between all nodes and therefore a maximum total length of the road network, and minimal costs for the network mean the shortest possible total length for the roads.

Considering only the first demand mentioned, the solution of the optimization process is given by a road network where every node is connected to the network by just one (the shortest possible) link, which leads to a *minimal link system* (see Fig. 7.2b), also known as a *minimal spanning tree*. On the other hand, considering only the second demand, the solution is simply given by a network where every node is connected to every other node by a direct link, which leads to a *direct link system* (see Fig. 7.2a).

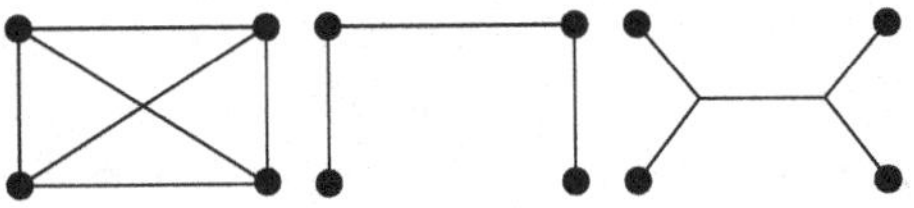

Fig. 7.2. (**a**) Direct link system; (**b**) minimal link system; (**c**) Steiner tree for a set of four nodes

Compared to these two idealized limiting cases, the road network in most real applications is a *compromise* among different requirements. Figure 7.3 shows an example of the town *Martina Franca* in upper Italy. The roads of different widths and the perimeter road that surrounds the old town are clearly shown in the aerial photograph. To reach arbitrary destinations from arbitrary locations within the perimeter, one usually has a choice between numerous different combinations of roads. The left part of Fig. 7.3 shows the connections which would be chosen to minimize the *length* of each possible path on the *existing road network.* The thickness of the lines indicates how much a given road would be used. The right part, on the other hand, shows the connections chosen to minimize the *topological distance* on the existing road network, i.e., the connections that cross the smallest possible number of *intersections.* The patterns resulting from these two different optimization criteria display considerable differences; in the latter case, the perimeter road is used a lot more because it avoids possible intersections, but it increases the distance traveled. The real road network is a compromise between these different optimized solutions.

An evaluative perspective that considers only requirements such as "detours" and construction costs to optimize a road network, of course, neglects

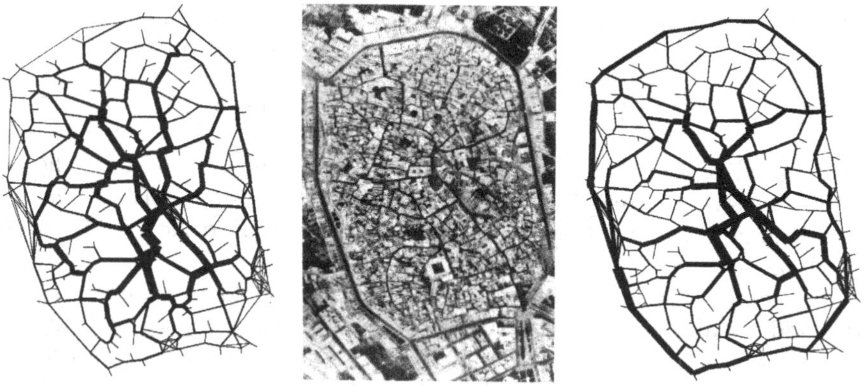

Fig. 7.3. (*middle*) Aerial photograph of the town of Martina Franca, (*left*) usage pattern resulting from an optimization of the distances (shortest connections on the existing road network), (*right*) usage pattern resulting from an optimization of the topological distances (minimum number of intersections crossed) [424]

all problems that usually occur in using roads, such as traffic density, flow capacity, etc. Here, we focus mainly on the *optimized establishment of links*; thus the definition of "costs" takes into account only the length of the road, not the width or equipment, and "detour" takes into account only the traveling distance, not the traveling time or traveling velocity. Further, the evaluation does not consider any requirements related to the use of the area covered by the road network. In cities or urban areas, optimization of road networks usually has to take into account that many subareas are occupied or preserved, and therefore cannot be divided by an "optimized" road. Thus, for applications of the model to cities, the evaluation should also include a term expressing the optimization of subareas as well. So far, our considerations may be applied to road networks that link different cities at certain distances. In this range, the terms of a minimal and a direct link system and the compromise between these limit cases make sense for geographic applications.

7.2.2 The Evaluation Function

Because the minimization of both "detour" and "costs" cannot be satisfied at the same time, the road optimization problem in the form considered belongs to the class of frustrated optimization problems, mentioned in Sect. 7.1.1. To find optimized solutions, we want to apply a special class of evolutionary strategies, the Boltzmann and Darwinian strategies, as well as a mix of both, which are introduced in Sects. 7.1.2 and 7.1.3. In a first step, we have to define a potential function (or *fitness function*) $U(\boldsymbol{x})$ that evaluates a given road network.

Let us consider a set of nodes, $p_1 \dots p_N$, which are connected by straight lines representing the roads. The configuration space is defined by the number of all possible graphs g to connect the nodes and is of the order $2^{N(N-1)/2}$. Each graph should be evaluated due to the following potential function [452]:

$$U(g, \lambda) = (1 - \lambda)D(g) + \lambda C(g) \, , \qquad 0 \leq \lambda \leq 1 \, . \tag{7.49}$$

Here, $D(g)$ represents the mean detour to reach different nodes on the existing network, whereas $C(g)$ represents the total costs for constructing and maintaining the network, which should be proportional to the total length of the links.

To minimize the potential $U(g)$, both terms should be minimized. The demand for a minimized detour between two points represents a *local constraint* on the network. Hence, a minimized mean detour between *every* two nodes is considered an averaged local constraint, whereas the demand for a minimized total length of the network is a *global constraint*. For the example of the road network considered local constraints represent the interests of users who don't like detours, and global constraints are given by the interests of the government which has to pay for the road construction and therefore tries to minimize it.

The parameter λ is introduced to weight between the two contradicting demands. The case $\lambda \to 0$ leads to a potential function, which only minimizes the detour regardless of the costs of the network, and finally results in a *direct link system* (Fig. 7.2a). In this case, any two nodes are connected by the shortest possible link (equal to the length of the distance); the total length of the network, however, reaches a maximum.

In the opposite case, $\lambda \to 1$, only the costs of the network will be minimized, which finally leads to a *minimal link system* (Fig. 7.2b). This implies a very long connection between arbitrary nodes, which includes large detours, because the connection on the existing network exceeds the distance by far. Moreover, a minimal link system might be susceptible to breakdown because every node has only one connection to the net. We note that in the node system considered here, the minimal link system will be different from the known Steiner *tree* (Fig. 7.2c) because we assume that any link between nodes is a straight line and no additional subnodes are constructed to connect given nodes. To estimate the difference, we note that the ratio of the length of the Steiner tree and the length of the minimal spanning tree is always $\geq \sqrt{3}/2$ [173, 562].

For practical applications, both limiting cases, $\lambda = \{0, 1\}$, could be sufficient under certain circumstances. A minimal link system is appropriate, e.g., if the connection distance can be passed at very high speed. Then the time to reach a given node does not count, and a detour on the network could be easily accepted. On the other hand, a direct link system will be appropriate if the costs of establishing a network do not count compared to the traveling time which should be as short as possible. For road networks, however, λ should be different from zero or one; then it can be a measure of the *degree of frustration* of the problem or a measure of accepting compromises.

Here, the direct link system will be used as a *reference state*, indicated by the symbol $\star$. The direct link system has the advantage that the values for both the detour and the costs of the road network are known. The total length of the network is simply given by all direct distances, $l_{i,j}$, between the nodes and gets its maximum value $L^\star$:

$$L^\star = \frac{1}{2} \sum_{i,j=1}^{N} \sqrt{(x_i - x_j)^2 + (y_i - y_j)^2} = \frac{1}{2} \sum_{i,j=1}^{N} l_{i,j} \,. \tag{7.50}$$

x_i, y_i are the coordinates of node i in a two-dimensional plane (see also Table 7.1). For values of $0 \le \lambda \le 1$, the mean detour $D(g)$ is defined as follows:

$$D(g) = \frac{1}{2} \sum_{i,j=1}^{N} h_{i,j} - l_{i,j} \;, \quad D(g^\star) = 0 \,, \tag{7.51}$$

where $h_{i,j}$ is the length of the *shortest route* that connects nodes i and j *on the existing graph.* Obviously, for a graph representing the direct link system, $h_{i,j} = l_{i,j}$, holds; hence the detour $D(g^\star)$ is zero.

The expression $D(g)$ can be normalized by the total length of the direct link system, which is a known constant for a given set of nodes:

$$d(g) = \frac{D(g)}{L^\star} = \frac{1}{2} \sum_{i,j=1}^{N} \frac{h_{i,j} - l_{i,j}}{L^\star} \,. \tag{7.52}$$

Another possibility, proposed in [424, p. 82], reads

$$d(g) = \frac{1}{2} \sum_{i,j=1}^{N} \frac{h_{i,j} - l_{i,j}}{l_{i,j}} \,. \tag{7.53}$$

Here, the detours related to shorter distances are weighted higher than those for larger distances. We restrict ourselves to (7.52).

To specify the term representing the costs, we assume that the costs are simply proportional to the total length of the graph:

$$C(g) = \frac{1}{2} \sum_{i,j=1}^{N} \vartheta_{ij} l_{i,j} \,. \tag{7.54}$$

Here, ϑ_{ij} expresses the connectivity of the nodes: $\vartheta_{ij} = 1$, if there is a *direct* connection between nodes i and j, and $\vartheta_{ij} = 0$ else. After normalization of $C(g)$, similar to (7.52),

$$c(g) = \frac{C(g)}{L^\star} = \frac{1}{2} \sum_{i,j=1}^{N} \frac{\vartheta_{ij} l_{i,j}}{L^\star} \,, \tag{7.55}$$

the potential function describing the road network reads finally [452],

$$u(g,\lambda) = \frac{U(g)}{2L^\star} = \frac{1}{2L^\star}\sum_{i,j=1}^{N}(1-\lambda)\,(h_{i,j}-l_{i,j}) + \lambda\,\vartheta_{ij}l_{ij}\,. \tag{7.56}$$

For the reference state (direct link system), the values are $d(g^\star) = d_{\text{min}} = 0$ for the mean detour, $c(g^\star) = c_{\text{max}} = 1$ for the costs, and $u(g^\star) = \lambda$ for the potential. Based on (7.56), a hypothetical network will be optimized with respect to the mean detour and the total length of the network.

7.2.3 Results of Computer Simulations

We carried out our computer simulations of the evolutionary optimization of a road network for a system of 39 nodes [451, 452]. The coordinates of the nodes are listed in Table 7.1.

The initial state for the simulations is given by the graph presented in Fig. 7.4 for $t = 0$. Here, a very dense connection exists among the nodes. There is no need to start the simulations exactly with a direct link system because for $\lambda \neq 0$, this is is not an optimized state of the problem. Basically, the simulations could start with an arbitrary graph. During every time step of the simulation, the graph is first mutated by adding, removing, or exchanging one link between nodes, and then evaluated by using (7.56).

Figure 7.4 shows a time series displaying different stages of the optimization process for $\lambda = 0.975$. The decrease of temperature is due to

$$T(t+1) = \frac{T_0}{1+at}\,, \tag{7.57}$$

where $T_0 = 0.01$ is the initial temperature and $a = 0.002$ is the cooling rate. The thickness of the lines indicates how much a given road is used for the shortest possible connection of any node to any other. These considerations can also be used to estimate *travel demands* from a *static* perspective.

When we start the simulations with a dense graph, the optimization process occurs in *two stages.* During the first stage, the network is strictly thinned out, whereas during the second and much longer stage, the links between the different nodes are balanced with respect to the costs and the mean detour. How long the second stage takes compared to the first, depends, of course, on the value of λ. For $\lambda \to 0$, the thinning out of the network does not play an important role in optimization; therefore, the first stage should be rather short.

The two stages of the optimization process are, for the given initial situation, a rather typical scenario, because for a *dense graph,* the removal of links results in a larger "gain," whereas the exchange of links does not affect the value of the optimization function significantly. On the other hand, if a thinned-out network already exists, the strict removal of links eventually

Table 7.1. Coordinates of the nodes in the (x, y) plane, used for the simulations [452]

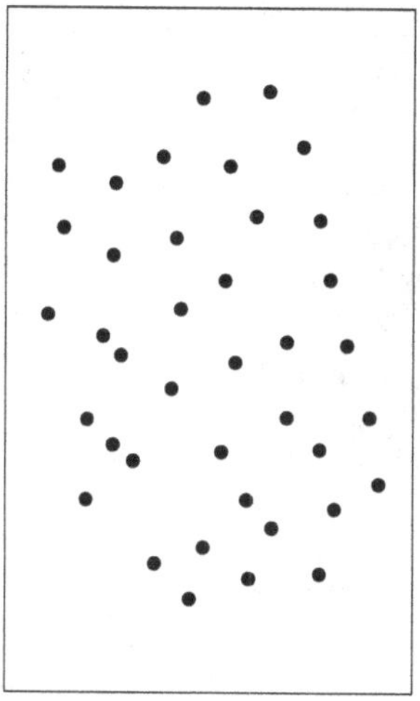

Node	x	y
1	0.121667	0.388333
2	0.131667	0.291667
3	0.203333	0.438333
4	0.218333	0.233333
5	0.231667	0.078333
6	0.231667	0.331667
7	0.256667	0.163333
8	0.305000	0.370000
9	0.310000	0.463333
10	0.321667	0.086667
11	0.336667	0.253333
12	0.361667	0.160000
13	0.396667	0.478333
14	0.398333	0.325000
15	0.440000	0.260000
16	0.448333	0.063333
17	0.480000	0.145000
18	0.488333	0.415000
19	0.493333	0.503333
20	0.508333	0.171667

Node	x	y
21	0.518333	0.340000
22	0.556667	0.246667
23	0.598333	0.415000
24	0.598333	0.536667
25	0.601667	0.121667
26	0.638333	0.160000
27	0.645000	0.463333
28	0.648333	0.320000
29	0.661667	0.190000
30	0.695000	0.550000
31	0.718333	0.120000
32	0.718333	0.356667
33	0.731667	0.485000
34	0.760000	0.393333
35	0.788333	0.293333
36	0.811667	0.221667
37	0.826667	0.463333
38	0.833333	0.360000
39	0.863333	0.273333

leads again to increasing values in the optimization function, which are rarely accepted, but the exchange of links may still result in improvements of the graph.

These two stages are investigated in more detail in Fig. 7.5. As we see in Fig. 7.5a, the rearrangements in the network during the first stage are characterized by rather large fluctuations in the evaluation function but do

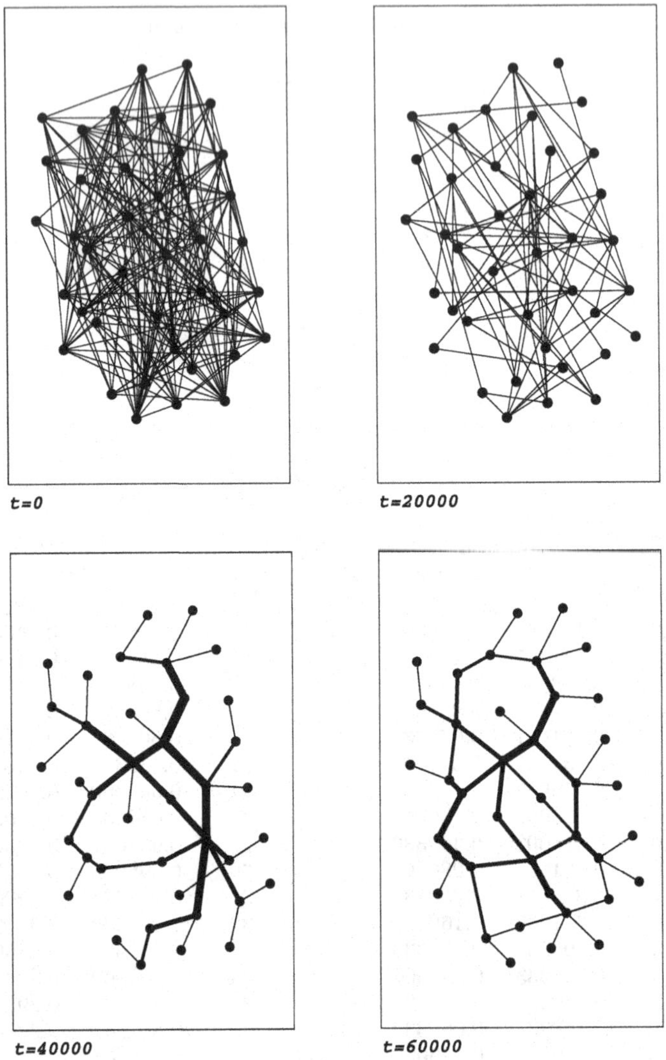

Fig. 7.4. Optimization of a network of 39 nodes ($\lambda = 0.975$). The graph is shown after different time steps [451, 452]

not affect the mean detour significantly. The second stage starts with a steep decrease in the potential function, which is related to a remarkable increase in the mean detour. This increase is reduced only during a long period of time, indicating the painstaking process of balancing the contradictory requirements, costs and detour, to optimize the network.

Figure 7.5b displays the ratio between costs and mean detour during the optimization process. With respect to time, the curve starts in the lower right

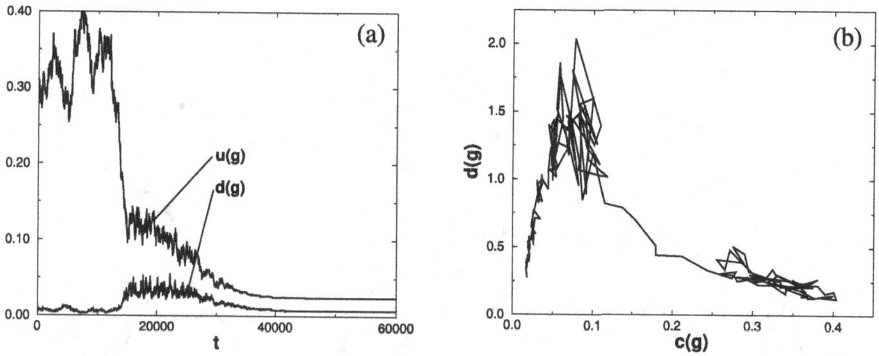

Fig. 7.5. (**a**) Time dependence of the potential $u(g)$ and the mean detour $d(g)$, (**b**) mean detour $d(g)$ versus costs $c(g)$ during optimization of the network. The data points are picked from a single run every 200 time steps ($\lambda = 0.975$) [451, 452]

corner and ends in the lower left corner. Again, the first stage (which is related to the right part of the curve) is characterized by a steep decrease in costs and a moderate change in the detour. The transition into the second stage is indicated by a large increase in the detour, and the crossover is marked by the maximum region of the curve. During the second stage, a slow decrease of both detour and costs occurs which eventually results in the quasi-stationary state of a balanced network.

With respect to practical applications, it is very remarkable that during the last stage, a considerable decrease in the detour occurs, whereas costs remain nearly constant. That means that global constraints (e.g., the costs paid by the "government") are already satisfied, but regarding local constraints (e.g., the interests of users), further improvements can be achieved if the optimization proceeds long enough. As mentioned above, this should be rather typical of the final stage of simulations because removal of links (which mostly affect costs) is rarely accepted, but exchange of links (which mostly affect the detour) still results in better solutions. In Sect. 7.3.2, we give an analytical approximation of whether more effort in optimization may eventually yield still better solutions for the graph and we estimate the computer time needed for additional improvements.

The optimized final state, of course, depends on the parameter λ which influences the density of the graph. Figure 7.6 presents results for the optimized network for different values of λ. In agreement with the discussion in Sect. 7.2.2 for small λ, we find networks, where most of the nodes are connected to all neighboring nodes and for large λ, networks that are close to the minimal link system.

Finally, we would like to compare the results of the Boltzmann strategy and the mixed strategy which also includes Darwinian elements. As shown in the simulations, the Boltzmann strategy finds suitable results in the asymptotic limit (about 60,000 simulation steps). However, the mixed Boltzmann–

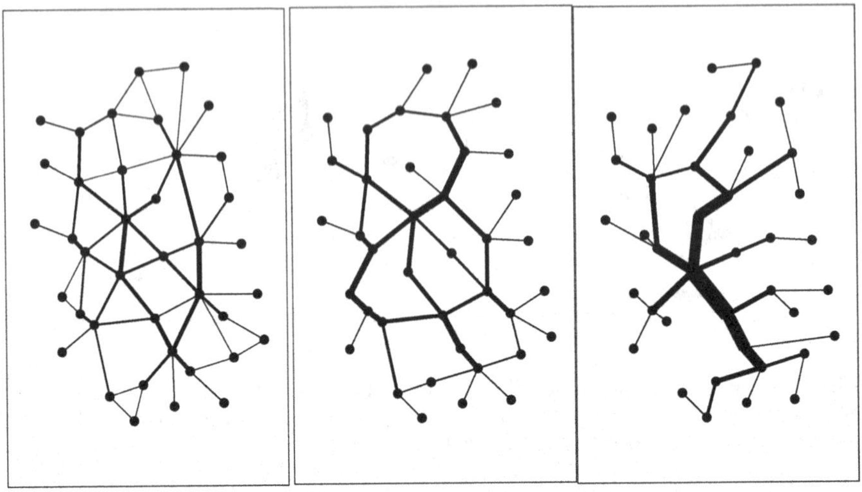

Fig. 7.6. Optimized networks after $t = 60,000$ simulation steps for different values of parameter λ: (**a**) $\lambda = 0.900$; (**b**) $\lambda = 0.975$; (**c**) $\lambda = 0.990$ [451, 452]

Darwinian strategy already finds optimal graphs in a much shorter simulation time, as shown in Fig. 7.7 for 10,000 simulation steps (obtained for the same number of searchers in both simulations).

The optimization function relaxes very fast compared to the Boltzmann curve. With respect to the networks obtained after 10,000 time steps, we find already balanced graphs with the mixed optimization strategy, whereas the graphs obtained from the Boltzmann strategy still display failures in optimization. For larger values of parameter λ, which means for networks closer to the minimal link system, this advantage becomes even more remarkable, as shown in the comparison between $\lambda = 0.6$ and $\lambda = 0.8$ in Fig. 7.7. Hence, we can conclude from the above computer simulations, that Boltzmann and Darwinian strategies provide a tool suitable for finding optimized solutions. However, the consideration of biological elements in the strategies allows us to reduce the number of time steps in the simulation and enhance the performance of the optimization, compared to Boltzmann-like strategies.

We want to add that the advantage of the mixed strategy compared to the Boltzmann strategy can also be substantiated by calculating the convergence velocity for the optimization process (see Sect. 7.1.3). The analytical results are obtained, however, only for specific simple potentials [18, 20] and are by now out of reach for the complex problem considered here.

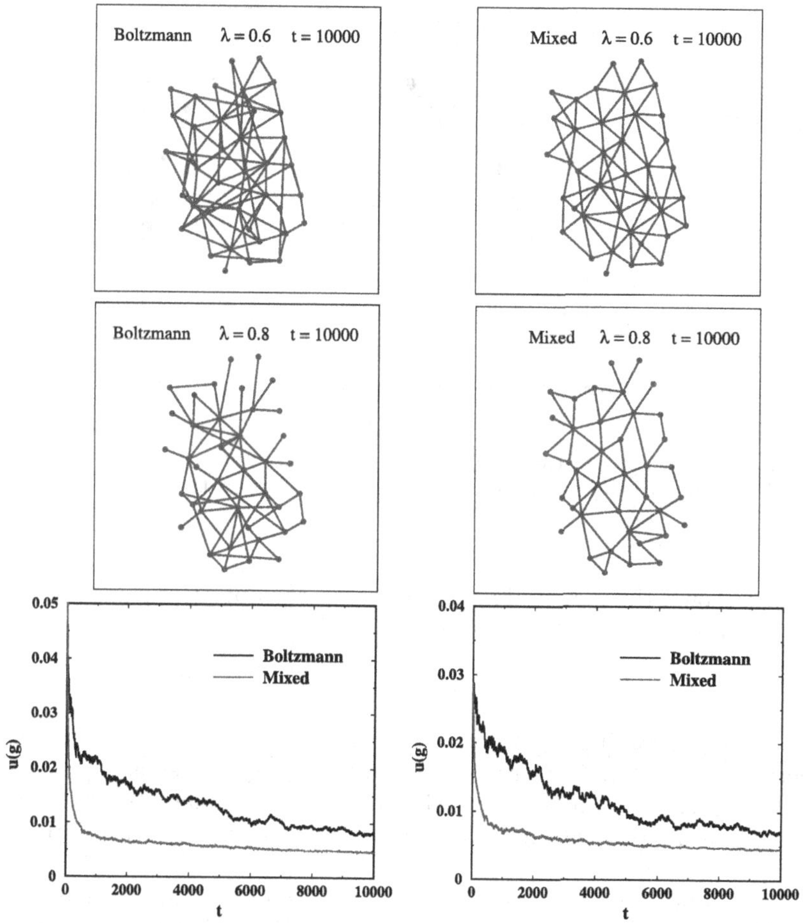

Fig. 7.7. Comparison of Boltzmann and mixed strategies for network optimization. (*left*) $\lambda = 0.6$; (*right*) $\lambda = 0.8$. The networks presented are obtained after 10,000 simulation steps; the related potential is displayed below. The ensemble consists of 16 searchers for both strategies. For the selection process, tournament selection is used, with a probability of 0.3 per searcher and time step [451, 452]

7.3 Asymptotic Results on the Optimization Landscape

7.3.1 Optimization Values in the Asymptotic Limit

To characterize the optimization landscape of road networks in more detail, we finally investigate some asymptotic features, obtained from the Boltzmann strategy [452]. First, the dependence of the optimized state on the parameter λ is discussed. Whereas Fig. 7.6 shows some realizations of the network obtained for different λ, in Fig. 7.8, the potential values for the optimized network, u^{opt}, obtained *asymptotically* are plotted versus parameter λ. As

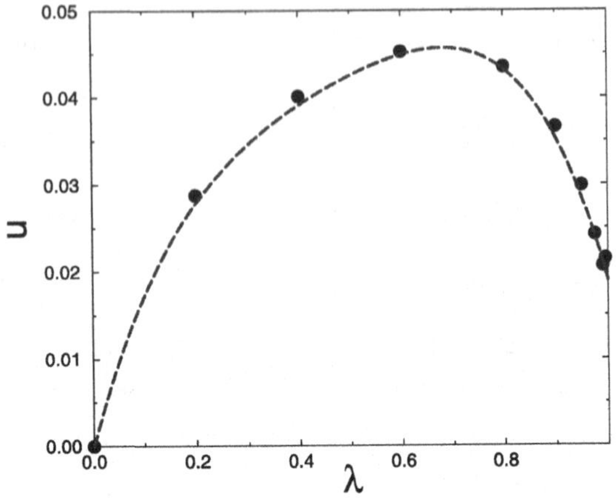

Fig. 7.8. Dependence of the asymptotic potential minimum on parameter λ. *Dots* mark the results of the simulations, averaged over 50 runs; the *dashed line* is given by (7.58) [451, 452]

the plot shows, the potential minimum in the asymptotic regime can be well approximated by a fourth-order power function of λ:

$$u^{\mathrm{opt}}(\lambda) = \lambda\ \left(0.2155 - 0.4949\,\lambda + 0.6380\,\lambda^2 - 0.3395\,\lambda^3\right)\ . \qquad (7.58)$$

This allows predicting asymptotic value of $u(g)$ and can be used to estimate whether a simulation has already reached the asymptotic stage. On the other hand, this relation can also be used to test different functions for a decrease in temperature $T(t)$, which finally have to yield the known asymptotic value.

For an optimized network, (7.58) does not completely determine the best possible cost and the best affordable detour. From (7.49), (7.58), a linear relation between the asymptotic values of $d(g)$ and $c(g)$ results:

$$d^{\mathrm{opt}}\left(c^{\mathrm{opt}}\right) = \frac{u^{\mathrm{opt}}(\lambda)}{1-\lambda} - \frac{\lambda}{1-\lambda}\,c^{\mathrm{opt}}\ . \qquad (7.59)$$

Equation (7.59) characterizes the *range of compromises* that can be found asymptotically for a fixed value of λ. For $\lambda < 1$, c^{opt} has to be larger than the costs for the minimal link system, $c_{\min}$, which defines the lower boundary for $c(g)$, and less than $c(g^\star) = c_{\max} = 1$, which is the upper boundary, obtained for the direct link system. For a medium range of $0 \leq \lambda \leq 1$ and an appropriate large number of nodes, there should be many sets of $c^{\mathrm{opt}}, d^{\mathrm{opt}}$ that fulfill (7.59), as also indicated by Fig. 7.9.

This means that an optimized road network for a given λ allows different realizations for costs and mean detours. Within the range of possible compromises defined by (7.59), we obtain different graphs to connect nodes. Here, those sets of $c^{\mathrm{opt}}, d^{\mathrm{opt}}$ that are related to lower costs, lead to graphs

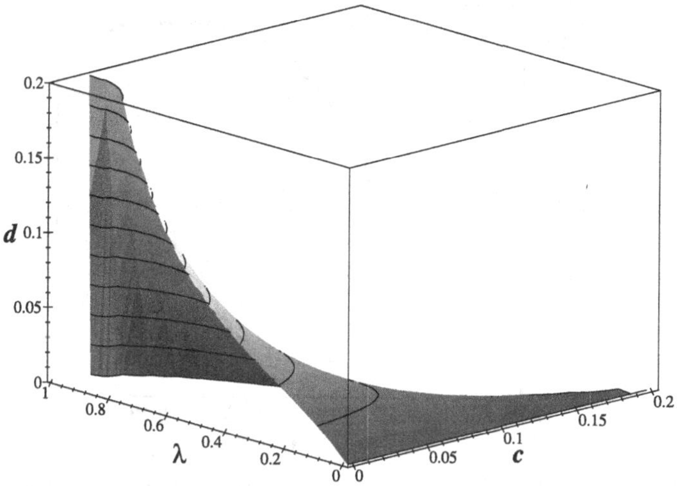

Fig. 7.9. Range of compromises, (7.59), between detour, d^{opt}, and costs, c^{opt}, for an optimized network, depending on λ [452]

with dominating trunk routes (such as in Fig. 7.6c), whereas sets related to smaller mean detours lead to graphs with ring routes. This interesting feature should be investigated by subsequent computer simulations.

7.3.2 Density of States in the Asymptotic Limit

From the Boltzmann distribution reached in the asymptotic limit, $p_i^0 \sim \exp(-U_i/T)$, (7.15), we obtain the probability $P^0(U)$ of finding a certain value of U in the interval dU:

$$P^0(U)\, dU = n(U) \exp(-U/T)\, dU\,. \tag{7.60}$$

The function $n(U)$ is called the degeneracy function or *density of states*, which is widely used in statistical physics and thermodynamic optimization processes [12]. It contains important information about the structure of the optimization landscape [415] because it tells us how sparsely states of a given quality are distributed. The density of states, however, measures only the number and the order of possible solutions but not the geometry and topology of the search space itself.

Usually, the best available optimization states are found in the tail of the distribution $n(U)$, corresponding to a very small density of state. Thus, in the limit $n(U) \to 0$, the best possible potential value of a given problem can be estimated. Formally, the density of states may be found from the equilibrium solution,

$$n(U) = P^0(U) \exp(U/T)\,. \tag{7.61}$$

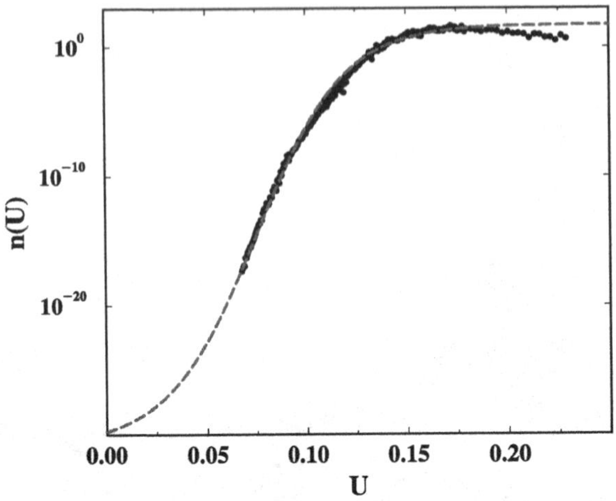

Fig. 7.10. Density of states $n(U)$ in the asymptotic limit ($t = 100,000$ time steps). The probability distribution $P^0(U)$ was obtained from a simulation of 1000 graphs, carried out for different temperatures: $T = 0.001, 0.002, 0.005, 0.1$ ($\lambda = 0.950$). *Dots* mark the results of the simulations; the *dashed line* results from (7.62) [452]

After determining the equilibrium probability distribution $P^0(U)$ for the optimization of a road network by a Boltzmann strategy, we find the density of states, (7.61), presented in Fig. 7.10.

Note that the density of states provides useful information for those optimization problems that can be characterized by a smooth function $n(U)$, with $n(U) \to 0$ both in the limit of $U \to U_{\text{min}}$ and $U \to U_{\text{max}}$. For problems that deal with a singular solution, such as the "needle-in-a-haystack" problem, $n(U)$ is not smooth, and therefore the concept is less appropriate.

From the density of states, information can be derived about the *effort* of optimization for a given problem. Optimization problems characterized by very steep decay in $n(U)$ for good values of U, in general, do not need too much effort in simulations because the number of good solutions is very limited. Once a good solution is found, one can almost be sure that more effort does not necessarily result in better optimization values.

More interesting, however, are optimization problems that display long tails in the density of state, $n(U)$, in the region of optimal solutions, $U \to U_{\text{min}}$. Here, the number of better solutions is figuratively "nonexhaustible", which means that more effort in optimization always has a good chance of yielding better solutions. As Fig. 7.10 indicates, the network problem discussed here, belongs to this class of problems. The existence of long tails in the distribution of the density of states in the range of better potential values corresponds to the fact that during the final stage of the simulation, global constraints are already satisfied, but further improvements can be made to match local constraints. Hence, the search for more comfort in the solutions is worthwhile.

For the network considered with $\lambda = 0.950$, we find that the frequency of better solutions can be estimated by

$$\begin{aligned} n(U) &= c_0 \exp\{c_1 \tanh[c_2(U - c_3)]\}\,, \\ c_0 = 2.8306 \times 10^{-15}\,, c_1 &= 37.4982\,, c_2 = 22.6000\,, c_3 = 0.0750\,, \end{aligned} \tag{7.62}$$

which, in the limit $U \to 0$, can be approximated by

$$\begin{aligned} n(U) &\sim \exp(-b + aU) \\ \text{with } a &= c_1 c_2\,,\ b = c_1 c_2 c_3\,. \end{aligned} \tag{7.63}$$

The non-Gaussian distribution of $n(U)$ is a notable difference from other optimization problems that yield either Gaussian distribution for $n(U)$ [415], or display very steep decay, such as the LABS problem [57]. Compared to these cases, the frustrated network problem is characterized by rather weak decay in $n(U)$. The best values we have obtained so far, $U \approx 0.07$ [452], are not yet in the region of saturation, which indicates a good chance of finding better solutions (e.g., $U \approx 0.03 - 0.05$) in the long run.

Let us estimate the effort needed to improve a given solution U_0 by the amount ΔU to obtain $U_0 - \Delta U$. Assuming that the time to find a better solution is proportional to the reciprocal density of states,

$$\frac{\tau}{\tau_0} \simeq \frac{n(U_0)}{n(U_0 - \Delta U)}\,, \tag{7.64}$$

we obtain with (7.63) for the network problem,

$$\frac{\tau}{\tau_0} \simeq \exp(a\Delta U)\,. \tag{7.65}$$

This gives a useful estimate for the computer time (or effort, respectively) needed to improve a given optimal solution by a certain amount, which depends on only one characteristic parameter, a.

Finally, note that the network problem discussed here is a very graphic example for elucidating frustrated optimization problems; therefore, it may serve as a toy model for testing different approaches.

8. Analysis and Simulation of Urban Aggregation

8.1 Spatial Structure of Urban Aggregates

8.1.1 Urban Growth and Population Distribution

The previous chapter demonstrated an application of the Brownian agent approach to optimization of a road network, which is an *urban structure*. In the optimization procedure, Brownian agents played their role as searchers that could reproduce, depending on their success. In this chapter, we deal with another urban structure – specific urban aggregations, such as (mega)cities and their suburbs. Despite its complex dependence on various influences, urban growth also seems to follow certain laws that we would like to understand by means of an agent-based model. At the end of this chapter, Brownian agents will represent *urban growth units* that tend to aggregate with respect to specific rules. However, to understand the aim of the model, first it will be useful to look at the *problems behind* urban aggregation.

The vast urbanization of land is among the major challenges facing us at the beginning of the twenty-first century and also confronts the natural sciences on their way to new understanding of complex phenomena. Estimates tell us that today the increase of the *urban world population* is about 40 to 50 million people *per year*. Figure 8.1 shows that, despite the many reasons that cause the growth of the world and urban population, the increase is almost exponential. Even if we assume saturation effects in growth, by the year 2050, 10,000 million people may be alive, almost all of them in urban areas.

These circumstances raise questions about *land consumption* for building homes for these people, about the optimal or maximal tolerable *size of megacities*, and about efficient *transportation structures*, and about possible *disaggregation* processes in urban growth. Therefore, in this and the next chapter, we develop models based on the framework introduced so far, which can reflect certain features of these problems. These models do not claim to give a complete picture; however, they might shed some new light on established ideas about hierarchical planning of urban settlements.

Usually, one is convinced that the development of urban areas is determined by numerous factors, difficult to grasp, such as cultural, sociological, economic, political, and ecological ingredients. On the other hand, it has been

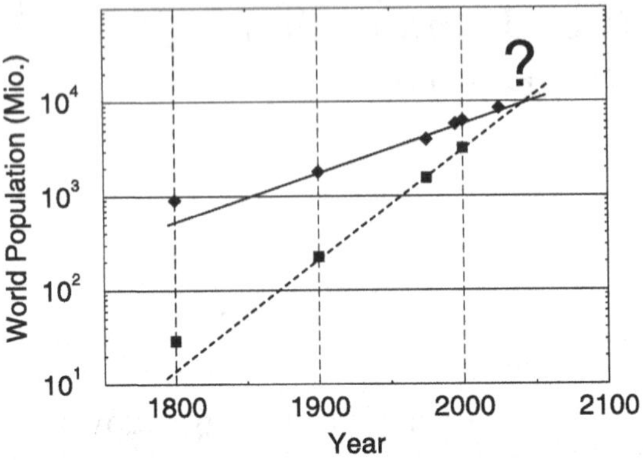

Fig. 8.1. Growth of the world population (*diamonds*) and the urban population (*squares*) in a logarithmic plot. The *dotted* and *dashed lines* indicate the approximation by an exponential growth law, which may result in a fictive urban catastrophe by the year of 2050 [463]

proved that even the rather complex process of urban growth on a certain level of abstraction may be described by simple rules.

One example which was already discussed by Auerbach in 1913 [25], is *population distribution* among different urban settlements, which indicates a hierarchical structure in the set of settlements. If cities are ranked by their numbers of inhabitants (i.e., the city with the largest population is assigned rank 1, the second largest rank 2, etc.), then the *rank–size distribution* follows a power law, also known as Zipf-distribution or Pareto-distribution:

$$n_k(t) = n_1(t)\, k^{q(t)}\,, \quad q(t) < 0\,. \tag{8.1}$$

$n_k(t)$ is the *population* of the settlement with rank k, $n_1(t)$ is the population of the largest city, and $q(t)$ is the Pareto coefficient. Note that n_k is often interpreted as the *size* of the settlement, but this is not completely correct, although there are, of course, relations between the size of a city and its population. In Sect. 8.2.1, we will investigate the real size (area) distribution of urban aggregates, instead of the population.

Interestingly, the rank–size distribution, (8.1), of settlements remains stable over a long period of time, even if the total population number increases remarkably (see Fig. 8.2). However, for the Pareto coefficient $q(t)$, a slight decrease in the course of time can be found [152].

Due to Zipf [571], rank–size distribution is considered an *equilibrium state* between *two counteractive forces* that influence urban growth: (i) a "force of concentration" to avoid transportation costs during production and (ii) a "force of diversification" resulting from the fact that production occurs in the vicinity of natural resources, which are rather dispersed. Hence, the facilities for production and consumption are also spatially distributed.

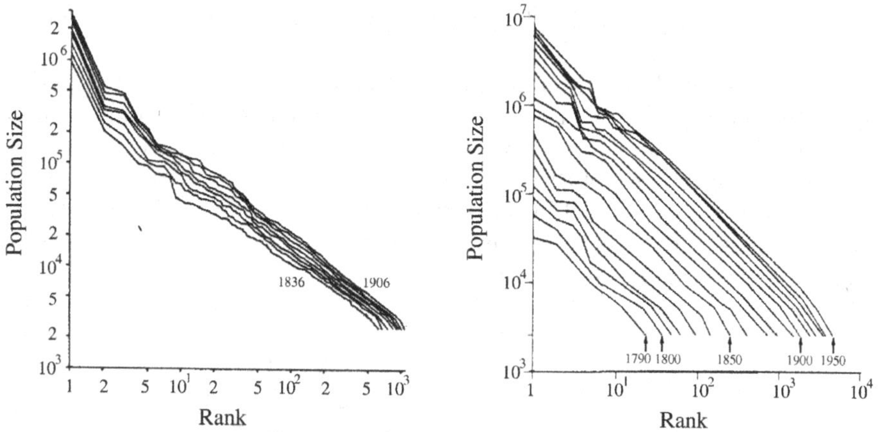

Fig. 8.2. Rank–size distribution of the population of cities in different countries in the course of time: (*left*) France (years 1836–1906) [152, 393], (*right*) USA (years 1790–1950) [152, 571]

Recent stochastic models derive the population distribution of cities within a migration model [193–195, 545], which is based on a certain "attractivity" of settlements. This attractivity depends on the population number in a nonlinear manner. Under specific assumptions for the attractivity, the occurrence of a Pareto distribution in the population distribution has been shown [192], however, with the restriction of a *constant* total urban population.

The hierarchy of cities, expressed in a stable rank–size distribution, is also found on a *spatial scale*. In 1933, Christaller [85] developed a *central place theory* that tries to explain the spatial hierarchical distribution of cities by using economic arguments. He already pointed to the fact that certain characteristic distances exist between settlements of the same hierarchical level. Recently, the central place problem was also investigated using methods of nonlinear dynamics to describe spatial interactions [3, 550, 551, 560]. For further details, we refer to Sect. 9.1.1.

In this chapter, the investigation of urban aggregates is restricted to a very basic level, which considers only *structural* or *morphological features* of single urban areas, instead of dealing with nationwide sets of settlements.

From the perspective of physics, urban aggregations, like cities with surrounding satellite towns and commercial areas, or large agglomerations of megacities, can be described as special kinds of *clusters* on a two-dimensional surface. Using a black-and-white reduction, these clusters represent the *built-up area* (black pixel = occupied site, built up, white pixel = empty site, no building). Here the different types of buildings are *not considered*. Restricted to these "black-and-white" maps, the structural features of urban clusters display analogies to growth phenomena in physics that can be investigated by quantitative analysis.

One of the similarities to physical growth phenomena is the *morphological transition* in urban form. In the course of time, the growth of population at a specific location has led to growth of the built-up area and also to changes in the *morphological form* of the urban aggregate. In Fig. 8.3, we see already, on the phenomenological level, a *morphological transition* from the condensed, "container-like" form of a medieval town to the scattered morphology of today's larger urban aggregations.

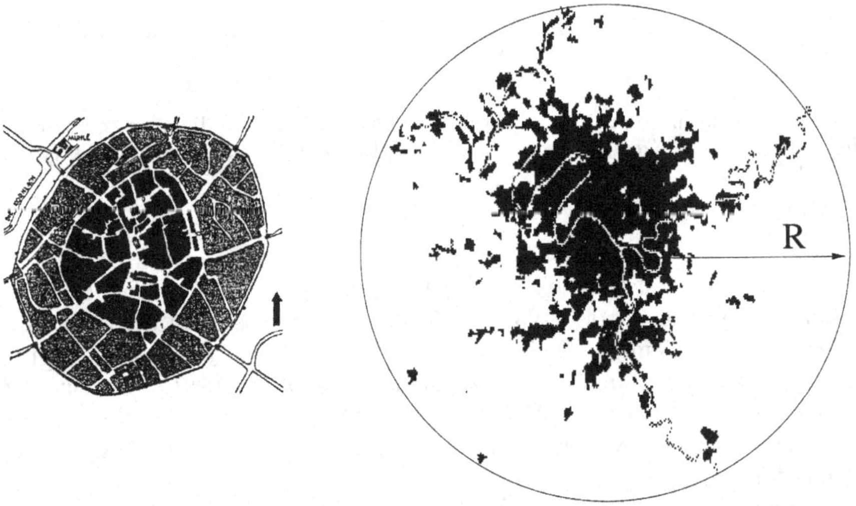

Fig. 8.3. Morphological transition from compact to fractal urban form: (*left*: a) Nördlingen (Germany) [250], (*right*: b) Paris (France) [250]. Both maps are not on the same scale, but the morphological change becomes obvious

The shape of the medieval town tries to mimimize the ratio between the length of the town wall and the area inside the walls, as can be seen, for example, for the German town of Nördlingen (see Fig. 8.3a). If we assume that the urban aggregate is embedded in a circle of radius R, then for the medieval town, the built-up area A inside this circle is almost equal to the circle's area: $A \sim R^2$. The evolution of Paris (see Fig. 8.3b) also started out with such a condensed form. During further growth, however, the urban form broke up to a fractal-like morphology, where the built-up area A is related to the spatial extension R by the relation:

$$A(R) \sim R^{D_\mathrm{f}}\,, \quad 1 < D_\mathrm{f} < 2\,. \tag{8.2}$$

Here, D_f is the fractal dimension of the urban aggregate, discussed later in Sect. 8.1.3. We do not want to discuss the sociological, economical, military

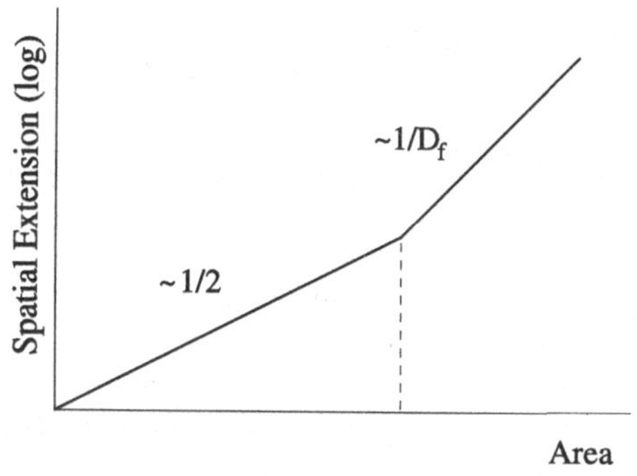

Fig. 8.4. Morphological transition from compact to fractal urban form as a nonequilibrium second-order phase transition. The total urban built-up area A is plotted vs. the logarithm of the spatial extension R. The transition is indicated by the change in the slope

etc. reasons which may have led to such a morphological change, but on the structural level considered here, we mention analogies to a *nonequilibrium second-order phase transition* known in physics (see Fig. 8.4).

Figure 8.3b gives an example of the urban aggregates investigated in the following sections. For our investigations, we use digitized "black-and-white" maps, where the length scale (between 1:50,000 and 1:1,000,000) defines the resolution. Note again that our approach does not consider the internal structures of the urban area; in particular, there is *no* valuation of the types of buildings (garage or skyscraper) and *no* valuation of the importance of the area (downtown or suburb).

Of course, this 2-bit characteristic of the urban area is a major reduction of the complex structure. Recently, spatial urban models based on *cellular automata* have been suggested, which allow a multidimensional description of a single site. After some fundamental discussion about the value of these models in urban and regional science [39, 89, 255, 507], today different types of cellular models are used to simulate complex urban growth [29, 40, 41, 45, 97, 388, 390].

White et al. [552, 554, 556] developed a cellular automaton, where every cell can be in one of four states, characterizing its current use: V (vacant), H (housing), I (industrial), C (commercial). Further, each cell can change its current state, depending on both the state of other cells in its vicinity and some developmental rules. However, to obtain rather realistic *urban land-use patterns*, extensive transition matrices for the possible transitions have to be used [552]. Thus, for integrated dynamic regional modeling, the cellular automaton should be linked to other components, such as geographical information systems (GIS) or economic and demographic models [38, 553, 555].

In the next sections, however, we neglect the interdependences between urban land-use patterns and urban growth, while concentrating only on the spatial structures of urban aggregates. We further neglect interdependences between urban growth and *architectural* issues.

8.1.2 Mass Distribution of Urban Aggregates: Berlin

In this section, the analysis of the spatial distribution of a built-up area is based mainly on a time series representing the evolution of Berlin. Figure 8.5 shows that the city of Berlin is clearly distinguishable from the surrounding area only by the year 1800. During its further urban evolution, Berlin seems to merge more and more with settlements in the neighborhood. The resulting metropolitan area is now denoted as the urban aggregate of "Berlin" regardless of independent cities like Potsdam or Oranienburg, which are part of this cluster, too.

After defining a coordinate system for the digitized map, the *center of gravity* (center of mass) of the urban cluster can be calculated as follows:

$$\boldsymbol{R}_m = \frac{1}{N} \sum_{i=1}^{N} \boldsymbol{r}_i \, . \tag{8.3}$$

$\boldsymbol{r}_i$ is the space vector of the black pixels ($i = 1, \ldots, N$) with mass unit $m_i = 1$. Note the problem that the coordinates of the center of gravity $\boldsymbol{R}_m$ are strongly affected by the sector of the map used for the calculations. For example, in urban agglomerations with large settlement areas in the outer regions or with distinct asymmetries in the built-up area, the center of gravity can be shifted far from the urban center thought.

The *radial expansion* of an urban cluster is not easily defined. Due to the frayed structure typical of fractals, it is hard to distinguish between the city and the surrounding area. However, an estimate can be obtained by calculating the *radius of gyration* R_g [139, 523], which is a weighted measure of the distance of the different pixels from the center of gravity $\boldsymbol{R}_m$:

$$R_g^2 = \frac{1}{N} \sum_{i=1}^{N} (\boldsymbol{r}_i - \boldsymbol{R}_m)^2 \, . \tag{8.4}$$

In Fig. 8.6, the increase in the radius of gyration for the urban area of Berlin is shown. The plot indicates that the radial expansion of Berlin between 1875 and 1945 can be approximated by an exponential growth law.

In addition to the spatial extension, another quantity of interest is the *maximum border distance* [247, 251], which is the maximum distance between an arbitrary mass point within the aggregate and the nearest inner or outer area of free space. To estimate the urban aggregate of Berlin, the settlement area of Fig. 8.5e is circumscribed consecutively by isochrones measuring 0.25 km wide, proceeding from the outside toward the center. As can be obtained from Fig. 8.7, the distance to the nearest inner or outer border is *2 km* at maximum, which is a relatively small value for an extended city like Berlin.

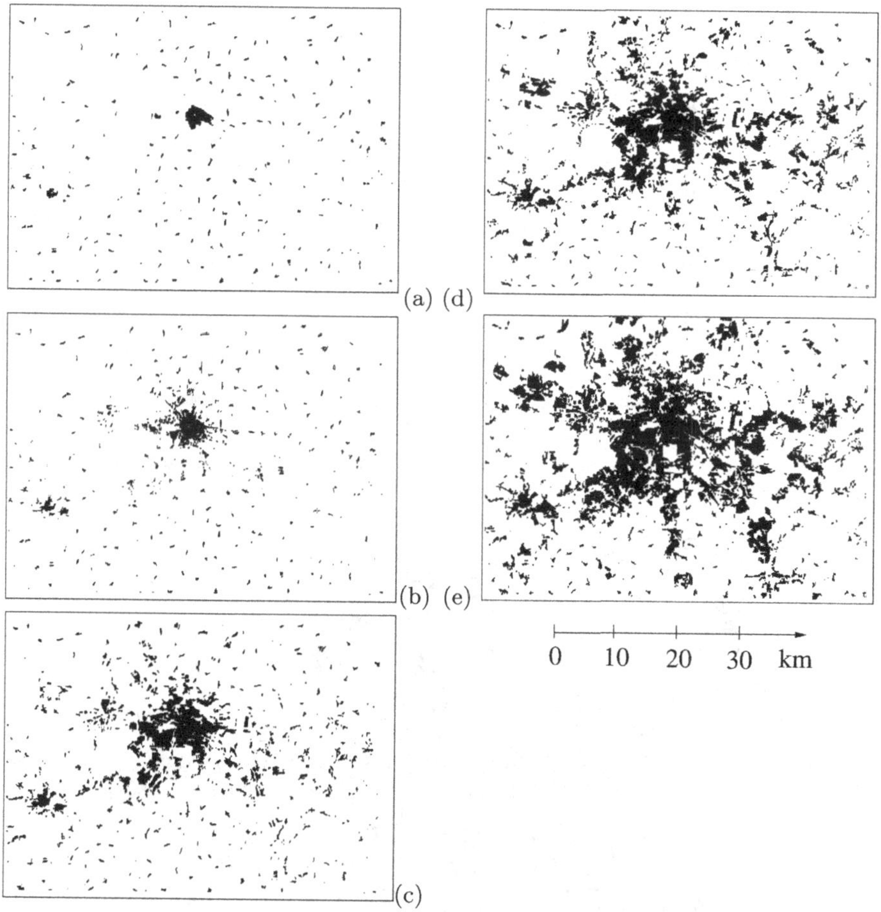

Fig. 8.5. Region of Berlin in different years: (**a**) 1800; (**b**) 1875; (**c**) 1910; (**d**) 1920; (**e**) 1945 [249][1]

The maximum border distance provides a measure of urban living quality because people usually want to live both close to the urban center and close to natural areas, such as parks or natural environments. On the structural level, these two contradicting demands lead to fractal organization of the settlement, as discussed in the next section.

---

[1] Parts of this Berlin series have been recently reprinted and/or analyzed in different publications, e.g. [43, 156, 320], without ever quoting the *correct* source. Noteworthy, credit for this series should be given to K. Humpert and co-workers at the *Städtebauliches Institut*, University of Stuttgart, who redrew it from historical maps. The series first appeared in [249]; for the material available, see [248].

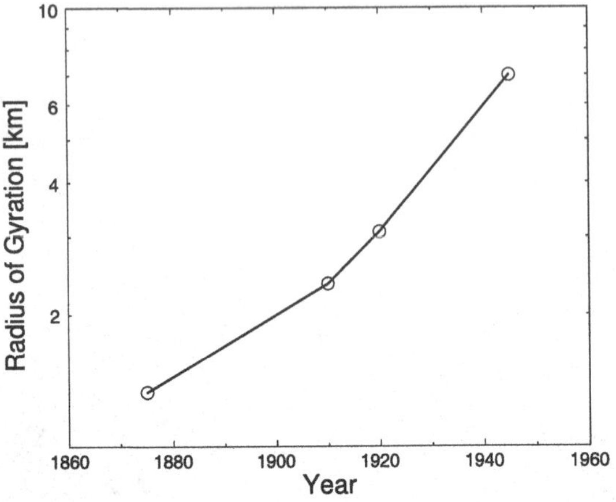

Fig. 8.6. Radius of gyration in km (logarithmic scale) for the urban cluster of Berlin [465]

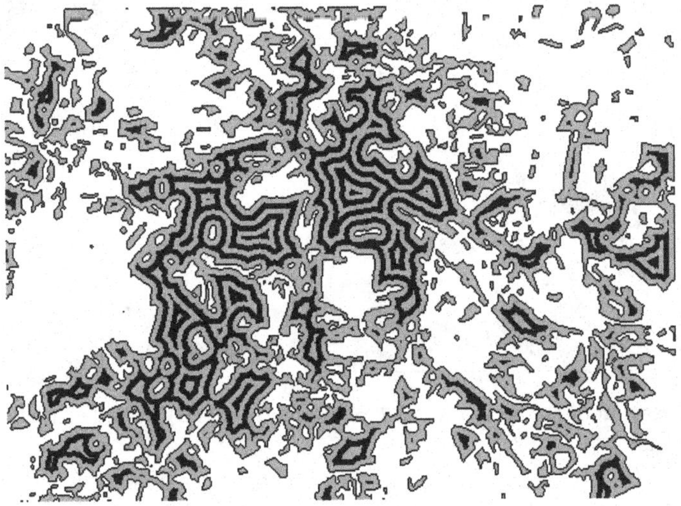

Fig. 8.7. Estimate of the maximum border distance of Berlin (1945) by isochrones measuring 0.25 km wide [462, 465]

In the following, we want to characterize the mass distribution of the urban aggregate of Berlin in dependence on the distance from the center. Therefore, around the center of gravity R_m, concentric circles are drawn with a radius R which is increased by dR for every new circle. Then, dN is the number of pixels in a ring of area dA:

$$\begin{aligned} dN &= N(R + dR) - N(R)\,, \\ dA &= A(R + dR) - A(R) \;=\; 2\pi R dR\,. \end{aligned} \tag{8.5}$$

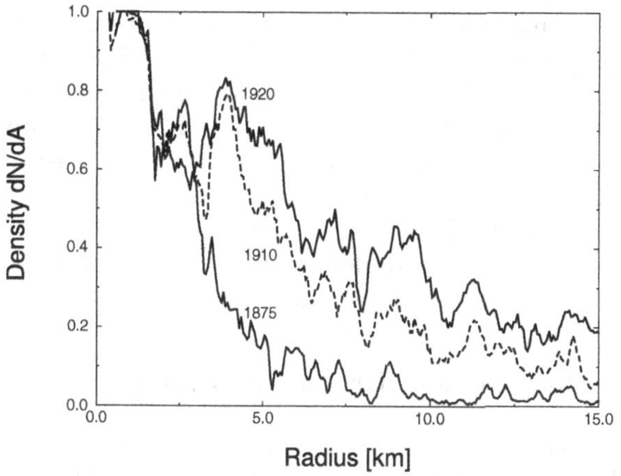

Fig. 8.8. Settlement density (dN/dA) for Berlin vs. distance from the center of gravity in different years: 1875, 1910, 1920 [462, 465]

Figure 8.8 shows the change of the mass density, dN/dA, in dependence on the distance R from the center. The values represent the ratio of the built-up area compared to the total area of the ring; an amount of one therefore indicates a completely built-up area.

For the years 1910 and 1920, we find a second ring of very dense settlement appearing at a distance of 4–6 km from the center of Berlin. This indicates the suburbs built for the mass of factory workers during that time. Despite that insight, the plot of Fig. 8.8 does not seem to be useful for further

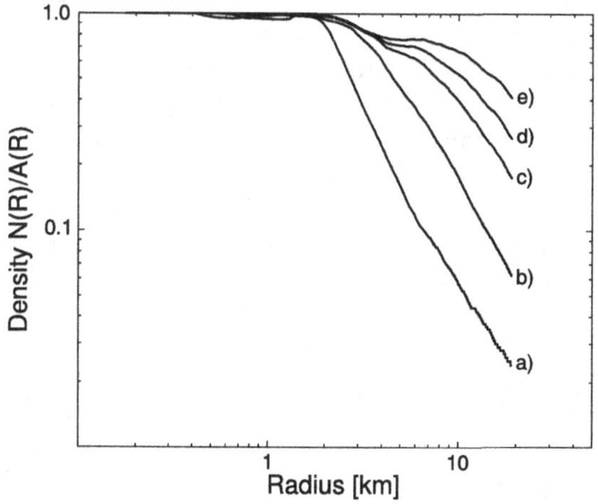

Fig. 8.9. Settlement density $N(R)/A(R)$ for Berlin vs. distance from the center of gravity in a double-logarithmic plot for different years: (a) 1800; (b) 1875; (c) 1910; (d) 1920; (e) 1945 [465]

evaluation. More appropriate is an investigation of the *total built-up area* $N(R)$ in dependence on the distance. Thus, instead of the settlement density, dN/dA, we calculate now the value $N(R)/A(R)$ which is the total built-up area normalized by the total area $A(R)$ of the circle. The double-logarithmic plot in Fig. 8.9 shows the compact core of the city (high density) with a radius of about 2 km. Outside this core, the density continuously decreases with a descent that is related to the *fractal dimension* D_{f} of the settlement area. To discuss this relation in more detail, we investigate the fractal properties of urban aggregates in the next section.

8.1.3 Fractal Properties of Urban Aggregates

Recently, *fractal concepts* have been used to describe the structural features and the kinetics of urban clusters [36, 42–44, 150, 151, 154–157, 179, 320, 552, 554]. The existence of fractal properties over several spatial scales indicates a *hierarchical organization* of the urban aggregate [157]. The fractal dimension is one possible measure for describing the spatial distribution of the built-up area, as already mentioned with respect to Fig. 8.9. The continuous descent of the density $N(R)/A(R)$ is related to the fractal dimension D_{f} as follows:

$$\frac{N(R)}{4\pi R^2} \sim R^{D_{\mathrm{f}}-2} \,. \tag{8.6}$$

As shown in Fig. 8.9c–e, some of the regions within the descending parts of the plots can be characterized by a different slope and hence, by a different fractal dimension. The so-called *multifractality* is a phenomenon typical of inhomogeneous clusters [138, 523], indicating that, within an urban cluster, more or less compact built-up areas exist.

There are different numerical treatments for calculating the fractal dimension, and the results are slightly different depending on the method used [156, 523]. In the following, we indicate the method used by an index of the dimension D. The *correlation dimension* D_{c} should be appropriate for describing the fractal dimension of inhomogeneous fractals. D_{c} is related to the correlation function $c(r)$, which gives a measure whether a certain pixel belongs to the structure:

$$c(r) \sim r^{D_{\mathrm{c}}-2} \,. \tag{8.7}$$

$c(r)$ can be calculated by counting the number of pixels in a given ring, as discussed in Sect. 8.1.2. However, now the center of the ring is not only the center of gravity, but different pixels are used as centers, and the results are averaged. Figure 8.10 gives the correlation function for the urban cluster of Berlin. From Fig. 8.10, the *fractal dimension of Berlin* is calulated from (8.7) as $D_{\mathrm{c}} = 1.79 \pm 0.01$.

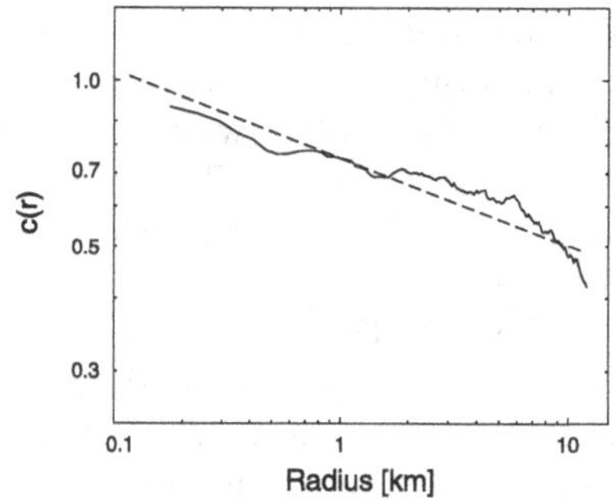

Fig. 8.10. Correlation function $c(r)$ vs. radius in a double-logarithmic plot (Berlin, 1945) [465]

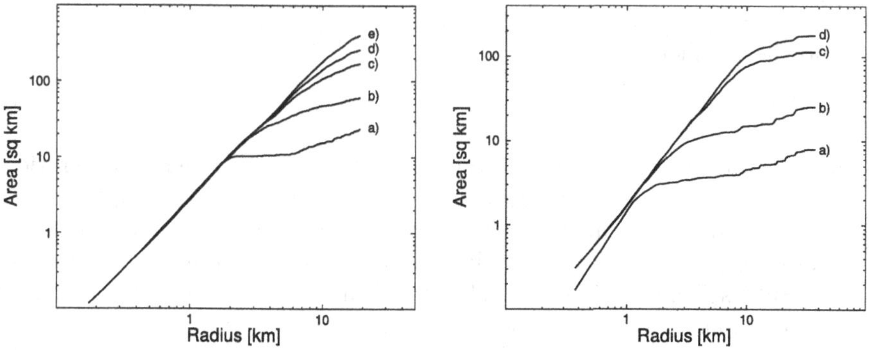

Fig. 8.11. Built-up area $N(R)$ vs. radius in a double-logarithmic plot: (*left*) Berlin (a: 1800; b: 1875; c: 1910; d: 1920; e: 1945) [462, 465] (*right*) Munich (a: 1850; b: 1900; c: 1950; d: 1965) [465]

Whereas Figs. 8.8 and 8.9 show the built-up density, Fig. 8.11 presents the built-up area (number of pixels) of Berlin, in dependence on the radius, instead.

In Fig. 8.11, both of the plots for Berlin and for Munich indicate that the spatial distribution of the settlement area approaches a characteristic power function, $\log A \propto \log R$, in the course of time. This distribution, again, is related to a fractal dimension:

$$N(R) \sim R^{D_r}\,, \tag{8.8}$$

where D_r now is the radial dimension to indicate the different computation compared to D_c. The following values can be obtained from Fig. 8.11:

$$\begin{aligned} \text{Berlin (1945):} \quad D_r &= 1.897 \pm 0.006\,, \\ \text{Munich(1965):} \quad D_r &= 1.812 \pm 0.004\,. \end{aligned}$$

For Berlin, a comparison of both D_r and D_c shows a slightly larger value for D_r, which results from the different definitions.

In Fig. 8.11, the plots for different years are nearly identical for small distances from the center, indicating that almost no additional buildup occurred in the central area, but the settlement grew mainly in the outer regions, near the border of the surrounding free space. The value of the radius at which the different plots deviate from the power law, turning into a horizontal line, can give an estimate for an *effective settlement radius* R_{eff}, characterizing the radial expansion of the city. For Berlin, R_{eff} was about 2 km in 1875, but the plots approach the power law more and more in the course of time, which means that the settlement continuously grew and merged with the surrounding area. The final value approached, $N(R_{\text{max}})/4\pi R_{\text{max}}^2$, gives an estimate of the *mean density* of the urban cluster.

Another possibility of characterizing the fractal properties of urban aggregates is to calculate the *area–perimeter relation* [157]. For a *compact* two-dimensional geometric object, there is a quadratic relation between the area a and the perimeter p:

$$a = f\, p^2 \,. \tag{8.9}$$

The prefactor f depends on the type of regular object under consideration; e.g., for a circle, $f = 1/(4\pi)$. However, it can be proved [156, 323] that (8.9) does not hold for fractals and a *linear* relation results instead. This can also be confirmed by quantitative analysis of 60 metropolitan areas from all over the world, as Fig. 8.12 shows. Frankhauser et al. [156, 157] obtained an almost linear relation between settlement area and total perimeter:

$$a \sim p^D \;, \quad D = 1.05 \,, \tag{8.10}$$

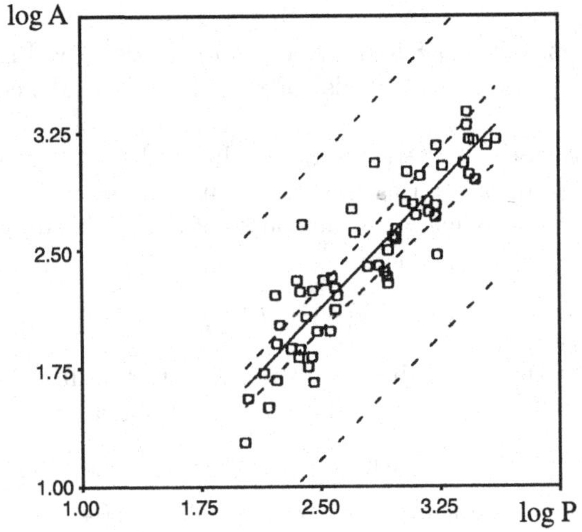

Fig. 8.12. Double-logarithmic representation of the surface of the built-up area of 60 agglomerations in relation to their perimeter [156, 157]

which also agrees with estimates by Humpert et al. [247, 251]. So, the linear relation between the total built-up area and the total perimeter of an urban agglomeration seems to be a universal feature of these aggregates.

8.2 Rank–Size Distribution of Urban Aggregates

8.2.1 Analysis of the Rank–Size Distribution

As the different pictures of urban aggregates indicate, these are not just one merged cluster, but consist of a large number of separated clusters of different sizes. To analyze the cluster distribution, we can do a cluster analysis that gives the size and the number of different clusters forming the urban aggregate [462, 464]. First, we have to determine whether or not a certain cluster is connected to other clusters. Each of the separated clusters gets a number (label). Afterward, the number of pixels in each labeled cluster is determined, and the clusters are sorted with respect to their size, thus determining their rank number (the largest cluster gets rank 1, etc.)

Figure 8.13 shows the development of the rank–size cluster distribution for Berlin and Munich. In particular for the development of Berlin, the rank–size distribution approaches a power law in the course of time, i.e., during the continuous differentiation of the settlement area. In Fig. 8.13, this power law is indicated by a straight slope, which can be described by the following equation:

$$\log n_k = \log n_1 + q \log k$$

or

$$n_k = n_1\, k^q\,, \quad q < 0\,, \tag{8.11}$$

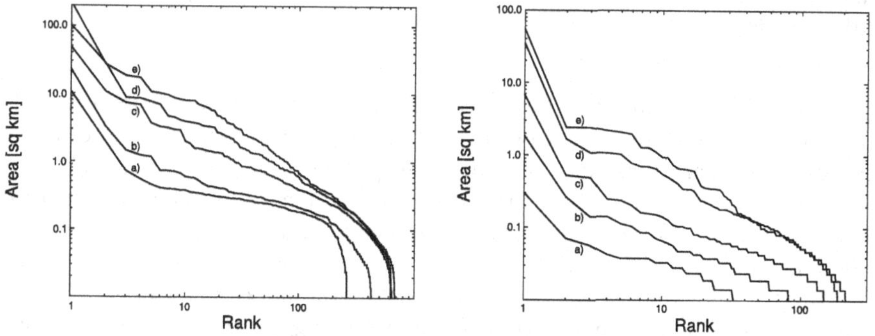

Fig. 8.13. Evolution of the rank–size distribution of connected built-up areas (clusters) in a double-logarithmic plot: (*left*) Berlin (a: 1800; b: 1875; c: 1910; d: 1920; e: 1945) [462–465], (right) Munich (a: 1800; b: 1850; c: 1900; d: 1950; e: 1965) [464, 465]

where n_k is the size of the cluster with rank k and $n_1(t)$ is the size of the largest cluster that serves as normalization. Equation (8.11) agrees with the Pareto–Zipf distribution, (8.1), already mentioned in Sect. 8.1.1 for population distribution. The Pareto–Zipf distribution is a characteristic feature of hierarchically organized systems; this means for the case considered that the urban settlement area is formed by clusters of all sizes in the course of time.

If the evolution of urban aggregates occurs so that the cluster distribution approaches a Pareto distribution in time, we may conclude that the establishment of this distribution could be a measure of a fully developed urban aggregate – and, on the other hand, the deviation from this distribution could be a measure of the *developmental potentiality* with respect to the morphology of the settlement. Looking at the rank–size distributions of different urban aggregations (see Fig. 8.14), we find that this developed stage is sometimes already reached (see Fig. 8.14d – Philadelphia).

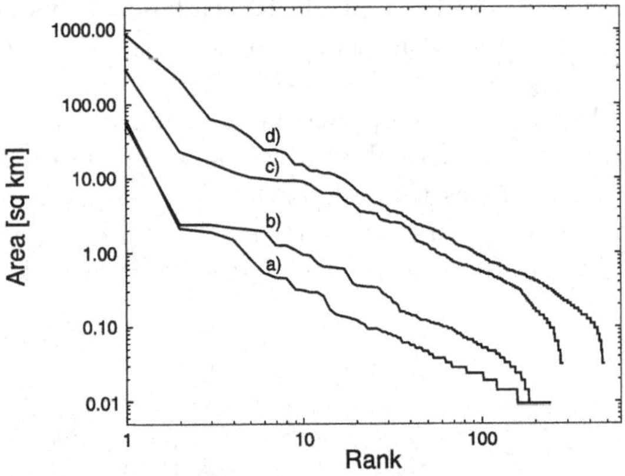

Fig. 8.14. Rank–size distributions of connected built-up areas in a double-logarithmic plot: (a) Daegu (1988); (b) Munich (1965); (c) Moscow (1980); (d) Philadelphia (1980) [463–465]

Table 8.1 contains the Pareto exponents q which result from Fig. 8.14 due to (8.11). Our investigations show that the Pareto exponents of different urban agglomerations have similar values. This indicates that the Pareto exponent could serve, within the scope of certain error limits, as a *structural measure of developed urban aggregates*, which may complement the fractal dimension.

In the following, we want to derive a kinetic model to describe the evolution of the rank–size distribution of urban aggregates toward a Pareto distribution.

Table 8.1. Pareto exponent q for different urban agglomerations. The values are obtained from Fig. 8.14 by linear regression of the first 100 ranks [463–465]

City	Pareto Exponent
Munich 1965	-1.23 ± 0.02
Daegu 1988	-1.31 ± 0.03
Moscow 1980	-1.15 ± 0.02
Philadelphia 1980	-1.32 ± 0.01

8.2.2 Master Equation Approach to Urban Growth

Because urban aggregates consist of clusters of different sizes, we introduce the cluster–size distribution $\boldsymbol{n}$ of the aggregate as follows: $\boldsymbol{n} = \{n_1, n_2, \ldots, n_k, \ldots, n_A\}$. Here, n_k is the size (the number of pixels) of the cluster with number k, and a total number of A clusters exist ($k = 1, \ldots, A$). The total number of pixels is obtained by summation:

$$N_{\text{tot}}(t) = \sum_{k=1}^{A} n_k \,, \tag{8.12}$$

and should, for an *evolving settlement*, increase in the course of time. Figure 8.15 shows the increase in the urban cluster of Berlin, as obtained from the time series of Fig. 8.5. As shown, the total increase in the urban area of Berlin follows an *exponential growth* law for the years 1870 to 1945.

The cluster distribution $\boldsymbol{n}$ can change due to two fundamental processes: (i) the formation of new clusters and (ii) the growth of already existing clusters. In principle, we should also consider the shrinkage and disappearance

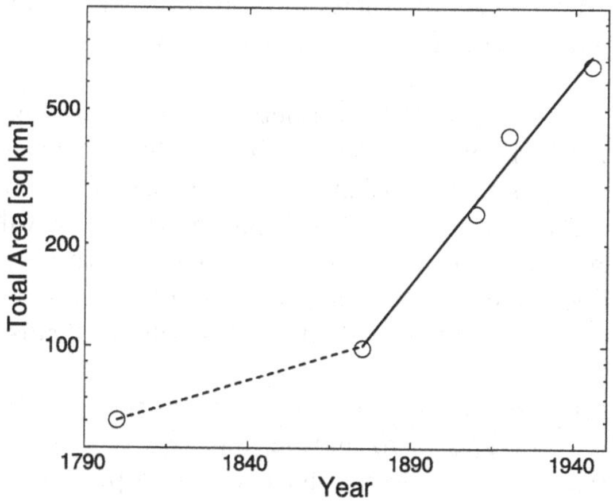

Fig. 8.15. Growth of the settlement area of Berlin in a logarithmic plot [463–465]

of already existing clusters. However, the probability of these processes is rather small for growing urban areas, and therefore they are neglected in the following.

The *formation of new clusters* can be described by a symbolic reaction

$$A \xrightarrow{w_1} A+1\,, \tag{8.13}$$

which means that the total number of clusters A increases by one at that time. The probability of forming a new cluster during the next time interval, w_1, should certainly depend on the total growth of the urban settlement, $N_{\text{tot}}(t)$, which can be approximated using empirical data (see Fig. 8.15). Thus, we find the ansatz,

$$w_1 = w(A+1, t+1|A, t) = c(N_{\text{tot}})\,. \tag{8.14}$$

However, for simplicity we assume a constant probability for the formation of new clusters, i.e., $c = \text{const}$ in the following.

The *growth of already existing clusters* can be described by the symbolic reaction

$$n_k \xrightarrow{w_k} n_k + 1\,, \tag{8.15}$$

which means that the size n of cluster k increases by one at that time. The probability that this process occurs during the next time interval, w_k, should depend on both the already existing size of the cluster and on the existing cluster distribution. Hence, we find the following ansatz:

$$w_k = w(n_k+1, t+1|n_k, t) = g\frac{n_k}{N_{\text{tot}}}\,, \qquad g = 1 - c(N_{\text{tot}})\,. \tag{8.16}$$

The dependence on $N_{\text{tot}}(t)$ represents *global coupling* between the growth processes of all separated clusters. Therefore, the growth probability of a specific cluster within the urban aggregate is not independent of the other existing clusters.

By means of the factor g, the ratio between the formation of new clusters and the growth of existing clusters can be weighted. Hence, g depends on the probability of cluster formation, c. If we consider the total increase in the urban area, $N_{\text{tot}}(t)$, as a known function, then the value of g in the computer simulations determines, whether urban growth results mainly from the formation of new clusters or from the growth of already existing clusters. Usually, we chose values of $g = 0.90\ldots0.99$, which means $c = 0.1\ldots0.01$ accordingly.

In a given time interval, the morphology of an urban aggregate can change due to numerous processes of formation and growth of clusters that occur simultanously. All possible processes have a certain probability, so the result of the evolution is not clearly determined. Instead of a deterministic approach, we have to use a *probabilistic approach* considering the uncertainity of the future and the limited predictability of evolution. Therefore, we introduce a probability distribution $P(\mathbf{n},t) = P(n_1, n_2, \dots, n_k, \dots, n_A, t)$ that gives the probability that a particular of all possible cluster distributions $\{n_1, n_2, \dots, n_k, \dots, n_A\}$ will be found after a certain time. The change in this probability distribution in the course of time can be described by a *master equation* (see Sect. 1.3.3) that considers all possible processes leading to a change in a particular distribution:

$$
\begin{aligned}
\frac{\partial P(\boldsymbol{n},t)}{\partial t} = & -\sum_i w_k(n_k,t) P(n_1,\dots,n_k,\dots,t) \\
& +\sum_{i\neq 1} w_k(n_k-1,t) P(n_1,\dots,n_k-1,\dots,t) \\
& -w_1(A,t) P(n_1,\dots,n_A,t) \\
& +w_1(A-1,t) P(n_1,\dots,n_{A-1},t) \,.
\end{aligned}
\tag{8.17}
$$

Due to global coupling of the kinetic processes given by (8.12), it is rather complicated to derive analytical results for (8.17). However, this coupling can be neglected by assuming that the total number of clusters, A, is indeed large, but much less then the total number of pixels, N_{tot}. This is basically a restriction of the probability of forming new clusters, c, which has to be small enough. With the assumption $1 \ll A \ll N_{\mathrm{tot}}$, the master equation can be resolved:

$$
P(n_1,\dots,n_k,\dots,t) \approx \prod_{k=1}^{A} P(n_k,t) \,. \tag{8.18}
$$

Equation (8.18) means that instead of the probability of the whole cluster distribution, $P(\boldsymbol{n},t)$, only the probability of finding clusters of a certain size, $P(n_k,t)$ has to be considered now. With this probability, the mean size of a cluster k at a given time can be calculated:

$$
\langle n_k(t)\rangle = \sum_{n_k=1}^{N_{\mathrm{tot}}} n_k \, P(n_k,t) \,, \tag{8.19}
$$

where the summation in (8.19) is over all possible values of n_k, each of which may have a certain probability $P(n_k,t)$. As a result, we obtain [190]

$$
\langle n_k(t)\rangle \sim N_{\mathrm{tot}} \, k^{-(1-c)} \,. \tag{8.20}
$$

Equation (8.1) indicates an exponential relation between the size of a cluster, n_k, and its number (rank), k, as already discussed for the Pareto–Zipf distribution, (8.11). Here, the Pareto exponent q is given by

$$
q = c - 1 < 0 \,. \tag{8.21}
$$

Thus, we conclude that the change in the cluster distribution, described by the master equation, (8.17), occurs so that eventually, on average, clusters of all sizes exist, which leads to a hierarchical composition of an urban aggregate.

The descent of the Pareto–Zipf distribution expressed by q depends remarkably on the probability of the formation of new clusters, c. For large values of c, the distribution becomes flat; for smaller values of c, on the other hand, the distribution becomes steeper, indicating a stronger hierarchical organization that is determined mainly by growth of the largest cluster.

8.2.3 Simulation of the Rank–Size Distribution: Berlin

To show a real example of the evolution of the rank–size cluster distribution, we have simulated the development of Berlin for the years between 1910 and 1945 [463, 464]. The initial state for the simulation is given by the urban cluster of Berlin (1910) (see also Fig. 8.5c). The simulations are carried out for two time intervals: (i) from 1910 to 1920 and (ii) from 1910 to 1945. For those times, the increase of the total built-up area, $N_{\text{tot}}(t)$, is known from empirical data (see also Fig. 8.15).

Stochastic simulations of urban growth proceed as follows:

1. At time t, all possible probabilities of the formation of new clusters and the growth of existing clusters are calculated.
2. The probability w for every process is compared with a random number drawn from the interval $(0, 1)$. Only if w is larger than the random number is the related process carried out, i.e., either cluster growth: $n_k \to n_k + \Delta n$, or cluster formation: $\Delta n \to n_{A+1}$ occur. Here, Δn is the the amount of pixels which is just optically recognized for the given resolution of the map (in the simulation of the growth of Berlin, these are 40 pixels). A random place is chosen for the formation of new clusters; for cluster growth, the number of pixels is randomly distributed on the cluster.
3. Afterward, the simulation continues with the next time interval unless the total increase in the built-up area is reached: $\sum \Delta n = N_{\text{tot}}(t_{\text{fin}})$.

The first simulations which were carried out that way, however, proved that the empirical rank–size distribution for Berlin could not be reproduced. The reason for that is obvious. During the growth of urban clusters, in addition to the global coupling mentioned, local coupling also exists due to the spatial neighborhood of the clusters. This means that during growth, the clusters increase their spatial extension and therefore can merge with neighboring clusters – a phenomenon known from physics as coagulation. In this way, two small clusters can create one larger cluster very fast, which affects the Pareto distribution remarkably. Eventually, if all clusters merge – which is known as percolation – the cluster distribution collapses because only one large cluster exists.

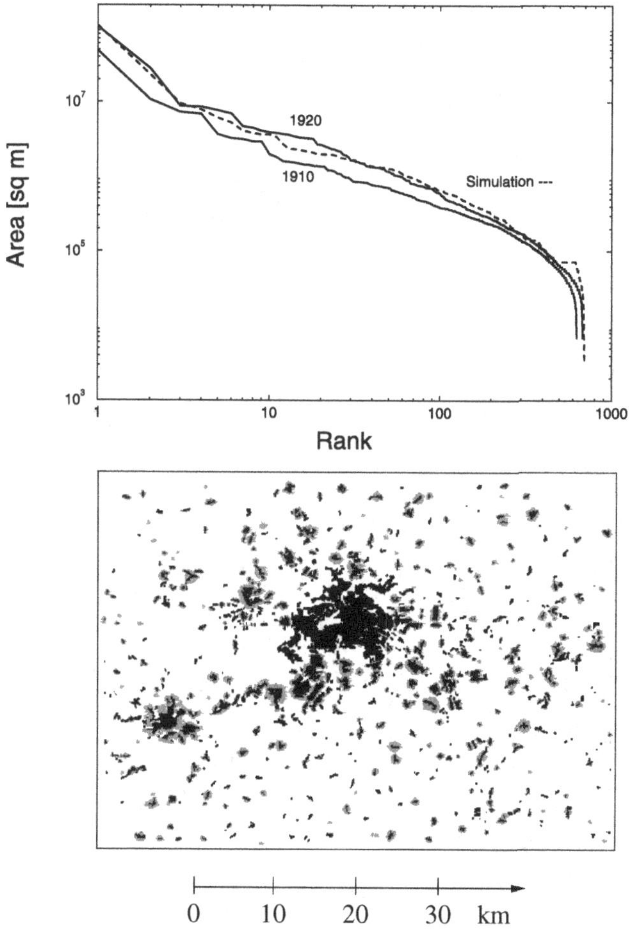

Fig. 8.16. (*top*) Simulation of the rank–size cluster distribution of Berlin (1910–1920); (*bottom*) simulated spatial distribution of urban clusters in Berlin (1920), *black*: initial state (Berlin 1910), *gray*: simulated growth area. Parameter: $c = 0.09$ [463–465]

With respect to the growth probabilities of (8.16), during the simulations, the largest cluster (rank 1) dominates the whole growth process. During its growth, which is proportional to its size, this cluster also merges with the smaller clusters surrounding it, which leads again to a jump in its size. Eventually, the whole cluster distribution collapses. However, such an evolution was not found empirically. Instead, separate clusters are also found in large urban agglomerations, which leads us to the conclusion that the growth probabilities have to be modified at least for clusters with small rank.

Our simulations [463–465] proved that the Pareto distribution for the urban area of Berlin can be reproduced with the *additional assumption* that

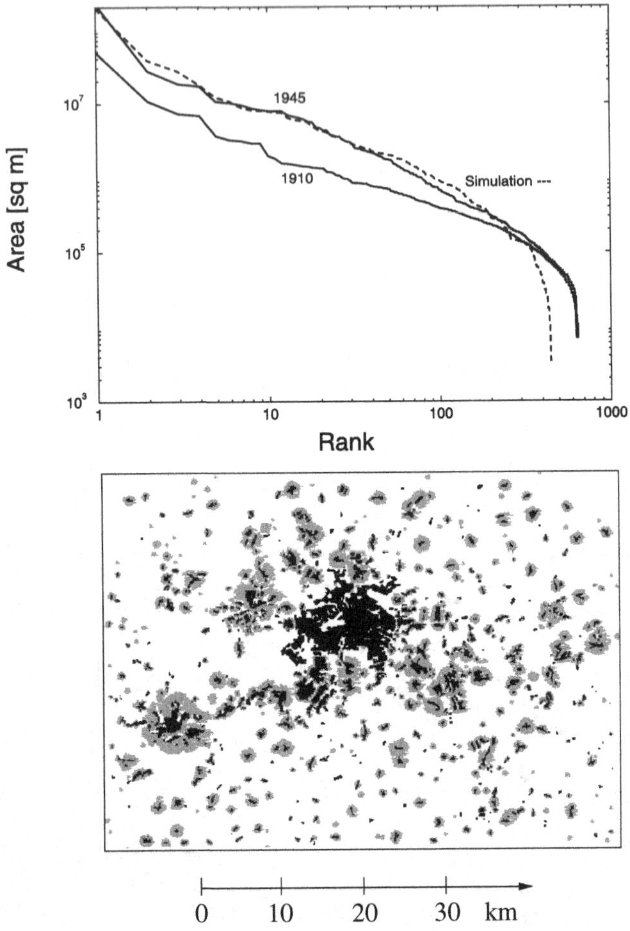

Fig. 8.17. (*top*) Simulation of the rank–size cluster distribution of Berlin (1910–1945), (*bottom*) simulated spatial distribution of urban clusters in Berlin (1945), *black*: initial state (Berlin 1910), *gray*: simulated growth area. Parameter: $c = 0.04$ [463–465]

the cluster with rank 1 does not grow automatically proportionally to its size but only due to coagulation with neighboring clusters; whereas clusters with rank 2 or higher grow again proportionally to their sizes, due to (8.16), and of course due to coagulation. Substantiation for this additional assumption will be given in Sect. 8.3.2.

The results of computer simulations are shown in Figs. 8.16 and 8.17 for two different time intervals. The results clearly indicate that the rank–size cluster distribution of Berlin can be well reproduced for the year 1920, and even better for the year 1945, which confirms our kinetic assumptions for the simulation. The simulations also indicate that the probability of the formation of new clusters, c, should have decreased in the course of time because good agreement with the empirical distribution for the year 1945 was found with a value for c, which is less than half that used for the simulation of 1920.

We want to point out again that information about the total increase in the built-up area has been drawn only from empirical data; this could have also been calculated from an exponential growth law. We did not use any information about which of the different clusters grew and by what amount – this process was entirely simulated by the master equation.

The agreement between the simulations and the real rank–size cluster distribution demonstrates only that the evolution of the size distribution of different clusters has been described correctly. This means that one can estimate how much a cluster of a particular size should grow during the time interval considered – independent of its location within the urban aggregate. These simulations cannot reproduce the *real* spatial distribution of different clusters because no information about spatial coordinates is considered in the kinetic assumptions. With respect to Berlin, we do not draw conclusions whether "Lichtenrade" or "Zehlendorf" (two specific suburbs) will grow; we give estimates only about the development of a cluster of a certain size. However, these estimates agree sufficiently with the real evolution of an urban aggregate, as the simulations have shown.

8.2.4 Forecast of Future Evolution: Daegu

The good agreement between the real and the simulated rank–size distribution for Berlin gives rise to simulating also the future evolution of urban aggregates which, with respect to their structural morphology, today are similar to an early stage in the evolution of Berlin. As one example, we have simulated the evolution of the South Korean metropolis Daegu. Despite the fact that there are millions of inhabitants today, the distribution of the built-up area of Daegu in 1988 (see Fig. 8.18) resembles that of Berlin in 1800 or 1875 (see Fig. 8.5). In both cases, one large cluster dominates the whole structure, whereas in the surrounding area only very small clusters exist. Hence, from a structural perspective, the developmental resources discussed in Sect. 8.2.1 are rather large.

From a time series[2], showing the development of the settlement area of Daegu, Fig. 8.18, we know the growth rate in the past. As for the simulation of Berlin, we assume exponential growth that allows extrapolating future growth. Thus we find an estimate for the increase in the total area, $N_{\text{tot}}(t)$, for Daegu (see Fig. 8.19).

For the simulations, we have used the assumptions for the formation of new clusters and the growth of existing clusters, as discussed above, (8.14), (8.16). As for the simulation of the growth of Berlin (1910–1945), it is further

[2] Courtesy of K. Humpert and co-workers at the *Städtebauliches Institut*, University of Stuttgart. For the material available, see [248].

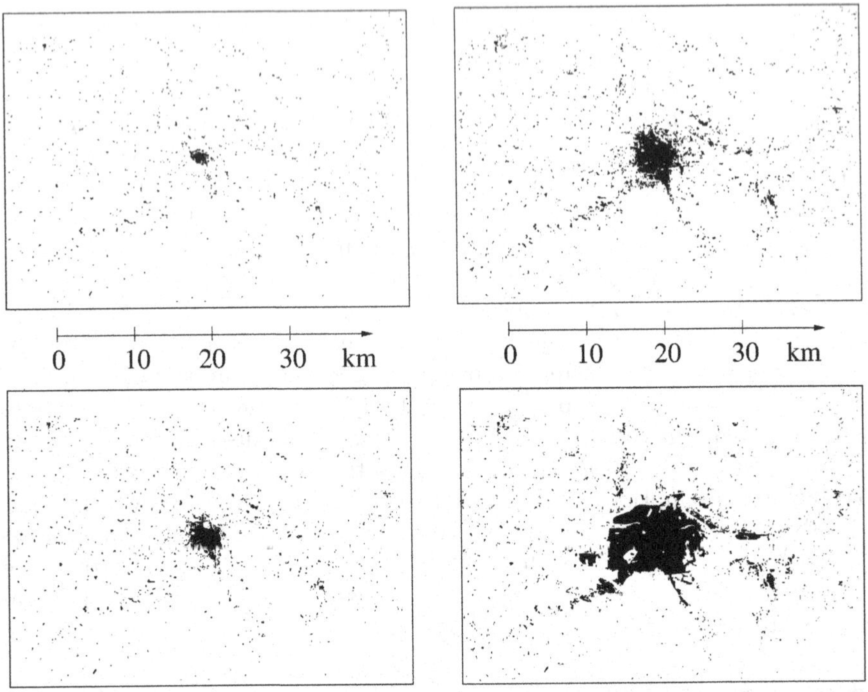

Fig. 8.18. Region of Daegu (South Korea) in different years: (*upper left*) 1915; (*lower left*) 1954; (*upper right*) 1974; (*lower right*) 1988[2]

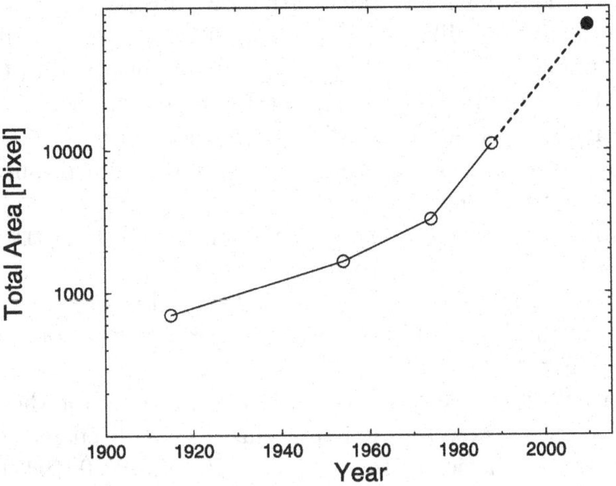

Fig. 8.19. Increase in the built-up area of Daegu in a logarithmic plot. The black dot indicates the estimated growth up to the year 2010 obtained from an exponential growth law [464, 465]

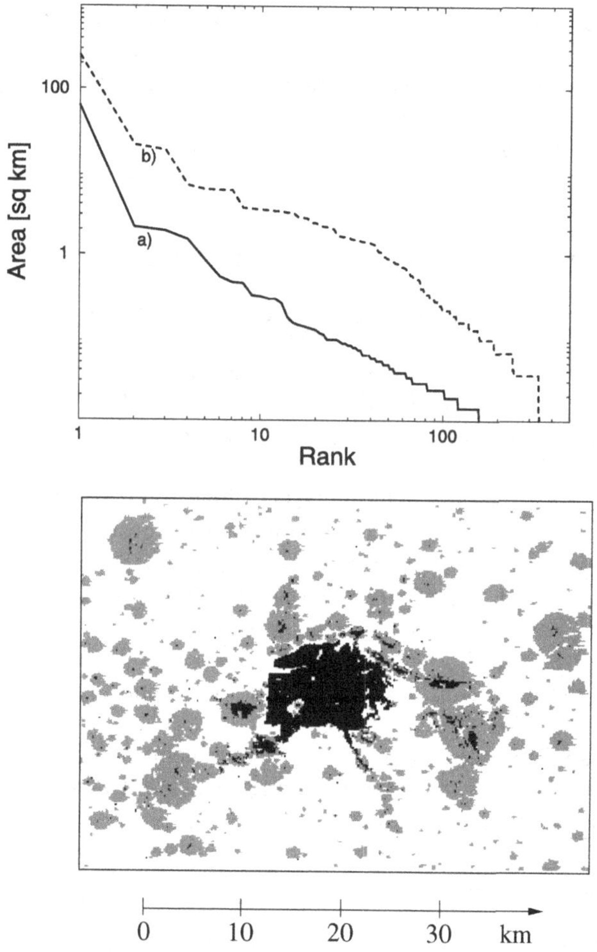

Fig. 8.20. (*top*) Simulation of the rank–size cluster distribution of Daegu (1988–2010); (*bottom*) simulated spatial distribution of the urban clusters of Daegu (2010), *black*: initial state (Daegu 1988); *gray*: simulated growth area. Parameter: $c = 0.01$ [464, 465]

assumed that a cluster with rank 1 does not grow proportionally to its size but only by coagulation with neighboring clusters. Starting with the known spatial distribution of the settlement area of Daegu in 1988, we have simulated the possible growth of the settlement up to the year 2010, which is, as in the case of Berlin, a time interval of about 35 years. The result of the simulations may serve as a first attempt at a prognosis of future development and is shown in Fig. 8.20.

The simulations indicate again that, for future evolution, the rank–size cluster distribution of the urban aggregate may approach the characteristic

Pareto distribution. With respect to the spatial distribution, this evolution is accompanied by forced growth of the rather underdeveloped suburbs in the region, which may now dominate the growth of the central cluster. Eventually, this process results in a hierarchical spatial organization typical of a fully developed urban aggregate containing clusters of all sizes.

8.3 Kinetic Models of Urban Growth

8.3.1 Fractal Growth and Correlated Growth Models

The discovery of the fractal properties of urban agglomerations raised the idea of applying physical growth models that generate fractal structures to simulate also the growth of urban clusters. Previous attempts [36, 37, 44] were mainly based on fractal aggregation mechanisms, such as DLA (diffusion limited aggregation) [138, 139, 523] or DBM (dielectrical breakdown model) [366, 523, 524]. In these models, particles for cluster growth are released at a longer distance from the cluster, then diffuse and aggregate with certain probabilities to the existing cluster, if they hit it by chance. In the DLA model, constant probabilities of the attachment of particles exist, which lead to dendritic growth typical of fractals. In the DBM model, these probabilities can be determined due to an additional potential condition that basically considers the occupation of neighboring sites.

To be more specific (see Fig. 8.21), the attachment probability at a given site p_{xy} at time t depends on a potential Φ_{xy}, which is the average over the potentials of the neighboring sites:

$$\Phi_{xy} = \frac{1}{4}\Big\{\Phi_{x+1,y} + \Phi_{x,y+1} + \Phi_{x-1,y} + \Phi_{x,y-1}\Big\}\,. \qquad (8.22)$$

It yields $\Phi_{x,y} = 0$, if the site (x, y) belongs to the cluster. The simulated "breakdown" occurs from the points of discharge, $\Phi_{x,y} = 0$, to the highest potential in the field, which is $\Phi_{x,y} = 1$ far away (at a distance R) from the cluster. With the assumed boundary conditions, the probability that a site adjacent to the existing pattern is the next point of discharge is given by

$$p(x, y, t) = \frac{\Phi^{\eta}_{xy}}{\sum\limits_{x,y\in C} \Phi^{\eta}_{xy}}\,. \qquad (8.23)$$

The summation is over all sites (x, y) that are part of C, the interface to the pattern of discharge at time t.

Further, an additional parameter η occurs in (8.23), which weights the influence of the potential field [366], thus changing the compactness of the clusters, which leads to patterns with *different fractal dimensions.* For $\eta = 1$,

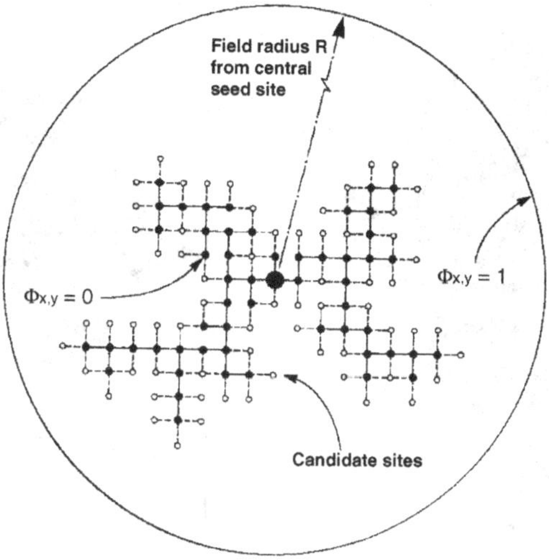

Fig. 8.21. Sketch of the DBM aggregation model on a square lattice [37, 366]

the usual diffusion-limited aggregation pattern occurs, with a fractal dimension $D_\mathrm{f} = 1.7$. In the limit $\eta \to 0$, more compact forms arise, with $D_\mathrm{f} \to 2$, whereas in the limit $\eta \to \infty$, more linear forms occur, with $D_\mathrm{f} \to 1$.

The relation between DBM and urban growth has been suggested by Batty [37, 43]. The potential Φ^{η}_{xy} is interpreted to reflect the available space in the immediate vicinity of a site (x, y), but is also influenced by the cluster. The parameter η, on the other hand, might be regarded as a measure of *planning control*. The variation of η then allows us to generate a continuum of different urban forms with different fractal dimensions (see Fig. 8.22).

To obtain a closer match between the patterns shown in Fig. 8.22 and real urban forms, Batty [37] further suggested considering a confined space, that allows the cluster to grow only in certain angular sectors of the lattice. Hence, with the two adjustable parameters, $0 < \theta < 2\pi$ and $0 < \eta < 5$, which consider topological and planning contraints, urban patterns have been generated, which show at least with respect to their fractal dimensions, analogies to existing urban aggregates.

However, in addition to the visual mismatch between real and simulated urban aggregates, the adaptation of fractal growth models to urban growth has several theoretical drawbacks. DLA and DBM models only produce connected clusters, whereas we have shown in the previous section that urban aggregates rather consist of individually growing separated clusters. Regarding the urban growth zones, DLA and DBM models have the largest growth potential at the tips of the cluster, far away from the center. There is no influence or feedback of an existing urban core to future growth. Moreover, phenomena specific for urban agglomerations, such as the burn-out of urban

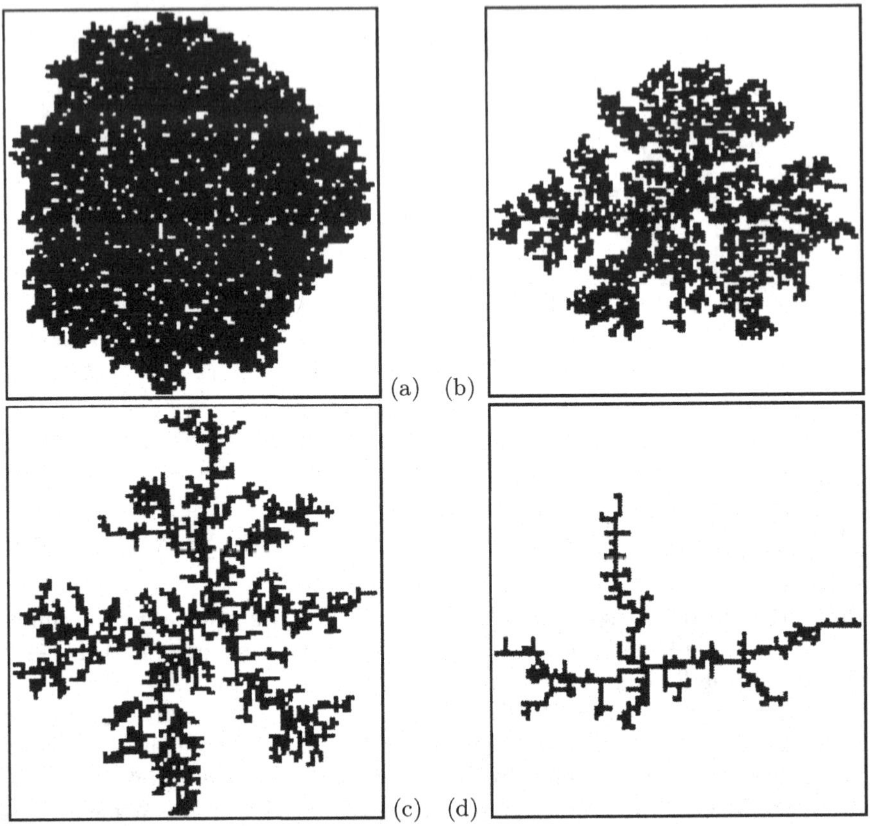

Fig. 8.22. Urban forms generated by systematic distortions of the DBM field. The different values of η result in different fractal dimensions D_{f}: (**a**) $\eta = 0$, $D_{\mathrm{f}} = 1.971$; (**b**) $\eta = 0.5$, $D_{\mathrm{f}} = 1.858$; (**c**) $\eta = 1$, $D_{\mathrm{f}} = 1.701$; (**d**) $\eta = 2$, $D_{\mathrm{f}} = 1.409$ (redrawn from [37, 43])

growth, the emergence of new growth zones, or the coexistence of urban clusters, cannot be sufficiently described within this approach.

The compactness of an urban cluster can be adjusted in the simulation, but it is not thought that the compactness will change in the course of time. As pointed out in Sect. 8.1.1, there is a morphological change during evolution that resembles a second-order phase transition. To put this morphological transition into the framework of aggregation models, urban clusters during an early stage are rather compact, similar to Eden clusters [129, 139, 523]; only in later stages do they change their form to the dendritic morphology of diffusion-like aggregates.

To include this morphological change in urban growth models, additional correlations between the existing aggregate and its further growth are needed. Frankhauser [156] suggested a simulation model that takes into account the interaction between the occupation of a site and the blocking of neighboring sites for further growth.

Recently, Makse et al. [320] used a correlated percolation model [319, 321] to model urban growth by taking into account correlations between development units and the existing aggregation (see Fig. 8.23). They assume (i) that the development units are positioned at a distance r from the center with an occupancy probability,

$$p(r) = \frac{\varrho(r)}{\varrho_0}\,, \quad \varrho(r) = \varrho_0 e^{-\lambda(t) r}\,. \tag{8.24}$$

Here, $\varrho(r)$ is the population density at distance r, and $\lambda(t)$ is the density gradient, both known from empirical data from actual urban systems. Noteworthy, the population density profile of cities shows a tendency to decentralization [87, 344] that is quantified by the decrease in $\lambda(t)$ in the course of time.

Occupation occurs if the occupancy variable $u(r)$, which is an uncorrelated random number, is smaller than the occupation probability, $p(r)$, at site r. Because $\lambda(t)$ decreases in the course of time, the occupation probability in the outer regions increases with time.

Further, it is assumed (ii) that the development units are not positioned at random, but the occupation probability increases, if the neighborhood is also occupied. To introduce correlations among the variables, the (uncorrelated) occupancy variables $u(r)$ are transformed [320, 321] into a new set of random variables $n(r)$ with long-range power law correlations that decay as $r^{-\alpha}$. The *correlation exponent* α has then to be adjusted to match the simulated and the actual urban pattern. Qualitative comparison showed the best agreement for the strongly correlated case ($\alpha = 0.05$).

Additional correlations have the effect of agglomerating the units around the urban area. However, as $\varrho(r)$ decreases with the distance from the center, the probability that the central cluster remains connected during its further growth also decreases with r, and eventually separated clusters occur. The mean radius for the spatial extension of the central (connected) cluster has been estimated at $r_{\mathrm{f}} \sim \lambda^{-1}$.

The quest for additional correlations in urban growth models can, of course, be satisfied in many ways. In Sect. 8.3.3, we want to apply our model of *Brownian agents* to introduce these additional correlations by means of the field $h(r,t)$, which now plays the role of an attraction field for potential growth units. But first, we need to discuss some observations of urban agglomeration.

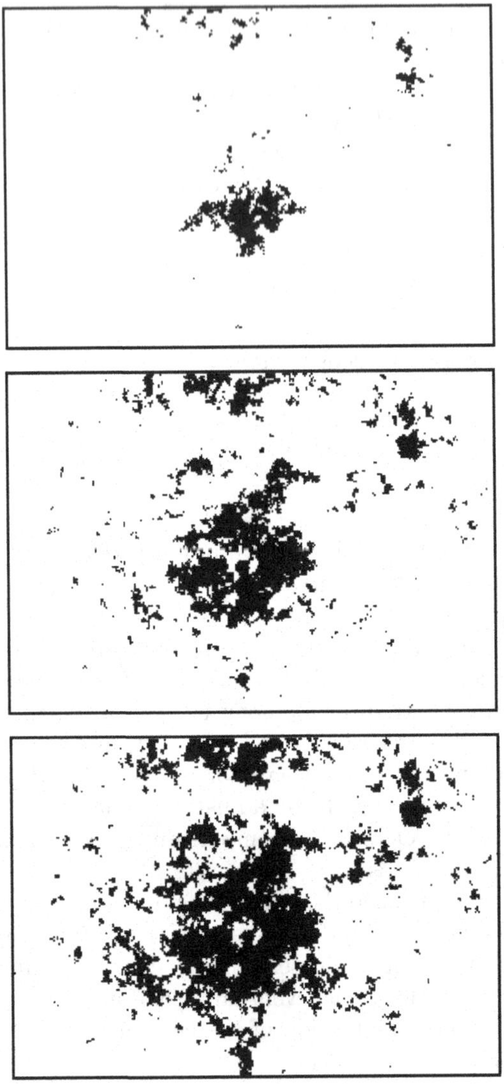

(a)

(b)

(c)

Fig. 8.23. Dynamic urban simulations redrawn from [320] which claim qualitative relation to the growth of Berlin for the years 1875, 1920, and 1945 (see Fig. 8.5). The correlation exponent was fixed at $\alpha = 0.05$; the occupation probability $p(r)$ was chosen to correspond to the density profiles obtained from the real urban pattern of Berlin in the respective years

8.3.2 Shift of Growth Zones

From empirical observations of urban growth, two typical scenarios for the late stage of urban evolution can be obtained. One scenario, which can be observed, for example, for the growth of Berlin (cf. Fig. 8.24), is characterized

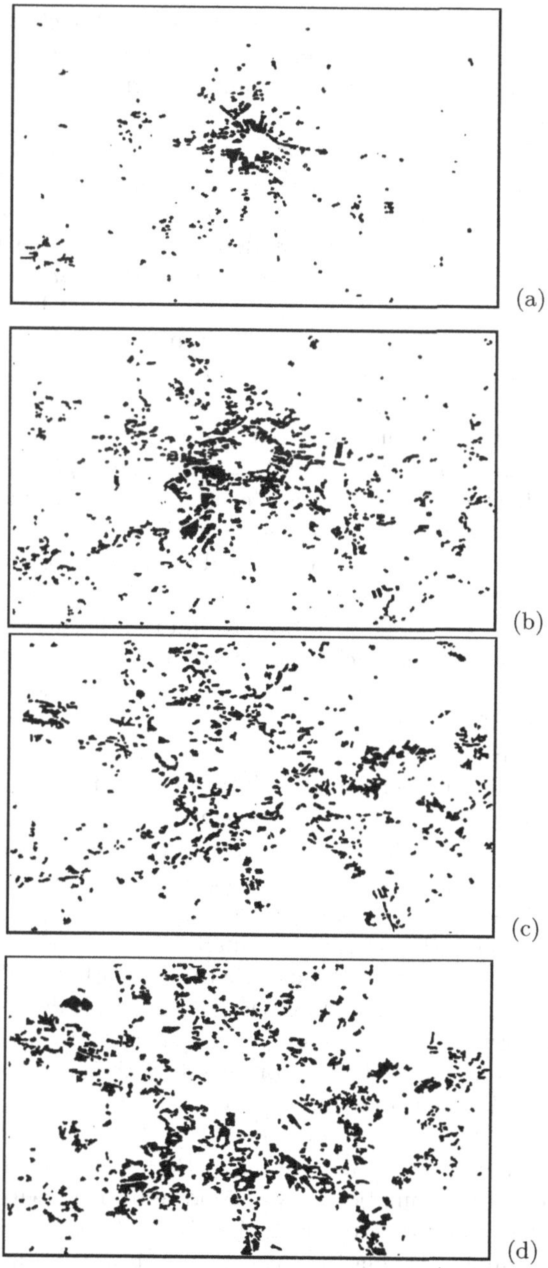

Fig. 8.24. Shift of growth zones into outer regions shown for the evolution of Berlin from 1800 to 1945 (see Fig. 8.5). The sequence shows only the new urban sites occupied during the given time interval: (**a**) 1800–1875; (**b**) 1875–1910; (**c**) 1910–1920; (**d**) 1920–1945 [249]

by a *moving growth front* toward the outer regions of the cluster. The shift of growth zones is accompanied by *slowing down* of growth in the *inner parts* of the agglomeration in the course of time until the growth in the inner regions stops, eventually.

This scenario does *not* mean that there are no construction activities in the center any longer. The center of Berlin, for example, has seen vast construction in recent years. But, with respect to the *built-up* area, which is only considered here, these construction activities just mean that *former* built-up areas such as the "Potsdamer Platz" are *rebuilt* now, after some time lag. It does not mean, for instance, that empty (not built-up) spaces in the center, such as the "Tiergarten," a large park, become available as new construction areas. The real new growth, if any, takes place only in the outer regions that still have free space.

The first scenario is typical of *central growth* governed by *one* predominating center. *Local inhibition*, resulting from the *depletion of free space*, prevents growth in the center; thus the growth front is moving outward. Similar dynamic behavior has already been discussed in Sect. 3.3.1, where particles injected in the center, precipitated at a certain distance from their origin after a chemical reaction, $A + B \rightarrow C$. There, the depletion of B near the center caused the shift of the precipitation front. We mention that activator–inhibitor systems, such as discussed in Sect. 3.3.3, also display similar behavior. In the example of (3.75), the inhibitor reduces the replication or the "growth" of the activator and therefore restricts growth to regions where no inhibitor exists yet.

The second scenario mentioned is different from the first mainly in that there are *multiple growth centers* in a *close vicinity* now, instead of one predominating center. Each of these centers originates its own growth front, but the close distance between them does not allow the establishment of major growth zones, as in central growth. As the example of the German Rhine–Ruhr area shows (cf. Fig. 8.25), simultaneous growth of different urban centers leads to a merger with those of their neighborhood. The competition for free space among the different centers very soon results in a deadlock for further growth. Eventually, the only growth capacity is in the outer region again. Hence, this scenario may be continued by the first scenario, with the newly formed agglomeration as the "center."

These considerations can be used to explain the ad hoc assumption in the simulation of the *rank–size distribution* of Berlin, discussed in Sect. 8.2.3. We recall that these simulations show good agreement with the empirical distribution, if the additional assumption is used that the largest cluster (rank 1) grows only by coagulating with smaller clusters in the neighborhood. This can be considered *indirect* growth. *Direct growth* of the central cluster proportional to its size, due to (8.16), might have occurred during an early stage of evolution. But in the later stage, this direct growth process burns out, whereas smaller clusters near the central cluster grow further and eventually

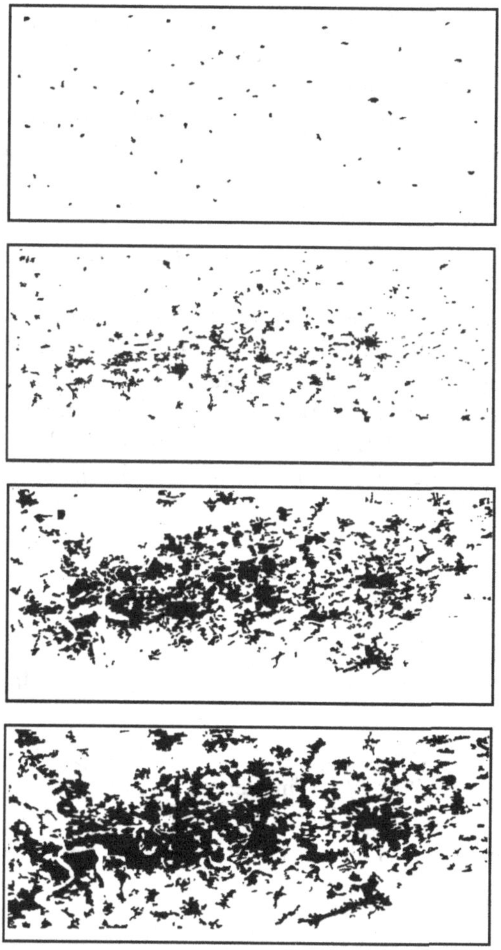

Fig. 8.25. German Rhine–Ruhr area in different years: (a) 1857; (b) 1890; (c) 1957; (d) 1980 [249]

coagulate with it. Obviously, direct growth of the central cluster decreases in the course of time.

The question *why* the cluster of rank 1 stops its direct growth, and *when* does it stop this growth, can be answered by considering the shift of growth zones due to the depletion of free space. To model this shift, we will apply reaction–diffusion dynamics similar to that discussed in Sect. 3.3.1. A specific three-component system, denoted as the *A-B-C model* [463] describes some specific features of the urban evolution; the *spatial dependence* of growth rates, which is neglected in the simulation of the rank–size distribution. In addition to the *global coupling* of urban cluster growth resulting from (8.12), the A-B-C model introduced in the following section, also considers *local cou-*

pling by a self-consistent field, generated by existing urban clusters. This field reflects the *attraction* of the metropolitan area and influences local growth probabilities, thus producing nonlinear feedback between existing clusters and further urban growth. The microscopic dynamics of this structure formation process will be described again by means of Brownian agents, introduced in Sect. 3.1.

8.3.3 Simulating Urban Growth with Brownian Agents

In this section, we introduce the basic features of the A-B-C model for urban growth. This is an agent-based model that symbolically considers the different influences resulting in decentralized urban agglomeration, the tendency of forced growth near attractive locations, on the one hand, and the availability of free space, on the other hand. The relation between the different components is shown in Fig. 8.26 and will be elucidated in the following.

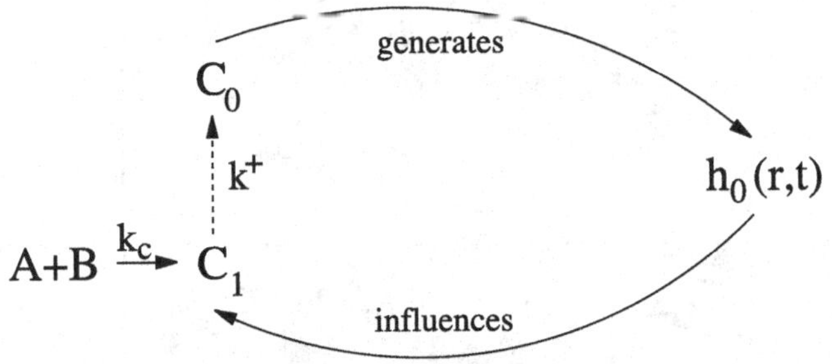

Fig. 8.26. Circular causation among the three components of the urban growth model

(1) Local Aggregation of Agents

The basic assumption of our model is nonlinear feedback between the existing urban aggregate and its further growth, which is mediated by an attraction field created by the settlement. The aggregation process should be described by *Brownian agents* C that exist in two different internal states, $\theta \in \{0, 1\}$. Here, C_0 represents agents that are already "aggregated" (no motion); hence, the spatial distribution $n_0(r,t)$ describes the *built-up area* at a given time. C_1, on the other hand, represents *growth units* that are not yet aggregated; $n_1(r,t)$ is the corresponding distribution. The transformation of a moving growth unit into a precipitated (nonmoving) built-up unit can be expressed by the symbolic reaction: $C_1 \xrightarrow{k^+} C_0$; k^+ is the reaction rate, or the *growth rate* of the urban cluster.

We further assume that the *existing* urban aggregate creates a specific spatiotemporal field $h_0(r,t)$ which represents the *attraction* of the urban area. We are convinced that this attraction exists; however, we are not specific about the reasons for this attraction; there could be economic, political, or cultural reasons, or just a certain way of life that makes the urban area attractive. Within our model, we simply assume that the attraction field follows the same reaction–diffusion dynamics, as discussed in Sect. 3.1.2:

$$\frac{\partial h_0(\boldsymbol{r},t)}{\partial t} = -k_0 h_0(\boldsymbol{r},t) + s_0 \sum_i \delta_{0,\theta_i}\, \delta(\boldsymbol{r}-\boldsymbol{r}_i) + D_0 \Delta h_0(\boldsymbol{r},t)\,. \qquad (8.25)$$

Equation (8.25) means that the attraction field is steadily produced by the existing built-up area ($\theta_i = 0$) at rate s_0. On the other hand, an attraction that is not maintained can also fade out in the course of time, expressed by rate k_0. Finally, we consider that the attraction of an urban area can spread out into the surrounding area, expressed by the diffusion constant D_0.

Nonlinear feedback between the existing urban aggregate and its further growth is given by the assumption that C_1 agents, i.e., growth units that are not aggregated so far, are affected by the attraction field of the existing aggregation (i) in their motion and (ii) in their "precipitation," i.e., in their transition into a built-up unit. The motion of the growth units C_1 at position r_i is influenced by the gradient of the attraction field, so that they tend to come close to the maxima of the attraction (see also Sect. 3.2.1). The resulting Langevin equation in the overdamped limit reads

$$\frac{d\boldsymbol{r}_i}{dt} = \mu \left.\frac{\partial h_0(\boldsymbol{r},t)}{\partial \boldsymbol{r}}\right|_{r_i} + \sqrt{2\,D}\,\boldsymbol{\xi}_i(t)\,, \qquad (8.26)$$

where $\mu = \alpha/\gamma_0$ represents the strength of the response to the gradient. The second term on the right-hand side is the stochastic force that keeps the units moving; D is the diffusion coefficient.

In dependence on the local value of the attraction field, a growth unit can precipitate and transform into a built-up unit – either by attachment to an existing cluster or by formation of a new one. The transition rate of the growth unit, k^+, depends on the *normalized local attraction*:

$$\frac{\partial n_0(\boldsymbol{r},t)}{\partial t} = k^+ n_1(\boldsymbol{r},t) \quad , \quad k^+ \equiv k^+(\boldsymbol{r},t) = \frac{h_0(\boldsymbol{r},t)}{h_0^{max}(t)}\,. \qquad (8.27)$$

To summarize nonlinear feedback, we conclude that agents representing built-up units do not move, but create an attraction field, which affects the movement of agents representing growth units, which do not create an attraction field. Growth units can be transformed into built-up units, thus further increasing the attraction of the urban aggregation.

(2) Local Depletion of Free Space

The demand for new built-up areas cannot always be satisfied in desired places because of local depletion of free space, which is the second important feature affecting urban growth. Let component A represent free space, whose density is $a(\boldsymbol{r},t)$. Initially, for $t = 0$, we shall assume that free space is equally distributed: $a(\boldsymbol{r},t_1) = a_0 = A_0/S$, where S is surface area. However, for $t > 0$, the decrease in free space leads to the emergence of depletion zones in A.

Further, we assume that component B represents the *demand* for built-up areas, with spatial density $b(\boldsymbol{r},t)$. This demand could result from different reasons, which are not discussed here, and could be concentrated at different locations. Hence, the production of demand, expressed by the symbolic reaction, $\xrightarrow{\beta} B$, could reflect different situations:

$$
\begin{array}{ll}
\beta = b_0 = B_0\,/\Delta t & \Rightarrow \text{demand equally distributed}\,,\\
\beta = b_0\,\delta(\boldsymbol{r}-\boldsymbol{r}_0) & \Rightarrow \text{demand only in main center } \boldsymbol{r}_0\,,\\
\beta = \sum_k b_0^k \delta(\boldsymbol{r}-\boldsymbol{r}_k)\,,\ \sum_i b_0^k = B_0\,/\Delta t & \Rightarrow \text{demand in different centers } \boldsymbol{r}_k\,.
\end{array}
\tag{8.28}
$$

Here, B_0 is the *total demand* in the time–space interval considered. In addition to linear production of demand, nonlinear production could also be considered, e.g., $B \xrightarrow{\beta} 2\,B$.

As pointed out, not every demand for a built-up area can be satisfied where it was initiated; it has to match some free space available. Therefore, we assume that component B diffuses at a constant D_b from the centers, until it meets agents of the A component, which represent the free space and which do not diffuse. Only B and A, both demand and free space, can create growth units C_1 which account for urban growth, expressed by the symbolic reaction: $A + B \xrightarrow{k_c} C_1$. Here, k_c describes the reaction rate at which free space disappears. Due to local depletion of A, component B has to move further away from the centers of demand to create some growth units C_1 in outer regions, which in turn try to move back to the urban centers with respect to the attraction field $h_0(\boldsymbol{r},t)$. Along their way, moving growth units pass different aggregations, such as suburbs, which create local attraction. This increases the probability that the growth units precipitate before they have reached the very centers of the urban aggregate with the highest attraction.

The dynamics for the densities of A, B and C_1 can be described by the following set of coupled reaction–diffusion equations for $a(\boldsymbol{r},t)$, $b(\boldsymbol{r},t)$ and a Fokker–Planck equation for $n_1(\boldsymbol{r},t)$:

$$
\frac{\partial a(\boldsymbol{r},t)}{\partial t} = -k_c\, a(\boldsymbol{r},t)\, b(\boldsymbol{r},t)\,, \tag{8.29}
$$

$$
\frac{\partial b(\boldsymbol{r},t)}{\partial t} = \beta - k_c\, a(\boldsymbol{r},t)\, b(\boldsymbol{r},t) + D_b\, \Delta b(\boldsymbol{r},t)\,, \tag{8.30}
$$

$$
\begin{aligned}
\frac{\partial n_1(\boldsymbol{r},t)}{\partial t} &= -\frac{\partial}{\partial \boldsymbol{r}}\left[\mu \frac{\partial h_0(\boldsymbol{r},t)}{\partial \boldsymbol{r}} - D_c \frac{\partial n_1}{\partial \boldsymbol{r}}\right] \\
&\quad + \beta\, a(\boldsymbol{r},t)\, b(\boldsymbol{r},t) - k^+\, n_1(\boldsymbol{r},t)\,.
\end{aligned}
\tag{8.31}
$$

Together with (8.25), (8.27), (8.29)–(8.31) have to be solved simultaneously to describe the reinforced urban growth with respect to local attraction and local burnout. To demonstrate the applicability of the kinetic model outlined here, we have simulated the evolution of the urban area of Berlin [463].

8.3.4 Results of Computer Simulations: Berlin

For the computer simulation, three input parameters must be known: the free space A_0 available in the region at the start time t_{start}, the demand for built-up areas, B_0, during the simulated time interval $t_{\text{end}} - t_{\text{start}}$, and the existing urban aggregation, $n_0(\boldsymbol{r}, t_{\text{start}})$, at time t_{start}, to calculate the initial attraction field. Whereas the cluster distribution of the urban aggregate is given from empirical data (maps), the values for A_0, B_0 can be obtained from a plot which shows the evolution of the total built-up area (see Fig. 8.27).

The straight slope of the curves in Fig. 8.27 indicates that the increase in the total built-up area approaches a characteristic power law, $N(R) \propto R^{D_{\text{f}}}$, in the course of time, which determines the fractal dimension, D_{f}, of the urban aggregate. It has been found that during urban evolution, the fractal dimension remains constant. We count on that fact when calculating the free space available in the region. If we consider a certain spatial extension (e.g., $R^\star = 10$ km), then the fractal dimension determines the *maximum value* of the built-up area within that area, $N_{\max}(R^\star)$, where N is the number of occupied pixels. The free space available in the region at a certain time is just the difference between the actual value of the built-up area, $N(R^\star, t)$, and $N_{\max}(R^\star)$ (see Fig. 8.27):

$$A_0\,(R^\star, t) = N_{\max}(R^\star) - N(R^\star, t)\,. \tag{8.32}$$

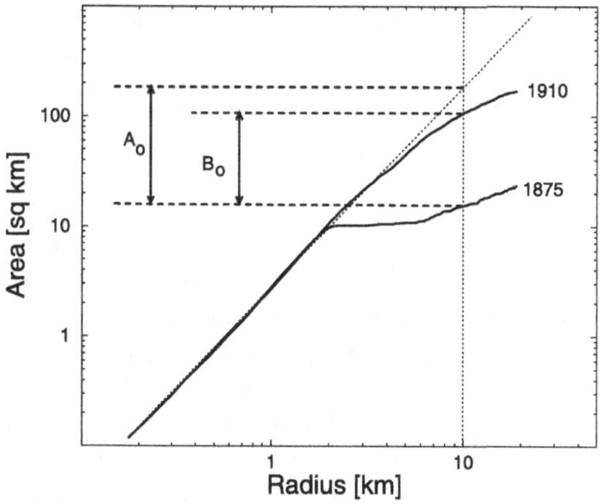

Fig. 8.27. Total built-up area $N(R)$ of Berlin vs. distance R from the center in a double-logarithmic plot for different years [463]

Assuming, that the free space in the region is initially equally distributed, the mean density of the free space is also given by

$$a_0 = \frac{A_0}{S} = \frac{N_{\max}(R^\star)}{4\pi R^{\star\,2}} . \tag{8.33}$$

For the area of Berlin, the mean density of free space is calculated as $a_0 = 0.43$ pixels/lattice unit.

The demand for built-up areas in a certain time interval can be derived from the time series of Fig. 8.27 by calculating the difference between the built-up areas at the beginning and at the end:

$$B_0 = N(R^\star, t_{\mathrm{end}}) - N(R^\star, t_{\mathrm{start}}) . \tag{8.34}$$

For future predictions, the value of $N(R^\star, t_{\mathrm{end}})$ is, of course, not known from empirical data; however, as proved by Fig. 8.15, the value can also be estimated by an exponential growth law. For our simulations of the urban area of Berlin, we have simply assumed a constant demand for built-up areas during the time interval, which is initiated only in the center of Berlin (r_0). Hence, the production of B is described by

$$\beta = b_0\, \delta(\boldsymbol{r} - \boldsymbol{r}_0) \ , \quad b_0 = B_0 \,/(t_{\mathrm{end}} - t_{\mathrm{start}}) . \tag{8.35}$$

Further simulations could also include major cities in the neighborhood as additional centers of demand.

Our simulations of the Berlin area are carried out for the time interval 1910 to 1920. Therefore, we have to take into account that, due to the previous evolution, depletion zones of the free space already exist, which surround the existing clusters. The total consumption of free space is, of course, equal to the total amount of occupied space, known from empirical data. Because we assume that component B, which reflects the demand, diffuses until it reaches component A, the existing depletion zones resulting from the $A+B$ reaction can be simply described by a circle that surrounds the existing clusters. The radii of these depletion circles, R_i^d, can be estimated by

$$R_i^d = \sqrt{\frac{N_i}{4\pi a_0}} , \tag{8.36}$$

where N_i is the pixel size of a particular urban cluster and a_0 is the mean density of the free space, given above. Outside these depletion zones, we assume again an equal distribution of free space: $a(r > R_i^d, t) = a_0$. The initial configuration for the simulation of the Berlin area is given in Fig. 8.28.

The attraction field $h_0(\boldsymbol{r}, t)$ is calculated according to (8.25) and is permanently updated during the simulated urban evolution. Figure 8.29 shows a spatial section of the initial attraction field; the location of the related urban clusters can be seen in Fig. 8.28. Obviously, the suburbs in the Berlin area create their own attraction field which may affect the probability of precipitation of incoming growth units.

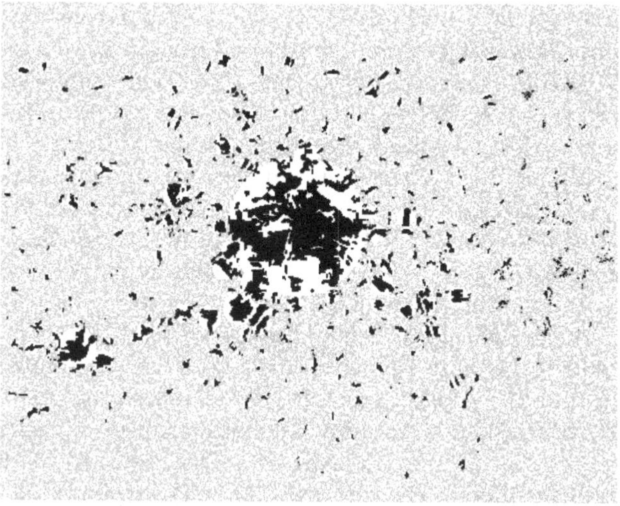

Fig. 8.28. Initial situation for the computer simulations: (*black*) built-up area of Berlin, 1910, (*gray*) density of free space, (*white*) depletion zones [463]

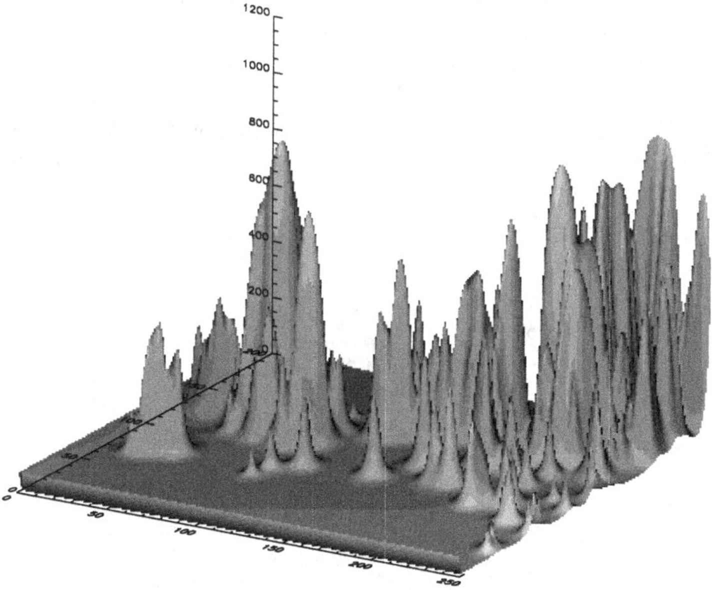

Fig. 8.29. Calculated attraction field $h_0(\boldsymbol{r}, t)$ for the initial situation: Berlin area 1910. The spatial section, which is identical with the lower left part in Fig. 8.28, shows, in the middle, the area of Potsdam with the adjacent suburban areas of Berlin on the right-hand side [463]

During the simulation, B agents are initiated at a constant rate in the center of Berlin. These agents have to diffuse through the depletion zones to meet some A agents to create growth units C_1. The consumption of A leads to an expansion of the depletion zones; hence the growth units are always released at greater distances from the center of demand. During their diffusion toward the maxima of the attraction field, there is an increasing probability that the growth units precipitate before they reach the inner urban centers. Eventually, this may result in stoppage of inner urban growth.

Figure 8.30 proves that, as a result of the dynamics described above, the inner part of the settlement area of Berlin does not grow remarkably, although the demand in the built-up area is initiated only in the center. Instead, we find major growth in those suburban areas that have a suitable "mixture" of attractivity *and* free space available. Compared to the real evolution in the time interval considered, the simulated growth of the Potsdam area (the *lower left part* of Fig. 8.30) is not sufficient. This is due to the fact that we considered only demand in the center of Berlin. As explained above, further simulations could also consider spatial distribution of demand in different centers.

To conclude our simulations, we want to point out that the A-B-C model of urban cluster growth can reflect major features of urban growth, such as (i) the depletion of free space, (ii) the burnout of urban growth in the inner parts of the settlement and the movement of the growth zones into outer regions, and (iii) the attraction of existing settlement areas which affects the further growth of the aggregate. These effects lead to a kind of "decentralized agglomeration" characterized by the formation of new clusters and the merger of neighboring urban clusters, on one hand, and the spatial coexistence of

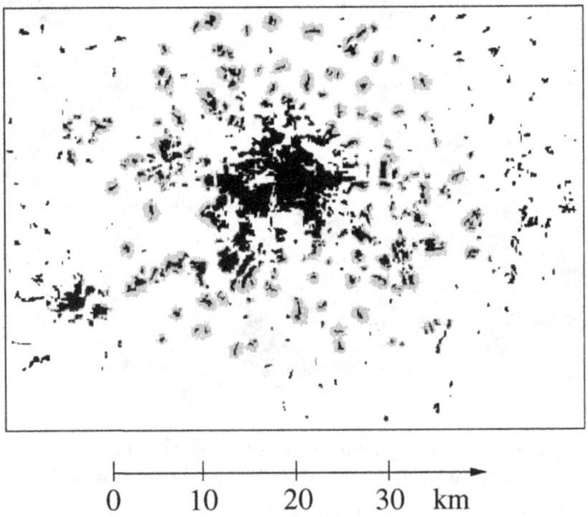

Fig. 8.30. Simulated growth of the Berlin area 1910–1920. (*black*) built-up area 1910, (*gray*) simulated growth area [463]

separated urban clusters instead of *one* merged mega-cluster, on the other hand, depending on the evolution of the depletion zones and the generated attraction of the larger clusters.

How much these features account for urban evolution depends on a couple of parameters which have to be adjusted accordingly. This could be possible, e.g., by estimates based on time series of the previous urban growth. However, a theory for most of these parameters and a suitable explanation, e.g., for the creation and the spread of urban attraction, can be found only in collaboration with town planners, regional developers, economists, etc. Finally, we note again that the kinetic models derived in this chapter focus on the *structural* properties of urban aggregations. This restriction, on the other hand, allows us to derive simple methods suitable for estimating future urban growth. This forecast, of course, is based on the conviction that self-organization plays an important role in urban evolution and that the structure generating forces can be featured within a formal approach.

9. Economic Agglomeration

9.1 Migration and Agglomeration of Workers

9.1.1 Spatial Economic Patterns

In this chapter, Brownian agents play their role as *economic agents*, namely, employed and unemployed *workers*. Depending on the local economic situation that changes in time, employees can be fired, or the unemployed can be hired. This in turn leads to changes in the supply and demand of products and, together with complex local and global feedback processes, to the rise and fall of whole economic regions. In a world of fast growing population and intercontinental trade relations, the *local* emergence of new economic centers, on the one hand, and the *global* competition of existing economic centers, on the other hand, has a major impact on the socioeconomic and political stability of our future world [281].

Hence, not only economists have paid a lot of attention to the problem of economic agglomeration and *economic geography* [162, 224, 284, 394, 396, 448]. Spatiotemporal pattern formation in urban (see Chapter 8) and economic systems has been investigated for quite a long time [3, 225, 226, 282, 285]. There is also a well-known German tradition in economic location theory, associated with names like von Thünen [504], Weber [527], Christaller [85], and Lösch [306]. The *central-place theory*, already mentioned in Sect. 8.1.1, tried to explain both the hierarchical and the spatial organization of economic locations. In his famous hexagonal scheme (see Fig. 9.1), Christaller elucidated the spatial distribution of locations with different hierarchies. Here, every location on a certain level of the hierarchy is surrounded by a number of locations on the next lower level, and so forth. From a spatial perspective, this implies that locations on the *same* hierarchical level have a certain characteristic *distance* from each other, or, as Christaller pointed out, they have a certain "complement area" or "attraction area," that increases with the economic importance of the location.

Recent approaches developed by Weidlich et al. and Haag et al. investigate the problem of settlement formation, location theory, and city size distribution within a stochastic dynamic theory, based on *master equations* and economically motivated *utility functions* [99, 192–194, 536, 537, 539, 540, 544, 545]. One of these elaborated spatial models [360, 546] combines,

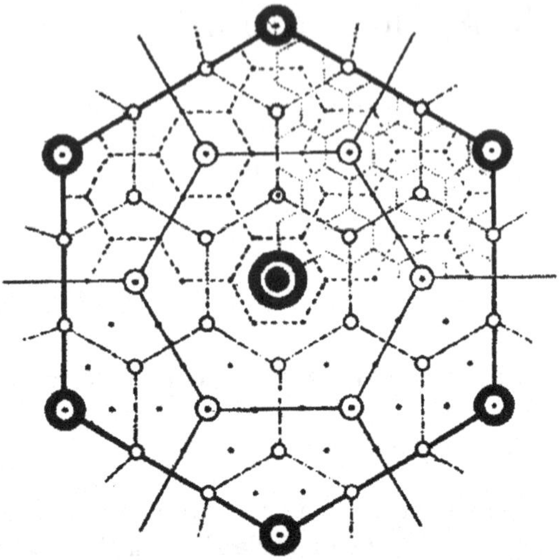

Fig. 9.1. Scheme of the central place theory, after Christaller [85, p. 71], which indicates the hierarchical and spatial organization of locations

for example, the economic activities (production, consumption, transport) of different subpopulations, i.e., *productive* populations, such as "craftsmen" and "peasants," and *service* populations, such as "landowners" and "transporters." The "economic sector" of the model considers fixed costs for production, which depend on population density and transportation costs. Moreover, a "migratory sector" considers migration of the productive populations because they try to maximize their net income, which decreases with fixed costs and transportation costs. The inclusion of fixed costs in the model causes a saturation effect in the density of craftsmen, which therefore tend to distribute in different "towns." Further inclusion of transportation costs causes a tendency to form "towns" close together. This is shown in computer simulations (see Fig. 9.2) that are based on a unit rhombus of 12×12 lattice cells and periodic boundary conditions.

Although the above model has been outlined as a stochastic model, the computer simulations in Fig. 9.2 are not carried out as stochastic simulations (see Sect. 3.1.4). Rather, they are based on related mean value equations derived from master equations. Hence, for the "migratory sector," a set of $(A+2)I^2$ coupled ordinary nonlinear differential equations has to be solved, where A is the number of migrating subpopulations (which is two in the case considered) and I^2 is the size of the unit rhombus.

Different from these investigations, in this chapter, we focus on an *agent-based* simulation of the dynamics of economic concentration. To apply the methods developed in the previous chapters, *economic agents* are now described by means of *Brownian agents.* In the following, we will outline a model which, from a minimalistic point of view, shows the establishment of distinct

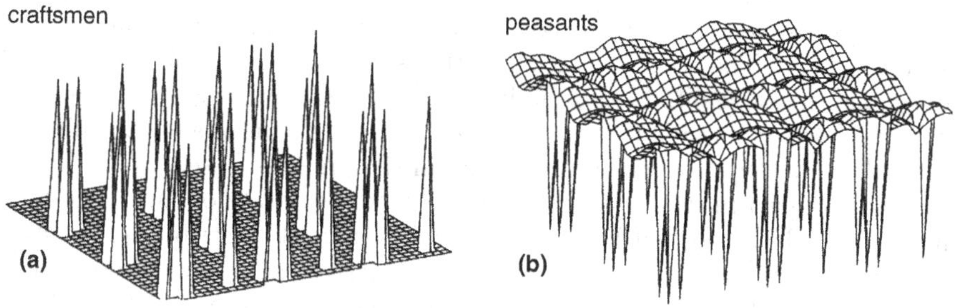

Fig. 9.2. (**a**) Stationary formation of differentiated "town structures" inhabited by "craftsmen" in each unit area. (**b**) Differentiated ring-shaped rural settlements of "peasants" around "town structures." The model considers fixed costs and transportation costs and a joint probability of production and consumption corresponding to competititve minimization of transportation costs [360, 536, 537]

economic centers and their stable coexistence at a certain critical distance from each other, in accordance with central-place theory.

9.1.2 Model Equations for Migration and Employment

Let us consider a two-dimensional system with $i = 1, \dots, N$ *economic agents*, which are represented by *Brownian agents.* These agents are again characterized by two variables: their current location, given by the space coordinate $\boldsymbol{r}_i$; and an internal state θ_i, which can be either one or zero: $\theta \in \{0, 1\}$. Agents with the internal state $\theta = 0$ are considered *employed agents*, C_0; agents with $\theta = 1$ are considered *unemployed agents*, C_1. At a certain rate k^+ (hiring rate), an unemployed agent becomes employed, and at a rate k^- (firing rate), an employed agent becomes unemployed; this can be expressed by the symbolic reaction,

$$C_0 \underset{k^+}{\overset{k^-}{\rightleftarrows}} C_1 \,. \tag{9.1}$$

Employed agents are considered immobile, whereas unemployed agents can *migrate.* The movement of a migrant may depend on both erratic circumstances and deterministic forces that attract him to a certain place. Within a stochastic approach, this movement can be described by the following overdamped Langevin equation:

$$\frac{d\boldsymbol{r}_i}{dt} = \boldsymbol{f}(\boldsymbol{r}_i) + \sqrt{2D}\, \boldsymbol{\xi}_i(t) \,. \tag{9.2}$$

$\boldsymbol{f}(\boldsymbol{r}_i)$ describes the local value of a deterministic force that influences the motion of an agent. We note here again, that agents are *not* subject to long-range forces, which may attract them over large distances, but only to *local* force. That means that migrants do *not* count on global information to guide their movements, but respond only to local information, which will be specified later. The second term in (9.2) describes random influences on the movement of an agent. As (9.2) indicates, an unemployed agent will move in a very predictable way if the guiding force is large and the stochastic influence is small, and it will act rather randomly in the opposite case.

As discussed in detail in Sect. 3.1.1, the current state of the agent community can be described by the canonical N-particle distribution function $P(\theta_1, \boldsymbol{r}_1, \ldots, \theta_N, \boldsymbol{r}_N, t)$, from which we can derive the spatiotemporal densities $n_\theta(\boldsymbol{r}, t)$ of unemployed and employed agents by means of (3.9). With $\theta \in \{0, 1\}$, we obtain the spatiotemporal density of employed agents, $n_0(\boldsymbol{r}, t)$, and of unemployed agents, $n_1(\boldsymbol{r}, t)$. For simplicity, we omit the index θ by defining

$$n_0(\boldsymbol{r}, t) = l(\boldsymbol{r}, t) \;, \quad n_1(\boldsymbol{r}, t) = n(\boldsymbol{r}, t) \,. \tag{9.3}$$

We assume only that the total number of agents is constant, whereas the density of employed and unemployed agents can change in space and time:

$$N_0(t) = \int_A l(\boldsymbol{r}, t)\, d\boldsymbol{r} \;, \quad N_1(t) = \int_A n(\boldsymbol{r}, t)\, d\boldsymbol{r} \;, \quad N_0(t) + N_1(t) = N \,. \tag{9.4}$$

The density of *employed* agents can be changed only by local "hiring" and "firing," which with respect to (9.1), can be described by the reaction equation:

$$\frac{\partial}{\partial t} l(\boldsymbol{r}, t) = k^+ \, n(\boldsymbol{r}, t) \; - \; k^- \, l(\boldsymbol{r}, t) \,. \tag{9.5}$$

The density of *unemployed* agents can be changed by two processes, (i) migration and (ii) hiring of unemployed and firing of employed agents. With respect to the overdamped Langevin equation, (9.2), which describes the movement, we can derive the following reaction–diffusion equation (see also Sect. 3.1.1):

$$\begin{aligned} \frac{\partial}{\partial t} n(\boldsymbol{r}, t) \; &= \; - \frac{\partial}{\partial \boldsymbol{r}} \boldsymbol{f}(\boldsymbol{r}, t)\, n(\boldsymbol{r}, t) + D_n \frac{\partial^2}{\partial \boldsymbol{r}^2} n(\boldsymbol{r}, t) \\ & \quad - k^+ \, n(\boldsymbol{r}, t) + k^- \, l(\boldsymbol{r}, t) \,. \end{aligned} \tag{9.6}$$

The first term of the r.h.s. of (9.6) describes the change in local density due to the force $\boldsymbol{f}(\boldsymbol{r}, t)$, the second term describes the migration of unemployed agents in terms of a diffusion process with D_n the diffusion coefficient, and the third and the fourth terms describe local changes in density due to "hiring" or "firing" of agents.

By now, we have a complete dynamic model that describes local changes of employment and unemployment, as well as the migration of unemployed agents. However, so far, some of the important features of this dynamic model are not specified, namely, (i) the deterministic influences on a single migrant, expressed by $\boldsymbol{f}(\boldsymbol{r},t)$; and (ii) the hiring and firing rates, k^+, k^-, which locally change employment density. These variables depend, of course, on local economic conditions; hence we will need additional economic assumptions.

To determine $\boldsymbol{f}(\boldsymbol{r},t)$, we assume that an unemployed agent, who can migrate due to (9.2), responds to *local gradients* in real *wages* (per capita). The migrant tries to move to places with a higher wage but again counts only on information in the vicinity. Hence, the guiding force $\boldsymbol{f}(\boldsymbol{r}_i)$ is determined as follows:

$$\boldsymbol{f}(\boldsymbol{r},t) = \frac{\partial}{\partial \boldsymbol{r}}\omega(\boldsymbol{r},t)\,. \tag{9.7}$$

For further discussion, we need some assumptions about the local distribution of real wages, $\omega(\boldsymbol{r},t)$. It is reasonable to assume that *local wages* may be a function of the local density of *employed agents*, $\omega(\boldsymbol{r},t) = \omega[l(\boldsymbol{r},t)]$. Some specific assumptions about this dependence will be discussed in Sect. 9.2.1. Because of the functional dependence, $\omega[l(\boldsymbol{r},t)]$, we now have *nonlinear feedback* in our model of migration and employment, which is also shown in Fig. 9.3. Due to the local production of the employees, denoted by C_0 in (9.1), a local wage is generated, which influences the migration of unemployed agents, i.e., they are attracted by positive gradients in the local wage. Moreover, unemployed agents can be employed, and vice versa, which further affects the local wage.

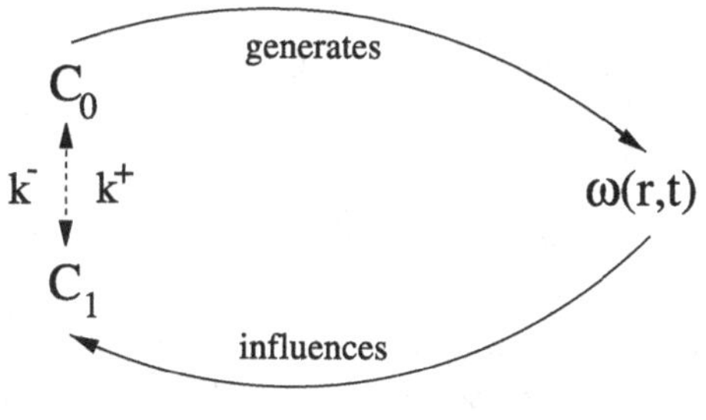

Fig. 9.3. Circular causation between unemployed agents (C_1) and employed agents (C_0). The employees generate a field $\omega(\boldsymbol{r},t)$, which may represent their incomes and locally attracts migrating unemployed agents. Further, the "hiring" of unemployed and the "firing" of employed agents are considered [447]

In our model, we can distinguish between two different timescales: one describes the migration of unemployed agents and the second the processes of "hiring" and "firing." In the next section, we will derive a competition equation, which is based on the assumption that the timescale for hiring and firing of workers is more determining than the timescale of migration. If we explicitly consider unemployment in our model, this assumption would imply that there are always enough unemployed workers who can be hired on demand. A *growing economy*, however, might be determined just by the opposite limiting case: It is important to *attract* workers/consumers to a certain area before the output of production can be increased. Then, the timescale for the dynamics will be determined mostly by *migration* processes. This means that the spatial distribution of employed agents can be assumed as a *quasi-stationary equilibrium* compared to the spatial distribution of unemployed agents:

$$\frac{\partial}{\partial t} l(\boldsymbol{r},t) = 0 \quad \Rightarrow \quad l^{\text{stat}}(\boldsymbol{r},t) = \frac{k^+}{k^-}\, n(\boldsymbol{r},t)\,. \tag{9.8}$$

Hence, the local density of employed agents can be expressed as a function of the local density of unemployed agents available, which itself changes due to migration on a slower timescale. With (9.8), we can then rewrite the spatial derivative for wages as follows:

$$\frac{\partial \omega(\boldsymbol{r},t)}{\partial \boldsymbol{r}} = \frac{\delta \omega[l(\boldsymbol{r},t)]}{\delta l}\,\frac{\partial l(\boldsymbol{r},t)}{\partial \boldsymbol{r}} = \frac{\delta \omega[l(\boldsymbol{r},t)]}{\delta l}\,\frac{k^+}{k^-}\,\frac{\partial n(\boldsymbol{r},t)}{\partial \boldsymbol{r}}\,, \tag{9.9}$$

where δ denotes the functional derivative. Using (9.7)–(9.9), the reaction–diffusion equation for the change of the density of unemployed agents, (9.6), can now be rewritten as follows:

$$\frac{\partial}{\partial t} n(\boldsymbol{r},t) = \frac{\partial}{\partial \boldsymbol{r}}\left[D_{\text{eff}}\frac{\partial n(\boldsymbol{r},t)}{\partial \boldsymbol{r}}\right]. \tag{9.10}$$

Similar to the investigations in Sect. 3.2.3, the r.h.s. of (9.10) now has the form of a usual diffusion equation, and D_{eff} is an *effective diffusion coefficient*:

$$D_{\text{eff}} = D_n - \frac{k^+}{k^-}\,\frac{\delta \omega}{\delta l}\, n(\boldsymbol{r},t)\,. \tag{9.11}$$

Here, D_n is the "normal" diffusion coefficient of unemployed agents, (9.6). The additional terms reflect the fact that unbiased diffusion is changed because of the response of migrants to local differences in wage distribution. As we see from (9.11), there are two contradicting forces determining the effective diffusion coefficient: normal diffusion, which keeps unemployed agents moving and the response to the wage gradient.

If the local wage decreases with the number of employed agents, then $\delta\omega/\delta l < 0$, and effective diffusion increases. That means that unemployed agents migrate away from regions where employment may result in an effective decrease of marginal output. However, if $\delta\omega/\delta l > 0$, then the wage effectively *increases* in regions with a larger number of employed agents, and unemployed agents are *attracted* to these regions. As we see in (9.11), for a certain *positive feedback* between the local wage and the employment density, the effective diffusion coefficient can be *locally negative*, and unemployed agents do not leave these areas once they are there. Economically speaking, these agents stay there because they may profit from the local increase in employment density. We want to emphasize that this interesting dynamic behavior has been derived *without* any *explicit* economic assumptions.

Before specifying the economic functions in Sect. 9.2.1, we first want to discuss the case that the timescale of transitions is the most influential, which allows the derivation of a selection equation from the current model.

9.1.3 Derivation of Competition Dynamics

In the previous section, we suggested specific dynamics for the migration of agents, which assumes that only *unemployed* agents *move*, whereas employed agents are immobile. Other models in economic geography, however, assume migration dynamics for employees. Krugman [283] (see also [284] for the results) discusses a dynamic spatial model, where "workers are assumed to move toward locations that offer them higher real wages." Krugman makes no attempt "to model the moving decisions explicitly," but he assumes the following "law of motion of the economy":

$$\frac{d\,\lambda_j}{dt} = \varrho\,\lambda_j\,(\omega_j - \bar{\omega})\,. \tag{9.12}$$

Here, λ_j is the "share of the manufactoring labor force in location j", ϱ is a scaling constant, and ω_j is the real wage at location j. $\bar{\omega}$ is the "average real wage," which is defined as

$$\bar{\omega} = \frac{\sum_j \lambda_j\,\omega_j}{\sum_j \lambda_j}\;, \quad \sum_j \lambda_j = 1\,. \tag{9.13}$$

Here, the sum is over all different regions, where j serves as a space coordinate, and it is assumed that "at any point in time there will be location-by-location full employment." The assumed *law of motion of the economy*, (9.12), then means that "workers move away from locations with below-average real wages and towards sites with above-average real wages."

From a dynamic perspective, (9.12) is identical with a selection equation of the Fisher–Eigen type, (3.62) [130, 145], which is repeated here:

$$\frac{dx_i}{dt} = x_i\Big(E_i - \langle E_i\rangle\Big)\;, \quad \langle E_i\rangle = \frac{\sum_i E_i\,x_i}{\sum_i x_i}\,. \tag{9.14}$$

Here, $x_i = N_i/N$ is the fraction of individuals of subspecies i, E_i is the fitness of species i, and $\langle E_i\rangle$ is the mean fitness. N denotes the total population size. It can be proved [142] that the Fisher–Eigen equation describes a competition process which finally leads to a stable state with only *one* surviving species. The competition process may occur on a very long timescale, but asymptotically, stable coexistence of many different species is impossible.

The Fisher–Eigen equation, (9.14), can be derived from quite general assumptions about dynamics. Suppose that dynamics for species i is proportional to the current concentration x_i:

$$\frac{dx_i}{dt} = \alpha_i(\dots, x_i, x_j, \dots)\, x_i\,. \tag{9.15}$$

The prefactor $\alpha_i(\dots, x_i, x_j, \dots)$ describes the replication/extinction rate of species i and may also consider couplings between different species. Further, we assume that the *total number* of individuals in *all* subspecies is *constant*:

$$\sum_{i=1}^{N} x_i = \text{const} = 1\,. \tag{9.16}$$

This condition prevents unlimited replication of all species. Hence, the prefactor α_i in (9.15) should consist of at least two parts:

$$\alpha_i(\dots, x_i, x_j, \dots) = E_i - k\,, \tag{9.17}$$

where E_i is the replication rate, assumed proportional to the fitness of species i, and k is the extinction rate or death rate, which is assumed to be the same constant for all species. Equation (9.16), combined with (9.15), (9.17), requires that

$$\sum_{i=1}^{N} \frac{dx_i}{dt} = 0\ , \quad k = \frac{\sum_i E_i\, x_i(t)}{\sum_i x_i(t)} = \langle E_i(t)\rangle\ , \tag{9.18}$$

which then yields the dynamics of (9.14).

On the other hand, we could also assume that replication occurs via interaction between two species, i and j, and consider the "advantage," $E_i - E_j$, for the fitness:

$$\alpha_i(\dots, x_i, x_j, \dots) = \sum_{j\neq i} \Big(E_i - E_j\Big)\, x_j\,. \tag{9.19}$$

This ansatz would also yield the dynamics of (9.14) because of the definition of the mean value and the condition, (9.16). Note that we have used (9.19) in Sect. 7.1.3 to describe the "difference selection" among searchers.

Fisher–Eigen dynamics describes a competition process between different subspecies i, where the mean value, $\langle E_i(t)\rangle$, decides about winners and losers: for fitness $E_i > \langle E_i(t)\rangle$, the species continues to grow, on average, whereas for $E_i < \langle E_i(t)\rangle$, the species most likely becomes extinct. If the same dynamics is applied to the migration of workers, as (9.12) suggests, it describes competition of the manufacturing labor force at different locations. However, in an economic context, this assumed "law of motion of the economy" involves some deficiencies:

1. A constant total number of employed workers is assumed; a change of the total number or unemployment is not discussed.
2. Employed workers move. In fact, if workers want to move to a place that offers higher wages, they first have to be free; that means an unemployed worker moves and then has to be reemployed at the new location. This process is completely neglected (or else, it is assumed that it occurs at infinite velocity).
3. Workers move immediately if their wages are below the average, regardless of the migration distance to places with higher wages and with no doubt about their reemployment.

Further, the competition equation, (9.12), assumes coupling between *all* fractions of workers at different locations due to $\bar{\omega}$. The more refined dynamic model introduced in the previous section considers different types of agents, namely, unemployed and employed agents, and therefore does not assume *direct* coupling. However, it is also possible to derive the proposed competition equation (9.12) from the general dynamic model presented in Sect. 9.1.2. This will be done in the following to elucidate the implicit assumptions that lead to (9.12). The derivation is very similar to the procedure discussed in Sect. 3.2.5. With respect to the economic context discussed here, the derivation is based on four approximations:

(i) In [283], unemployment is not discussed. But, with respect to the dynamic model presented in Sect. 9.1.2, unemployed agents exist in an overwhelmingly large number. Thus, the first assumption for deriving Krugman's "law of motion" is that the local change in the density of unemployed agents due to hiring and firing can simply be neglected.

(ii) Further, it is assumed that the spatial distribution of unemployed agents is in a *quasi-stationary* state. This does not mean that the distribution does not change, but that the distribution relaxes fast into quasi-stationary equilibrium, compared to the distribution of employed agents.

With assumptions (i) and (ii), (9.6) for the density of unemployed agents reduces to

$$\frac{\partial}{\partial t} n(\boldsymbol{r},t) \;=\; -\frac{\partial}{\partial r} f(\boldsymbol{r},t)\, n(\boldsymbol{r},t) + D_n \,\frac{\partial^2}{\partial r^2} n(\boldsymbol{r},t) \;=\; 0\,. \tag{9.20}$$

Integration of (9.20) leads to the known canonical distribution,

$$\begin{aligned} n^{\mathrm{stat}}(\boldsymbol{r},t) &= \bar{n}\frac{\exp\left[\int f(\boldsymbol{r},t)/D_n\, dr\right]}{\left\langle \exp\left[\int f(\boldsymbol{r},t)/D_n\, dr\right]\right\rangle}\,, \\ \left\langle \exp\left[\int f(\boldsymbol{r},t)/D_n\, dr\right]\right\rangle &= \frac{1}{A}\int_A \exp\left[\int f(\boldsymbol{r}',t)/D_n\, dr'\right] dr\,, \end{aligned} \tag{9.21}$$

where the expression $\langle \ldots \rangle$ describes the mean value and $\bar{n} = N_1/A$ is the mean density of unemployed agents.

(iii) For a derivation of (9.12), we now have to specify $f(\boldsymbol{r},t)$ in (9.21). Here, it is assumed that a single unemployed agent who migrates due to (9.2), responds to the *total local income* in a specific way:

$$f(\boldsymbol{r},t) = \frac{\partial}{\partial r} \ln \left[\omega(\boldsymbol{r})\, l(\boldsymbol{r},t)\right]. \tag{9.22}$$

Equation (9.22) means that the migrant is guided by *local gradients* in *total income*, where $\omega(\boldsymbol{r})$ is local income and $l(\boldsymbol{r},t)$ the local density of employed workers. We note here again that migrants do not count on information about the highest global income; they only "know" about their vicinity. With assumption (9.22), we can rewrite (9.21) using the discrete notation,

$$n_j^{\text{stat}}(t) = \bar{n}\, \frac{\omega_j\, l_j(t)}{\sum_j \omega_j\, l_j(t) \,/\, \sum_j}. \tag{9.23}$$

The corresponding equation (9.5) for $l_j(t)$ reads, in discrete notation,

$$\frac{\partial}{\partial t} l_j(t) = k^+ n_j(t) \; - \; k \quad l_j(t). \tag{9.24}$$

Because (9.12) deals with fractions instead of densities, we have to divide (9.24) by $\sum_j l_j(t)$, which leads to

$$\frac{d}{dt}\lambda_j(t) = k^- \lambda_j \left(\frac{k^+}{k^-} \frac{n_j}{l_j} - 1 \right). \tag{9.25}$$

(iv) Krugman assumes that the total number of employed workers is constant, so we use this assumption to replace the relation between the hiring and the firing rate in (9.25). Equation (9.24) yields

$$\sum_j \frac{d}{dt}\lambda_j(t) = 0 \quad \Rightarrow \quad \frac{k^+}{k^-} = \frac{\sum_j l_j}{\sum_j n_j}. \tag{9.26}$$

n_j in (9.25) is now replaced by the quasi-stationary value, (9.23). Inserting (9.26) into (9.25) and using the definitions of average wage $\bar{\omega}$, (9.13), and $\bar{n} = \sum_j n_j / \sum_j$, we finally arrive at

$$\frac{d}{dt}\lambda_j(t) = \frac{k^-}{\bar{\omega}}\, \lambda_j\, (\omega_j - \bar{\omega}). \tag{9.27}$$

Equation (9.27) is identical with Krugman's law of motion of the economy, (9.12), if the prefactor ϱ in (9.12) is identified as $\varrho = k^-/\bar{\omega}$. Hence, ϱ is a slowly varying parameter that depends on both the firing rate, which determines the timescale, and on the average wage $\bar{\omega}$, which may change in the course of time.

It is an interesting question whether assumptions (i)–(iv) which lead to Krugman's law of motion in the economy, have some practical value in an economic context. Equation (9.27) implies that finally *all* workers are located in *one* region j, where the local real wage ω_j is equal to the mean real wage $\bar{\omega}$. In [283, 284], computer simulations for 12 locations (on a torus) are discussed that show the stable coexistence of two (sometimes three) centers, roughly evenly spaced across the torus, with exactly the same number of workers. The coexistence of these centers might result from some specific economic assumptions in Krugman's model of spatial aggregation, which includes complex issues such as consideration of transportation costs, distinction between agricultural and manifactured goods, and price index. On the other hand, coexistence might also result from the lack of fluctuations that could have revealed the instability of the (deterministic) stationary state. Apart from a random initial configuration, stochastic influences are not considered in [283, 284].

In the following, we will focus on the more interesting question whether the general dynamic model of migration and employment introduced in Sect. 9.1.2, can produce stable coexistence of different economic centers in the presence of fluctuations. This may allow us to overcome some of the deficiencies in (9.12).

9.2 Dynamic Model of Economic Concentration

9.2.1 Production Function and Transition Rates

In Sect. 9.1.2, we introduced the general dynamics of migration and employment without specific economic assumptions. But for a further discussion of the model, the three remaining functions which are unspecified until now, (i) $\delta\omega/\delta l$, (ii) k^{+}, (iii) k^{-}, have to be specified, and that is, of course, where economics comes into play.

To determine the economic functions, we refer to a *perfectly competitive industry* (where "perfect" means complete or total). In this standard model [80], the economic system is composed of many firms, each small relative to the size of the industry. These firms represent separate economies sharing common pools of labor and capital. New competitors can freely enter/exit the market; hence the number of production centers is not limited or fixed. Further, it is assumed that every firm produces the same (one) product and every firm uses only *one* variable input. Then, the *maximum profit condition* tells us that "firms will add inputs as long as the marginal revenue product of that input exceeds its market price." When labor is a variable input, the price of labor is the wage, and a profit-maximizing firm will hire workers as long as the marginal revenue product exceeds the wage of the added worker.

The marginal revenue product MRP = MP × P is the additional revenue a firm earns by employing one additional unit of input, where P is the price of output and MP is the marginal product. In a perfectly competitive industry, however, no single firm has any control over prices. Instead, the price results from the interaction of many suppliers and many consumers. In a perfectly competitive industry, every firm sells its output at the market equilibrium price, which is simply normalized to one, hence the marginal revenue product is determined by MP.

The marginal product, MP, can be derived from a *production function*

$$Y(\boldsymbol{r},t) = A(\boldsymbol{r},t)\, g(\boldsymbol{r},t)\,, \tag{9.28}$$

which describes the relationship between inputs and outputs (i.e., the technology of production). Usually, the production function may include the effect of different inputs, such as capital, public goods, and natural resources. Here, we concentrate only on one variable input, labor. Thus, $g(\boldsymbol{r},t)$, it is assumed, is a Cobb–Douglas production function

$$g(\boldsymbol{r},t) = l^{\beta}(\boldsymbol{r},t) \tag{9.29}$$

with a common (across regions) exponent β; $l(\boldsymbol{r},t)$ is the local density of employees. The exponent β describes how a firm's output depends on the scale of operations. Increasing "returns to scale" is characterized by a lower average cost with an increasing scale of production; hence $\beta > 1$ in this case. On the other hand, if an increasing scale of production leads to higher average costs (or lower average output), we have the situation of decreasing "returns to scale", with $\beta < 1$. In the following, we will restrict the discussion to $\beta < 1$, common to all regions.

The prefactor $A(\boldsymbol{r},t)$ represents the economic details of the level of productivity. We assume that these influences can be described by two terms [447]:

$$A(\boldsymbol{r},t) = A_{\mathrm{c}} + A_{\mathrm{u}}\,. \tag{9.30}$$

$A_{\mathrm{c}} = \mathrm{const}$ summarizes those output dependences on capital, resources, etc. which are not explicitly discussed here. The new second term, A_{u}, considers *cooperative effects* that result from *interactions* among the workers. Because all cooperative effects are *nonlinear*, A_{u} is a nonlinear function of $l(\boldsymbol{r},t)$: $A_{\mathrm{u}} = A_{\mathrm{u}}\left[l(\boldsymbol{r},t)\right]$. Hence, the production function depends explicitly only on $l(\boldsymbol{r},t)$:

$$Y\left[l(\boldsymbol{r},t)\right] = \Big\{A_{\mathrm{c}} + A_{\mathrm{u}}[l(\boldsymbol{r},t)]\Big\}\, l^{\beta}(\boldsymbol{r},t)\,. \tag{9.31}$$

The marginal product (MP) is the additional output created by adding one more unit of input, i.e., MP $= dY/dl$, if labor is input. The wage of a potential worker will be the marginal product of labor:

$$w[l(\boldsymbol{r},t)] = \frac{\delta Y\left[l(\boldsymbol{r},t)\right]}{\delta l}\,. \tag{9.32}$$

If a firm faces a market wage rate of $\omega^\star$ (which could be required, e.g., by minimum wage laws), then, in accordance with the maximum profit condition, a firm will hire workers as long as

$$\frac{\delta Y[l(\boldsymbol{r},t)]}{\delta l} > \omega^\star . \tag{9.33}$$

To complete our setup, we need to discuss how the prefactor A_u depends on the density of employees, $l(\boldsymbol{r},t)$. Here, we assume that the cooperative effects will have an effect only in an intermediate range of l. For small production centers, the *synergetic effect* resulting from mutual stimulation among workers is too low. On the other hand, for very large production centers, the advantages of cooperation might be overshadowed by the disadvantages of massing agents. Thus we will assure, that $Y \sim l^\beta$ in both the limit $l \to 0$ and $l \to \infty$. These assumptions are concluded in the ansatz,

$$A_\mathrm{u}\left[l(\boldsymbol{r},t)\right] \sim \exp\Big\{u\left[l(\boldsymbol{r},t)\right]\Big\}, \tag{9.34}$$

where the *utility function* $u(l)$ describes mutual stimulation among workers in powers of l:

$$u(l) = a_0 + a_1 l + a_2 l^2 + \dots . \tag{9.35}$$

The series will be truncated after the second order. The constants a_i characterize the effect of cooperation, with $a_0 > 0$. The case $a_1 > 0,\ a_2 < 0$ especially considers saturation effects in cooperation, i.e., the advantages of cooperation will be overshadowed by the disadvantages of crowding. This idea implies that there is an optimal size for taking advantage of the cooperative effect, which is determined by the ratio of a_1 and a_2. If one believes that cooperative effects are always an increasing function in l, then simply $a_2 > 0$ can be assumed. On the other hand, if cooperative effects are neglected, i.e., $a_1 = a_2 = 0$, then we obtain the "normal" production function,

$$Y(l) = \bar{A}\, l^\beta . \tag{9.36}$$

The constant $\bar{A}$ is determined by a relation between A_c and a_0 which results from (9.31), (9.34), (9.35). Within our approach, constant A_c is not specified. If we, without restrictions of the general case, choose

$$\bar{A} = A_\mathrm{c} + \exp(a_0) \equiv\ 2A_\mathrm{c}\,, \tag{9.37}$$

then the production function, $Y(l)$, (9.31), with respect to cooperative effects can be expressed as follows [447]:

$$Y\left[l(\boldsymbol{r},t)\right] = \frac{\bar{A}}{2}\left[1 + \exp\left(a_1 l + a_2 l^2\right)\right]\, l^\beta . \tag{9.38}$$

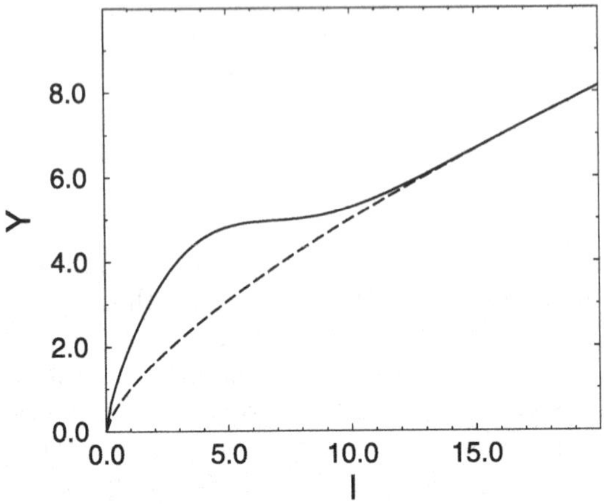

Fig. 9.4. Production function $Y(l)$ vs. density of employed workers, l. The *solid line* results from (9.38). Parameters: $\bar{A} = 2$, $a_1 = 0.06$, $a_2 = -0.035$, $\beta = 0.7$. The *dashed line* shows the production function *without* cooperative effects ($a_1 = a_2 = 0$) [447]

Figure 9.4 presents the production function, (9.38), depending on the density of employees, which can be compared with the "normal" production function, $Y(l)$, (9.36). Clearly, we see an increase in the total output due to cooperative effects among workers. Because $a_1 > 0, a_2 < 0$, this increase has a remarkable effect only in an intermediate range in l. The cooperative effect vanishes for both $l \to 0$ and $l \to \infty$.

Once the production function is determined, we also have determined the local wage $\omega[l(\boldsymbol{r},t)]$, (9.32), as a function of the density of employees:

$$\begin{aligned} \omega\left[l(\boldsymbol{r},t)\right] = {} & \frac{\bar{A}}{2}\left[1+\exp\left(a_1\,l+a_2\,l^2\right)\right]\;\beta\,l^{\beta-1} \\ & +\frac{\bar{A}}{2}\;\exp\left(a_1\,l+a_2\,l^2\right)\;(a_1+2a_2 l)\;l^{\beta}\,. \end{aligned} \tag{9.39}$$

Hence, the derivative $\delta\omega/\delta l$, used for the effective diffusion coefficient D_{eff}, (9.11), is also determined. Figure 9.5 shows both functions dependent on the density of employees.

Figure 9.5b indicates that, within a certain range of l, the derivative $\delta\omega/\delta l$ can be positive due to cooperative effects. With respect to the discussion in Sect. 9.1.2, this means that the effective diffusion coefficient, (9.11), can possibly be negative, i.e., unemployed workers will stay in these regions.

The transition rate k^+ for hiring unemployed workers is implicitly already given by the conclusion that firms hire workers as long as the marginal revenue product exceeds the wage of the worker. In accordance with (9.33), we define

$$k^+ = k^+[l(\boldsymbol{r},t)] = \eta\,\exp\left\{\frac{\delta Y\,[l(\boldsymbol{r},t)]}{\delta l} - \omega^\star\right\}. \tag{9.40}$$

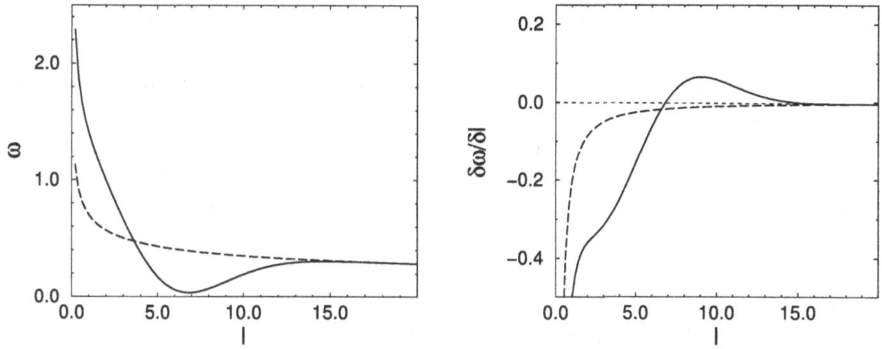

Fig. 9.5. (*left*: a) Wage ω, (9.39); (*right*: b) derivative $\delta\omega/\delta l$ vs. density of employed workers, l. The *solid lines* show the functions *with*; the *dashed lines without* cooperative effects. For parameters, see Fig. 9.4 [447]

Here, η determines the timescale of the transitions. k^+, which is now a function of the local economic situation, is significantly larger than the level of random transitions, represented by η, only if the maximum profit condition allows hiring. Otherwise, the hiring rate tends to zero.

The firing rate k^- can be simply determined opposite to k^+. Then, from the perspective of an employee, firing is caused by the local situation of the economy, i.e., by *external reasons*: workers are fired if $\delta Y/\delta l < \omega^\star$. A more refined description, however, should also consider *internal reasons*: workers cannot only *lose* their job, they can also *quit* their jobs themselves for better opportunities, e.g., because they want to move to a place where they earn higher wages.

It is reasonable to assume that the internal reasons depend again on spatial gradients in wages. Due to (9.2) the unemployed agent migrates while guided by gradients in wages. An employed agent at the same location may have the same information. Noteworthy again, this is only local information about differences in wages. If the local gradients in wages are small, internal reasons to quit the job vanish, and firing depends entirely on the (external) economic situation. However, if these differences are large, the employee might quit his job for better opportunities.

We note that the latter process was already considered in Krugman's "law of motion of the economy", Sect. 9.1.3. Different from the assumptions involved in (9.12), here the process: *employment* $\to$ *unemployment* $\to$ *migration* $\to$ *reemployment* is *explicitly* modeled. It does not occur at infinite velocity, and there is no guarantee of reemployment.

Hence, we define the "firing rate," which describes the transition from an employed to an unemployed agent, as follows:

$$k^- = k^-[l(\boldsymbol{r},t)] = \eta\, \exp\left(-\left\{\frac{\delta Y\,[l(\boldsymbol{r},t)]}{\delta l} - \omega^\star\right\} + c\,\frac{\partial\omega(\boldsymbol{r})}{\partial r}\right). \tag{9.41}$$

The additional parameter c can be used to weight the influence of spatial gradients on the employee. $c = 0$ simply means that internal reasons are neglected, and the agent is fired only for external reasons. This might also be a reasonable assumption; however, it includes a pitfall. In a growing economy, eventually almost all unemployed agents might be hired at some locations. But, as long as $dY/dl > \omega^{\star}$, employed agents are almost never fired; thus there is a shortage of free agents for further growth. Eventually, the dynamics sticks in a deadlock because agents cannot move even if there are locations in their neighborhoods that may offer them higher wages. Therefore, it is reasonable to choose $c > 0$.

Finally, in this section, we have determined the variables $f(\boldsymbol{r}, t)$, k^{+}, and k^{-} via a production function $Y[l(\boldsymbol{r}, t)]$, which represents certain economic assumptions. Now, we can turn back to the dynamic model described in Sect. 9.1.2, which can be solved now by computer simulations.

9.2.2 Simulation of Spatial Economic Agglomeration

The computer simulation of the economic model has to deal with three different processes: (i) movement of Brownian agents with $\theta_i = 1$ due to the overdamped Langevin equation, (9.2); (ii) transition of agents due to rates, k^{+}, (9.40), k^{-}, (9.41); and (iii) generation of the field $\omega(\boldsymbol{r}, t)$. In Sect. 3.1.4, we already discussed how the equations of motion have to be discretized in time. For the overdamped case considered here, we find from (9.2),

$$x_i(t + \Delta t) = x_i(t) + \left.\frac{\partial \omega}{\partial x}\right|_{x_i} \Delta t + \sqrt{2 D_n \Delta t}\,\mathrm{GRND}\,. \tag{9.42}$$

The equation for the y position reads accordingly. Here, GRND is a Gaussian random number whose mean equals zero and standard deviation equals unity.

To calculate the spatial gradient of the wage field ω, (9.39), we have to consider its dependence on $l(\boldsymbol{r}, t)$, which is a local density. The density of employed agents is calculated assuming that the surface is divided into boxes with spatial (discrete) indexes u, v and unit length Δs. Then, the local density is given by the number of agents with $\theta_i = 0$ inside a box of size $(\Delta s)^2$:

$$l(\boldsymbol{r}, t) \;\Rightarrow\; l_{uv}(t) = \frac{1}{(\Delta s)^2} \int_{x_u}^{x_u + \Delta s} \int_{y_v}^{y_v + \Delta s} dx'\, dy'\, C_0(x', y')\,. \tag{9.43}$$

We note that $\Delta s \gg \Delta x$, where $\Delta x = x(t + \Delta t) - x(t)$ is the spatial move of a migrating agent in the x direction during time step Δt, (9.42). That means that the migration process is really simulated as a motion of agents on a two-dimensional plane, rather than a hopping process between boxes.

Using the box coordinates u, v, the production function and the wage field can be rewritten as follows:

$$Y\left[l(\boldsymbol{r}, t)\right] \;\Rightarrow\; Y\left[l_{uv}(t)\right] = Y_{uv}(t)\,, \quad \omega(\boldsymbol{r}) \;\Rightarrow\; \omega_{uv} = \frac{\delta Y_{uv}(t)}{\delta l}\,. \tag{9.44}$$

The spatial gradient is then defined as

$$\frac{\partial \omega_{uv}}{\partial x} = \frac{\omega_{u+1,v} - \omega_{u-1,v}}{2\Delta s}\,, \quad \frac{\partial \omega_{uv}}{\partial y} = \frac{\omega_{u,v+1} - \omega_{u,v-1}}{2\Delta s}\,, \tag{9.45}$$

where the indexes $u-1$, $u+1$, $v-1$, $v+1$ refer to the left, right, lower, and upper boxes adjacent to box uv. We further note that *periodic boundary conditions* are used for the simulations; therefore, the neighboring box is always specified.

Using the discretized versions, (9.43)–(9.45), transition rates can be reformulated as k^{+}_{uv}, k^{-}_{uv} accordingly. They determine the *average number* of transitions of an agent in the internal state θ_i, located in box uv, during the next time step Δt. In Sect. 3.1.4, we already explained how the actual number of reactions, which is a *stochastic* variable, can be determined and how to choose the appropriate time step $\Delta t = \tau$, (3.22). These considerations will be applied here. Hence, we can conclude the procedure to simulate the movement and the transition of agents as follows:

1. Calculate the density of agents $l_{uv}(t)$, (9.43).
2. Calculate the production function $Y_{uv}(t)$ and the wage field $w_{uv}(t)$, (9.44).
3. Calculate the sum over all possible transitions, which determines the mean lifetime t_m, (3.21), of the current system state.
4. Calculate from the mean lifetime the *actual* time step $\Delta t = \tau$, (3.22) by drawing a random number.
5. Move all agents with $\theta_i = 1$ according to the discretized Langevin equation, (9.42), using the time step τ.
6. Calculate which one of the agents undergoes a transition by drawing a second random number, (3.24).
7. Update the system time: $t = t + \tau$, and continue with step 1.

In the following, we want to discuss some features of the dynamics by means of a computer with 500 Brownian agents. Initially, every agent is randomly assigned an internal parameter (either 0 or 1) and a position on the surface, which has been divided into 10×10 boxes of unit length $\Delta s = 1$. The diffusion coefficient, which describes the mobility of migrants, is set to 0.01 (in arbitrary units). So, a simple Brownian particle would need approximately a time of $t = 25$ for a mean spatial displacement of 1 (which is the spatial extension of a box). The minimum wage $\omega^{\star}$ is set to 0.1. Further, $\eta = 1$ and $c = 1$; for the remaining parameters, see Fig. 9.4.

In the following, we will restrict the discussion to the *spatiotemporal evolution* of the densities of employed and unemployed agents, shown in Figs. 9.6 and 9.7, and to the production function, shown in Fig. 9.9. Figure 9.8 presents the evolution of the total number of employed and unemployed agents in terms of the fraction $x_\theta = N_\theta/N$ with respect to (9.4). Other quantities of interest, such as the spatiotemporal wage distribution and the local values of "hiring" and "firing" rates have, of course, been calculated in the simulation.

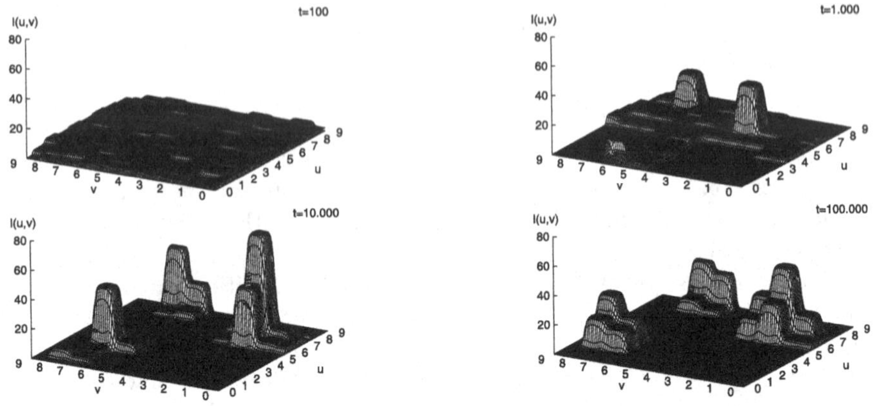

Fig. 9.6. Spatial density of employed agents, $l(u, v)$, for different times t. For parameters and initial conditions of the simulation, see text [447, 448]

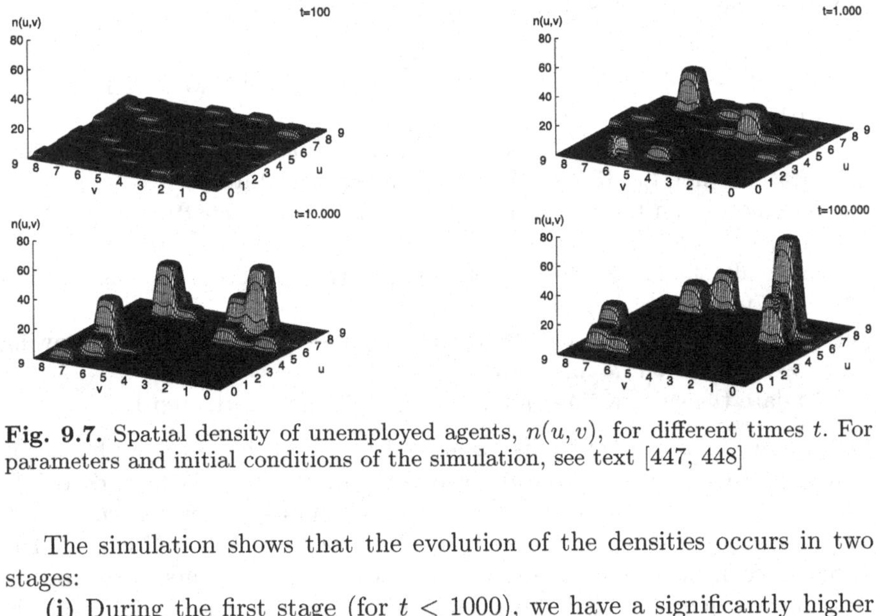

Fig. 9.7. Spatial density of unemployed agents, $n(u, v)$, for different times t. For parameters and initial conditions of the simulation, see text [447, 448]

The simulation shows that the evolution of the densities occurs in two stages:

(i) During the first stage (for $t < 1000$), we have a significantly higher fraction of employed agents (up to 70% of the agent community). They are broadly distributed in numerous small economic centers, which have an employment density of about 5. This can be understood using Figs. 9.4, 9.5. The increase in output due to cooperative effects allows fast establishment of numerous small firms, which count on mutual stimulation among their workers and at the beginning offer higher wages (compared to the model without cooperative effects).

This growth strategy, however, is not sufficient for output on larger scales because the marginal product drastically decreases (before it may increase

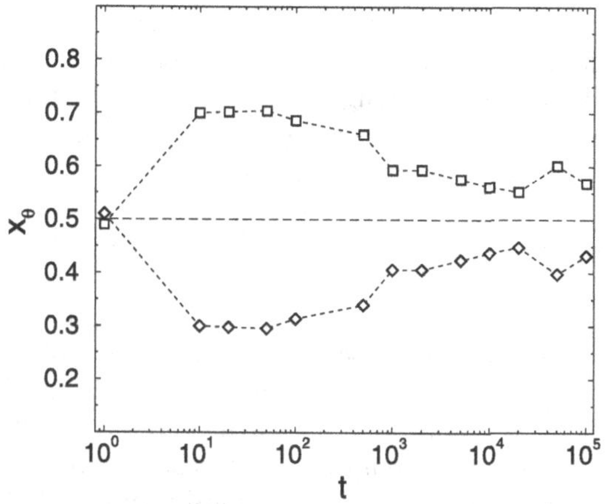

Fig. 9.8. Fraction x_θ of employed agents ($\square$) and unemployed agents ($\diamond$) vs. time t. The *marks* indicate simulation data that correspond to Figs. 9.6, 9.7 [447, 448]

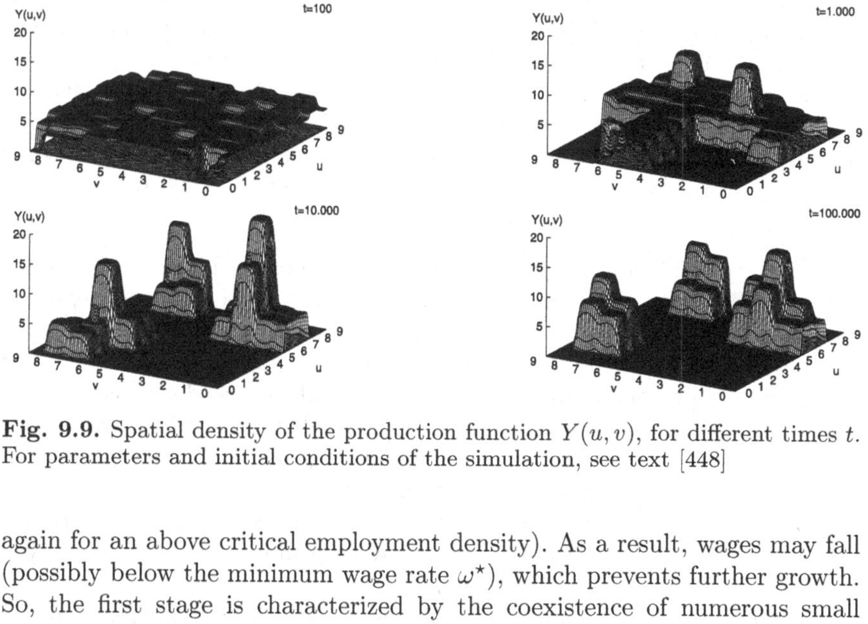

Fig. 9.9. Spatial density of the production function $Y(u, v)$, for different times t. For parameters and initial conditions of the simulation, see text [448]

again for an above critical employment density). As a result, wages may fall (possibly below the minimum wage rate $\omega^\star$), which prevents further growth. So, the first stage is characterized by the coexistence of numerous small economic centers.

(ii) During the second and much longer stage (for $t > 1000$), some of these small centers have overcome this economic bottleneck, which allows them to grow further. In the simulations, this crossover occurred between $t = 500$ and $t = 800$. In the model considered here, we have only one variable input, labor, so the crossover is mainly due to stochastic fluctuations in $l(\boldsymbol{r}, t)$.

After the bottleneck, the marginal output increases again, and also wages increase with the density of employment (see Figs. 9.4, 9.5). This in turn affects the migration of unemployed agents, which, due to (9.2), is determined by two forces: local attraction to areas with higher wages and stochastic influences that keep them moving. When $\partial\omega/\partial l > 0$, the attraction may exceed the stochastic forces, so unemployed agents are bound to these regions, once they got there. This, in turn, is important for further growth of these economic centers, which need agents to hire. As a result, we observe on the spatial level the concentration of employed *and* unemployed agents in the *same* regions.

For their further growth, the centers that overcome the economic bottleneck attract the labor force *locally* at the expense of the former small economic centers. This leads to the interesting situation that economic growth and decline occur *at the same time* in our model, but at different locations, which results in specific *spatiotemporal* patterns. As a result of the competition process, the small centers, which previously employed about 70% of the agents, disappear. In the new (larger) centers, however, only 60% of the agents can be employed (see Fig. 9.8), so the competition and the resulting large-scale production effectively also results in an *increase in unemployment.* But the employment rate of 60% seems to be a stable value for the given parameter setup because it is also kept in the long run with certain fluctuations.

As the result of our dynamic model, we find the establishment of distinct *extended major economic regions* out of a random initial distribution, as shown in Figs. 9.6, 9.7, 9.9. Each of these economic regions consists of some *subregions* (in terms of boxes). In the long run (up to $t = 100,000$), we find some remarkable features of the dynamics:

(a) *Stable coexistence* of the major economic regions, which keep a certain distance from each other. This *critical distance* – which is a *self-organized phenomenon* – prevents the regions from affecting each other. In fact, finally, there is no force between these regions which would pull away employed or unemployed agents and guide them over long distances to other regions. Thus, coexistence is really stable, even in the presence of *fluctuations.* Each of the major economic regions has established its own attraction/supply area, a result which proves – from the *dynamic* model – predictions of the *central-place theory* (see Sect. 9.1.1). Noteworthy, the centers do not necessarily have to have the same number of employed agents to coexist.

(b) A *quasi-stationary nonequilibrium* within the major economic centers. As we see, even in the long run, the local densities of employed and unemployed agents, or production, do not reach a fixed stationary value. The dynamics does not simply converge into an equilibrium state, but still follows *stochastic eigendynamics.* Within the major regions, there are still exchange processes among the participating boxes, hiring and firing, attraction, and migration. Hence, the fraction of employed agents continues to fluctuate around the mean value of 60% (cf. Fig. 9.8).

We conclude our economic model of migration and employment and the computer simulations. We have considered two types of Brownian agents: employed agents, which are immobile, but (as the result of their work) generate a wage field, and unemployed agents which migrate by responding to local gradients in the wage field. Further, a transition between employed and unemployed agents (hiring and firing) is considered.

The economic assumptions used in the model have been derived from a production function, $Y(l)$, which describes the output of production, depending on the variable input of labor. Our ansatz for $Y(l)$ specifically counts on the influence of cooperative interactions among agents on a certain scale of production. We have seen in the simulations that cooperative effects can initiate an increase in economic output, which leads to the establishment of many small firms. But the final situation, which is characterized by the stable coexistence of major economic regions, does not include any of those small firms and seems to be independent of this intermediate state. This is mainly due to the fact that we considered only *one* product. A more complex production function with *different* outputs, however, should also result in the coexistence of small (innovative) firms and large-scale production. So, we conclude that our model may serve as a experimental model for simulating the influence of different social and economic assumptions on the spatiotemporal patterns in migration and economic aggregation.

10. Spatial Opinion Structures in Social Systems

10.1 Quantitative Sociodynamics

10.1.1 Socioconfiguration

In this chapter, *Brownian agents* will represent *individuals* of a spatially distributed "community" who are also allowed to *migrate*. The different *opinions* of the agents on a specific subject may be described by the *internal parameter* θ_i. The agents' *decisions* to migrate and to keep or to change their opinions may be influenced by direct or indirect interactions with other agents, such as *persuasion*, *support*, or *imitation*. We try to capture these rather complex processes in a formal model that reflects some *basic* aspects of social dynamics, although it still can be treated by analytical methods of statistical physics. In fact, in recent years there has been a lot of interest in applications of physical paradigms for *quantitative* description of social processes [100, 174, 210, 449, 518, 536, 541, 543]. These attempts usually raise controversial discussion. From the perspective of life and social sciences, one is afraid of an unjustified reduction of complex relations in social systems to fit them into a rather "mechanical" description [206]. From the perspective of physics, on the other hand, it is claimed that the description of social processes "evidently lies outside of the realm of physics" (to quote an unknown referee).

Despite these objections, the development of the interdisciplinary field, "science of complexity," has led to the insight that complex dynamic processes may also result from simple interactions, and even *social* structure formation can be well described by a mathematical approach. This is not an artifact. Statistical mechanics is meant to comprise any phenomena where the relationship between microscopic properties and macroscopic behavior plays a role. The problem, however, is to understand carefully the reductions in system elements and their interactions, when turning to social systems. Here, we are usually confronted with individuals capable of mental reflection and purposeful actions, creating their own reality, and the question how this interferes with a rather autonomous or "self-organized" social dynamics is far from being solved.

Nevertheless, a broad range of *dynamical* methods originally developed in a *physical* context, for instance, methods of synergetics [198, 533], stochastic processes [209], deterministic chaos [100, 240, 241, 305], and lattice gas models [261, 300], have been successfully applied to describe *social phenomena.* In particular, Weidlich et al. [536, 538, 541, 543] and Helbing et al. [209, 210, 223] developed *quantitative sociodynamics* based on the framework of master equations and Boltzmann equations.

The new quest for quantitative methods in sociology is also aided by the extended capabilities of numerical computation. In addition to "conventional" computer simulations of social processes [174, 207, 354], nowadays *artificial societies* [135] and *agent-based modeling* also play important roles [26, 175, 176, 471, 495, 503, 513].

Among the recent fields of interest for a quantitative description of social processes, we mention *interregional migration* [193, 195, 447, 545], *chaotic behavior* in social systems [100, 305], and in particular the problem of *collective opinion formation* [30, 262, 296, 370, 450, 538]. The formation of public opinion is among the most challenging problems in social science because it reveals a complex dynamics that may depend on different internal and external influences. This includes, e.g., the influence of political leaders, the biasing effect of mass media, and individual features, such as persuasion and support for other opinions. Synergetic modeling concepts for sociodynamics with application to collective political opinion formation can be found in [536, 538, 541]. The transition from private attitude to public opinion is discussed within the *social impact theory* [296, 370], which can be reformulated in terms of the *lattice gas model* of statistical mechanics [261, 262, 300] (see also Sect. 10.1.2). Further, social paradoxes of majority rule voting [164, 165], consensus and attitude change in groups [167], and polarization phenomena in social systems [534, 535] have been investigated by physical methods.

The models presented in the following provide a *stochastic approach* to the problem of opinion formation. We will also consider migration of individuals as an additional possibility of action. As in previous chapters, we will discuss different levels for the dynamics description, namely, master equations for individuals and subpopulations and mean-value equations. As a first step, we need to specify our quantitative description of the social system, and this is necessarily accompanied by a certain *reduction* of the problem, which may neglect some specific sociological, psychological, and political interdependences.

Let us consider a two-dimensional spatial system of total area A, where a community of N agents (individuals of a social group) exists. To allow a spatial description, we assume that the system is divided into z boxes of equal size $A^\star = A/z$. Then N^k is the number of agents at location (that means, in a box) k, where $k = 1, 2, \dots, z$ refers to a spatial variable. Each agent is characterized by an individual parameter θ_i, representing its *opinion* (with respect to a definite aspect or problem). This should be an *internal degree of freedom*, which can have discrete values. That means that different opinions are assigned a number on the scale,

$$\boldsymbol{M} = (-m, \dots, -2, -1, 0, 1, 2, \dots, m) \;, \quad \theta_i \in \boldsymbol{M} \,. \tag{10.1}$$

Here, it is assumed that opinions can be sorted so that the extreme left spectrum of opinions is characterized by large negative values of θ, whereas the extreme right spectrum of opinions is identified by large positive numbers of θ. Opinions with small (positive or negative) values of θ indicate the indifferent middle region of opinions (with respect to a definite problem). We note that most investigations restrict themselves to the case of *two opposite opinions* [261, 262, 534].

We now have two different levels of description: (i) the (microscopic) level of *agent* i characterized by an opinion θ_i and a spatial coordinate k; and (ii) the (macroscopic) level of *subpopulations*, which can be defined either with respect to their location k or with respect to their opinion θ:

$$N^k = \sum_{\theta=-m}^{m} N_\theta^k \,, \quad N_\theta = \sum_{k=1}^{z} N_\theta^k \,. \tag{10.2}$$

Here, N_θ^k is the number of agents with opinion θ at location k. Assuming that the total number of agents is conserved,

$$N = \sum_{k=1}^{z} N^k = \sum_{\theta=-m}^{m} N_\theta = \sum_{k=1}^{z} \sum_{\theta=-m}^{m} N_\theta^k = \text{const.} \tag{10.3}$$

The spatial distribution of opinions can be described by the *socio-configuration* $\boldsymbol{N}$. This is a discrete vector, which, it has been shown, is the appropriate macrovariable to express the state of the "society" considered [536, 541, 543]:

$$\boldsymbol{N} = \left[\ldots (N_\theta^1 \, N_\theta^2, \ldots, N_\theta^z) \ldots \right] \,, \quad \theta = -m, \ldots m \,. \tag{10.4}$$

The given socioconfiguration $\boldsymbol{N}$, (10.4), can be changed on the microscopic level by two different processes: (i) individual opinion changes and (ii) location changes due to migration of agents. To describe these processes within a stochastic approach, we first derive the appropriate master equations and then specify possible transition rates.

10.1.2 Stochastic Changes and Transition Rates

Let $p_i(\theta_i, t)$ be the probability density of finding agent i with opinion θ_i. This probability changes in the course of time due to the following master equation [see (1.42)]:

$$\frac{\partial}{\partial t} p_i(\theta_i, t) = \sum_{\theta'} w(\theta_i|\theta_i') p(\theta_i', t) - p(\theta_i, t) \sum_{\theta'} w(\theta_i'|\theta_i) \,. \tag{10.5}$$

$w(\theta_i'|\theta_i)$ means the transition rate to change opinion θ_i into one of the possible θ_i' during the next time step, with $w(\theta_i|\theta_i) = 0$.

Further, let $p(N_\theta, t)$ denote the probability density of finding N_θ agents in the community that share opinion θ:

$$p(N_\theta, t) = \sum_{i=1}^{N} \delta_{\theta,\theta_i}\, p(\theta_i, t) \,. \tag{10.6}$$

The master equation for $p(N_\theta, t)$ reads explicitly

$$\begin{aligned} \frac{\partial}{\partial t} p(N_\theta, t) &= w(N_\theta|N_\theta - 1)\, p(N_\theta - 1, t) + w(N_\theta|N_\theta + 1)\, p(N_\theta + 1, t) \\ &\quad - p(N_\theta, t) \Big[w(N_\theta + 1|N_\theta) + w(N_\theta - 1|N_\theta) \Big] \,. \end{aligned} \tag{10.7}$$

If the transition depends only on opinions θ, θ' but does not vary with agents i, we may write $w(\theta'|\theta)$ instead if $w(\theta'_i|\theta_i)$.

The transition rates $w(N'_\theta|N_\theta)$ can then be assumed to be proportional to the rate $w(\theta'|\theta)$ to change a given opinion and to the number of agents that can change their opinions in the given direction:

$$\begin{aligned} w(N_\theta + 1|N_\theta) &= \sum_{\theta' \neq \theta} N_{\theta'}\, w(\theta|\theta') \,, \\ w(N_\theta - 1|N_\theta) &= N_\theta \sum_{\theta' \neq \theta} w(\theta'|\theta) \,. \end{aligned} \tag{10.8}$$

In the following section, it is assumed that the opinion change results from *indirect* interactions among agents, which are mediated either via a utility function, a social impact, or a communication field. Possible *direct* interactions among agents are discussed in Sect. 10.3.

For the transition rates of the opinion change, $w(\theta'_i|\theta_i)$, a general form has been suggested by Weidlich et al. [223, 536, 538, 543], which also assumes variations among agents i:

$$w(\theta'_i|\theta_i) = \eta \exp\left[u_i(\theta'_i) - u_i(\theta_i)\right] . \tag{10.9}$$

Here η [1/s] defines the timescale of the transitions. The function $u_i(\theta)$ represents a *dynamic utility*, which could be a very complex function, describing the "gain" or "loss," if agent i changes its opinion θ_i. The "payoff" function $\Delta u_i = u_i(\theta'_i) - u_i(\theta_i)$ is an *individual* function, which may also consider psychological measures, such as "satisfaction," "attitude," and "affirmative latent trends" [518].

Another quantitative approach to the dynamics of opinion formation is given by the concept of *social impact* [300, 370], which is based on methods similar to the cellular automata approach [564]. Social impact describes the force on an agent to keep or to change its current opinion. It is assumed that the change of opinions depends on social impact I_i and a "social temperature" T [261, 262] that is a measure of randomness in social interaction.

A possible ansatz for the transition rate that has the same form as (10.9) reads

$$w(\theta_i'|\theta_i) = \eta \exp(I_i/T)\,. \tag{10.10}$$

T represents the erratic circumstances of opinion change: in the limit $T \to 0$, opinion change is more determined by I_i, leading to the deterministic case. As (10.10) indicates, the probability of changing the opinion is rather small, if $I_i < 0$. Hence, a negative social impact on agent i represents a condition for *stability*. To be specific, the social impact may consist of three parts:

$$I_i = I_i^{\mathrm{p}} + I_i^{\mathrm{s}} + I_i^{\mathrm{ex}}\,. \tag{10.11}$$

I_i^{p} represents influences imposed on the agent by other members of the group, e.g., to change or to keep its opinion. I_i^{s}, on the other hand, is self-support for one's own opinion, $I_i^{\mathrm{s}} < 0$, and I_i^{ex} represents external influences, e.g., from government policy, or mass media, which may also support a certain opinion.

Within the simplified approach of the social impact theory, only *two opposite opinions* $\theta_i = \pm 1$ are considered. Further, every agent can be ascribed a single parameter, the "strength", s_i, and a social distance d_{ij}^n is defined, which measures the distance between each two agents (i, j) in an n-dimensional *social space* that does not necessarily coincide with the physical space. It is assumed that the impact between two agents decreases with the social distance in a nonlinear manner. The above assumptions are embodied in the following ansatz [261, 262]:

$$I_i = -\theta_i \sum_{j=1, j\neq i}^{N} s_j \theta_j / d_{ij}^n \; - \sigma s_i \; + e_i\,. \tag{10.12}$$

σ is the so-called self-support parameter. The external influence e_i may be regarded as a global preference toward one of the opinions. A negative social impact on agent i is obtained (i) if most of the opinions in its social vicinity match its own opinion; (ii) if the impact resulting from opposite opinions is at least not large enough to compensate for its self-support; or (iii) if the external influences do not force the agent to change its opinion, regardless of self-support or the impact of the community.

In the current form, the social impact model displays analogies to *lattice gas models*. The equilibrium statistical mechanics of the social impact model was formulated in [300], and in [261, 262], the occurrence of phase transitions and bistability in a social system in the presence of a strong leader or an external impact have been analyzed. Despite these extensive studies, there are several basic disadvantages in the concept. In particular, the social impact theory assumes that the impact on an agent is *instantaneously* updated if some opinions are changed in the group (which basically means communication at infinite velocity). Moreover, no memory effects are considered in the social impact; the community is affected only by the current state of the opinion distribution, regardless of its history and past experience. Spatial effects in a physical space are not considered here; any "spatial" distributions of opinions refer to the social space.

Finally, the agents are not allowed to move. The migration effect is often neglected in models dealing with collective opinion formation, although there are also stochastic models of migration based on utility functions [193, 195, 536]. However, in a spatial model, it seems realistic to consider that agents sometimes prefer to *migrate* to places where their opinion is supported rather than change their opinion. If $p(N^k,t)$ denotes the probability of finding N^k agents at location k (in box k) at time t, the change in $p(N^k,t)$ due to migration processes can be described by the master equation:

$$\begin{aligned}\frac{\partial}{\partial t}p(N^k,t) &= w(N^k|N^k-1)\,p(N^k-1,t) + w(N^k|N^k+1)\,p(N^k+1,t) \\ &\quad -p(N^k,t)\left[w(N^k+1|N^k) + w(N^k-1|N^k)\right] . \end{aligned} \tag{10.13}$$

The transition probabilities are again assumed proportional to the probability $w(l|k)$ for an agent to move from location k to location l and to the number of agents that migrate in the given direction:

$$\begin{aligned} w(N^k+1|N^k) &= \sum_{l=1,l\neq k}^{z} N_l\, w(k|l) , \\ w(N^k-1|N^k) &= N^k \sum_{l=1,l\neq k} w(l|k) . \end{aligned} \tag{10.14}$$

Equation (10.14) describes migration between adjacent boxes and also "hopping" between boxes at greater distances. To specify the transition rates $w(l|k)$, an ansatz similar to (10.9) can be chosen, where the "payoff" function $\Delta u = u(l) - u(k)$ now describes the advantages for an agent to move from location k to l, while keeping its opinion. Note that $w(l|k)$ are assumed to depend only on the location k, l and do not vary for different agents i.

To summarize the possible changes in the socioconfiguration, we define $P(\boldsymbol{N},t)$ as the probability of finding the configuration $\boldsymbol{N}$, (10.4), at time t. The time-dependent evolution of $P(\boldsymbol{N},t)$ can be described by a *multivariate master equation* [169], which reads

$$\begin{aligned}\frac{\partial P(\boldsymbol{N},t)}{\partial t} = &\sum_{\theta'=-m}^{m}\sum_{k=1}^{z} \Big[w(\theta|\theta')\,(N^k_{\theta'}+1)\,P(N^k_\theta-1, N^k_{\theta'}+1, \boldsymbol{N}^\star, t) \\ &\qquad - w(\theta'|\theta)\,N^k_\theta\,P(\boldsymbol{N},t)\Big] \\ + &\sum_{\theta=-m}^{m}\sum_{k,l=1,k\neq l}^{z} \Big[w(k|l)\,(N^l_\theta+1)\,P(N^k_\theta-1, N^l_\theta+1, \boldsymbol{N}^\star, t) \\ &\qquad - w(l|k)\,N^k_\theta\,P(\boldsymbol{N},t)\Big] . \end{aligned} \tag{10.15}$$

The first summation terms describe all possible *opinion changes* that may either lead to the assumed socio-configuration $\boldsymbol{N}$ or change it; the second summation describes all possible *migration processes*, respectively. Here, $\boldsymbol{N}^\star$ denotes those elements of vector $\boldsymbol{N}$, (10.4), that are not changed by the transition; the changed elements are explicitly written.

The multivariate master equation, (10.15), is a *discrete* description of the possible dynamics processes, such as migration or opinion change. The crucial problem is determining the *transition rates*, $w(l|k)$, $w(\theta'|\theta)$, that may depend on payoff or social impact. Here, the freedom of choice often turns out to be a pitfall, which can be avoided only by a profound sociological theory. Before we investigate the discrete dynamics in more detail in Sect. 10.3, we will first turn to a simpler description of the problem that is based on the model of Brownian agents. It has the advantage of overcoming some of the drawbacks mentioned with respect to the social impact theory, but, on the other hand, the model will consider only indirect interactions between agents and only diffusion-like migration.

10.2 Collective Opinion Formation of Brownian Agents

10.2.1 Dynamic Equations

In this book, we have extensively studied the indirect interaction of Brownian agents "communicating" via a self-consistent field created by them, which in turns influences their further behavior. In the context of collective opinion formation, this field can be interpreted as a *spatiotemporal multicomponent communication field*, $h_\theta(\boldsymbol{r},t)$ (see also Sect. 1.2.2). Every agent contributes permanently to this field with its opinion, θ_i and with its personal "strength" of information s_i at its current spatial location $\boldsymbol{r}_i$. The information generated in this way, has a certain lifetime, $1/k_\theta$; further, it can spread throughout the system in a diffusion-like process, where D_θ represents the diffusion constant for information exchange. In this section, we want to restrict our discussion to the case of only *two opposite opinions*, $\theta_i \in \{-1,+1\}$; hence the communication field should also consist of two components, each representing one opinion. For simplicity, it is assumed that the information resulting from the different opinions has the same lifetime, $k_{-1} = k_{+1} = k_h$ and the same way of spatial distribution,

$$k_{-1} = k_{+1} = k_h\,, \quad D_{-1} = D_{+1} = D_h\,. \tag{10.16}$$

More complex cases have been considered as well [470, 471]. Consequently, the spatiotemporal change of the communication field can be embodied again in the following equation [see (3.3)]:

$$\frac{\partial}{\partial t} h_\theta(\boldsymbol{r},t) = \sum_{i=1}^{N} s_i\; \delta_{\theta,\theta_i}\; \delta(\boldsymbol{r}-\boldsymbol{r}_i) \;-\; k_h h_\theta(\boldsymbol{r},t) \;+\; D_h \Delta h_\theta(\boldsymbol{r},t)\,. \tag{10.17}$$

Instead of social impact, (10.12), the communication field $h_\theta(\boldsymbol{r},t)$ influences the agent i as follows. At a certain location $\boldsymbol{r}_i$, the agent with opinion $\theta_i = +1$ is affected by two kinds of information: the information resulting from agents who share its opinion, $h_{\theta=+1}(\boldsymbol{r}_i,t)$, and the information resulting from the opponents $h_{\theta=-1}(\boldsymbol{r}_i,t)$. The diffusion constant D_h determines how fast it will receive any information, and the decay rate k_h determines how long generated information will exist. Depending on *local* information, the agent has two opportunities to act: (i) it can *change its opinion*, and (ii) it can *migrate* toward locations that provide larger support for its current opinion.

For the change of opinions, we can adopt the transition rate, (10.10), by replacing the influence of the social impact I_i with the influence of the local communication field. A possible ansatz reads

$$w(\theta_i'|\theta_i) = \eta \exp\left[\frac{h_{\theta'}(\boldsymbol{r}_i,t) - h_\theta(\boldsymbol{r}_i,t)}{T}\right], \quad w(\theta_i|\theta_i) = 0\,. \tag{10.18}$$

T that has been interpreted in terms of "social temperature" in the social impact theory [261, 262] is a parameter that represents the *erratic circumstances of decisions* based on incomplete or incorrect transmission of information. Note that T is measured in units of the communication field. In the limit $T \to 0$, the opinion change rests only on the difference $\Delta h(\boldsymbol{r}_i,t) = h_\theta(\boldsymbol{r}_i,t) - h_{\theta'}(\boldsymbol{r}_i,t)$, leading to "rational" decisions (see also [166]), i.e., decisions that are totally determined by external information. In the limit $T \to \infty$, on the other hand, the influence of the information received is attenuated, leading to "random" decisions. Again, for a fixed T, the probability of changing opinion θ_i is rather small, if the local field $h_\theta(\boldsymbol{r}_i,t)$, which is related to the support for opinion θ_i, overcomes the local influence of the opposite opinion.

The movement of an agent may depend on both erratic circumstances represented by T and on the influence of the communication field. Within a stochastic approach, this movement can be described by the following overdamped Langevin equation [cf. (3.2)]:

$$\frac{d\boldsymbol{r}_i}{dt} = \frac{\alpha_i}{\gamma_0}\left.\frac{\partial h_e(\boldsymbol{r},t)}{\partial \boldsymbol{r}}\right|_{\boldsymbol{r}_i} + \sqrt{2\,D_n}\,\xi(t)\,. \tag{10.19}$$

D_n is the spatial diffusion coefficient of the agents. $h^e(\boldsymbol{r},t)$ means an *effective* communication field (see Sect. 3.1.1) which results from $h_\theta(\boldsymbol{r},t)$, as specified below. The Langevin equation considers the response of an agent to the *gradient* of the field $h^e(\boldsymbol{r},t)$, where α_i is the individual response parameter weighting the importance of the information received. We mention that α_i may consider different types of responses as discussed in Sect. 3.1.1, for instance:

1. Agents try to move toward locations that provide the most support for their current opinion θ_i. In this case, they only count on the information which matches their opinion, $h^e(\boldsymbol{r},t) \rightarrow h_\theta(\boldsymbol{r},t)$ and follow the local ascent of the field ($\alpha_i > 0$).
2. Agents try to move away from locations that provide any negative pressure on their current opinion θ_i. In this case, they count on the information resulting from opposite opinions (θ'): $h^e(\boldsymbol{r},t) \rightarrow h_{\theta'}(\boldsymbol{r},t)$ and follow the local descent of the field ($\alpha_i < 0$).
3. Agents try to move away from locations, if they are forced to change their current opinion θ_i, but they can accept a vicinity of opposite opinions, as long as these are not dominating. In this case, they count on the information resulting from both supporting and opposite opinions, and the local difference between them is important: $h^e(\boldsymbol{r},t) \rightarrow [h_\theta(\boldsymbol{r},t) - h_{\theta'}(\boldsymbol{r},t)]$ with $\alpha_i > 0$.

For further discussion, we assume $h^e(\boldsymbol{r},t) \rightarrow h_\theta(\boldsymbol{r},t)$ for the effective communication field (case 1), and

$$\frac{\alpha_i}{\gamma_0} := \alpha > 0 \tag{10.20}$$

is treated as a constant independent of agent i that describes a *positive* response to the field. The decision process, however, shall depend on both components, $h_{-1}(\boldsymbol{r},t)$, $h_{+1}(\boldsymbol{r},t)$, as described by the transtion rates, (10.18). The nonlinear feedback between agents and the two-component communication field is summarized in Fig. 10.1.

In terms of synergetics, the communication field plays the role of an order parameter, which couples the individual actions, and in this way originates spatial structures and coherent behavior within a social group. We conclude that, compared to the social impact theory, the suggested approach has several advantages:

1. Consideration of the *spatial* distribution of the individual opinions in terms of physical space, which means that *spatial distances* between agents are also considered.

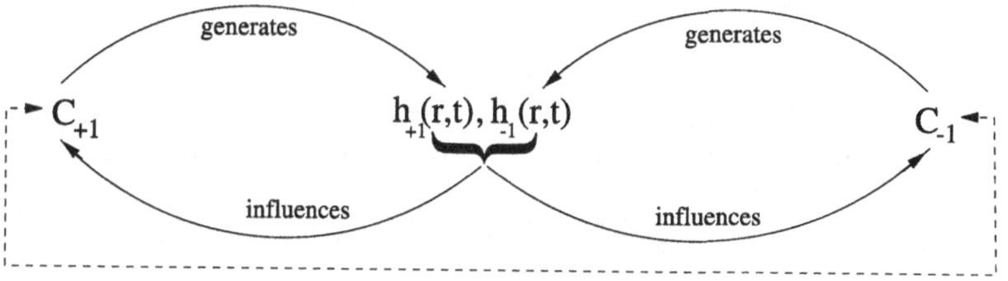

Fig. 10.1. Circular causation between agents with different opinions, C_{-1}, C_{+1} and the two-component communication field, $h_\theta(\boldsymbol{r},t)$.

2. Consideration of collective *memory effects*, which reflect the past experience, in terms of a finite lifetime of the information generated.
3. *Exchange of information* in the community at a *finite* velocity, modeled as a diffusion process. In most cases, agents are not instantaneously affected by the opinions of others, especially if they are not in the vicinity.
4. Possibility of *spatial migration* for agents. It seems more realistic to us that agents have the chance to migrate to places where their opinion is supported rather than change their opinion.

The complete dynamics of the community can be formulated again in terms of the canonical N-particle distribution function $P(\underline{\theta}, \underline{r}, t)$, (3.5), which gives the probability density of finding the N agents with opinions $\theta_1, ..., \theta_N$ at positions $\boldsymbol{r}_1, ..., \boldsymbol{r}_N$ on surface A at time t. This probability density can change in the course of time from both *opinion changes* and *migration* of agents. In Sect. 3.1.3, we derived two types of master equations, (3.6), (3.7), to describe the dynamics of $P(\underline{\theta}, \underline{r}, t)$. For the investigations in Sect. 10.2.4, we will consider (3.7), which holds for a continuous space and an overdamped motion of agents, (10.19). But first, we want to discuss a special case, which allows us to reduce the description to the level of subpopulations.

10.2.2 Subpopulation Sizes in a System with Fast Communication

For further discussion of the model, let us restrict our discussion to the case of very fast exchange of information in the system. Then spatial inhomogenities are equalized immediately; hence, the communication field $h_\theta(\boldsymbol{r}, t)$ can be approximated by a mean field $\bar{h}_\theta(t)$:

$$\bar{h}_\theta(t) = \frac{1}{A} \int\limits_A h_\theta(\boldsymbol{r}, t)\, dr^2 \,, \tag{10.21}$$

where A means system size. The equation for the mean field $\bar{h}_\theta(t)$ results from (10.17):

$$\frac{\partial \bar{h}_\theta(t)}{\partial t} = -k_h \bar{h}_\theta(t) + s\bar{n}_\theta \,, \tag{10.22}$$

where $s_i \equiv s$ and the mean density

$$\bar{n}_\theta = \frac{N_\theta}{A} \,, \quad \bar{n} = \frac{N}{A} \,, \tag{10.23}$$

where the number of agents with a given opinion θ fulfils the condition,

$$\sum_\theta N_\theta = N_{+1} + N_{-1} = N = \text{const.} \tag{10.24}$$

We note that in the mean–field approximation, no spatial gradients exist in the communication field. Hence, there is no additional driving force for agents to move, as assumed in (10.19). Such a situation can be imagined for communities existing in very small systems with small distances between different groups. In particular, in such small communities, the assumption of fast information exchange also holds. Thus, in this section, we restrict our discussion to *subpopulations* with a certain opinion rather than to agents at particular locations.

Due to (10.6), the probability density of finding N_θ agents in the community which shares opinion θ, is denoted by $p(N_\theta, t)$, which changes in the course of time due to the master equation, (10.7). The corresponding transition rates, (10.8), now read explicitly,

$$\begin{aligned} w(N_{+1}+1|N_{+1}) &= N_{-1}\,\eta\,\exp\left[(\bar{h}_{+1}-\bar{h}_{-1})/T\right], \\ w(N_{+1}-1|N_{+1}) &= N_{+1}\,\eta\,\exp\left[-(\bar{h}_{+1}-\bar{h}_{-1})/T\right]. \end{aligned} \tag{10.25}$$

The mean values for the number of agents with a certain opinion can be derived from the master equation, (10.7):

$$\langle N_\theta(t)\rangle = \sum_{\{N_\theta\}} N_\theta(t)\, p(N_\theta, t)\,, \tag{10.26}$$

where the summation is over all possible numbers of N_θ that obey condition (10.24). From (10.26), the deterministic equation for the change in N_θ can be derived in a first approximation, as follows [459]:

$$\frac{d}{dt}\langle N_\theta\rangle = \big\langle w(N_\theta+1|N_\theta) - w(N_\theta-1|N_\theta)\big\rangle\,. \tag{10.27}$$

For N_{+1}, this equation reads explicitly,

$$\begin{aligned} \frac{d}{dt}\langle N_{+1}\rangle = \Bigg\langle & N_{-1}\,\eta\,\exp\left[\frac{\bar{h}_{+1}(t)-\bar{h}_{-1}(t)}{T}\right] \\ & -N_{+1}\,\eta\,\exp\left[-\frac{\bar{h}_{+1}(t)-\bar{h}_{-1}(t)}{T}\right]\Bigg\rangle\,. \end{aligned} \tag{10.28}$$

Introducing now the fraction of a *subpopulation* with opinion θ: $x_\theta = \langle N_\theta\rangle/N$ and using the standard approximation to factorize (10.28), we can write it as

$$\begin{aligned} \dot{x}_{+1} &= (1-x_{+1})\,\eta\,\exp(a) - x_{+1}\,\eta\,\exp(-a)\,, \\ a &= \left[\bar{h}_{+1}(t)-\bar{h}_{-1}(t)\right]/T\,. \end{aligned} \tag{10.29}$$

Via $\Delta\bar{h}(t) = \bar{h}_{+1} - \bar{h}_{-1}$, this equation is coupled with the equation,

$$\Delta\dot{\bar{h}} = -k_h\,\Delta\bar{h} + s\,\bar{n}\,(2x_{+1}-1)\,, \tag{10.30}$$

which results from (10.22) for the two field components.

Within a quasi-stationary approximation, we can assume that the communication field *relaxes faster* than the distribution of opinions into a stationary state. Hence, with $\dot{\bar{h}}_\theta = 0$, we find from (10.22),

$$\bar{h}_{+1}^{\text{stat}} = \frac{s\,\bar{n}}{k_h} x_{+1}\,, \quad \bar{h}_{-1}^{\text{stat}} = \frac{s\,\bar{n}}{k_h}(1 - x_{+1})\,, \qquad a = \varrho\left(x_{+1} - \frac{1}{2}\right) \quad \text{with}\ \varrho = \frac{2s\,\bar{n}}{k_h T}\,. \tag{10.31}$$

Here, the parameter ϱ includes the specific *internal conditions* within the community, such as the total population size, the "social temperature," the individual strength of the opinions, and the lifetime of the information generated. Inserting a from (10.31) in (10.29), a closed equation for $\dot{x}_\theta$ is obtained, which can be integrated with respect to time (see Fig. 10.2a). We find that, depending on ϱ, different stationary values exist for the fraction of the subpopulations. For the critical value, $\varrho^c = 2$, the stationary state can be reached only asymptotically. Figure 10.2(b) shows the stationary solutions, $\dot{x}_\theta = 0$, resulting from the equation for x_{+1}:

$$(1 - x_{+1})\,\exp\left(\varrho\, x_{+1}\right) = x_{+1}\,\exp\left[\varrho\,(1 - x_{+1})\right]\,. \tag{10.32}$$

For $\varrho < 2$, $x_{+1} = 0.5$ is the only stationary solution, which means a stable community where opposite opinions have the same influence. However, for $\varrho > 2$, the equal distribution of opinions becomes unstable, and a separation process toward a preferred opinion is obtained, where $x_{\pm 1} = 0.5$ plays the role of a separatrix. Now we find two stable solutions where both opinions coexist at different fractions in the community, as shown in Fig. 10.2a,b. Hence, the subpopulation, e.g., with opinion +1 can exist either as a *majority* or as a *minority* within the community. Which of these two possible situations is realized depends, in a deterministic approach, on the initial fraction of the

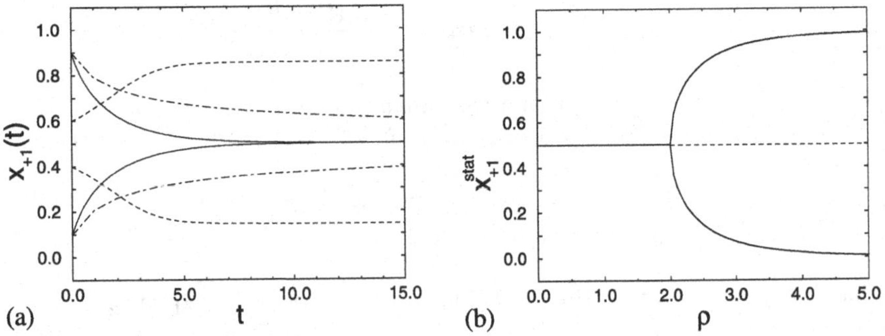

Fig. 10.2. (**a**) Time dependence of the fraction $x_{+1}(t)$ (10.29) of the subpopulation with opinion +1 for three different values of ϱ: 1.0 (*solid line*); 2.0 (*dot-dashed line*), 3.0 (*dashed line*). (**b**) Stationary solutions for x_{+1} (10.32) for different values of ϱ. The bifurcation at the critical value $\varrho^c = 2$ is clearly indicated [455]

subpopulation. For initial values of x_{+1} below the separatrix 0.5, the minority status will most likely be the stable situation, as Fig. 10.2a shows.

The bifurcation occurs at $\varrho^c = 2$, where the former stable solution $x_{+1} = 0.5$ becomes unstable. From the condition $\varrho = 2$, we can derive a *critical population size* [455],

$$N^{c} = k_h A T / s \,, \tag{10.33}$$

where for larger populations, an equal fraction of opposite opinions is certainly unstable. If we consider, e.g., a *growing community* with fast communication, then contradicting opinions are balanced, as long as the population number is small. However, for $N > N^c$, i.e., after a certain population growth, the community tends toward one of these opinions, thus *necessarily separating* into a *majority* and *minority*. Which of these opinions would be dominating, depends on small fluctuations in the bifurcation point and has to be investigated within a stochastic approach. We note that (10.33) for the critical population size can also be interpreted in terms of a critical social temperature, which leads to an opinion separation in the community. This will be discussed in more detail in Sect. 10.2.4.

From Fig. 10.2b, we see further that the stable coexistence between a majority and minority breaks down at a certain value of ϱ, where almost the whole community shares the same opinion. From (10.32), it is easy to find that, e.g., $\varrho \approx 4.7$ yields $x_{+1} = \{0.01; 0.99\}$, which means that about 99% of the community shares either opinion $+1$ or -1.

10.2.3 Influence of External Support

Now, we discuss the situation when the symmetry between the two opinions is broken due to external influences on agents. We may consider two similar cases: (i) the existence of a *strong leader* in the community, who supports one of the opinions with a strength s_l which is much larger than the usual strength s of other agents; (ii) the existence of an external field, which may result from government policy, mass media, etc. that support a certain opinion with a strength s_{m}. The additional influence $s_{\mathrm{ext}} := (s_l/A, s_{\mathrm{m}}/A)$ mainly affects the communication field, (10.17), due to an extra contribution, normalized by the system size A. If we assume an external support of opinion $\theta = +1$, the corresponding field equation in the mean–field limit (10.30) and the stationary solution (10.31) are changed as follows:

$$\begin{aligned} \dot{\bar{h}}_{+1} &= -k_h \bar{h}_{+1}(t) + s\,\bar{n} x_{+1} + s_{\mathrm{ext}} \,, \\ \bar{h}_{+1}^{\mathrm{stat}} &= \frac{s\,\bar{n}}{k_h} x_{+1} + \frac{s_{\mathrm{ext}}}{k_h} \,, \\ a &= \varrho\left(x_{+1} - \frac{1}{2}\right) + \frac{s_{\mathrm{ext}}}{k_h T} \,. \end{aligned} \tag{10.34}$$

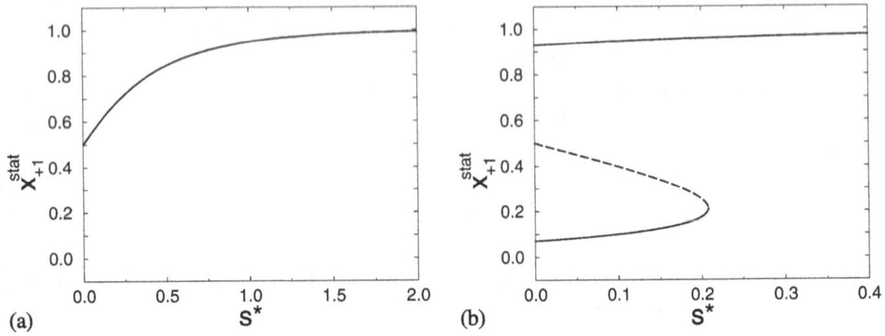

Fig. 10.3. Stable fraction of the subpopulation, $x_{+1}^{\rm stat}$, depending on the strength, $s^\star = s_{\rm ext}/k_hT$ of the external support. (**a**) $\varrho = 1$; (**b**) $\varrho = 3$. The *dashed line* in (**b**) represents the separation line for the initial conditions, which lead either to a minority or a majority status of the subpopulation [455]

Hence, in (10.32), which determines the stationary solutions, the arguments are shifted by a certain value,

$$(1 - x_{+1}) \exp\left(\varrho\, x_{+1} + \frac{s_{\rm ext}}{k_h T}\right) = x_{+1} \exp\left[\varrho\,(1 - x_{+1}) - \frac{s_{\rm ext}}{k_h T}\right]. \qquad (10.35)$$

Figure 10.3a,b shows how the critical and stable subpopulation sizes change for subcritical and supercritical values of ϱ, depending on the strength of the external support.

For $\varrho < \varrho^{\rm c}$ (see Fig. 10.3a), we see that there is still only one stable solution, but with an increasing value of $s_{\rm ext}$, the supported subpopulation exists as a majority. For $\varrho > \varrho^{\rm c}$ (see Fig. 10.3b), we observe again two possible stable situations for the supported subpopulation, either a minority or a majority status. But, compared to Fig. 10.2b, the symmetry between these possibilities is now broken due to the external support, which increases the region of initial conditions and leads to majority status.

Interestingly, at a critical value of $s_{\rm ext}$, the possibility of minority status completely vanishes. Hence, for a certain supercritical external support, the supported subpopulation will grow toward a majority, regardless of their initial population size, with no chance for the opposite opinion to be considered. This situation is quite often realized in communities with one strong political or religious leader ("fundamentalist dictatorships"), or in communities driven by external forces, such as financial or military power ("banana republics").

The value of the critical external support, $s_{\rm ext}^{\rm c}$, of course, depends on ϱ, which summarizes the internal situation in terms of the social temperature or the population size. From (10.35), we can derive a condition, where two of the three possible solutions coincide, thus determining the relation between $s_{\rm ext}^{\rm c}$ and ϱ as follows:

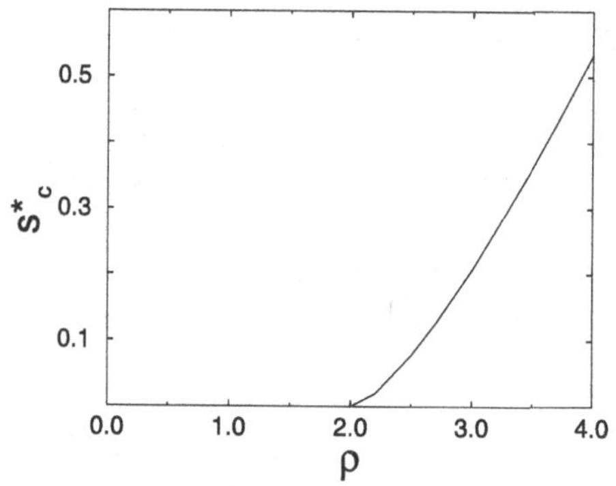

Fig. 10.4. Critical external support $s_c^\star$ (10.36) dependent on ϱ [455]

$$s_c^\star = \frac{s_{\text{ext}}^c}{k_h T} = \frac{1}{2}\ln\left[\frac{1-\sqrt{1-\frac{2}{\varrho}}}{1+\sqrt{1-\frac{2}{\varrho}}}\right] + \frac{1}{2}\varrho\sqrt{1-\frac{2}{\varrho}}\,. \tag{10.36}$$

Figure 10.4 shows how much external support is needed to paralyze a community with a given internal situation (ϱ) by one ruling opinion. As one can see, critical external support is an increasing function of parameter ϱ, which means that it is more difficult to paralyze a society with strong interpersonal interactions.

10.2.4 Critical Conditions for Spatial Opinion Separation

In the previous section, the existence of critical parameters, such as ϱ^c or s_{ext}^c, has been proven for a community with fast communication, where no inhomogeneities in the communication field can exist. In the more realistic case, however, we have finite diffusion coefficients for the information, and the mean field approximation, (10.22), is no longer valid. Instead of focusing on subpopulation sizes, we now need to consider the *spatial distribution* of agents with opposite opinions.

As mentioned in Sect. 10.2.1, the complete dynamics of the community with respect to both *opinion changes* and *migration* of agents can be described by the change in the probability density $P(\theta_1, \boldsymbol{r}_1, ..., \theta_N, \boldsymbol{r}_N, t)$. However, instead of investigating the stochastic dynamics, in this section we are interested in the critical conditions for spatial opinion separation, which can be best derived from density equations. In Sect. 3.1.3, we showed that the spatiotemporal density $n_\theta(\boldsymbol{r}, t)$ of agents with opinion θ can be obtained from $P(\underline{\theta}, \underline{r}, t)$ by means of (3.9). Considering the master equation, (3.7), for the dynamics of $P(\underline{\theta}, \underline{r}, t)$, we find the following equation for $n_\theta(\boldsymbol{r}, t)$ [see (3.11)]:

$$\begin{aligned}\frac{\partial}{\partial t} n_\theta(\boldsymbol{r}, t) = &- \boldsymbol{\nabla}\, [n_\theta(\boldsymbol{r}, t)\, \alpha\, \boldsymbol{\nabla} h_\theta(\boldsymbol{r}, t)] + D_n\, \Delta n_\theta(\boldsymbol{r}, t) \\ &- w(\theta'|\theta)\, n_\theta(\boldsymbol{r}, t) + w(\theta|\theta')\, n_{\theta'}(\boldsymbol{r}, t)\,, \end{aligned} \tag{10.37}$$

with transition rates obtained from (10.18):

$$w(\theta'|\theta) = \eta \exp\{[h_{\theta'}(\boldsymbol{r},t) - h_\theta(\boldsymbol{r},t)]/T\}. \tag{10.38}$$

With $\theta \in \{-1,+1\}$, (10.37) is a set of two reaction–diffusion equations coupled both via $n_\theta(\boldsymbol{r},t)$ and $h_\theta(\boldsymbol{r},t)$, (10.17). If we neglect any external support, assume again $s_i \equiv s$, and further insert the densities $n_\theta(\boldsymbol{r},t)$, (10.17) for the spatial communication field can be transformed into the linear deterministic equation,

$$\frac{\partial}{\partial t} h_\theta(\boldsymbol{r},t) = s\, n_\theta(\boldsymbol{r},t) \;-\; k_h h_\theta(\boldsymbol{r},t) \;+\; D_h \Delta h_\theta(\boldsymbol{r},t). \tag{10.39}$$

Solutions for spatiotemporal distributions of agents and opinions are now determined by four coupled equations, (10.37) and (10.39). For further discussions, we assume again that the spatiotemporal communication field *relaxes faster* than the related distribution of agents into a quasistationary equilibrium. The field $h_\theta(\boldsymbol{r},t)$ should still depend on time and space coordinates, but, due to the fast relaxation, there is a fixed relation to the spatiotemporal distribution of agents. Further, we neglect the independent diffusion of information, assuming that the spreading of opinions is due to the migration of agents. From (10.39), we find with $\dot{h}_\theta(\boldsymbol{r},t) = 0$ and $D_h = 0$,

$$h_\theta(\boldsymbol{r},t) = \frac{s}{k_h}\, n_\theta(\boldsymbol{r},t), \tag{10.40}$$

which can now be inserted into (10.37), thus reducing the set of coupled equations to two equations.

The homogeneous solution for $n_\theta(\boldsymbol{r},t)$ is given by the mean densities:

$$\bar{n}_\theta = \langle n_\theta(\boldsymbol{r},t)\rangle = \frac{\bar{n}}{2}. \tag{10.41}$$

Under certain conditions, however, the homogeneous state becomes unstable, and a spatial separation of opinions occurs. To investigate these critical conditions, we allow small fluctuations around the homogeneous state $\bar{n}_\theta$:

$$n_\theta(\boldsymbol{r},t) = \bar{n}_\theta + \delta n_\theta\,; \qquad \left|\frac{\delta n_\theta}{\bar{n}_\theta}\right| \ll 1. \tag{10.42}$$

Inserting (10.42) into (10.37), linearization gives

$$\frac{\partial \delta n_\theta}{\partial t} = \left(D_n - \frac{\alpha s \bar{n}}{2k_h}\right) \Delta \delta n_\theta + \left(\frac{\eta s \bar{n}}{k_h T} - \eta\right) (\delta n_\theta - \delta n_{-\theta}). \tag{10.43}$$

With the ansatz

$$\delta n_\theta \sim \exp\left(\lambda t + i\boldsymbol{\kappa}\boldsymbol{r}\right), \tag{10.44}$$

we find from (10.43) the dispersion relation $\lambda(\boldsymbol{\kappa})$ for small inhomogeneous fluctuations with wave vector $\boldsymbol{\kappa}$. This relation yields two solutions:

$$\begin{aligned} \lambda_1(\boldsymbol{\kappa}) &= -\kappa^2 C + 2B\,, \quad \lambda_2(\boldsymbol{\kappa}) = -\kappa^2 C\,, \\ B &= \frac{\eta\, s\, \bar{n}}{k_h T} - \eta\,, \quad C = D_n - \frac{\alpha s\, \bar{n}}{2k_h}\,. \end{aligned} \tag{10.45}$$

For homogeneous fluctuations, we obtain from (10.45),

$$\lambda_1 = \frac{2\eta\, s\, \bar{n}}{k_h T} - 2\eta\,, \quad \lambda_2 = 0\,, \quad \text{for } \boldsymbol{\kappa} = 0\,, \tag{10.46}$$

which means that the homogeneous system is marginally stable, as long as $\lambda_1 < 0$, or $s\,\bar{n}/k_h T < 1$. This result agrees with the condition $\varrho < 2$ obtained from the previous mean-field investigations in Sect. 10.2.2. The condition $\varrho = 2$ or $B = 0$, respectively, defines a *critical social temperature* [455]

$$T_1^{\rm c} = \frac{s\,\bar{n}}{k_h}\,. \tag{10.47}$$

For temperatures $T < T_1^{\rm c}$, the homogeneous state $n_\theta(\boldsymbol{r}, t) = \bar{n}/2$, where agents of both opinions are equally distributed, becomes unstable, and the spatial separation process occurs. This is analogous, e.g., to the *phase transition* in the Ising model of a ferromagnet. Here, the state with $\varrho < 2$ or $T > T_1^{\rm c}$, respectively, corresponds to the *paramagnetic* or disordered phase, and the state with $\varrho > 2$ or $T < T_1^{\rm c}$, respectively, corresponds to the *ferromagnetic* ordered phase.

The conditions of (10.46) denote a stability condition for the homogeneous case. To obtain stability to inhomogeneous fluctuations of wave vector $\boldsymbol{\kappa}$, the two conditions $\lambda_1(\boldsymbol{\kappa}) \leq 0$ and $\lambda_2(\boldsymbol{\kappa}) \leq 0$ have to be satisfied. Taking into account the critical temperature $T_1^{\rm c}$, (10.47), we can rewrite these conditions, (10.45), as follows:

$$\begin{aligned} \kappa^2\,(D_n - D_n^{\rm c}) - 2\,\eta \left(\frac{T_1^{\rm c}}{T} - 1\right) &\geq 0\,, \\ \kappa^2\,(D_n - D_n^{\rm c}) &\geq 0\,. \end{aligned} \tag{10.48}$$

Here, a *critical diffusion coefficient* $D_n^{\rm c}$ for agents appears, which results from the condition $C = 0$:

$$D_n^{\rm c} = \frac{\alpha}{2}\,\frac{s\,\bar{n}}{k_h}\,. \tag{10.49}$$

Hence, the condition

$$D_n > D_n^{\rm c} \tag{10.50}$$

denotes a second stability condition. To explain its meaning, let us consider that the diffusion coefficient of agents, D_n, may be a function of the social temperature T. This sounds reasonable because the social temperature is a measure of randomness in social interaction, and an increase in such a randomness leads to an increase in random spatial migration. The simplest relation for function $D_n(T)$ is the linear one, $D_n = \mu T$. By assuming this, we may rewrite (10.48) using a *second critical temperature* T_2^c, instead of the critical diffusion coefficient D_n^c:

$$\begin{aligned} \boldsymbol{\kappa}^2 \, \mu \, (T - T_2^c) - 2\eta \left(\frac{T_1^c}{T} - 1 \right) &\geq 0 \, , \\ \boldsymbol{\kappa}^2 \, \mu \, (T - T_2^c) &\geq 0 \, . \end{aligned} \tag{10.51}$$

The second critical temperature T_2^c reads as follows [455]:

$$T_2^c = \frac{\alpha}{2\mu} \frac{s \, \bar{n}}{k_h} = \frac{\alpha}{2\mu} \, T_1^c \, . \tag{10.52}$$

The occurrence of two critical social temperatures T_1^c, T_2^c allows a more detailed discussion of the stability conditions. Therefore, we have to consider two separate cases of (10.52): (1) $T_1^c > T_2^c$ and (2) $T_1^c < T_2^c$, which correspond either to the condition $\alpha < 2\mu$, or $\alpha > 2\mu$, respectively.

In the first case, $T_1^c > T_2^c$, we can discuss three ranges of temperature T:

1. For $T > T_1^c$, both eigenvalues $\lambda_1(\boldsymbol{\kappa})$ and $\lambda_2(\boldsymbol{\kappa})$, (4.28), are nonpositive for all wave vectors $\boldsymbol{\kappa}$, and the homogenous solution $\bar{n}/2$ is *completely stable.*
2. For $T_1^c > T > T_2^c$, the eigenvalue $\lambda_2(\boldsymbol{\kappa})$ is still nonpositive for all values of $\boldsymbol{\kappa}$, but the eigenvalue $\lambda_1(\boldsymbol{\kappa})$ is negative only for wave vectors that are larger than some critical value $\boldsymbol{\kappa}^2 > \boldsymbol{\kappa}_c^2$:

$$\boldsymbol{\kappa}_c^2 = \frac{2\eta}{\mu T} \frac{T_1^c - T}{T - T_2^c} \, . \tag{10.53}$$

   This means that, in the given range of temperatures, the homogeneous solution $\bar{n}/2$ is *metastable* in an infinite system because it is stable only to fluctuations with large wave numbers, i.e., to small-scale fluctuations. Large-scale fluctuations destroy the homogeneous state and result in a spatial separation process, i.e., instead of a homogenous distribution of opinions, individuals with the same opinion form separated *spatial domains* that coexist. The range of the metastable region is especially determined by the value of $\alpha < 2\mu$, which defines the difference between T_1^c and T_2^c.
3. For $T < T_2^c$, both eigenvalues $\lambda_1(\boldsymbol{\kappa})$ and $\lambda_2(\boldsymbol{\kappa})$ are positive for all wave vectors $\boldsymbol{\kappa}$ (except $\boldsymbol{\kappa} = 0$, for which $\lambda_2 = 0$ holds), which means that the homogeneous solution $\bar{n}/2$ is *completely unstable.* On the other hand, all systems with spatial dimension $L < 2\pi/\kappa_c$ are stable in this temperature region.

For case (2), $T_1^c < T_2^c$, which corresponds to $\alpha > 2\mu$, small inhomogeneous fluctuations already result in instability of the homogenous state for $T < T_2^c$, i.e., we have a direct transition from a completely stable to an completely unstable regime at the critical temperature $T = T_2^c$.

That means that the second critical temperature T_2^c marks the transition into complete instability. The metastable region, which exists for $\alpha < 2\mu$, is bounded by the two critical social temperatures, T_1^c and T_2^c. This allows us again to draw an analogy to the theory of phase transitions [516]. It is well known from phase diagrams (see Fig. 3.7) that the density-dependent *coexistence* curve $T_1^c(\bar{n})$ divides stable and metastable regions; therefore, we can name the critical temperature T_1^c, (10.47), as the *coexistence* temperature, which marks the transition into the metastable regime. On the other hand, the metastable region is separated from the completely unstable region by a second curve $T_2^c(\bar{n})$, known as the spinodal curve, which defines the region of *spinodal decomposition*. Hence, we can identify the second critical temperature T_2^c, (10.52), as the *instability* temperature.

We note that similar investigations of critical system behavior can be performed by discussing the dependence of stability conditions on the "social strength" s or on the total population number $N = A\bar{n}$. These investigations allow calculating a phase diagram for the opinion change in the model discussed, where we can derive critical *population densities* for the spatial opinion separation within a community.

10.3 Spatial Opinion Patterns in a Model of Direct Interactions

10.3.1 Transition Rates and Mean Value Equations

In the previous sections, we have assumed that there is only an *indirect* interaction between agents, mediated by the communication field $h_\theta(\boldsymbol{r}, t)$. In social systems, however, the opinion change of agents may also result from *direct local interactions* between two agents, which involves, e.g., *persuasion* or *supportiveness*.

To include these effects, we turn back to the discrete description of the socioconfiguration $\boldsymbol{N}$, (10.4), already described in Sect. 10.1. Further, we consider now a spectrum of different opinions, given by vector $\boldsymbol{M}$, (10.1). If C_θ^k denotes an agent with opinion θ located in box k, then we have to consider now the following possible transitions that lead to changes in configuration $\boldsymbol{N}$:

$$C_\theta^k \stackrel{w(\theta'|\theta)}{\longrightarrow} C_{\theta'}^k\,, \quad C_\theta^k + C_{\theta'}^k \stackrel{v(\theta'|\theta)}{\longrightarrow} 2C_{\theta'}^k\,, \quad C_\theta^k \stackrel{w(l|k)}{\longrightarrow} C_\theta^l\,. \tag{10.54}$$

The first symbolic reaction again describes opinion changes that result from indirect interactions or occur *spontaneously*. The second reaction describes opinion changes that result from *direct interactions* between two agents. Here, it is assumed that one agent has convinced or persuaded the other to adopt the opinion of the persuader. The third reaction, finally, describes the *migration* of an agent to another location, while keeping its current opinion.

For the individual transition rates $w(\theta'|\theta)$ and $w(l|k)$, an ansatz similar to those discussed in the previous sections could be chosen. However, here we use some different assumptions [450], which, on one hand, consider the "distance" between possible opinions, but, on the other hand, try to simplify the global couplings resulting from internal and external influences. This means, in particular, that instead of a multicomponent field $h_\theta(\boldsymbol{r},t)$, the effect of existing opinions is now simply summarized in a function Q which characterizes the *internal* social "climate":

$$Q = \frac{1}{z}\sum_{k=1}^{z} Q^k\,, \quad Q^k = \frac{1}{N^k}\sum_{\theta=-m}^{m} \tau_\theta\,\theta\,N_\theta^k\,. \qquad (10.55)$$

Q^k considers the opinions of all agents in box k with the weight τ_θ that ensures that, e.g., minorities also can have a certain influence on the local "climate." The value Q, which does not necessarily have to be an integer, describes a commonly accepted opinion. The external support of a certain opinion, on the other hand, is now simply expressed by a value E, which has one of the possible values of $\boldsymbol{M}$.

The first reaction in (10.54) considers that an agent can change its opinion *spontaneously*, that means without reasons that can be determined from outside. The change of opinion from θ to θ', however, depends on the difference between the opinions and on their distance to the "global" preferences, expressed in terms of Q and E. Thus, we choose the following ansatz:

$$\begin{aligned} w(\theta'|\theta) = \eta\,\exp\big[&-\alpha\,|\theta-\theta'| - \beta\,(|\theta'-E| - |\theta-E|) \\ &- \gamma\,(|\theta'-Q| - |\theta-Q|)\big]\,. \end{aligned} \qquad (10.56)$$

Equation (10.56) means that the rate of opinion change decreases with the difference between the old and the new opinion because it is unlikely that drastic opinion changes on a given subject occur spontaneously. Further, the rate decreases if the new opinion has a larger difference from the global preferences, i.e., the supported opinion E or the common opinion Q.

The influence of the three terms in (10.56) are weighted by small constants α, β, γ, which are normally positive values. But for agents with extreme opinions, β or γ could also be negative, which means that these agents always prefer an opinion opposite to the commonly accepted opinion Q, or to the external support E, for instance.

The direct interaction between agents that are at the same location k is described by agent transition rates $v(\theta'|\theta)$. In (10.54), only interactions between *two* agents at the same time are considered; further, it is assumed that only one agent changes its opinion, convinced by the other one. For this transition, the following ansatz is used:

$$v(\theta'|\theta) = \eta \exp\left[-\sigma\left(N_\theta^k - N_{\theta'}^k\right)\right] . \tag{10.57}$$

Equation (10.57) means that the transition rate due to persuasion increases if the new opinion is shared by a *larger number* of agents. The small parameter σ is positive, but it could also be negative for agents that are always opposed to the majority.

A simple diffusion process has been assumed in the previous sections for the migration process of agents. However, individual transition rates for migration, $w(l|k)$, can also consider jumps between far distant regions instead of continuous migration. Here, we restrict our assumptions again to a diffusion-like process, that means that migration occurs only between adjacent boxes, where d is related to the diffusion coefficient. However, instead of a constant diffusion coefficient, we now assume that the transition rate of migration depends on the opinion θ as follows:

$$w(l|k) = \begin{cases} d \exp[-\varepsilon\,|\theta|\,N_\theta^k] & \text{if } l \text{ adjacent to } k \\ 0 & \text{otherwise} . \end{cases} \tag{10.58}$$

Similar to biological species, which do not diffuse in a simple way [478], the spatial spread in social systems might also be related to repulsive or attractive forces. In (10.58), for instance, the migration rate *decreases* if a large number of agents sharing the same opinion θ, is present at the current location k, in this way modeling the fact that these subpopulations try to stay together. This holds, of course, only for $\varepsilon > 0$; in the opposite case, $\varepsilon < 0$, an increasing number in their own subpopulation even enforces the migration.

This behavior may further depend on where θ ranges on the scale of opinions. The indifferent (or "nonpolitical") opinions, which correspond to $|\theta| \approx 0$, are more or less equally distributed, and no specific aggregation occurs. On the other hand, agents with extreme opinions, i.e., with larger values of $|\theta|$, mostly try to stay together to support each other's opinions (e.g., in "autonomous subcultures"). Therefore, their migration rate might be rather small, *if* the number of like-minded individuals at their location is rather large.

We note that the assumptions for transition rates, (10.56), (10.57), (10.58), include only some of the many interdependences that may influence the change of opinion or the opinion-related migration process. Here, they rather serve as examples, which should elucidate the general dynamics of social pattern formation.

For the change in socioconfiguration $\boldsymbol{N}$, we assume again that the transition rates are proportional to the individual transition rates and to the number of agents involved. With respect to the direct interactions considered, the multivariate master equation for the configuration $\boldsymbol{N}$, (10.15), changes now to

$$\frac{\partial P(\boldsymbol{N},t)}{\partial t} = \sum_{\theta'=-m}^{m} \sum_{k=1}^{z} \Big\{ \big[w(\theta|\theta') + v(\theta|\theta')\,(N_\theta^k - 1)\big]\,(N_{\theta'}^k + 1) \times P(N_\theta^k - 1, N_{\theta'}^k + 1, \boldsymbol{N}^\star, t) - \mathrm{d}\big[w(\theta'|\theta) + v(\theta|\theta')\,N_\theta^k\big]\,N_\theta^k\,P(\boldsymbol{N},t)\Big\}$$
$$+ \sum_{\theta=-m}^{m} \sum_{k,l=1,k\neq l}^{z} \big[w(k|l)\,(N_\theta^l + 1)\,P(N_\theta^k - 1, N_\theta^l + 1, \boldsymbol{N}^\star, t) - w(l|k)\,N_\theta^k\,P(\boldsymbol{N},t)\big]\,. \tag{10.59}$$

The multivariate master equation can be solved by stochastic computer simulations (see Sect. 3.1.4) that focus on the probabilistic aspect of evolution. But here we restrict the discussion to mean values:

$$\langle N_\theta^k(t)\rangle = \sum_{\{\boldsymbol{N}\}} N_\theta(t)\,P(\boldsymbol{N},t)\,. \tag{10.60}$$

The summation is over all possible configurations of $\boldsymbol{N}$, (10.4), that obey the condition, (10.3). From (10.60), the deterministic equation for the change in N_θ^k can be derived in a first approximation, which neglects higher correlations:

$$\frac{d}{dt}\langle N_\theta^k\rangle = \sum_{\theta'=-m}^{m} \langle w(\theta|\theta')\,N_{\theta'}^k - w(\theta'|\theta)\,N_\theta^k\rangle + \Big\langle\Big[v(\theta|\theta') - v(\theta'|\theta)\Big]\,N_{\theta'}^k\,N_\theta^k\Big\rangle$$
$$+\sum_{l=1}^{z} \langle w(k|l)\,N_\theta^l \;-\; w(l|k)\,N_\theta^k\rangle\,. \tag{10.61}$$

If we introduce the abbreviations,

$$\mathcal{A}_{\theta\theta'} = \begin{cases} -w(\theta|\theta') & \text{for } \theta \neq \theta' \\ \sum\limits_{\theta'=-m}^{m} w(\theta|\theta') & \text{for } \theta = \theta'\,, \end{cases}$$
$$\mathcal{B}_{\theta\theta'} = v(\theta|\theta') - v(\theta'|\theta) = -\mathcal{B}_{\theta'\theta}\,, \tag{10.62}$$
$$\mathcal{C}_{lk} = \begin{cases} -w(k|l) & \text{for } l \neq k \\ \sum\limits_{l=1}^{z} w(l|k) & \text{for } l = k,\,, \end{cases}$$

(10.61) can be rewritten in the compact form,

$$\frac{d}{dt}\langle N_\theta^k\rangle = \sum_{\theta'=-m}^{m} \left(\langle\mathcal{A}_{\theta\theta'}\,N_{\theta'}^k\rangle + \langle\mathcal{B}_{\theta\theta'}N_{\theta'}^k\,N_\theta^k\rangle\right) + \sum_{l=1}^{z}\langle\mathcal{C}_{kl}\,N_\theta^l\rangle\,. \tag{10.63}$$

Equation (10.63) is a set of coupled equations. The conservation of the total number of agents yields

$$\sum_{\theta=-m}^{m}\sum_{k=1}^{z}\frac{d}{dt}\langle N_\theta^k\rangle = 0\,, \tag{10.64}$$

which finally results in

$$\sum_{\theta=-m}^{m} \mathcal{A}_{\theta\theta'} = 0\,, \quad \sum_{\theta=-m}^{m} \mathcal{B}_{\theta\theta'} = 0\,, \quad \sum_{k=1}^{z} \langle \mathcal{C}_{kl}\, N_{\theta}^{l} \rangle = 0\,. \tag{10.65}$$

In the following, we want to discuss the stationary and time-dependent solutions of (10.63) for different specific cases.

10.3.2 Stationary Solutions for a Single Box

Of particular interest is the question whether the dynamic system, (10.63), allows the existence of multiple steady states. In the context of social systems, steady-state solutions with more than one subpopulation θ indicate the possibility of (stable) *coexistence of different opinions* at the same location. Multiple steady states would correspond to different stable situations in a society with different fractions of coexisting opinions.

To estimate multiple steady states, we apply a method outlined in [140, 244, 245, 473] for investigating *multicell reaction systems*. These are systems that combine chemical reactions and transport, as are found, for example, in biological systems. Here, we have a similar dynamic situation in spite of different "chemical" species. Central to the investigation is the "cell mechanism," i.e., the reactions occurring within a cell or a box, respectively. If we concentrate on one box, neglecting the box index, then the reactions, (10.54), can be summarized in the following way:

$$\sum_{\theta} a_{\theta j} C_{\theta} \longrightarrow \sum_{\theta} b_{\theta j} C_{\theta}\,, \tag{10.66}$$

where θ describes the number of "species", i.e., opinions, and j the number of possible reaction types. Let us for simplicity consider only three different opinions, $\theta \in \{-1, 0, +1\}$. Due to (10.54), we have two different reaction types in each box, $j = 2$. A theorem by Shapiro and Horn [473] states that system dynamics may not exhibit sustained oscillations, multiple steady states, or asymmetrical steady states if the *deficiency* $\mathcal{D}$ of a mechanism is zero and the cell mechanism is weakly reversible. Here, the deficiency is defined as

$$\mathcal{D} = n - l - c\,, \tag{10.67}$$

where n is the number of distinct complexes in the mechanism, l is the number of linkage classes, and c is the rank of the stoichiometric matrix induced by the mechanism. A complex is of the form $a_{\theta j} C_{\theta}$; thus we have nine different complexes in the reaction system considered, which is illustrated in Fig. 10.5. The number of linkage classes is 2, as can also be seen from Fig. 10.5, and the rank of the stoichiometric matrix is 2. Hence, the deficiency for the considered change of opinions is $\mathcal{D} = 5$, which means that sustained oscillations, multiple steady states, or asymmetrical steady states are possible in the system. Notably, this conclusion holds independently of the specific assumptions for transition rates, i.e., (10.56), (10.57).

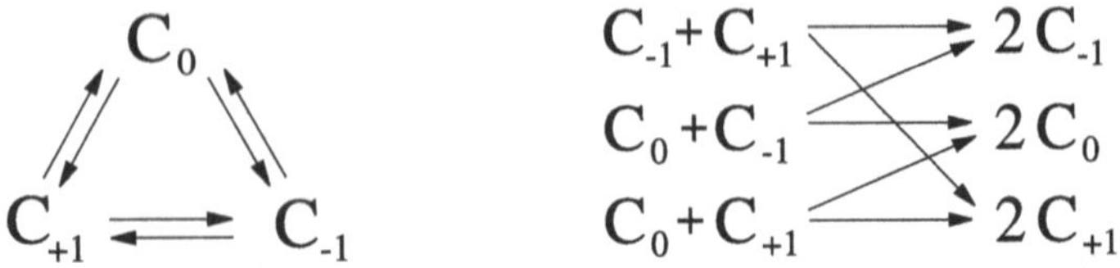

Fig. 10.5. Possible transitions among three opinions $\theta = \{-1, 0, +1\}$: (*left*) reversible transitions resulting from spontaneous opinion changes; (*right*) irreversible transitions resulting from direct interactions (persuasion)

A mechanism is denoted as weakly reversible if, when it is seen as a directed graph, every complex is contained in a directed cycle [473]. As Fig. 10.5 shows, one part of the mechanism is reversible, i.e., every reaction is reversible, whereas the other part is irreversible. Considering only spontaneous transitions of opinions, as assumed, e.g., in (10.56), the mechanism would be only reversible, $\mathcal{D}$ would be zero, and no nontrivial steady-state solutions exist.

For the interaction scheme in Fig. 10.5, the trajectories and the stationary solutions for three possible opinions in a single box are shown in Fig. 10.6. Here, we have used the fractions of the different opinions:

$$x_\theta = \frac{N_\theta}{N}, \quad \sum_\theta x_\theta = 1. \tag{10.68}$$

As the left part of Fig. 10.6 shows, the consideration of only individual decisions (spontaneous opinion changes) would result in the stable solution,

$$x_{-1} = x_{+1} = x_0 = 1/3, \tag{10.69}$$

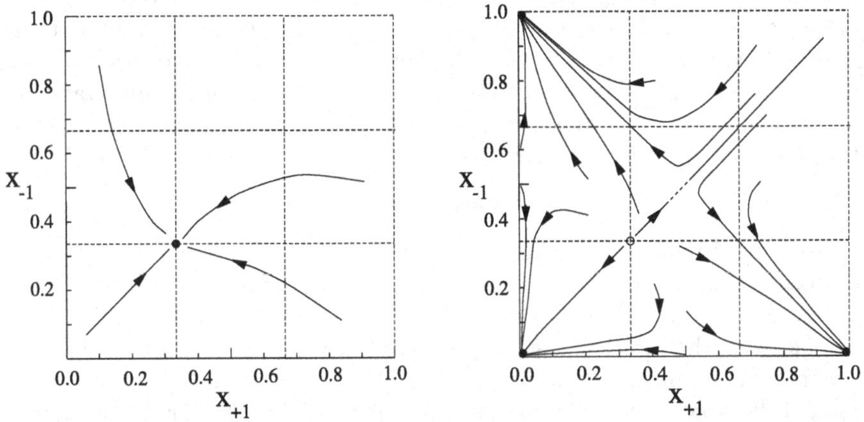

Fig. 10.6. Trajectories and stationary solutions in the state space x_{+1}, x_{-1}, $x_0 = 1 - x_{+1} - x_{-1}$. (*left*: a) only spontaneous opinion changes, ($\alpha = 0.3$, $\beta = \gamma = \sigma = 0$, $\eta = 1$); (*right*: b) direct interactions among all three opinions $\theta \in \{-1, 0, +1\}$. Parameters: $\alpha = 0.5$, $\beta = 0.5$, $\gamma = 0$, $\sigma = 0.01$, $E = 0$, $\eta = 1$

in the absence of external influences. However, a significant influence of E may shift the ratio toward the supported opinion. If we consider, additionally, direct interactions among *all* opinions, we find that the former stable state, (10.69), becomes an *unstable node*, as shown in the right part of Fig. 10.6. Additionally, an (unstable) *saddle point* appears. Further, we observe three new stable states characterized by one dominating opinion $x_\theta \to 1$.

Let us now consider direct interactions only between adjacent opinions, which is based on the assumption that agents with completely opposite opinions will not communicate with each other. Then, the reaction scheme changes to Fig. 10.7. Even if the number of complexes is now reduced by one, the deficiency is still nonzero, and nontrivial steady-state solutions can be expected. Figure 10.8 shows the trajectories and the stationary solutions for the interaction scheme of Fig. 10.7.

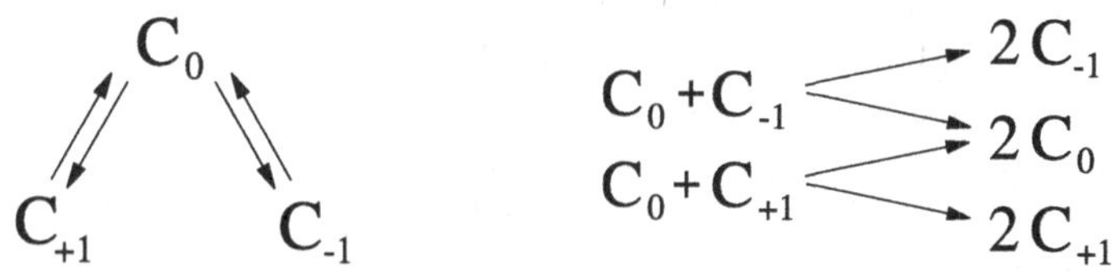

Fig. 10.7. Possible transitions between adjacent opinions $\theta = \{-1, 0, +1\}$: (*left*) reversible transitions resulting from spontaneous opinion changes; (*right*) irreversible transitions resulting from direct interactions (persuasion)

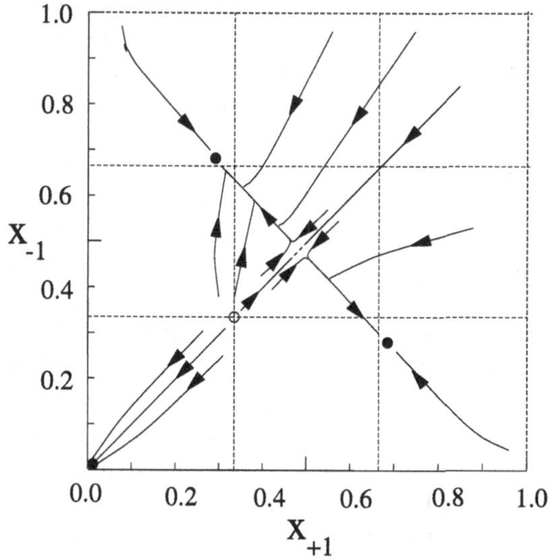

Fig. 10.8. Trajectories and stationary solutions in the state space x_{+1}, x_{-1}, $x_0 = 1 - x_{+1} - x_{-1}$ for direct interactions only between adjacent opinions $\theta = \{-1, 0, +1\}$. Parameters: $\alpha = 0.5$, $\beta = 0.5$, $\gamma = 0$, $\sigma = 0.01$, $E = 0$, $\eta = 1$

Considering only individual decisions and interactions between neighboring opinions, we find in Fig. 10.8 the same unstable node and the same saddle point as in Fig. 10.6, but the three stable states are now different: two stable states are characterized by the coexistence of two contradicting opinions, whereas the third stable state means one dominating opinion again.

To conclude the analysis, we note that for the model considered, multiple steady states of the homogeneous system exist only if we include direct interactions between agents. The fraction of respective opinions depends on whether direct interaction (communication, persuasion) among *all* opinions is considered or only between *adjacent* opinions.

10.3.3 Results of Computer Simulations

Eventually, we will solve the time-dependent mean-value equations, (10.61), (10.63), also with respect to the *spatial coordinate*, given by the box index k. Considering z boxes and M different opinions, with respect to conservation of the total number of agents, (10.3), this means a system of $z\,(M-1)$ coupled equations. Here, we assume a linear chain of $z = 50$ boxes arranged as a circle. Further, *five* different opinions are assumed for the spectrum of opinions $\boldsymbol{M}$, (10.1): $\theta \in \{-2, -1, 0, +1, +2\}$.

As an example, we discuss the spread of one extreme opinion [450] which is analogous to the *colonization* of an uninhabited area. As the initial condition, a central box ($j = 25$) with $N = 10{,}000$ inhabitants is assumed. These agents should all have the opinion $\theta = -2$, which means with respect to (10.58), their migration rate should be rather small, i.e., they are rather settled. Because all other boxes are uninhabited and the concentration of settled inhabitants is very high, a certain "pressure" for migration exists, expressed in a field E different from the leading opinion, for instance, $E = 0$.

The dynamics of colonization occurs in different stages. During the *first stage*, a new opinion, i.e., to migrate, is created in the mother box. This means a transition $\theta = -2 \to \theta = 0$. During the *second stage*, some of the inhabitants with $\theta = 0$ migrate into surrounding subsystems and, during a *third stage*, begin to resettle in the new subsystems by changing their opinion again from $\theta = 0 \to \theta = -2$. This is caused by their *cultural memory*, expressed by the internal field Q, (10.55), which is, especially at the beginning of the simulation, very close to $Q = -2$. If the concentration of inhabitants in the new subsystem increases again, a new cycle occurs.

The spatiotemporal evolution of the total number of inhabitants is shown in Fig. 10.9. In the course of time, a distinct spatial pattern appears at a certain distance from the mother box, which is characterized by regions of higher occupation, separated by regions of lower occupation.

We note the fact that the distribution of opinions in boxes with lower occupation of agents is different from those with higher occupation, which is also sketched in Fig. 10.10. In boxes with high occupation, we find a unimodal distribution dominated by agents with opinion $\theta = -2$, which are settled,

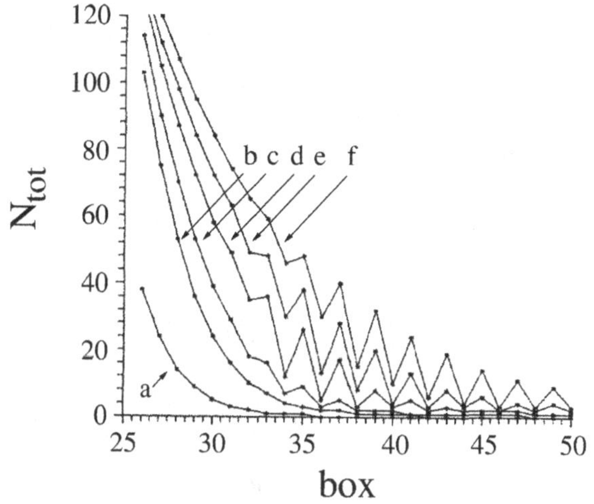

Fig. 10.9. Evolution of the total number of agents per box in a chain of boxes for different time steps: (a) 9; (b) 15; (c) 20; (d) 30; (e) 44; (f) 66. Initial state: $N_{-2} = 10,000$ in box 25; parameters: $\alpha = 0.2$, $\beta = 0.1$, $\gamma = 0.3$, $\sigma = 0.01$, $\varepsilon = 0.7$, $E = 0$, $\eta = 1$ [450]

whereas in boxes with low occupation, bimodal distribution is found. This can also be viewed in Fig. 10.11, which shows the spatial distribution of the opinions $\theta = 0$ and $\theta = -2$ at a given time. Here, we see a rather smooth distribution of agents with $\theta = 0$; their number is almost the same as the number of agents with $\theta = -2$ in boxes with low occupation, but much less in boxes with high occupation.

The example discussed has a certain analogy to precipitation phenomena already discussed in Sect. 3.3.1. In the so-called Liesegang *rings* (see Fig. 10.12), we find periodic precipitation at a certain distance from the origin, whereas near the origin, a rather continuous distribution exists. With respect to the model discussed here, the transition $\theta = 0 \rightarrow \theta = -2$ describes

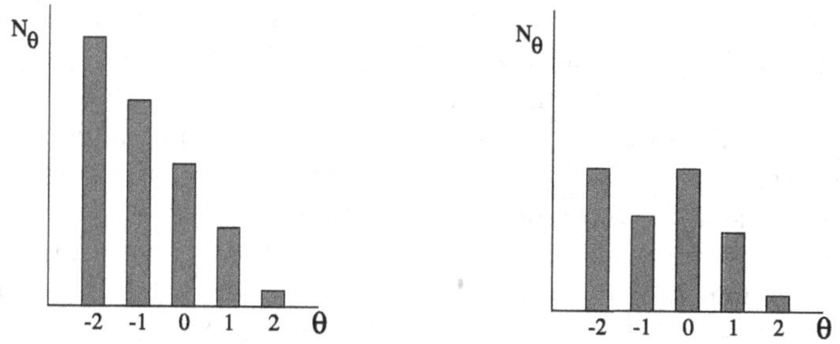

Fig. 10.10. Opinion distribution in boxes with high occupation (*left*) and low occupation (*right*)

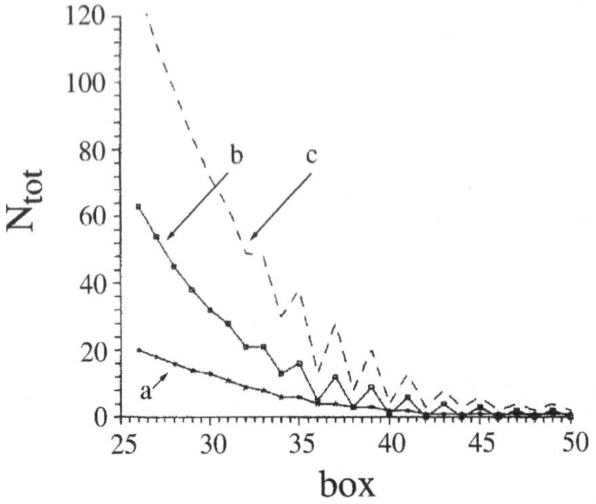

Fig. 10.11. Distribution of the number of agents per box for a particular time step $t = 44$: (a) N_0; (b) N_{-2}; (c) N_{tot} (see curve (e) in Fig. 10.9). For parameters, see Fig. 10.9

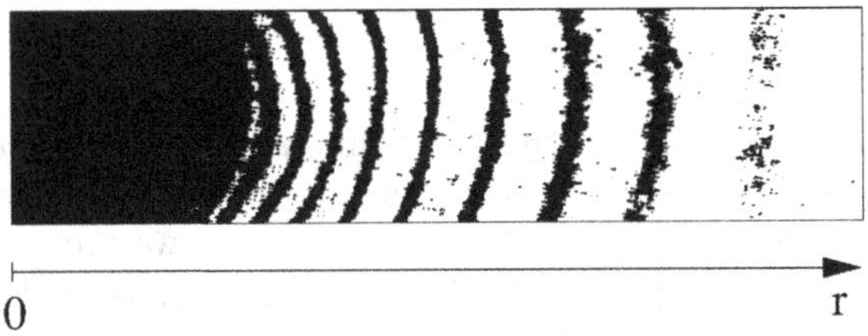

Fig. 10.12. Liesegang rings as an example of periodic precipitation

"precipitation" of agents who get settled. The internal field Q, which supports the opinion $\theta = -2$, can be regarded here as a measure of supersaturation. The nonlinear feedback that creates the local increase in opinion $\theta = -2$ results here from the direct interaction between agents, (10.57), which assumes that the opinion change is more likely in the direction of the majority opinion in each box. Hence, a slight increase of the respective opinion will be reinforced. On the other hand, the band-like structure that occurs at a distance from the center is the result of a depletion effect, as in the Liesegang rings. Because diffusion is slow, only a few agents per time interval appear in the "outer regions"; hence, the regions with higher occupation are necessarily separated by depleted zones of low occupation.

Finally, we would like to note that the different types of models for collective opinion formation derived in this chapter sketch only some *basic features* of structure formation in social systems. There is no doubt that in real human societies, more complex behavior among individuals occurs and that decision making and opinion formation may depend on numerous influences beyond quantitative classification. Here, we restricted ourselves to a simplified dynamic approach, which *purposely* stretches some analogies between physical and social systems. The results, however, display similarities to phenomena observed in social systems and allow interpretation within such a context. Examples of these analogies are the formation of minorities and majorities, the spatial separation of opinions for certain "social temperatures", as discussed in Sect. 10.2, and the spatial concentration of individuals with certain attitudes, as shown above. So, our approach may give rise to some further investigations in this direction.

Bibliography

[1] Akimov, V.; Soutchanski, M. (1994). Automata simulation of N-person social dilemma games. *J. Conflict Resolution* **38**, 138–148.

[2] Allen, P. (1998). Modelling complex economic evolution. In: F. Schweitzer; G. Silverberg (eds.), *Evolution und Selbstorganisation in der Ökonomie / Evolution and Self-Organization in Economics*, Berlin: Duncker & Humblot, vol. 9 of *Selbstorganisation. Jahrbuch für Komplexität in den Natur- Sozial- und Geisteswissenschaften*, pp. 47–75.

[3] Allen, P.; Sanglier, M. (1979). A dynamic model of growth in a central place system. *Geogr. Anal.* **11/3**, 256–272.

[4] Alt, W. (1980). Biased random walk models for chemotaxis and related diffusion approximations. *J. Math. Biol.* **9**, 147–177.

[5] Alt, W. (1988). Modelling of motility in biological systems. In: *ICIAM'87 Proceedings*, Philadelphia: SIAM, pp. 15–30.

[6] Alt, W. (1990). Correlation analysis of two-dimensional locomotion paths. In: W. Alt; G. Hoffmann (eds.), *Biological Motion*, Berlin: Springer, vol. 89 of *Lecture Notes in Biomathematics*, pp. 254–268.

[7] Alt, W. (1995). Elements of a systematic search in animal behavior and model simulations. *BioSystems* **34**, 11–26.

[8] Alt, W.; Hoffmann, G. (eds.) (1990). *Biological Motion*, vol. 89 of *Lecture Notes in Biomathematics*. Berlin: Springer.

[9] Andersen, P. W.; Arrow, K. J.; Pines, D. (eds.) (1988). *The Economy as an Evolving Complex System*. Reading, MA: Addison-Wesley.

[10] Anderson, T. L.; Donath, M. (1990). Animal behaviour as a paradigm for developing robot autonomy. In: P. Maes (ed.), *Designing Autonomous Agents*, Cambridge, MA: MIT Press, pp. 145–168.

[11] Andreazza, P.; Lefaucheux, F.; Mutaftschiev, B. (1988). Nucleation in confined space: Application to the crystallization in gels. *J. Cryst. Growth* **92**, 415–422.

[12] Andresen, B.; Hoffmann, K. H.; Mosegaard, K.; Nulton, J.; Pedersen, J. M.; Salamon, P. (1988). On lumped models for thermodynamic properties of simulated annealing problems. *J. Phys. (France)* **49**, 1485.

[13] Appleby, S. (1995). Estimating the cost of a telecommunications network using the fractal structure of the human population distribution. *IEEE Proc.-Commun.* **142**, 172–178.

[14] Arthur, W. B. (1993). On designing economic agents that behave like human agents. *J. Evol. Econ.* **3**, 1–22.

[15] Arthur, W. B. (1994). Inductive reasoning and bounded rationality. *Am. Econ. Assoc. Pap. Proc.* **84**, 406–411.

[16] Arthur, W. B.; Durlauf, S. N.; Lane, D. (eds.) (1997). *The Economy as an Evolving Complex System II*. Reading, MA: Addison-Wesley.

[17] Asselmeyer, T.; Ebeling, W. (1997). Mixing of thermodynamic and biological strategies in optimization. In: F. Schweitzer (ed.), *Self-Organization of Complex Structures: From Individual to Collective Dynamics*, London: Gordon and Breach, pp. 153–163.

[18] Asselmeyer, T.; Ebeling, W. (1997). Unified description of evolutionary strategies over continuous parameter spaces. *BioSystems* **41**, 167–178.

[19] Asselmeyer, T.; Ebeling, W.; Rosé, H. (1996). Smoothing representation of fitness landscapes – the genotype-phenotype map of evolution. *BioSystems* **39**, 63–76.

[20] Asselmeyer, T.; Ebeling, W.; Rosé, H. (1997). Evolutionary strategies of optimization. *Phys. Rev. E* **56**, 1171–1180.

[21] Astumian, R. D. (1997). Thermodynamics and kinetics of a Brownian motor. *Science* **276**, 917–922.

[22] Astumian, R. D.; Bier, M. (1994). Fluctuation driven ratchets: Molecular motors. *Phys. Rev. Lett.* **72/11**, 1766–1769.

[23] Attygalle, A. B.; Steghaus-Kovac, S.; Ahmed, V. U. (1991). *cis*-Isogeraniol, a recruitment pheromone of the ant *Leptogenys diminuta*. *Naturwissenschaften* **78**, 90–92.

[24] Attygalle, A. B.; Vostrowsky, O.; Bestmann, H. J.; Steghaus-Kovac, S.; Marschwitz, U. (1988). (3*R*,4*S*)-4-Methyl-3-heptanol, the trail pheromone of the ant *Leptogenys diminuta*. *Naturwissenschaften* **75**, 315–317.

[25] Auerbach, F. (1913). Das Gesetz der Bevölkerungskonzentration. *Petermanns Mitteilungen* **59/1**, 74–76.

[26] Axelrod, R. (1997). *The Complexity of Cooperation: Agent-Based Models of Competition and Collaboration.* Princeton, NJ: Princeton University Press.

[27] Bäck, T. (1994). Selective pressure in evolutionary algorithms: A characterization of selection mechanisms. In: R. K. Belew; L. B. Booker (eds.), *Proc. 1st IEEE Conf. Evol. Computation*, Piscataway, NJ: IEEE Press, pp. 57–62.

[28] Bäck, T. (1995). Generalized convergence models for tournament- and (5,l)-selection. In: L. Eshelman (ed.), *Proc. 6th Int. Conf. Genet. Algorithms*, San Francisco: Morgan Kaufmann, pp. 2–8.

[29] Bäck, T.; Doernemann, H.; Hammel, U.; Frankhauser, P. (1996). Modeling urban growth by cellular automata. In: H. M. Voigt; W. Ebeling; H. P. Schwefel; I. Rechenberg (eds.), *Parallel Problem Solving from Nature - PPSN IV*, Berlin: Springer, vol. 1141 of *Lecture Notes in Computer Science*, pp. 636–645.

[30] Banaerjee, A. V. (1992). A simple model of herd behavior. *Q. J. Econ.* **107**, 797–817.

[31] Bartels, J.; Schmelzer, J.; Schweitzer, F. (1990). The influence of depletion effects on homogeneous nucleation rates. *Zeitschrift für Physikalische Chemie (München)* **166**, 119–123.

[32] Bartussek, R.; Hänggi, P. (1995). Brownsche Motoren. Wie aus Brownscher Bewegung makroskopischer Transport wird. *Physikalische Blätter* **51/6**, 506–507.

[33] Bartussek, R.; Hänggi, P.; Kissner, J. G. (1994). *Europhysics Lett.* **28**, 459.

[34] Bartussek, R.; Hänggi, P.; Lindner, B.; Schimansky-Geier, L. (1997). Ratchets driven by harmonic and white noise. *Physica D* **109**, 17–23.

[35] Bartussek, R.; Reimann, P.; Hänggi, P. (1996). Precise numerics versus theory for correlation ratchets. *Phys. Rev. Lett.* **76/7**, 1166–1169.

[36] Batty, M. (1991). Cities as fractals: simulating growth and form. In: A. Crilly; R. Earnshaw; H. Jones (eds.), *Fractals and Chaos*, New York: Springer, pp. 43–69.

[37] Batty, M. (1991). Generating urban forms from diffusive growth. *Environ. Planning A* **23**, 511–544.
[38] Batty, M. (1992). Urban modeling in computer-graphic and geographic information system environments. *Environ. Planning B* **19**, 663–688.
[39] Batty, M. (1994). Urban models 25 years on. *Environ. Planning B* **21/5**, 515.
[40] Batty, M. (1997). Cellular automata and urban form: A primer. *J. Am. Planning Assoc.* **63/2**, 266.
[41] Batty, M.; Couclelis, H.; Eichen, M. (1997). Urban systems as cellular automata. *Environ. Planning B* **24/2**, 159.
[42] Batty, M.; Longley, P. A. (1987). Fractal-based description of urban form. *Environ. Planning B* **14**, 123–134.
[43] Batty, M.; Longley, P. A. (1994). *Fractal Cities.* London: Academic Press.
[44] Batty, M.; Longley, P. A.; Fotheringham, S. (1989). Urban growth and form: Scaling, fractal geometry, and diffusion-limited aggregation. *Environ. Planning A* **21/11**, 1447–1472.
[45] Batty, M.; Xie, Y. (1994). From cells to cities. *Environ. Planning B* **21**, 31–48.
[46] Becker, R. (1961). *Theorie der Wärme.* Berlin: Springer.
[47] Ben-Jacob, E.; Cohen, I.; Czirók, A. (1995). Smart bacterial colonies: From complex patterns to cooperative evolution. *Fractals* **3**, 849–868.
[48] Ben-Jacob, E.; Cohen, I.; Levine, H. (2000). Cooperative self-organization of microorganisms. *Adv. Phys.* **49(4)**, 395–554.
[49] Ben-Jacob, E.; Cohen, I.; Shochet, O.; Aranson, I.; Levine, H.; Tsimring, L. (1995). Complex bacterial patterns. *Nature* **373**, 566–567.
[50] Ben-Jacob, E.; Schochet, O.; Tenenbaum, A.; Cohen, I.; Czirók, A.; Vicsek, T. (1994). Generic modelling of cooperative growth patterns in bacterial colonies. *Nature* **368**, 46–49.
[51] Ben-Jacob, E.; Shochet, O.; Cohen, I.; Czirok, A.; Vicsek, T. (1995). Cooperative formation of chiral patterns during growth of bacterial colonies. *Phys. Rev. Lett.* **75/15**, 2899–2902.
[52] Ben-Jacob, E.; Shochet, O.; Tenenbaum, A.; Cohen, I.; Czirok, A.; Vicsek, T. (1994). Communication, regulation and control during complex patterning of bacterial colonies. *Fractals* **2/1**, 15–44.
[53] Benguigui, L. (1992). Some speculations on fractals and railway networks. *Physica A* **191**, 75–78.
[54] Berg, H. C. (1975). How bacteria swim. *Sci. Am.* **233**, 36–44.
[55] Berg, H. C. (1983). *Random Walks in Biology.* Princeton, NJ: Princeton University Press.
[56] Berg, H. C. (1990). Bacterial microprocessing. *Cold Spring Habor Symp. Quant. Biol.* **55**, 539–545.
[57] Bernasconi, J. (1987). Low autocorrelation binary sequences: Statistical mechanics and configuration space analysis. *J. Physique* **48**, 559.
[58] Biebricher, C. K.; Nicolis, G.; Schuster, P. (1995). *Self-Organization in the Physico-Chemical and Life Sciences*, vol. 16546 of *EU Report.*
[59] Biler, P. (1995). Growth and accretion of mass in an astrophysical model. *Appl. Math.* **23/2**, 179–189.
[60] Bloh, W. v.; Block, A.; Schellnhuber, H. J. (1997). Self-stabilization of the biosphere under global change: A tutorial geophysiological approach. *Tellus B* **49**, 249–262.
[61] Boisfleury-Chevance, A.; Rapp, B.; Gruler, H. (1989). Locomotion of white blood cells: A biophysical analysis. *Blood Cells* **15**, 315–333.

[62] Bonabeau, E.; Dorigo, M.; Théraulaz, G. (1999). *Swarm Intelligence: From Natural to Artificial Systems.* Santa Fe Institute Studies on the Sciences of Complexity, New York: Oxford University Press.

[63] Bonabeau, E.; Dorigo, M.; Theraulaz, G. (2000). Inspiration for optimization from social insect behaviour. *Nature* **406**, 39–42.

[64] Bonabeau, E.; Theraulaz, G.; Camazine, S. (1997). Self-organization in social insects. *Trends Ecol. Evol.* **12/5**, 188.

[65] Borgers, A.; Timmermans, H. J. P. (1986). City centre entry points, store location patterns and pedestrian route choice behaviour: A microlevel simulation model. *Socio-Economic Planning Sci.* **20**, 25–31.

[66] Boseniuk, T.; Ebeling, W. (1988). Optimization of NP-complete problems by Boltzmann-Darwin strategies including life-cycles. *Europhysics Lett.* **6**, 107.

[67] Boseniuk, T.; Ebeling, W.; Engel, A. (1987). Boltzmann and Darwin strategies in complex optimization. *Phys. Lett.* **125**, 307–310.

[68] Boyarsky, A. (1975). A Markov chain model for human granulocyte movement. *J. Math. Biol.* **2**, 69–78.

[69] Brandt, K. (1997). Regional dynamic processes in the economy. In: F. Schweitzer (ed.), *Self-Organization of Complex Structures: From Individual to Collective Dynamics*, London: Gordon and Breach, pp. 489–500.

[70] Brooks, R. A.; Maes, P. (eds.) (1994). *Artificial Life IV. Proceedings of the Fourth International Workshop on the Synthesis and Simulation of Living Systems.* Cambridge, MA: MIT Press.

[71] Bruckner, E.; Ebeling, W.; Jimenez Montano, M. A.; Scharnhorst, A. (1994). Hyperselection and innovation described by a stochastic model of technological change. In: L. Leydesdorff; P. van den Besselaar (eds.), *Evolutionary Economics and Chaos Theory: New Directions in Technology Studies*, London: Pinter, pp. 79–90.

[72] Bruckner, E.; Ebeling, W.; Jiménez-Montano, M. A.; Scharnhorst, A. (1996). Nonlinear effects of substitution – an evolutionary approach. *J. Evol. Econ.* **6**, 1–30.

[73] Buchholtz, V.; Pöschel, T. (1997). Adaptive evolutionary optimization of team work. *Int. J. Bifurcation Chaos* **7/3**, 751–757.

[74] Budrene, E. O.; Berg, H. C. (1991). Escherichia coli. *Nature* **349**, 630–633.

[75] Burchard, R. P. (1982). Trail following by gliding bacteria. *J. Bacteriol.* **152**, 495–501.

[76] Calenbuhr, V.; Deneubourg, J. L. (1990). A model for trail following in ants: Individual and collective behaviour. In: W. Alt; G. Hoffmann (eds.), *Biological Motion*, Berlin: Springer, pp. 453–469.

[77] Calenbuhr, V.; Deneubourg, J. L. (1991). Chemical communication and collective behaviour in social and gregarious insects. In: W. Ebeling; M. Peschel; W. Weidlich (eds.), *Models of Selforganization in Complex Systems - MOSES*, Berlin: Akademie-Verlag, vol. 64 of *Mathematical Research*, pp. 322–331.

[78] Caro, G. D.; Dorigo, M. (1998). An adaptive multi-agent routing algorithm inspired by ants behavior. In: *Proc. 5th Annu. Australasian Conf. Parallel Real-Time Syst. (PART98)*, Berlin: Springer, pp. 261–272.

[79] Carraway, R. L.; Morin, T. L.; Moskowitz, H. (1990). Generalized dynamic programming for multicriteria optimization. *Eur. J. Operational Res.* **44**, 95–104.

[80] Case, K. E.; Fair, R. C. (1992). *Principles of Economics.* Englewood Cliffs, NJ: Prentice-Hall.

[81] Challet, D.; Chessa, A.; Marsili, M.; Zhang, Y.-C. (2001). From minority games to real markets. *Quant. Finance* **1**, 168–176.

[82] Challet, D.; Zhang, Y.-C. (1998). On the minority game: Analytical and numerical studies. *Physica A* **256**, 514–532.

[83] Chopard, B.; Droz, M. (1998). *Cellular Automata Modeling of Physical Systems.* Collection Alea, Cambridge: Cambridge University Press.

[84] Chopard, B.; Luthi, P.; Droz, M. (1994). Microscopic approach to the formation of Liesegang patterns. *J. Stat. Phys.* **76/1–2**, 661–677.

[85] Christaller, W. (1933). *Die zentralen Orte in Süddeutschland.* Jena: Fischer. Eine ökonomisch-geographische Untersuchung über die Gesetzmäßigkeit der Verbreitung und Entwicklung der Siedlungen mit städtischen Funktionen. Reprints: Darmstadt: Wissenschaftliche Buchgesellschaft, 1980, *Central Places in Southern Germany*, (English translation by C.W. Baskin), London: Prentice-Hall, 1966.

[86] Cladis, P. E.; Palffy-Muhoray, P. (eds.) (1995). *Spatio-Temporal Patterns in Nonequilibrium Complex Systems.* Reading, MA: Addison-Wesley.

[87] Clark, C. (1951). Urban population densities. *J. R. Stat. Soc. A* **114**, 490–496.

[88] Cliff, D.; Miller, G. F. (1996). Co-evolution of pursuit and evasion 2: Simulation methods and results. In: P. Maes; M. Mataric; J.-A. Meyer; J. Pollack; S. Wilson (eds.), *From Animals to Animats*, Cambridge, MA: MIT Press, vol. 4, pp. 506–515.

[89] Couclelis, H. (1985). Cellular worlds: A framework for modeling micro-macro dynamics. *Environ. Planning A* **17**, 585–596.

[90] Crist, T. O.; Haefner, J. W. (1994). Spatial model of movement and foraging in harvester ants *(Pogonomyrmex)* (II): The roles of environment and seed dispersion. *J. Theor. Biol.* **166**, 315–323.

[91] Czirok, A.; Barabasi, A. L.; Vicsek, T. (1999). Collective motion of self-propelled particles: Kinetic phase transition in one dimension. *Phys. Rev. Lett.* **82(1)**, 209–212.

[92] Czirok, A.; Ben-Jacob, E.; Cohen, I.; Vicsek, T. (1996). Formation of complex bacterial colonies via self-generated vortices. *Phys. Rev. E* **54(2)**, 1791–1801.

[93] Czirok, A.; Vicsek, T. (2000). Collective behavior of interacting self-propelled particles. *Physica A* **281**, 17–29.

[94] Darley, V. (1994). Emergent phenomena and complexity. In: R. A. Brooks; P. Maes (eds.), *Artificial Life IV. Proceedings of the Fourth International Workshop on the Synthesis and Simulation of Living Systems*, Cambridge, MA: MIT Press, pp. 411–416.

[95] Darwen, J.; Yao, X. (1997). Speciation as automatic categorical modularization. *IEEE Trans. on Evol. Computation* **1(2)**, 101–108.

[96] Davis, B. (1990). Reinforced random walks. *Probl. Theor. Relat. Fields* **2**, 203–229.

[97] Deadman, P.; Brown, R.; Gimblet, R. (1993). Modelling rural residential settlement patterns with cellular automata. *J. Environ. Manage.* **37**, 147–160.

[98] DeAngelis, D. L.; Gross, L. J. (eds.) (1992). *Individual-Based Models and Approaches in Ecology: Populations, Communities, and Ecosystems.* New York: Chapman and Hall.

[99] Dendrinos, D. S.; Haag, G. (1984). Towards a stochastic theory of location: Empirical evidence. *Geogr. Anal.* **16**, 287–300.

[100] Dendrinos, D. S.; Sonis, M. (1990). *Chaos and Socio-Spatial Dynamics.* Berlin: Springer.

[101] Deneubourg, J. L.; Goss, S.; Franks, N.; Pasteels, J. M. (1989). The blind leading the blind: Modeling chemically mediated army ant raid patterns. *J. Insect Behav.* **2/5**, 719–725.

[102] Deneubourg, J. L.; Goss, S.; Franks, N.; Sendova-Franks, A.; Detrain, C.; Chretien, L. (1991). The dynamics of collective sorting: Robot-like ants and ant-like robots. In: J. A. Meyer; S. W. Wilson (eds.), *From Animals to Animats*, Cambridge, MA: MIT Press, pp. 356–363.

[103] Deneubourg, J. L.; Gregoire, J. C.; Le Fort, E. (1990). Kinetics of larval gregarious behavior in the bark beetle *Dendroctonus micans* (Coleoptera: Scolytidae). *J. Insect Behav.* **3/2**, 169–182.

[104] Derenyi, I.; Vicsek, T. (1995). Cooperative transport of Brownian particles. *Phys. Rev. Lett.* **75/3**, 374–377.

[105] Dickinson, R.; Tranquillo, R. T. (1993). A stochastic model for adhesion-mediated cell random motility and haptotaxis. *J. Math. Biol.* **31**, 563–600.

[106] Dorigo, M.; Bonabeau, E.; Theraulaz, G. (2000). Ant algorithms and stigmergy. *Future Generation Comput. Syst.* **16(8)**, 851–871.

[107] Dorigo, M.; Caro, G. D. (1999). The ant colony optimization meta-heuristic. In: D. Corne; M. Dorigo; F. Glover (eds.), *New Ideas in Optimization*, New York: McGraw-Hill, pp. 11–32.

[108] Dorigo, M.; Gambardella, L. M. (1997). Ant colonies for the travelling salesman problem. *BioSystems* **43**, 73–81.

[109] Dunn, G. A. (1983). Characterizing a kinesis response: Time averaged measures of cell speed and directional persistence. In: H. Keller; G. O. Till (eds.), *Leukocyte Locomotion und Chemotaxis*, Basel: Birkhäuser, pp. 14–33.

[110] Dunn, G. A.; Brown, A. F. (1987). A unified approach to analyzing cell motility. *J. Cell. Sci. Suppl.* **8**, 81–102.

[111] Durrett, R.; Levin, S. (1998). Spatial aspects of interspecific competition. *Theor. Population Biol.* **53**, 30–43.

[112] Dworkin, M.; Kaiser, D. (eds.) (1993). *Myxobacteria*, vol. II. Washington, DC: American Society for Microbiology.

[113] Ebeling, W. (1981). Structural stability of stochastic systems. In: H. Haken (ed.), *Chaos and Order in Nature*, Berlin: Springer, vol. 11 of *Springer Series in Synergetics*, pp. 188–195.

[114] Ebeling, W. (1990). Applications of evolutionary strategies. *Syst. Anal. Model. Simul.* **7**, 3–16.

[115] Ebeling, W. (1994). Self–organization, valuation and optimization. In: R. K. Mishra; D. Maaß; E. Zwierlein (eds.), *On Self–Organization*, Berlin: Springer, vol. 61 of *Springer Series in Synergetics*, pp. 185–196.

[116] Ebeling, W. (2000). Problems of a statistical ensemble theory for systems far from equilibrium. In: J. A. Freund; T. Pöschel (eds.), *Stochastic Processes in Physics, Chemistry and Biology*, Berlin: Springer, vol. 557 of *Lecture Notes in Physics*, pp. 390–399.

[117] Ebeling, W.; Engel, H.; Herzel, H. (1990). *Selbstorganisation in der Zeit.* Berlin: Akademie-Verlag.

[118] Ebeling, W.; Engel-Herbert, H. (1980). The influence of external fluctuations on self-sustained temporal oscillations. *Physica A* **104**, 378–396.

[119] Ebeling, W.; Feistel, R. (1982). *Physik der Selbstorganisation und Evolution.* Berlin: Akademie-Verlag.

[120] Ebeling, W.; Feistel, R. (1994). *Chaos und Kosmos. Prinzipien der Evolution.* Heidelberg: Spektrum Akademischer Verlag.

[121] Ebeling, W.; Freund, J.; Schweitzer, F. (1998). *Komplexe Strukturen: Entropie und Information.* Stuttgart: Teubner.

[122] Ebeling, W.; Rosé, H.; Schuchhardt, J. (1994). Evolutionary strategies for solving frustrated problems. In: R. K. Belew; L. B. Booker (eds.), *Proc. 1st IEEE Conf. Evol. Computation*, Piscataway, NJ: IEEE Press, pp. 79–81.

[123] Ebeling, W.; Schimansky-Geier, L.; Schweitzer, F. (1990). Stochastic theory of nucleation in open molecular systems. *Zeitschrift für physikalische Chemie (Neue Folge)* **169**, 1–10.

[124] Ebeling, W.; Schweitzer, F. (2001). Swarms of particle agents with harmonic interactions. *Theory Biosci.* **120(3–4)**, 207–224.

[125] Ebeling, W.; Schweitzer, F.; Schimansky-Geier, L.; Ulbricht, H. (1990). Stochastic approach to cluster formation in adiabatically expanding molecular beams. *Zeitschrift für physikalische Chemie (Leipzig)* **271**, 1113–1122.

[126] Ebeling, W.; Schweitzer, F.; Tilch, B. (1999). Active Brownian particles with energy depots modelling animal mobility. *BioSystems* **49**, 17–29.

[127] Edelstein-Keshet, L. (1994). Simple models for trail following behaviour: Trunk trails versus individual foragers. *J. Math. Biol.* **32**, 303–328.

[128] Edelstein-Keshet, L.; Watmough, J.; Ermentrout, G. B. (1995). Trail following in ants: Individual properties determine population behaviour. *Behav. Ecol. Sociobiol.* **36**, 119–133.

[129] Eden, M. (1961). A two-dimensional growth process. In: J. Neyman (ed.), *Proc. Fourth Berkeley Symposium on Mathematical Statistics and Probability, Vol. IV: Biology and Problems of Health*, Berkeley: University of California Press, pp. 223–239.

[130] Eigen, M. (1971). The self-organization of matter and the evolution of biological macromolecules. *Naturwissenschaften* **58**, 465.

[131] Eigen, M.; Schuster, P. (1979). *The Hypercycle.* Berlin: Springer.

[132] Einstein, A. (1905). Über die von der molekularkinetischen Theorie der Wärme geforderte Bewegung von in ruhenden Flüssigkeiten suspendierten Teilchen. *Annalen der Physik (Leipzig)* **17**, 549–560.

[133] Einstein, A. (1926). *Investigations on the Theory of the Brownian Motion.* London: Methuen (edited by R. Fürth).

[134] Engelmore, R.; Morgan, T. (eds.) (1988). *Blackboard Systems.* Insight Series in Artificial Intelligence. Reading, MA: Addison-Wesley.

[135] Epstein, J. M.; Axtell, R. (1996). *Growing Artificial Societies: Social Science from the Bottom Up.* Cambridge, MA: MIT Press/Brookings.

[136] Erdmann, U. (1997). *Ensembles von Van-der-Pol-Oszillatoren.* Master's Thesis, Humboldt University, Berlin.

[137] Erdmann, U.; Ebeling, W.; Schimansky-Geier, L.; Schweitzer, F. (2000). Brownian Particles far from Equilibrium. *Eur. Phys. J. B* **15(1)**, 105–113.

[138] Family, F. (1993). Fractal structures and dynamics of cluster growth. In: P. J. Reynolds (ed.), *On Clusters and Clustering. From Atoms to Fractals*, Amsterdam: North-Holland, pp. 323-344.

[139] Family, F.; Vicsek, T. (eds.) (1991). *Dynamics of Fractal Surfaces.* Singapore: World Scientific.

[140] Feinberg, M.; Horn, F. J. M. (1974). Dynamics of open chemical systems and the algebraic structure of the underlying reaction network. *Chem. Eng. Sci.* **29**, 775–789.

[141] Feistel, R.; Ebeling, W. (1982). Models of Darwin processes and evolutionary principles. *BioSystems* **15**, 291.

[142] Feistel, R.; Ebeling, W. (1989). *Evolution of Complex Systems. Self-Organization, Entropy and Development.* Dordrecht: Kluwer.

[143] Ferber, J. (1999). *Multi-Agent Systems: An Introduction to Distributed Artificial Intelligence.* Harlow: Addison-Wesley-Longman.

[144] Fischer, M. M.; Fröhlich, J. (eds.) (2001). *Knowledge, Complexity and Innovation Systems.* Advances in Spatial Sciences, Berlin: Springer.

[145] Fisher, R. A. (1930). *The Genetical Theory of Natural Selection.* Oxford: Oxford University Press.

[146] Fogel, D. B. (1995). *Evolutionary Computation – Towards a New Philosophy of Machine Intelligence.* Piscataway, NJ: IEEE Press.

[147] Föllmer, H. (1974). Random economies with many interacting agents. *J. Math. Econ.* **1**, 51–62.

[148] Fomin, F.; Petrov, N. (1996). Pursuit-evasion and search poblems on graphs. *Congressus Numerantium* **122**, 47–58.

[149] Franke, K.; Gruler, H. (1990). Galvanotaxis of human granulocytes: Electric field jump studies. *Eur. Biophys. J.* **18**, 335–346.

[150] Frankhauser, P. (1990). Fractal structures of urban systems. *Methods Operations Res.* **60**, 697–708.

[151] Frankhauser, P. (1991). Aspects fractals des structures urbaines. *L'Espace Géographique* **1**, 45–69.

[152] Frankhauser, P. (1991). *Beschreibung der Evolution urbaner Systeme mit der Mastergleichung.* Ph.D. Thesis, Universität Stuttgart.

[153] Frankhauser, P. (1991). The Pareto–Zipf distribution of urban systems as stochastic process. In: W. Ebeling; M. Peschel; W. Weidlich (eds.), *Models of Self-Organization in Complex Systems: MOSES*, Berlin: Akademie-Verlag, pp. 276–287.

[154] Frankhauser, P. (1992). Fractal properties of settlement structures. In: *Proc. 1st Int. Semin. Struct. Morphol.*, Montpellier-La Grande Motte, pp. 357–368.

[155] Frankhauser, P. (1994). Fractales, tissus urbains et réseaux de transport. *Rev. Écon. Pol.* **104 (2/3)**, 435–455.

[156] Frankhauser, P. (1994). *La Fractalité des Structures Urbaines.* Paris: Anthropos.

[157] Frankhauser, P.; Sadler, R. (1992). Fractal analysis of urban structures. In: *Natural Structures – Principles, Strategies and Models in Architecture and Nature. Proc. Int. Symp. SFB 230*, Stuttgart, vol. 4 of *Mitteilungen des SFB 230*, pp. 57–65.

[158] Franks, N.; Gomez, N.; Goss, S.; Deneubourg, J. L. (1991). The blind leading the blind in army ant raid patterns: Testing a model of self-organization *(Hymenoptera: Formicidae). J. Insect Behav.* **4/5**, 583–607.

[159] Freimuth, R. D.; Lam, L. (1992). Active walker models for filamentary growth patterns. In: L. Lam; V. Naroditsky (eds.), *Modeling Complex Phenomena*, New York: Springer, pp. 302–313.

[160] Fricke, T.; Schimansky-Geier, L. (1996). Moving spots in three dimensions in activator–inhibitor dynamics. In: H. Engel; F.-J. Niedernostheide; H.-G. Purwins; E. Schöll (eds.), *Self-Organization in Activator-Inhibitor Systems: Semiconductors, Gas-Discharges and Chemical Active Media-Dynamics*, Berlin: Wissenschaft & Technik, pp. 184–189.

[161] Fromherz, P.; Zeiler, A. (1994). Dissipative condensation of ion channels described by a Langevin–Kelvin equation. *Phys. Lett. A* **190**, 33–37.

[162] Fujita, M. (1989). *Urban Economic Theory.* Cambridge: Cambridge University Press.

[163] Fürth, R. (1920). Die Brownsche Bewegung bei Berücksichtigung einer Persistenz der Bewegungsrichtung. Mit Anwendungen auf die Bewegung lebender Infusorien. *Z. Phys.* **11**, 244–256.

[164] Galam, S. (1990). Social paradoxes of majority rule voting and renormalization group. *J. Stat. Phys.* **61(3–4)**, 943–951.

[165] Galam, S. (1991). Renormalisation group, political paradoxes, and hierarchies. In: W. Ebeling; M. Peschel; W. Weidlich (eds.), *Models of Self-Organization in Complex Systems: MOSES*, Berlin: Akademie-Verlag, pp. 53–59.

[166] Galam, S. (1997). Rational decision making: A random field Ising model at $T = 0$. *Physica A* **238**, 66–88.
[167] Galam, S.; Moscovici, S. (1991). Towards a theory of collective phenomena: Consensus and attitude change in groups. *Eur. J. Soc. Psychol.* **21**, 49–74.
[168] Galluccio, A.; Loebl, M.; Vondrak, J. (2000). A new algorithm for the Ising problem: Partition function for finite lattice graphs. *Phys. Rev. Lett.* **84**, 5924–5927.
[169] Gardiner, C. W. (1983). *Handbook of Stochastic Methods for Physics, Chemistry and the Natural Sciences.* Berlin: Springer.
[170] Gerhart, M.; Schuster, H. (1995). *Das Digitale Universum.* Braunschweig: Vieweg.
[171] Gervet, J.; Deneubourg, J. L. (1991). Task differentiation in *Polistes* wasp colonies: A model for self-organizing groups of robots. In: J. A. Meyer; S. W. Wilson (eds.), *From Animals to Animats*, Cambridge, MA: MIT Press, pp. 346–355.
[172] Ghent, A. W. (1960). A study of the group feeding behaviour of the larvae of the jack pine sawfly *Neodiprion pratti banksianae* Rho. *Behaviour* **16/1–2**, 110–148.
[173] Gilbert, E. N.; Pollack, H. O. (1968). Steiner minimal trees. *SIAM J. Appl. Math.* **16**, 1–29.
[174] Gilbert, N.; Doran, J. (eds.) (1994). *Simulating Societies: The Computer Simulation of Social Processes.* London: University College.
[175] Gilbert, N.; Terna, P. (2000). How to build and use agent-based models in social science. *Mind Soc.* **1**, 57–72.
[176] Gilbert, N.; Troitzsch, K. G. (1999). *Simulation for the Social Scientist.* Buckingham, Philadelphia: Open University Press.
[177] Gipps, P. G.; Marksjö, B. (1985). A micro-simulation model for pedestrian flows. *Math. Comput. Simulation* **27**, 95–105.
[178] Goldberg, D. E. (1989). *Genetic Algorithms in Search, Optimization and Machine Learning.* Reading, MA: Addison-Wesley.
[179] Goodchild, M.; Mark, D. M. (1987). The fractal nature of geographical phenomena. *Ann. Assoc. Am. Geogr.* **77**, 265–178.
[180] Goss, S.; Aron, S.; Deneubourg, J. L.; Pasteels, J. M. (1989). Self-organized shortcuts in the Argentine ant. *Naturwissenschaften* **76**, 579–581.
[181] Graham, R. (1973). Statistical theory of instabilities in stationary nonequilibrium systems with applications to lasers and nonlinear optics. In: G. Höhler (ed.), *Quantum Statistics in Optics and Solid State Physics*, Berlin: Springer, vol. 66 of *Springer Tracts in Modern Physics*, p. 111.
[182] Grassé, P. P. (1959). La reconstruction du nid et les coordinations interindividuelles chez *Bellicositermes Natalensis* et *Cubitermes sp.* La théorie de la stigmergie: Essai d'interprétation du comportement des termites constructeurs. *Insectes Sociaux* **6**, 41–81.
[183] Grimsehl, E. (1977). *Lehrbuch der Physik, Bd. 1: Mechanik, Akustik, Wärmelehre.* Leipzig: Teubner, 22nd ed.
[184] Gruler, H. (1995). Cell migration, molecular machines, and living liquid crystals. *J. Trace Microprobe Tech.* **13(3)**, 403–412.
[185] Gruler, H. (1995). New insights into directed cell migration: Characteristics and mechanisms. *Nouv. Rev. Fr. Hematol.* **37**, 255–265.
[186] Gruler, H.; Boisfleury-Chevance, A. d. (1994). Directed cell movement and cluster formation: Physical principles. *J. Physique I (France)* **4**, 1085–1105.
[187] Gruler, H.; Bültmann, B. (1984). Analysis of cell movement. *Blood Cells* **10**, 61–77.

[188] Gruler, H.; Nuccitelli, R. (1991). Neural crest cell galvanotaxis: New data and novel approach to the analysis of both galvanotaxis and chemotaxis. *Cell Mot. Cytoskel.* **19**, 121–133.

[189] Grünbaum, D.; Okubo, A. (1994). Modelling social animal aggregation. In: S. A. Levin (ed.), *Frontiers in Theoretical Biology*, New York: Springer, vol. 100 of *Lecture Notes in Biomathematics*.

[190] Günther, R.; Shapiro, B.; Wagner, P. (1992). Physical complexity and Zipf's law. *J. Theor. Phys.* **31**, 525–543.

[191] Guttowitz, H. (ed.) (1991). *Cellular Automata: Theory and Experiment*. Cambridge: MIT Press.

[192] Haag, G. (1994). The rank-size distribution of settlements as a dynamic multifractal phenomenon. *Chaos, Solitons & Fractals* **4**, 519–534.

[193] Haag, G.; Dendrinos, D. S. (1983). Towards a stochastic theory of location: A nonlinear migration process. *Geogr. Anal.* **15**, 269–286.

[194] Haag, G.; Munz, M.; Pumain, P.; Sanders, L.; Saint-Julien, T. (1992). Interurban migration and the dynamics of a system of cities: 1. The stochastic framework with an application to the French urban system. *Environ. Planning A* **24**, 181–198.

[195] Haag, G.; Weidlich, W. (1984). A stochastic theory of interregional migration. *Geogr. Anal.* **16**, 331–357.

[196] Haefner, J. W.; Crist, T. O. (1994). Spatial model of movement and foraging in harvester ants *(Pogonomyrmex)* (I): The roles of memory and communication. *J. Theor. Biol.* **166**, 299–313.

[197] Haken, H. (1973). *Zeitschrift für Physik* **273**, 267

[198] Haken, H. (1978). *Synergetics. An Introduction. Nonequilibrium Phase Transitions in Physics, Chemistry and Biology*. Berlin: Springer, 2nd enl. ed.

[199] Haken, H. (1983). *Advanced Synergetics – Instability Hierarchies of Self-Organizing Systems and Devices*. Berlin: Springer.

[200] Hall, R. L. (1977). Amoeboid movement as a correlated walk. *J. Math. Biol.* **4**, 327–335.

[201] Hall, R. L.; Peterson, S. C. (1979). Trajectories of human granulocyles. *Biophys. J.* **25**, 365–372.

[202] Halsey, T. C.; Leibig, M. (1990). Electrodeposition and diffusion-limited aggregation. *J. Chem. Phys.* **92/6**, 3756–3767.

[203] Hänggi, P.; Bartussek, R. (1996). Brownian rectifiers: How to convert Brownian motion into directed transport. In: J. Parisi; S. C. Müller; W. Zimmermann (eds.), *Nonlinear Physics of Complex Systems – Current Status and Future Trends*, Berlin: Springer, pp. 294–308.

[204] Harada, Y.; Iwasa, Y. (1994). Lattice population dynamics for plants with dispersing seeds and vegetative propagation. *Res. Population Ecol.* **36**, 237–249.

[205] Hegselmann, R.; Flache, A. (1998). Understanding complex social dynamics: A plea for cellular automata based modelling. *J. Artif. Soc. Soc. Simulation* **1(3)**. `http://www.soc.surrey.ac.uk/JASSS/1/3/1.html`.

[206] Hegselmann, R. H.; Mueller, U.; Troitzsch, K. G. (eds.) (1996). *Modeling and Simulation in the Social Sciences from the Philosophy of Science Point of View*. Dordrecht: Kluwer.

[207] Hegselmann, R. H.; Peitgen, H. O. (eds.) (1996). *Modelle sozialer Dynamiken: Ordnung, Chaos und Komplexität*. Wien: Hölder-Pichker-Tempsky.

[208] Helbing, D. (1992). A fluid-dynamic model for the movement of pedestrians. *Complex Syst.* **6**, 391–415.

[209] Helbing, D. (1993). Stochastic and Boltzmann-like models for behavioral changes, and their relation to game theory. *Physica A* **193**, 241–258.

[210] Helbing, D. (1995). *Quantitative Sociodynamics. Stochastic Methods and Models of Social Interaction Processes.* Dordrecht: Kluwer Academic.

[211] Helbing, D. (1997). *Verkehrsdynamik. Neue physikalische Modellierungskonzepte.* Berlin: Springer.

[212] Helbing, D. (2001). Traffic and related self-driven many-particle systems. *Rev. Mod. Phys.* **73(4)**, 1067–1141.

[213] Helbing, D.; Farkas, I.; Molnar, P.; Vicsek, T. (2002). Simulation of pedestrian crowds in normal and evacuation situations. In: M. Schreckenberg; S. D. Sharma (eds.), *Pedestrian and Evacuation Dynamics*, Berlin: Springer, pp. 21–58.

[214] Helbing, D.; Farkas, I.; Vicsek, T. (2000). Freezing by heating in a driven mesoscopic system. *Phys. Rev. Lett.* **84**, 1240–12.

[215] Helbing, D.; Farkas, I.; Vicsek, T. (2000). Simulating dynamical features of escape panic. *Nature* **407**, 487–490.

[216] Helbing, D.; Greiner, A. (1997). Modeling and simulation of multilane traffic flow. *Phys. Rev. E* **55**, 5498–5507.

[217] Helbing, D.; Keltsch, P.; Molnár, P. (1997). Modelling the evolution of human trail systems. *Nature* **388**, 47–50.

[218] Helbing, D.; Molnár, P. (1995). Social force model for pedestrian dynamics. *Phys. Rev. E* **51/5**, 4282–4286.

[219] Helbing, D.; Molnár, P. (1996). Fußgängerdynamik in der Stadt. In: K. Teichmann; J. Wilke (eds.), *Prozeß und Form natürlicher Konstruktionen*, Berlin: Ernst & Sohn, p. 217.

[220] Helbing, D.; Molnár, P. (1997). Self-organization phenomena in pedestrian crowds. In: F. Schweitzer (ed.), *Self-Organization of Complex Structures: From Individual to Collective Dynamics*, London: Gordon and Breach, pp. 569–578.

[221] Helbing, D.; Molnar, P.; Schweitzer, F. (1994). Computer simulations of pedestrian dynamics and trail formation. In: *Evolution of Natural Structures. Proc. 3rd Int. Symp. SFB 230*, Stuttgart, vol. 9 of *Mitteilungen des SFB 230*, pp. 229–234.

[222] Helbing, D.; Schweitzer, F.; Keltsch, J.; Molnár, P. (1997). Active walker model for the formation of human and animal trail systems. *Phys. Rev. E* **56/3**, 2527–2539.

[223] Helbing, D.; Weidlich, W. (1995). Quantitative Soziodynamik. Gegenstand, Methodik, Ergebnisse und Perspektiven. *Kölner Zeitschrift für Soziologie und Sozialpsychologie* **47**, 114–140.

[224] Henderson, J. V. (1979). *Economic Theory and the Cities.* New York: Academic Press.

[225] Henderson, J. V. (1988). *Urban Development. Theory, Fact, and Illusion.* Oxford: Oxford University Press.

[226] Henderson, J. V. (1996). Ways to think about urban concentration: Neoclassical urban systems versus the new economic geography. *Int. Reg. Sci. Rev.* **19**, 31–36.

[227] Henderson, L. F. (1971). The statistics of crowd fluids. *Nature* **229**, 381–383.

[228] Henderson, L. F. (1974). On the fluid mechanics of human crowd motion. *Transp. Res.* **8**, 509–515.

[229] Henderson, L. F.; Jenkins, D. M. (1974). Response of pedestrians to traffic challenge. *Transp. Res.* **8**, 71–74.

[230] Henderson, L. F.; Lyons, D. J. (1972). Sexual differences in human crowd motion. *Nature* **240**, 353–355.

[231] Henisch, H. K. (1988). *In Vitro Veritas. Crystals in Gels and Liesegang Rings.* Cambridge: Cambridge University Press.

[232] Herrero, M. A.; Velasquez, J. J. L. (1996). Singularity patterns in a chemotaxis model. *Math. Ann.* **306**, 583–623.

[233] Hespanha, P.; Kim, H. J.; Sastry, S. (1999). Multiple-agent probabilistic pursuit-evasion games. Technical report, Dept. Electrical Engineering and Computer Science, University of California at Berkeley.

[234] Höfer, T. (1999). Chemotaxis and aggregation in the cellular slime mould. In: S. C. Müller; J. Parisi; W. Zimmermann (eds.), *Transport and Structure. Their Competitive Roles in Biophysics and Chemistry*, Berlin: Springer, vol. 532 of *Lecture Notes in Physics*, pp. 137–150.

[235] Höfer, T.; Sherratt, J. A.; Maini, P. K. (1995). *Dictyostelium discoideum*: Cellular self-organization in an excitable biological medium. *Proc. R. Soc. London B* **259**, 249–257.

[236] Holland, J.; Miller, J. (1991). Adaptive agents in economic theory. *Am. Econ. Rev. Pap. Proc.* **81**, 365–370.

[237] Holland, J. H. (1975). *Adaptation in Natural and Artificial Systems.* Ann Arbor: University of Michigan Press.

[238] Hölldobler, B.; Möglich, M. (1980). The foraging system of *Pheidole militicida (Hymenoptera: Formicidae). Insectes Sociaux* **27/3**, 237–264.

[239] Hölldobler, B.; Wilson, E. O. (1990). *The Ants.* Cambridge, MA: Belknap.

[240] Hołyst, J. A.; Hagel, T.; Haag, G. (1997). Destructive role of competition and noise for control of microeconomical chaos. *Chaos, Solitons & Fractals* **8/9**, 1489–1505.

[241] Hołyst, J. A.; Hagel, T.; Haag, G.; Weidlich, W. (1996). How to control a chaotic economy? *J. Evol. Econ.* **6**, 31–42.

[242] Hongler, M. O.; Ryter, D. M. (1978). *Zeitschrift für Physik B* **31**, 333

[243] Hopfield, J. J.; Tank, D. W. (1985). Computing RC-networks. *Biol. Cybernetics* **52**, 141.

[244] Horn, F. J. M. (1973). On a connexion between stability and graphs in chemical kinetics. I. Stability and the reaction diagram. *Proc. R. Soc. London A* **334**, 299–312.

[245] Horn, F. J. M. (1973). On a connexion between stability and graphs in chemical kinetics. II. Stability and the complex graph. *Proc. R. Soc. London A* **334**, 313–330.

[246] Huberman, B.; Glance, N. (1993). Evolutionary games and computer simulations. *Proc. Natl. Acad. Sci. USA* **90(16)**, 7715–7718.

[247] Humpert, K.; Becker, S.; Brenner, K. (1996). Entwicklung großstädtischer Agglomerationen. In: K. Teichmann; J. Wilke (eds.), *Prozeß und Form natürlicher Konstruktionen*, Berlin: Ernst & Sohn, pp. 182–189.

[248] Humpert, K.; Brenner, K.; Becker, S. (eds.) (2002). *Fundamental Principles of Urban Growth.* Wuppertal: Müller und Busmann.

[249] Humpert, K.; Bohm, H. J.; Nagler, H. (1989). Natürliche Prozesse – Haus und Stadt. Die universellen Gestaltwerdungsprozesse menschlicher Siedlungen. In: *Beiträge z. 1. Int. Symp. SFB 230, Teil 2*, Stuttgart, vol. 3 of *Mitteilungen des SFB 230*, pp. 17–38.

[250] Humpert, K.; Brenner, K. (1992). Das Phänomen der Stadt als fraktale Struktur. In: *Das Phänomen der Stadt. Berichte aus Forschung und Lehre*, Stuttgart: Städtebauliches Institut, Universität Stuttgart, pp. 223–269.

[251] Humpert, K.; Brenner, K.; Becker, S. (1996). Von Nördlingen bis Los Angeles - fraktale Gesetzmäßigkeiten der Urbanisation. *Spektrum der Wissenschaft*, 18–22.

[252] Huston, M.; DeAngelis, D.; Post, W. (1988). New computer models unify ecological theory. *BioScience* **38**, 682–691.

[253] Ikegami, T. (1994). From genetic evolution to emergence of game strategies. *Physica D* **75**, 310–327.

[254] Ising, E. (1925). Beitrag zur Theorie des Ferromagnetismus. *Zeitschrift für Physik* **31**, 235–258.

[255] Itami, R. (1988). Cellular worlds – models for dynamic conceptions of landscape. *Landscape Architecture*, 52–57.

[256] Jäger, W.; Luckhaus, S. (1992). On explosions of solutions to a system of partial differential equations modelling chemotaxis. *Trans. AMS* **329**, 819–824.

[257] Jantsch, E. (1980). *The Self-Organizing Universe. Scientific and Human Implications of the Emerging Paradigm of Evolution.* Oxford: Pergamon.

[258] Jimenez Montano, M. A.; Ebeling, W. (1980). A stochastic evolutionary model of technological change. *Collective Phenomena* **3**, 107–114.

[259] Johnson, D. S.; Lenstra, J. K.; Rinooy Kan, A. H. G. (1978). The complexity of the network design problem. *Networks* **8**, 279–285.

[260] Jülicher, F.; Prost, J. (1995). Cooperative molecular motors. *Phys. Rev. Lett.* **75/13**, 2618–2621.

[261] Kacperski, K.; Hołyst, J. A. (1996). Phase transitions and hysteresis in a cellular automata-based model of opinion formation. *J. Stat. Phys.* **84**, 169–189.

[262] Kacperski, K.; Hołyst, J. A. (1997). Leaders and clusters in a social impact model of opinion formation. In: F. Schweitzer (ed.), *Self-Organization of Complex Structures: From Individual to Collective Dynamics*, London: Gordon and Breach, pp. 367–378.

[263] Kai, S. (ed.) (1992). *Pattern Formation in Complex Dissipative Systems.* Singapore: World Scientific.

[264] Kai, S.; Müller, S. C. (1985). Spatial and temporal macroscopic structures in chemical reaction systems: Precipitation patterns and interfacial motion. *Sci. Form* **1/1**, 9–39.

[265] Kaiser, D. (1979). Social gliding is correlated with the presence of pili in *Myxococcus xanthus. Proc. Natl. Acad. Sci. USA* **76/11**, 5952–5956.

[266] Kappler, O. (1931). *Annalen der Physik* **11**, 233.

[267] Kareiva, P.; Shigesada, N. (1983). Analyzing insect movement as a correlated random walk. *Oecologica* **56**, 234–238.

[268] Kauffman, S. A. (1993). *The Origins of Order: Self-Organization and Selection in Evolution.* Oxford: Oxford University Press.

[269] Kayser, D. R.; Aberle, L. K.; Pochy, R. D.; Lam, L. (1992). Active walker models: Tracks and landscapes. *Physica A* **191**, 17–24.

[270] Keller, E.; Segel, L. (1970). Initiation of slime mold aggregation viewed as an instability. *J. Theor. Biol.* **26**, 399–415.

[271] Keller, J. B.; Rubinow, S. I. (1981). Recurrent precipitation and Liesegang rings. *J. Chem. Phys.* **74/9**, 5000–5007.

[272] Kirkpatrick, S.; Gelatt, C. D.; Vecchi, M. P. (1983). Optimization by simulated annealing. *Science* **220**, 671.

[273] Kirman, A. (1992). Whom or what does the representative agent represent? *J. Econ. Perspect.* **6**, 126–139.

[274] Kirman, A. (1993). Ants, rationality, and recruitment. *Q. J. Econ.* **108**, 37–155.

[275] Kirman, A.; Zimmermann, J.-B. (eds.) (2001). *Economies with Heterogeneous Interacting Agents*, vol. 503 of *Lecture Notes in Economics and Mathematical Systems.* Berlin: Springer.

[276] Klimontovich, Y. L. (1994). Nonlinear Brownian motion. *Physics-Uspekhi* **37(8)**, 737–766.

[277] Kohring, G. A. (1996). Ising models of social impact: The role of cumulative advantage. *J. Physique I (France)* **6**, 301–308.

[278] Kooijman, S. A. L. M. (1993). *Dynamic Energy Budgets in Biological Systems.* Cambridge: Cambridge University Press.

[279] Krischer, K.; Mikhailov, A. S. (1994). Bifurcation to traveling spots in reaction-diffusion systems. *Phys. Rev. Lett.* **73**, 23.

[280] Kropp, J.; Block, A.; Bloh, W. v.; Klenke, T.; Schellnhuber, H. J. (1997). Multifractal characterization of microbially induced magnesian calcite formation in recent tidal flat sediments. *Sediment. Geol.* **109**, 37–51.

[281] Krugman, P. (1991). *Geography and Trade.* Cambridge, MA: MIT Press.

[282] Krugman, P. (1991). Increasing returns and economic geography. *J. Political Econ.* **99**, 483–499.

[283] Krugman, P. (1992). A dynamic spatial model. *National Bureau of Economic Research Working Paper* **4219**.

[284] Krugman, P. (1996). *The Self-Organizing Economy.* Oxford: Blackwell.

[285] Krugman, P. (1996). Urban concentration: The role of increasing returns and transportation costs. *Int. Reg. Sci. Rev.* **19**, 5–30.

[286] Kube, C. R.; Zhang, H. Z. (1994). Collective robotics: from social insects to robots. *Adaptive Behav.* **2(2)**, 189–218.

[287] Laarhoven, P. J. M.; Aarts, E. H. C. (1987). *Simulated Annealing: Theory and Applications.* Dordrecht: Reidel.

[288] Lam, L. (1995). Active walker models for complex systems. *Chaos, Solitons & Fractals* **6**, 267–285.

[289] Lam, L. (1995). Electrodeposition pattern formation: An overview. In: H. Merchant (ed.), *Defect Structure, Morphology and Properties of Deposits,* The Minerals, Metals & Materials Society, pp. 169–193.

[290] Lam, L.; Naroditsky, V. (eds.) (1992). *Modeling Complex Phenomena.* New York: Springer.

[291] Lam, L.; Pochy, R. (1993). Active walker models: Growth and form in nonequilibrium systems. *Comput. Phys.* **7**, 534–541.

[292] Lam, L.; Veinott, M. C.; Pochy, R. (1995). Abnormal spatiotemporal growth. In: P. E. Cladis; P. Palffy-Muhoray (eds.), *Spatio-Temporal Patterns in Nonequilibrium Complex Systems,* Reading, MA: Addison-Wesley, pp. 659–670.

[293] Lambrinos, D.; Maris, M.; Kobayashi, H.; Labhart, T.; Pfeifer, R.; Wehner, R. (1997). An autonomous agent navigating with a polarized light compass. *Adaptive Behav.* **6(1)**, 175–206.

[294] Lane, D. (1992). Artificial worlds and economics. *J. Evol. Econ.* **3**, 89–107.

[295] Langton, C. G. (ed.) (1994). *Artificial Life III. Proc. Workshop on Artificial Life,* June 1992. Reading, MA: Addison-Wesley.

[296] Latané, B.; Nowak, A.; Liu, J. M. (1994). Dynamism, polarization, and clustering as order parameters of social systems. *Behavioral Sci.* **39**, 1–24.

[297] Lauffenburger, D. A. (1985). Chemotaxis: Analysis for quantitative studies. *Biotech. Prog.* **1/3**, 151–160.

[298] Lee, K. J.; McCormick, W. D.; Pearson, J. E.; Swinney, H. L. (1994). Experimental observation of self-replicating spots in a reaction-diffusion system. *Nature* **369**, 215–218.

[299] Leven, R. W.; Koch, B. P.; Pompe, B. (1989). *Chaos in dissipativen Systemen.* Berlin: Akademie-Verlag.

[300] Lewenstein, M.; Nowak, A.; Latané, B. (1992). Statistical mechanics of social impact. *Phys. Rev. A* **45(2)**, 703–716.

[301] Lewin, K. (1951). *Field Theory in Social Science.* New York: Harper and Brothers.

[302] Lindgren, K.; Nordahl, M. G. (1994). Evolutionary dynamics of spatial games. *Physica D* **75**, 292–309.
[303] Lindner, B.; Schimansky-Geier, L.; Reimann, P.; Hänggi, P. (1997). Mass separation by ratchets. *Proc. Am. Phys. Soc.* **411**, 309–314.
[304] Lobo, J.; Schuler, R. (1997). Efficient organization, urban hierarchies and landscape criteria. In: F. Schweitzer (ed.), *Self-Organization of Complex Structures: From Individual to Collective Dynamics*, London: Gordon and Breach, pp. 547–558.
[305] Lorenz, H. W. (1993). *Nonlinear Dynamical Equations and Chaotic Economy.* Berlin: Springer.
[306] Lösch, A. (1940). *Die räumliche Ordnung der Wirtschaft.* Jena: Fischer. Reprint: *The Economics of Location* (English translation by W.G. Woglom), New Haven: Yale University Press, 1954.
[307] Lotka, A. J. (1941). The law of urban concentration. *Science* **94**.
[308] Lotka, A. J. (1945). The law of evolution as a maximum principle. *Hum. Biol.* **17/3**, 167–194.
[309] Løvås, G. G. (1994). Modelling and simulation of pedestrian traffic flow. *Transp. Res. B* **28/6**, 429–443.
[310] Lovely, P. S.; Dahlquist, F. W. (1975). Statistical measures of bacterial motility and chemotaxis. *J. Theor. Biol.* **50**, 477–496.
[311] Luczka, J.; Bartussek, R.; Hänggi, P. (1995). White noise induced transport in periodic structures. *Europhysics Lett.* **31**, 431.
[312] Luna, F.; Stefansson, B. (eds.) (2000). *Economic Simulations in Swarm: Agent-Based Modelling and Object Oriented Programming*, vol. 14 of *Advances in Computational Economics*. Dordrecht: Kluwer.
[313] Maddox, J. (1994). Directed motion from random noise. *Nature* **369**, 181.
[314] Maddox, J. (1994). More models of muscle movement. *Nature* **368**, 287.
[315] Maes, P. (ed.) (1991). *Designing Autonomous Agents. Theory and Practice From Biology to Engineering and Back.* Cambridge, MA: MIT Press.
[316] Magnasco, M. O. (1993). Forced thermal ratchets. *Phys. Rev. Lett.* **71/10**, 1477–1481.
[317] Magnasco, M. O. (1994). Molecular combustion motors. *Phys. Rev. Lett.* **72/16**, 2656–2659.
[318] Magnasco, M. O. (1996). Brownian combustion engines. In: M. Millonas (ed.), *Fluctuations and Order: The New Synthesis*, New York: Springer, pp. 307–320.
[319] Makse, H.; Havlin, S.; Schwartz, M.; Stanley, H. E. (1996). Method for generating long-range correlations for large systems. *Phys. Rev. E* **53/5**, 5445–5449.
[320] Makse, H.; Havlin, S.; Stanley, H. E. (1995). Modelling urban growth patterns. *Nature* **377**, 608–612.
[321] Makse, H.; Havlin, S.; Stanley, H. E.; Schwartz, M. (1995). Novel method for generating long-range correlations. *Chaos, Solitons & Fractals* **6**, 295–303.
[322] Malchow, H.; Schimansky-Geier, L. (1985). *Noise and Diffusion in Bistable Nonequilibrium Systems*, vol. 5 of *Teubner-Texte zur Physik*. Leipzig: Teubner.
[323] Mandelbrot, B. B. (1983). *The Fractal Geometry of Nature.* New York: Freeman.
[324] Mansilla, R. (1997). A new algorithmic approach to the minority game. *Complex Syst.* **11**, 387–401.
[325] Margolus, N. (1984). Physics-like models of computation. *Physica D* **10**, 81–95.

[326] Marimon, R.; McGrattan, E.; Sargent, T. J. (1990). Money as a medium of exchange in an economy with artificially intelligent agents. *J. Econ. Dynamics Control* **14**, 329–373.
[327] Maschwitz, U.; Mühlenberg, M. (1975). Zur Jagdstrategie einiger orientalischer *Leptogenys*-Arten *(Formicidae: Ponerinae)*. *Oecologia* **38**, 65–83.
[328] Matsuda, H.; Ogita, A.; A.Sasaki; Sato, K. (1992). Statistical mechanics of population: The lattice Lotka–Voltera model. *Prog. Theor. Phys.* **88**, 1035–1049.
[329] Matsuda, H.; Tamachi, N.; Ogita, A.; A.Sasaki (1987). A lattice model for population biology. In: E. Teramato; M. Yamaguti (eds.), *Mathematical Topics in Biology*, New York: Springer, vol. 71 of *Lecture Notes in Biomathematics*, pp. 154–161.
[330] Matsushima, M.; Ikegami, T. (1998). Evolution of strategies in the three person iterated Prisoner's Dilemma game. *J. Theor. Biol.* **195**, 53–67.
[331] Matsushita, M.; Sano, M.; Hayakawa, Y.; Honjo, H.; Sawada, Y. (1984). Fractal structures of zinc metal leaves grown by electrodeposition. *Phys. Rev. Lett.* **53/3**, 286–289.
[332] Mayne, A. J. (1954). Some further results in the theory of pedestrians and road traffic. *Biometrica* **41**, 375–389.
[333] McBryan, O. A. (1994). An overview of message passing environments. *Parallel Computing* **20(4)**, 417–444.
[334] McCoy, B.; Wu, T. (1978). *The Two-Dimensional Ising Model.* Cambridge, MA: Harvard University Press.
[335] Metropolis, N.; Rosenbluth, A.; Rosenbluth, M.; Teller, A.; Teller, E. (1953). *J. Chem. Phys.* **21**, 1087.
[336] Meyer, J. A.; Wilson, S. W. (eds.) (1991). *From Animals to Animats. Proc. 1st Int. Conf. Simulation of Adaptive Behav.* Cambridge, MA: MIT Press.
[337] Mikhailov, A.; Zanette, D. H. (1999). Noise-induced breakdown of coherent collective motion in swarms. *Phys. Rev. E* **60**, 4571–4575.
[338] Mikhailov, A. S. (1990). *Foundations of Synergetics.* vol. I: *Distributed Active Systems.* Berlin: Springer.
[339] Mikhailov, A. S. (1993). Collective dynamics in models of communicating populations. In: H. Haken; A. S. Mikhailov (eds.), *Interdisciplinary Approaches to Nonlinear Complex Systems*, Berlin: Springer.
[340] Mikhailov, A. S.; Meinköhn, D. (1997). Self-motion in physico-chemical systems far from thermal equilibrium. In: L. Schimansky-Geier; T. Pöschel (eds.), *Stochastic Dynamics*, Berlin: Springer, vol. 484 of *Lecture Notes in Physics*, pp. 334–345.
[341] Millonas, M. M. (1992). A connectionist type model of self-organized foraging and emergent behavior in ant swarm. *J. Theor. Biol.* **159**, 529–552.
[342] Millonas, M. M. (ed.) (1996). *Fluctuations and Order: The New Synthesis.* New York: Springer.
[343] Millonas, M. M.; Dykman, M. I. (1994). Transport and current reversal in stochastically driven ratchets. *Phys. Lett. A* **185**, 65–69.
[344] Mills, E. S.; Tan, J. P. (1980). A comparison of urban population density functions in developed and developing countries. *Urban Stud.* **17**, 313–321.
[345] Minar, M.; Burkhart, R.; Langton, C.; Askenazy, M. (1996). *The Swarm Simulation System: A Toolkit for Building Multi-Agent Simulations.* Santa Fe Institute.
[346] Minsky, M.; Papert, S. (1973). *Artificial Intelligence.* Eugene: Oregon State System of Higher Education.
[347] Minsky, M. L. (1986). *The Society of Mind.* New York: Simon & Schuster.

[348] Misra, K. B. (1991). Multicriteria redundancy optimization using an efficient search procedure. *Int. J. Syst. Sci.* **22**, 2171–2183.

[349] Moller, R.; Lambrinos, D.; Pfeifer, R.; Labhart, T.; Wehner, R. (1998). Modeling ant navigation with an autonomous agent. In: R. Pfeifer; B. Blumberg; J.-A. Meyer; S. Wilson (eds.), *From Animals To Animats*, Cambridge, MA: MIT Press, vol. 5, pp. 185–194.

[350] Molnár, P. (1996). *Modellierung und Simulation der Dynamik von Fußgängerströmen.* Aachen: Shaker.

[351] Molnar, P.; Starke, J. (2000). Communication fault tolerance in distributed robotic systems. In: L. E. Parker; G. Bekey; J. Barhen (eds.), *Distributed Autonomous Robotic Systems*, Berlin: Springer, vol. 4, pp. 99–108.

[352] Molofsky, J.; Durrett, R.; Dushoff, J.; Griffeath, D.; Levin, S. (1999). Local frequency dependence and global coexistence. *Theor. Population Biol.* **55**, 270–282.

[353] Moore, C. (1997). Majority-vote cellular automata, ising dynamics, and P-completeness. *J. Stat. Phys.* **88**, 795–805.

[354] Mueller, U.; Troitzsch, K. (eds.) (1996). *Social Science Microsimulation: A Challenge for Computer Science.* Berlin: Springer.

[355] Mühlenbein, H. (1989). Parallel genetic algorithm, population dynamics and combinatorial optimization. In: H. Schaffer (ed.), *Proc. 3rd Int. Conf. Genet. Algorithms*, San Francisco: Morgan Kaufmann, pp. 416–421.

[356] Mühlenbein, H.; Schlierkamp-Voosen, D. (1994). The science of breeding and its application to the breeder genetic alogorithm. *Evol. Computation* **1**, 335–360.

[357] Müller, J. P.; Wooldridge, M. J.; Jennings, N. R. (eds.) (1997). *Intelligent Agents III : Agent Theories, Architectures, and Languages.* Berlin: Springer.

[358] Müller, M.; Wehner, R. (1988). Path integration in desert ants, *Cataglyphis fortis. Proc. Natl. Acad. Sci. USA* **85**, 5287–5290.

[359] Müller, S.; Kai, S.; Ross, J. (1982). Curiosities in periodic precipitation patterns. *Science* **216**, 635–637.

[360] Munz, M.; Weidlich, W. (1990). Settlement formation, II. Numerical simulation. *Ann. Reg. Sci.* **24**, 177–196.

[361] Nagel, K.; Rasmussen, S.; Barrett, C. L. (1997). Network traffic as a self-organized critical phenomenon. In: F. Schweitzer (ed.), *Self-Organization of Complex Structures: From Individual to Collective Dynamics*, London: Gordon and Breach, pp. 579–592.

[362] Nakamura, H. (1980). Ecological studies of the European pine sawfly, *Neodiprion sertifer* (Geoffroy) (Hymenoptera: Diprionidae), I. The effect of larval aggregation and its form. *Jpn. J. Appl. Entomol. Zool.* **24**, 137–144.

[363] Navin, P. D.; Wheeler, R. J. (1969). Pedestrian flow characteristics. *Traffic Eng.* **39**, 31–36.

[364] Nicolis, G.; Prigogine, I. (1977). *Self-Organization in Non-Equilibrium Systems: From Dissipative Structures to Order Through Fluctuations.* New York: Wiley.

[365] Niedersen, U.; Schweitzer, F. (eds.) (1993). *Ästhetik und Selbstorganisation*, vol. 4 of *Selbstorganisation. Jahrbuch für Komplexität in den Natur- Sozial- und Geisteswissenschaften.* Berlin: Duncker & Humblot.

[366] Niemeyer, L.; Pietronero, L.; Wiesmann, H. J. (1984). Fractal dimension of dielectric breakdown. *Phys. Rev. Lett.* **52**, 1033–1036.

[367] Nolfi, S.; Floreano, D. (2000). *Evolutionary Robotics: The Biology, Intelligence, and Technology of Self-Organizing Machines.* Cambridge, MA: MIT Press.

[368] Nossal, R. (1983). Stochastic aspects of biological locomotion. *J. Stat. Phys.* **30**, 391–399.
[369] Nossal, R.; Weiss, G. H. (1974). A descriptive theory of cell migration on surfaces. *J. Theor. Biol.* **47**, 103–113.
[370] Nowak, A.; Szamrej, J.; Latané, B. (1990). From private attitude to public opinion: A dynamic theory of social impact. *Psychol. Rev.* **97**, 362–376.
[371] Nowak, M. A.; May, R. M. (1992). Evolutionary games and spatial chaos. *Nature* **359**, 826–829.
[372] Nowak, M. A.; May, R. M. (1993). The spatial dilemmas of evolution. *Int. J. Bifurcation Chaos* **3(1)**, 35–78.
[373] Nulton, J. D.; Salamon, P. (1988). Statistical mechanics of combinatorial optimization. *Phys. Rev. A* **37**, 1351.
[374] O'Connor, K.; Zusman, D. R. (1989). Patterns of cellular interactions during fruiting-body formation in *Myxococcus xanthus*. *J. Bacteriol.* **171/11**, 6013–6024.
[375] Okubo, A. (1986). Dynamic aspects of animal grouping: Swarms, schools, flocks, and herds. *Adv. Biophys.* **22**, 1–94.
[376] Osyczka, A.; Kundu, S. (1995). A new method to solve generalized multicriteria optimization problems using the simple genetic algorithm. *Structural Optimization* **10**, 94–99.
[377] Othmer, H. G.; Dunbar, S. R.; Alt, W. (1988). Models of dispersal in biological systems. *J. Math. Biol.* **26**, 263–298.
[378] Othmer, H. G.; Stevens, A. (1997). Aggregation, blowup and collapse: The ABC's of taxis in reinforced random walks. *SIAM J. Appl. Math.* **57/4**, 1044–1081.
[379] Parisi, J.; Müller, S. C.; Zimmermann, W. (eds.) (1996). *Nonlinear Physics of Complex Systems – Current Status and Future Trends.* Berlin: Springer.
[380] Parisi, J.; Müller, S. C.; Zimmermann, W. (eds.) (1998). *A Perspective Look at Nonlinear Media – From Physics to Biology and Social Sciences.* Berlin: Springer.
[381] Parrish, J. K.; Hamner, W. (eds.) (1997). *Animal Groups in Three Dimensions.* Cambridge: Cambridge University Press.
[382] Parunak, H. V. D. (1997). Go to the ant: Engineering principles from natural agent systems. *Ann. Operations Res.* **75**, 69–101.
[383] Pasteels, J. M.; Deneubourg, J. L. (eds.) (1987). *From Individual To Collective Behavior in Social Insects*, vol. 54 of *Experientia Supplementum.* Basel: Birkhäuser.
[384] Pasteels, J. M.; Gregoire, J. C.; Rowell-Rahier, M. (1983). The chemical ecology of defense in arthropods. *Annu. Rev. Entomol.* **28**, 263–289.
[385] Patlak, C. S. (1953). Random walk with persistence and external bias. *Bull. Math. Biol.* **15**, 311–338.
[386] Payne, T. L.; Birch, M. C.; Kennedy, C. E. (eds.) (1986). *Mechanisms in Insect Olfaction.* Oxford: Clarendon Press.
[387] Perrin, J. B. (1908). *Comptes Rendus, Paris* **146**, 967.
[388] Phipps, M. (1989). Dynamical behavior of cellular automata under the constraint of neighborhood coherence. *Geogr. Anal.* **21/3**, 197–215.
[389] Pochy, R. D.; Kayser, D. R.; Aberle, L. K.; Lam, L. (1993). Boltzmann active walkers and rough surfaces. *Physica D* **66**, 166–171.
[390] Portugali, J. (2000). *Self-Organization and the City.* Berlin: Springer.
[391] Portugali, J.; Benenson, I. (1997). Human agents between local and global forces in a self-organizing city. In: F. Schweitzer (ed.), *Self-Organization of Complex Structures: From Individual to Collective Dynamics*, London: Gordon and Breach, pp. 537–546.

[392] Pricer, J. L. (1908). The life history of the carpenter ant. *Biol. Bull. Mar. Biol. Lab. Woods Hole* **14**, 177–218.

[393] Pumain, D. (1982). *La Dynamique des Villes.* Paris: Economica.

[394] Pumain, D. (1993). Geography, physics and synergetics. In: G. Haag; U. Mueller; K. G. Troitzsch (eds.), *Economic Evolution and Demographic Change*, Berlin: Springer, vol. 395 of *Lecture Notes in Economics and Mathematical Systems*, pp. 157–175.

[395] Pumain, D.; Haag, G. (1991). Urban and regional dynamics – Towards an integrated approach. *Environ. Planning A* **23**, 1301–1313.

[396] Puu, T. (1993). Pattern formation in spatial economics. *Chaos, Solitons & Fractals* **3**, 99–129.

[397] Racz, Z.; Plischke, M. (1985). Active zone of growing clusters: Diffusion-limited aggregation and the Eden model in two and three dimensions. *Phys. Rev. A* **31/2**, 985–994.

[398] Rapp, B.; de Boisfleury-Chevance, A.; Gruler, H. (1988). Galvanotaxis of human granulocytes. Dose-response curve. *Eur. Biophys. J.* **16**, 313–319.

[399] Rascle, M.; Ziti, C. (1995). Finite time blow-up in some models of chemotaxis. *J. Math. Biol.* **33**, 388–414.

[400] Rateitschak, K.; Klages, R.; Hoover, W. G. (2000). The Nosé–Hoover thermostatted Lorentz gas. *J. Stat. Phys.* **101**, 61–77.

[401] Rayleigh, J. W. S. (1877). *The Theory of Sound.* London: Macmillan. Second, revised and enlarged edition 1894–96, reprinted by Dover, 1976.

[402] Rechenberg, I. (1994). *Evolutionsstrategie '94.* Stuttgart: Frommann-Holzboog.

[403] Reichenbach, H. (1966). *Myxococcus spp. (Myxobacerales)*: Schwarmentwicklung und Bildung von Protocysten. *Publikationen zu wissenschaftlichen Filmen* **1A**, 557–578.

[404] Reichenbach, H. (1974). *Chondromyces apiculatus (Myxobacerales)*: Schwarmentwicklung und Bildung von Protocysten. *Publikationen zu wissenschaftlichen Filmen. Sektion Biologie* **7/3**, 245–263.

[405] Reichl, L. E. (1992). *The Transition to Chaos.* New York: Springer.

[406] Reimann, P.; Bartussek, R.; Häußler, R.; Hänggi, P. (1996). *Phys. Lett. A* **215**, 26.

[407] Reimann, P.; Grifoni, M.; Hänggi, P. (1997). Quantum ratchets. *Phys. Rev. Lett.* **79/1**, 10–13.

[408] Rettenmeyer, C. W. (1963). Behavioral studies of army ants. *Univ. Kans. Sci. Bull.* **44**, 281–465.

[409] Richter, G. (1999). *Flip-Tick Architecture: A design paradigm for cycle-oriented distributed systems. GMD Report 84*, GMD, Sankt Augustin.

[410] Richter, G.; Schmitz, A.; Veit, H. (2000). Towards more design flexibility for architectures of autonomous robot control system. In: E. Pagiello; F. Groen; T. Arai; R. Dillmann; A. Stentz (eds.), *Intelligent Autonomous Systems 6*, IOS Press, vol. 6, pp. 777–784.

[411] Riethmüller, T.; Rosenkranz, D.; Schimansky-Geier, L. (1996). Granular flow modelled by Brownian particles. In: D. E. Wolf; M. Schreckenberg; A. Bachem (eds.), *Traffic and Granular Flow*, Singapore: World Scientific, p. 293.

[412] Risken, H. (1984). *The Fokker–Planck Equation.* Springer, Berlin.

[413] Ritz, D. A. (1994). Social aggregation in pelagic invertebrates. *Adv. Mar. Biol.* **30**, 155–216.

[414] Röpke, G. (1987). *Statistische Mechanik für das Nichtgleichgewicht.* Berlin: Deutscher Verlag der Wissenschaften.

[415] Rosé, H.; Ebeling, W.; Asselmeyer, T. (1996). The density of states – A measure of the difficulty of optimization problems. In: H. M. Voigt; W. Ebeling;

H. P. Schwefel; I. Rechenberg (eds.), *Parallel Problem Solving from Nature – PPSN IV*, Berlin: Springer, vol. 1141 of *Lecture Notes in Computer Science*, pp. 208–217.

[416] Rosé, H.; Hempel, H.; Schimansky-Geier, L. (1994). Stochastic dynamics of catalytic CO oxidation on Pt(100). *Physica A* **206**, 421.

[417] Rousselet, J.; Salome, L.; Adjari, A.; Prost, J. (1994). Directional motion of Brownian particles induced by a periodic asymmetric potential. *Nature* **370**, 446–448.

[418] Rovinsky, A. B.; Menzinger, M. (1992). Chemical instability induced by a differential flow. *Phys. Rev. Lett.* **69**, 1193–1196.

[419] Rustem, B.; Velupillai, K. (1990). Rationality, computability and complexity. *J. Econ. Dynamics Control* **14**, 419–432.

[420] Sakoda, J. M. (1971). The checkerboard model of social interaction. *J. Math. Sociol.* **1**, 119–132.

[421] Salamon, P.; Nulton, J.; Robinson, J.; Pedersen, J. M.; Ruppeiner, G.; Liao, L. (1988). Simulated annealing with constant thermodynamic speed. *Comput. Phys. Commun.* **49**, 423.

[422] Sargent, T. J. (1993). *Bounded Rationality in Macroeconomics.* Oxford: Clarendon Press.

[423] Schaaf, R. (1985). Stationary solutions of chemotaxis systems. *Trans. AMS* **292**, 531–556.

[424] Schaur, E. (1991). *Non-Planned Settlements: Characteristic Features – Path Systems, Surface Subdivision*, vol. 39 of *IL*. Stuttgart: Universtität Stuttgart.

[425] Schelling, T. (1969). Models of segregation. *Am. Econ. Rev.* **59**, 488–493.

[426] Schellnhuber, H. J. (1996). Towards an integrated model of the Earth system. *EGS Newsl.* **59**, 8.

[427] Schienbein, M.; Gruler, H. (1993). Langevin equation, Fokker–Planck equation and cell migration. *Bull. Math. Biol.* **55**, 585–608.

[428] Schimansky-Geier, L.; Kschischo, M.; Fricke, T. (1997). Flux of particles in sawtooth media. *Phys. Rev. Lett.* **79/18**, 3335–3338.

[429] Schimansky-Geier, L.; Mieth, M.; Rosé, H.; Malchow, H. (1995). Structure formation by active Brownian particles. *Phys. Lett. A* **207**, 140–146.

[430] Schimansky-Geier, L.; Pöschel, T. (eds.) (1997). *Stochastic Dynamics*, vol. 484 of *Lecture Notes in Physics*. Berlin: Springer.

[431] Schimansky-Geier, L.; Schweitzer, F.; Mieth, M. (1997). Interactive structure formation with Brownian particles. In: F. Schweitzer (ed.), *Self-Organization of Complex Structures: From Individual to Collective Dynamics*, London: Gordon and Breach, pp. 101–118.

[432] Schimansky-Geier, L.; Zülicke, C.; Schöll, E. (1991). Domain formation due to Ostwald ripening in bistable systems far from equilibrium. *Z. Physik B* **84**, 433–441.

[433] Schimansky-Geier, L.; Zülicke, C.; Schöll, E. (1992). Growth of domains under global constraints. *Physica A* **188**, 436–442.

[434] Schmelzer, J.; Schweitzer, F. (1987). Ostwald ripening of bubbles in liquid-gas solutions. *J. Non-Equilibrium Thermodynamics* **12**, 255–270.

[435] Schnaars, S. P. (1994). *Managing Imitation Strategies: How Later Entrants Seize Markets from Pioneers.* New York: Free Press.

[436] Schneirla, T. C. (1933). Studies of army ants in Panama. *J. Comp. Psychol.* **15**, 267–299.

[437] Schneirla, T. C. (1940). Further studies on the army ant behavior pattern: Mass organization in the swarm-raiders. *J. Comp. Psychol.* **29**, 401–461.

[438] Schreckenberg, M.; Sharma, S. D. (eds.) (2002). *Pedestrian and Evacuation Dynamics.* Berlin: Springer.

[439] Schuster, H. G. (1984). *Deterministic Chaos – An Introduction.* Weinheim: Physik-Verlag.

[440] Schwefel, H.-P. (1981). *Numerical Optimization of Computer Models.* New York: Wiley.

[441] Schweitzer, F. (1992). Simulation of cluster growth in pores with diffusion interaction. *Surf. Sci.* **272**, 235–239.

[442] Schweitzer, F. (1996). Selbstorganisation von Wege- und Transportsystemen. In: K. Teichmann; J. Wilke (eds.), *Prozeß und Form natürlicher Konstruktionen*, Berlin: Ernst & Sohn, pp. 163–169.

[443] Schweitzer, F. (1996). Self-organization of trail networks using active Brownian particles. In: R. Hofestädt; T. Lengauer; M. Löffler; D. Schomburg (eds.), *Computer Science and Biology*, IMISE Report No. 1, Leipzig, pp. 299–301.

[444] Schweitzer, F. (1997). Active Brownian particles: Artificial agents in physics. In: L. Schimansky-Geier; T. Pöschel (eds.), *Stochastic Dynamics*, Berlin: Springer, vol. 484 of *Lecture Notes in Physics*, pp. 358–371.

[445] Schweitzer, F. (1997). Structural and functional information – an evolutionary approach to pragmatic information. *World Futures: J. Gen. Evol.* **50**, 533–550.

[446] Schweitzer, F. (1997). Wege und Agenten: Reduktion und Konstruktion in der Selbstorganisationstheorie. In: H. J. Krug; L. Pohlmann (eds.), *Evolution und Irreversibilität*, Berlin: Duncker & Humblot, vol. 8 of *Selbstorganisation. Jahrbuch für Komplexität in den Natur- Sozial- und Geisteswissenschaften*, pp. 113–135.

[447] Schweitzer, F. (1998). Modelling migration and economic agglomeration with active Brownian particles. *Adv. Complex Syst.* **1/1**, 11–37.

[448] Schweitzer, F. (1998). Ökonomische Geographie: Räumliche Selbstorganisation in der Standortverteilung. In: F. Schweitzer; G. Silverberg (eds.), *Evolution und Selbstorganisation in der Ökonomie/Evolution and Self-Organization in Economics*, Berlin: Duncker & Humblot, vol. 9 of *Selbstorganisation. Jahrbuch für Komplexität in den Natur- Sozial- und Geisteswissenschaften*, pp. 97–125.

[449] Schweitzer, F. (ed.) (2002). *Modeling Complexity in Economic and Social Systems.* Singapore: World Scientific.

[450] Schweitzer, F.; Bartels, J.; Pohlmann, L. (1991). Simulation of opinion structures in social systems. In: W. Ebeling; M. Peschel; W. Weidlich (eds.), *Models of Self-Organization in Complex Systems: MOSES*, Berlin: Akademie-Verlag, pp. 236–243.

[451] Schweitzer, F.; Ebeling, W.; Rosé, H.; Weiss, O. (1996). Network optimization using evolutionary strategies. In: H. M. Voigt; W. Ebeling; H. P. Schwefel; I. Rechenberg (eds.), *Parallel Problem Solving from Nature – PPSN IV*, Berlin: Springer, vol. 1141 of *Lecture Notes in Computer Science*, pp. 940–949.

[452] Schweitzer, F.; Ebeling, W.; Rosé, H.; Weiss, O. (1998). Optimization of road networks using evolutionary strategies. *Evol. Computation* **5/4**, 419–438.

[453] Schweitzer, F.; Ebeling, W.; Tilch, B. (1998). Complex motion of Brownian particles with energy depots. *Phys. Rev. Lett.* **80/23**, 5044–5047.

[454] Schweitzer, F.; Ebeling, W.; Tilch, B. (2001). Statistical mechanics of canonical-dissipative systems and applications to swarm dynamics. *Phys. Rev. E* **64**, 021110-(1–12).

[455] Schweitzer, F.; Hołyst, J. (2000). Modelling collective opinion formation by means of active Brownian particles. *Eur. Phys. J. B* **15(4)**, 723–732.

[456] Schweitzer, F.; Lao, K.; Family, F. (1997). Active random walkers simulate trunk trail formation by ants. *BioSystems* **41**, 153–166.

[457] Schweitzer, F.; Schimansky-Geier, L. (1994). Clustering of active walkers in a two-component system. *Physica A* **206**, 359–379.

[458] Schweitzer, F.; Schimansky-Geier, L. (1996). Clustering of active walkers: Phase transitions from local interactions. In: M. Millonas (ed.), *Fluctuations and Order: The New Synthesis*, New York: Springer, pp. 293–305.

[459] Schweitzer, F.; Schimansky-Geier, L.; Ebeling, W.; Ulbricht, H. (1988). A stochastic approach to nucleation in finite systems. Theory and computer simulations. *Physica A* **150**, 261–278.

[460] Schweitzer, F.; Schimansky-Geier, L.; Ebeling, W.; Ulbricht, H. (1988). Stochastics of nucleation in isolated gases including carrier molecules. *Physica A* **153**, 573–591.

[461] Schweitzer, F.; Silverberg, G. (eds.) (1998). *Evolution und Selbstorganisation in der Ökonomie/Evolution and Self-Organization in Economics*, vol. 9 of *Selbstorganisation. Jahrbuch für Komplexität in den Natur-, Sozial- und Geisteswissenschaften*. Berlin: Duncker & Humblot.

[462] Schweitzer, F.; Steinbrink, J. (1994). Die Berliner Bebauungsstruktur. *ARCH+* **122**, 34.

[463] Schweitzer, F.; Steinbrink, J. (1997). Urban cluster growth: Analysis and computer simulation of urban aggregations. In: F. Schweitzer (ed.), *Self-Organization of Complex Structures: From Individual to Collective Dynamics*, London: Gordon and Breach, pp. 501–518.

[464] Schweitzer, F.; Steinbrink, J. (1998). Estimation of megacity growth: Simple rules versus complex phenomena. *Appl. Geogr.* **18/1**, 69–81.

[465] Schweitzer, F.; Steinbrink, J. (2002). Analysis and computer simulation of urban cluster distributions. In: K. Humpert; K. Brenner; S. Becker (eds.), *Fundamental Principles of Urban Growth*, Wuppertal: Müller und Busmann, pp. 142–157.

[466] Schweitzer, F.; Tilch, B. (1999). Modelling a dynamic switch by means of active Brownian particles. Preprint, to be submitted.

[467] Schweitzer, F.; Tilch, B. (1999). Stochastic approach to network formation in an active walker model. Preprint, to be submitted.

[468] Schweitzer, F.; Tilch, B. (2002). Self-assembling of networks in an agent-based model. *Phys. Rev. E* **66**, 026113-(1–9).

[469] Schweitzer, F.; Tilch, B.; Ebeling, W. (2000). Uphill motion of active Brownian particles in piecewise linear potentials. *Eur. Phys. J. B* **14(1)**, 157–168.

[470] Schweitzer, F.; Zimmermann, J. (2001). Communication and self-organization in complex systems: A basic approach. In: M. M. Fischer; J. Fröhlich (eds.), *Knowledge, Complexity and Innovation Systems*, Advances in Spatial Sciences, Berlin: Springer, chap. 14, pp. 275–296.

[471] Schweitzer, F.; Zimmermann, J.; Mühlenbein, H. (2002). Coordination of decisions in a spatial agent model. *Physica A* **303(1-2)**, 189–216.

[472] Segel, L. A. (1977). A theoretical study of receptor mechanisms in bacterial chemotaxis. *SIAM J. Appl. Math.* **32/3**, 653–665.

[473] Shapiro, A.; Horn, F. J. M. (1979). On the possiblity of sustained oscillations, multiple steady states, and asymmetric steady states in multicell reaction systems. *Math. Biosci.* **44**, 19–39.

[474] Sibiani, P.; Pedersen, K. M.; Hoffmann, K. H.; Salamon, P. (1990). Monte Carlo dynamics of optimization: A scaling description. *Phys. Rev. A* **42**, 7080.

[475] Silverberg, G.; Verspagen, B. (1994). Collective learning, innovation and growth in a boundedly rational, evolutionary world. *J. Evol. Econ.* **4**, 207–226.

[476] Simon, H. (1992). *Economics, Bounded Rationality and the Cognitive Revolution.* Brookfield, VT: Edward Elgar.
[477] Sinai, Y. G. (1970). Dynamical systems with elastic reflections. *Russ. Math. Surv.* **25**, 137.
[478] Skellam, J. G. (1973). The formulation and interpretation of mathematical models of diffusionary processes in population biology. In: M. S. Bartlett; R. W. Hiorns (eds.), *The Mathematical Theory of the Dynamics of Biological Populations*, New York: Academic Press, pp. 63–85.
[479] Slanina, F. (2000). Social organization in the minority game model. *Physica A* **286(1–2)**, 367–376.
[480] Starke, J.; Schanz, M.; Haken, H. (1998). Self-organized behaviour of distributed autonomous mobile robotic systems by pattern formation principles. In: R. Dillmann; T. Lüth; P. Dario; H. Wörn (eds.), *Distributed Autonomous Robotic Systems*, Berlin: Springer, vol. 3, pp. 89–100.
[481] Stauffer, D.; Stanley, H. E. (1990). *From Newton to Mandelbrot. A Primer in Theoretical Physics.* Berlin: Springer.
[482] Steels, L. (ed.) (1995). *The Biology and Technology of Intelligent Autonomous Agents.* Berlin: Springer.
[483] Steiglitz, K.; Weiner, P.; Kleitman, D. J. (1969). The design of minimum-cost survivable networks. *IEEE Trans. Circuit Theory* **CT-16/4**, 455–460.
[484] Steinbock, O. (1998). Path optimization in chemical and biological systems on the basis of excitation waves. In: J. Parisi; S. C. Müller; W. Zimmermann (eds.), *A Perspective Look at Nonlinear Media – From Physics to Biology and Social Sciences*, Berlin: Springer, pp. 179–191.
[485] Steuer, R. E. (1989). *Multiple Criteria Optimization: Theory, Computation, and Application.* Malabar, FL: Krieger.
[486] Steuernagel, O.; Ebeling, W.; Calenbuhr, V. (1994). An elementary model for directed active motion. *Chaos, Solitons & Fractals* **4**, 1917–1930.
[487] Stevens, A. (1990). Simulations of the aggregation and gliding behavior of Myxobacteria. In: W. Alt; G. Hoffmann (eds.), *Biological Motion*, Berlin: Springer, vol. 89 of *Lecture Notes in Biomathematics*, pp. 548–555.
[488] Stevens, A. (1991). A model for gliding and aggregation of Myxobacteria. In: A. Holden; M. Markus; H. G. Othmer (eds.), *Nonlinear Wave Processes in Excitable Media*, New York: Plenum Press, pp. 269–276.
[489] Stevens, A. (1993). Aggregation of Myxobacteria – a many particle system. In: *Proc. First Eur. Conf. Math. Appl. Biol. Med.*, Winnipeg: Wuerz, pp. 519–524.
[490] Stevens, A. (1995). Trail following and aggregation of Myxobacteria. *J. Biol. Syst.* **3**, 1059–1068.
[491] Stevens, A. (1996). Simulation of chemotaxis-equations in two space dimensions. In: J. Parisi; S. C. Müller; W. Zimmermann (eds.), *Nonlinear Physics of Complex Systems – Current Status and Future Trends*, Berlin: Springer.
[492] Stevens, A.; Schweitzer, F. (1997). Aggregation induced by diffusing and nondiffusing media. In: W. Alt; A. Deutsch; G. Dunn (eds.), *Dynamics of Cell and Tissue Motion*, Basel: Birkhäuser, pp. 183–192.
[493] Stratonovich, R. L. (1967). *Topics in the Theory of Random Noise.* New York: Gordon and Breach, Vol. 2.
[494] Sudd, J. H.; Franks, N. R. (1987). *The Behavioural Ecology of Ants.* Glasgow: Blackie.
[495] Suleiman, R.; Troitzsch, K. G.; Gilbert, N. (eds.) (2000). *Tools and Techniques for Social Science Simulation.* Heidelberg: Physica Verlag.
[496] Sunderam, V. (1990). PVM. A framework for parallel distributed computing. *Concurrency Pract. Exper.* **3(4)**, 315–339.

[497] Sunderam, V. (1994). The PVM concurrent computing system: Evolution, experiences and trends. *Parallel Computing* **20(4)**, 531–545.

[498] Szabó, G.; Antal, T.; Szabó, P.; Droz, M. (2000). Spatial evolutionary prisoner's dilemma game with three strategies and external constraints. *Phys. Rev. E* **62**, 1095–1115.

[499] Szabó, G.; Csaba, T. (1998). Evolutionary prisoner's dilemma game on a square lattice. *Phys. Rev. E* **58(1)**, 69–73.

[500] Takayasu, H.; Inaoka, H. (1992). New type of self-organized criticality in a model of erosion. *Phys. Rev. Lett.* **68/7**, 966–969.

[501] Takayasu, H.; Inaoka, H. (1992). A simulation of pattern formation by erosion. In: S. Kai (ed.), *Pattern Formation in Complex Dissipative Systems*, Singapore: World Scientific, pp. 103–107.

[502] Teichmann, K.; Wilke, J. (eds.) (1996). *Prozeß und Form "Natürlicher Konstruktionen". Der Sonderforschungsbereich 230.* Berlin: Ernst & Sohn.

[503] Terna, P. (1998). Simulation tools for social scientists: Building agent based models with SWARM. *J. Artif. Soc. Soc. Simulation* **1(2)**, `http://www.soc.surrey.ac.uk/JASSS/1/2/4.html`.

[504] v. Thünen, J. H. (1826). *Der isolierte Staat in Beziehung auf Landwirtschaft und Nationalökonomie.* Rostock.

[505] Tilch, B. (1996). *Strukturbildung mit Aktiven Brownschen Teilchen.* Master's Thesis, Humboldt University, Berlin.

[506] Tilch, B.; Schweitzer, F.; Ebeling, W. (1999). Directed motion of Brownian particles with internal energy depot. *Physica A* **273(3–4)**, 294–314.

[507] Tobler, W. (1979). Cellular geography. In: S. Gale; G. Olsson (eds.), *Philosophy in Geography*, Dordrecht: Kluwer, pp. 379–386.

[508] Toffoli, T.; Margolus, N. (1987). *Cellular Automata Machines – A New Environment for Modeling.* Cambridge, MA: MIT Press.

[509] Tranquillo, R. T.; Alt, W. (1990). Glossary of terms concerning oriented movement. In: W. Alt; G. Hoffmann (eds.), *Biological Motion*, Berlin: Springer, vol. 89 of *Lecture Notes in Biomathematics*, pp. 510–517.

[510] Tranquillo, R. T.; Alt, W. (1994). Dynamic morphology of leukocytes: Statistical analysis and a stochastic model for receptor-mediated cell motion and orientation. In: N. Akkas (ed.), *Biomechanics of Active Motion and Division of Cells*, New York: Springer, pp. 437–443.

[511] Tranquillo, R. T.; Lauffenburger, D. (1987). Stochastic models of leukocyte chemosensory movement. *J. Math. Biol.* **25**, 229–262.

[512] Tranquillo, R. T.; Zigmond, S. H.; Lauffenburger, D. A. (1988). Measurement of the chemotaxis coefficient for human neutrophils in the under-agarose migration assay. *Cell Mot. Cytoskeleton* **11/1**, 1–15.

[513] Troitzsch, K. G.; Mueller, U.; Gilbert, G. N.; Doran, J. E. (eds.) (1996). *Social Science Microsimulation.* Berlin: Springer.

[514] Tsubaki, Y. (1981). Some beneficial effects of aggregation in young larvae of *Pryeria sinica* Moore (Lepidoptera: Zygaenidae). *Res. Population Ecol.* **23**, 156–167.

[515] Tsubaki, Y.; Shiotsu, Y. (1982). Group feeding as a strategy for exploiting food sources in the Burnet moth *Pryeria sinica. Oecologia (Berlin)* **55**, 12–20.

[516] Ulbricht, H.; Schmelzer, J.; Mahnke, R.; Schweitzer, F. (1988). *Thermodynamics of Finite Systems and the Kinetics of First-Order Phase Transitions.* Leipzig: Teubner.

[517] Urban, C. (ed.) (2000). *Agent-Based Simulations. Workshop 2000.* Ghent: SCS European Publishing House.

[518] Vallacher, R.; Nowak, A. (eds.) (1994). *Dynamical Systems in Social Psychology.* New York: Academic Press.

[519] Varela, F.; Bourgine, P. (1992). *Toward a Practice of Autonomous Systems. Proc. 1st Eur. Conf. Artif. Life.* Cambridge, MA: MIT Press.
[520] Veit, H.; Richter, G. (2000). The FTA design paradigm for distributed systems. *Future Generation Comput. Syst.* **16(6)**, 727–740.
[521] Venzl, G.; Ross, J. (1982). Nucleation and colloidal growth in concentration gradients (Liesegang rings). *J. Chem. Phys.* **77/3**, 1302–1307.
[522] Vichniac, G. Y. (1984). Simulating physics with cellular automata. *Physica D* **10**, 96–116.
[523] Vicsek, T. (1989). *Fractal Growth Phenomena.* Singapore: World Scientific.
[524] Vicsek, T.; Shlesinger, M.; Matsushita, M. (eds.) (1994). *Fractals in Natural Sciences.* Singapore: World Scientific.
[525] Wagner, G. P. (1995). Adaptation and the modular design of organisms. In: F. Moran; A. Moreno; J. Merelo; P. Chacon (eds.), *Advances in Artificial Life: Proc. 3rd Eur. Conf. Artif. Life*, Berlin: Springer, vol. 929 of *Lecture Notes in Artificial Intelligence*, pp. 317–328.
[526] Wagner, H.; Baake, E.; Gerisch, T. (1998). Ising quantum chain and sequence evolution. *J. Stat. Phys.* **92**, 1017–1052.
[527] Weber, A. (1909). *Über den Standort der Industrien, 1. Teil: Reine Theorie des Standortes.* Tübingen.
[528] Wehner, R. (1987). Spatial organization of foraging behavior in individually searching desert ants, *Cataglyphis* (Sahara Desert) and *Ocymyrmex* (Namib Desert). In: J. M. Pasteels; J. L. Deneubourg (eds.), *From Individual to Collective Behavior*, Basel: Birkhäuser, pp. 15–42.
[529] Wehner, R. (1992). Arthropods. In: F. Papi (ed.), *Animal Homing*, London: Chapman and Hall, pp. 45–144.
[530] Wehner, R.; Menzel, R. (1990). Do insects have cognitive maps? *Annu. Rev. Neurosci.* **13**, 403–414.
[531] Wehner, R.; Räber, R. (1979). Visual spatial memory in desert ants, *Cataglyphis bicolor (Hymenoptera, Formicidae). Experientia* **35**, 1569–1571.
[532] Wehner, R.; Wehner, S. (1990). Path integration in desert ants. Approaching a long-standing puzzle in insect navigation. *Monitore Zool. Ital. (NS)* **20**, 309–331.
[533] Wei-Bin, Z. (1991). *Synergetic Economics – Time and Change in Nonlinear Economics.* Berlin: Springer.
[534] Weidlich, W. (1971). The statistical description of polarization phenomena in society. *Brit. J. Math. Stat. Psychol.* **24**, 51.
[535] Weidlich, W. (1972). The use of statistical models in sociology. *Collective Phenomena* **1**, 51–59.
[536] Weidlich, W. (1991). Physics and social science – The approach of synergetics. *Phys. Rep.* **204**, 1–163.
[537] Weidlich, W. (1994). Settlement formation at the meso-scale. *Chaos, Solitons & Fractals* **4**, 507–518.
[538] Weidlich, W. (1994). Synergetic modelling concepts for sociodynamics with application to collective political opinion formation. *J. Math. Sociol.* **18**, 267–291.
[539] Weidlich, W. (1997). From fast to slow processes in the evolution of urban and regional settlement structures. In: F. Schweitzer (ed.), *Self-Organization of Complex Structures: From Individual to Collective Dynamics*, London: Gordon and Breach, pp. 475–488.
[540] Weidlich, W. (1997). Sociodynamics applied to the evolution of urban and regional structures. *Discrete Dynamics Nature Soc.* **1/2**, 85–98.
[541] Weidlich, W. (2000). *Sociodynamics. A Systematic Approach to Mathematical Modelling in the Social Sciences.* London: Harwood Academic Publishers.

[542] Weidlich, W.; Braun, M. (1992). The master equation approach to non-linear economics. *J. Evol. Econ.* **2**, 233–265.

[543] Weidlich, W.; Haag, G. (1983). *Concepts and Methods of a Quantitative Sociology: The Dynamics of Interacting Populations.* Berlin: Springer.

[544] Weidlich, W.; Haag, G. (1987). A dynamic phase transition model for spatial agglomeration processes. *J. Reg. Sci.* **27/4**, 529–569.

[545] Weidlich, W.; Haag, G. (eds.) (1988). *Interregional Migration – Dynamic Theory and Comparative Analysis.* Berlin: Springer.

[546] Weidlich, W.; Munz, M. (1990). Settlement formation, I. A dynamic theory. *Ann. Reg. Sci.* **24**, 83–106.

[547] Weiss, G. (ed.) (1999). *Multiagent Systems. A Modern Approach to Distributed Artificial Intelligence.* Cambridge, MA: MIT Press.

[548] Wenzel, J. W. (1991). Evolution of nest architecture. In: K. G. Ross; R. W. Matthews (eds.), *Social Biology of Wasps*, Ithaca, NY: Cornell University Press, pp. 480–521.

[549] White, D. J. (1982). The set of efficient solutions for multiple objective shortest path problems. *Comput. Operations Res.* **9/2**, 101–107.

[550] White, R. (1977). Dynamic central place theory: Results of a simulation approach. *Geogr. Anal.* **9**, 227–243.

[551] White, R. (1978). The simulation of central place dynamics: Two-sector systems and the rank–size distribution. *Geogr. Anal.* **10**, 201–208.

[552] White, R.; Engelen, G. (1993). Cellular automata and fractal urban form: A cellular modelling approach to the evolution of urban land-use patterns. *Environ. Planning A* **25**, 1175–1199.

[553] White, R.; Engelen, G. (1993). Cellular dynamics and GIS: Modelling spatial complexity. *Geogr. Syst.* **1**, 237–253.

[554] White, R.; Engelen, G. (1994). Urban system dynamics and cellular automata: Fractal structures between order and chaos. *Chaos, Solitons & Fractals* **4/4**, 563–583.

[555] White, R.; Engelen, G. (1997). Cellular automata as the basis of integrated dynamic regional modelling. *Environ. Planning B* **24/2**, 235.

[556] White, R.; Engelen, G. (1997). Multi-scale spatial modelling of self-organizing urban systems. In: F. Schweitzer (ed.), *Self-Organization of Complex Structures: From Individual to Collective Dynamics*, London: Gordon and Breach, pp. 519–535.

[557] Wilkinson, P. C. (1982). *Chemotaxis and Inflammation.* London: Churchill.

[558] Willebrand, H.; Niedernostheide, F. J.; Ammelt, E.; Dohmen, R.; Purwins, H. G. (1991). Spatio-temporal oscillations during filament splitting in gas discharge systems. *Phys. Lett. A* **152**, 437–445.

[559] Williams, H. C. W. L.; Wilson, A. G. (1980). Some comments on the theroetical and analytical structure of urban and regional models. *Sistemi Urbani* **2/3**, 203.

[560] Wilson, A. G. (1978). Spatial interaction and settlement structure: toward an explicit central place theory. In: A. Karqvist; L. Lundqvist; F. Snickars; J. Weibull (eds.), *Spatial Interaction, Theory and Planning Models*, Amsterdam: North Holland, pp. 137–156.

[561] Wilson, E. O. (1971). *The Insect Societies.* Cambridge, MA: Belknap.

[562] Winter, P. (1987). Steiner problems in networks: A survey. *Networks* **17**, 129–167.

[563] Wolfram, S. (1983). Statistical mechanics of cellular automata. *Rev. Mod. Phys.* **55(3)**, 601–644.

[564] Wolfram, S. (1986). *Theory and Application of Cellular Automata.* Singapore: World Scientific.

[565] Wooldridge, M.; Jennings, N. R. (1995). Intelligent agents: Theory and practice. *Knowledge Eng. Rev.* **10(2)**, 115–152.

[566] Yi-Der, C. (1997). Asymmetric cycling and biased movement of Brownian particles in fluctuating symmetric potentials. *Phys. Rev. Lett.* **79/17**, 3117–3120.

[567] Yakhnin, V. Z.; Rovinsky, A. B.; Menzinger, M. (1994). Differential-flow-induced pattern formation in the exothermic A→B reaction. *J. Phys. Chem.* **98**, 2116–2119.

[568] Yu, P. L. (1985). *Multiple-Criteria Decison Making.* New York: Plenum Press.

[569] Zhou, H.; Chen, Y. (1996). Chemically driven motility of Brownian particles. *Phys. Rev. Lett.* **77/1**, 194–197.

[570] Zigmond, S. H. (1977). Ability of polymorphonuclear leukocytes to orient in gradients of chemotactic factors. *J. Cell Biol.* **75**, 606–616.

[571] Zipf, G. K. (1949). *Human Behavior and the Principle of Least Effort.* Reading, MA: Addison-Wesley.

[572] Zürcher, U.; Doering, C. R. (1993). Thermally activated escape over fluctuating barriers. *Phys. Rev. E* **47**, 3862–3869.

Index

ERRATUM

Brownian Agents and Active Particles

Frank Schweitzer

ISBN 978-3-540-73844-2
E-ISBN 978-3-540-73845-9

The original version of this article unfortunately contained a mistake. The spelling of the booktitle was incorrect. Correct is: "Brownian Agents and Active Particles" (NOT Browning).